AF597483

HANDBOOK OF Media FOR Clinical Microbiology

RONALD M. ATLAS
University of Louisville

JAMES W. SNYDER
University of Louisville

CRC Press
Boca Raton New York London Tokyo

Library of Congress Cataloging-in-Publication Data

Catalog record is available from the Library of Congress.

Direct all inquiries to CRC Press, Inc., 2000 Corporate Blvd., N.W., Boca Raton, Florida 33431.

International Standard Book Number 0-8493-9497-X
Printed in the United States of America 1 2 3 4 5 6 7 8 9 0
Printed on acid-free paper

Preface

The *Handbook of Media for Clinical Microbiology* is a compilation of the formulations, methods of preparation, and applications for media used in the clinical microbiology laboratory. It is written in a consistent style with media listed alphabetically. Each listing includes medium composition, instruction for preparation commercial sources, and uses. The format of The *Handbook of Media for Clinical Microbiology* allows easy reference to the information needed to prepare media for the cultivation of microorganisms relevant to clinical diagnostics.

About the Authors:

Ronald M. Atlas is Professor of Biology at the University of Louisville. He received his B.S. from the State University of New York at Stony Brook in 1968 and his M.S. and Ph.D. from Rutgers in 1970 and 1972, respectively. Dr. Atlas has received a number of honors including: The University of Louisville Excellence in Research Award, Johnson and Johnson Fellowship for Biology, the ASM Award in Applied and Environmental Microbiology. He is listed in *American Men and Women of Science*.

He teaches a variety of courses in microbiology at the University of Louisville and has authored several textbooks in general microbiology and microbial ecology. He has written well over 100 research papers. His recent studies have been on the application of molecular techniques to detect pathogens for epidemiological investigations. He has served on the NIH Recombinant Advisory Committee (RAC), as well as on various advisory boards for the Environmental Protection Agency (EPA). He is Chairperson of the Environmental Committee of the Public and Scientific Affairs Board of the American Society for Microbiology. He has been a national lecturer for Sigma Xi, an American Society for Microbiology Foundation lecturer, and an Australian Society for Microbiology national lecturer. He has served on the editorial boards of *Applied and Environmental Microbiology, Binary, Advances in Microbial Ecology, BioScience, Biotechniques* and *Journal of Industrial Microbiology*, and is editor of *CRC Critical Reviews in Microbiology*.

James W. Snyder is an Associate Professor of Pathology in the Department of Pathology, University of Louisville School of Medicine and serves as the Chief of Microbiology, University of Louisville Hospital. He received his B.S. in 1969 and M.S. in 1970 from Eastern Kentucky University and Ph.D. in 1974 from the University of Dayton. Dr. Snyder is a diplomate of the American board of Medical Microbiology (ABMM) and, in 1989, was elected as a Fellow in the American Academy of Microbiology. He is an active member of the American Society of Microbiology (ASM) and the South Central Association for Clinical Microbiology (SCACM) for which he serves as chairman of the Program Planning committee, Public and Scientific Affairs, and Corporate Liaison. Dr. Snyder has received a number of honors including: The SCACM Outstanding Contribution to Microbiology Award, the Outstanding Educator Award from the University of Louisville Department of Pediatrics, the Spirit of Louisville Foundation Award, Outstanding Service Award as Chairman of the Committee on Continuing Education of the American Society for Microbiology, and election to the Alpha Omega Alpha Honor Medical Society.

He teaches medical microbiology at the University of Louisville School of Medicine, School of Allied Health, and School of Dentistry. He maintains an active research program in both applied and basic clinical microbiology including molecular techniques for the detection of pathogens, assessment and comparison of automated and nonautomated diagnostic systems and is an active member of the University of Louisville Hospital "Switch Therapy" Antibiotic Utilization Team. He has served on the ASM Board of Education (BET), the ASM Committee on Continuing Education, and is currently a member of the ASM Long Distance Education Subcommittee. He is the former Editor of *The Yearbook of Clinical Microbiology*, and *Certified Clinical Microbiology Data Base*. He is currently a member of the *CRC Critical Reviews in Microbiology* editorial board.

Table of Contents

Introduction

Organization

The media described in the *Handbook of Media for Clinical Microbiology* are organized alphabetically. The description of each medium includes its name(s), composition, instructions for preparation, commercial sources, safety precautions where needed, and uses.

Names of Media

Media often have numerous names. In many cases media with identical compositions produced by different companies have different names. For example, Trypticase Soy Agar produced by BBL Microbiology Systems, Tryptone Soy Agar produced by Oxoid Unipath, and Tryptic Soy Agar produced by Difco Laboratories have identical compositions. Many media also are known by acronyms. TSA, for example, is the common acronym for Trypticase Soy Agar. The *Handbook of Media for Clinical Microbiology* gives the various synonymous names in the index and directs the reader to see the entry where the information about that medium is given. In cases where modifications to a medium yield a new medium, such media generally are listed with the original medium name, followed by the term modified.

Trademarks

The names of some media, components of media, and other terms are registered trademarks. The trademarked items referred to in the *Handbook of Media for Clinical Microbiology* are listed below.

American Type Culture Collection® and ATCC® are trademarks of the American Type Culture Collection.

Bacto®, BiTek® and Difco® are trademarks of Difco Laboratories.

Oxoid® and Lab–Lemco® are trademarks of Unipath Ltd.

Acidase®, BBL®, Biosate®, CTA Medium®, DTA Medium®, DCLS Agar®, Desoxycholate®, Desoxycholate Agar®, Desoxycholate Citrate Agar®, Enterococcosel®, Eugonagar®, Eugonbroth®, GC-Lect®, Gelysate®, IsoVitaleX®, Mycobactosel®, Mycophil®, Mycosel®, Myosate®, Phytone®, Polypeptone®, Selenite-F Enrichment®, Thiotone®, Trichosel®, Trypticase®, TSA II®, and TSI Agar® are trademarks of Becton Dickinson and Co.

Composition of Media

Media for the cultivation of microorganisms contain the substances necessary to support the growth of microorganisms. Due to the diversity of microorganisms and their diverse metabolic pathways, there are numerous media. Even slight differences in the composition of a medium can result in dramatically different growth characteristics of microorganisms.

When methods for culturing microorganisms were first developed in the nineteenth century, largely by Robert Koch and his colleagues, animal and plant tissues were principally used as sources of nutrients to support microbial growth. One of the major discoveries of Fanny Hesse in Koch's laboratory was that agar could be used to form solidified culture media on which microorganisms would grow. Extracts of plants and animal tissues were prepared as broths or mixed with agar to form a variety of culture media. Virtually any plant, animal, or animal organ was considered for use in preparing media. Infusions were prepared from beef heart, calf brains, and beef liver, as a few examples. These classic infusions still form the primary components of many media that are widely used today, such as Brain Heart Infusion Agar and Liver Broth.

The composition section of each medium describes the ingredients that make up the medium, their amounts, and the pH. It lists those ingredients in order of decreasing amount. Solids are listed first showing the weights to be added, followed by liquids showing the volumes to be included in the medium.

The composition uses generic terms where these are applicable. For example, pancreatic digest of casein is marketed by various manufacturers as trypticase, tryptone, and other commercial product names. While there may well be differences between these products, such differences are undefined. Variations also occur between batches of products produced as digests of animal tissues.

Media for the cultivation of microorganisms have a source of carbon for incorporation into biomass. For autotrophs the carbon source most often is carbon dioxide which may be supplied as bicarbonate within the medium. Carbohydrates, such as glucose, or other organic compounds, such as acetate, various lipids, proteins, and other organic compounds are included in media as sources of carbon for heterotrophs. These carbon sources may also serve as the supply of energy. Other compounds, such as ammonium ions, nitrite ions, elemental sulfur and reduced iron may be used as the sources of energy for the cultivation of autotrophs. Nitrogen also is required for microbial growth. It may be supplied as inorganic nitrogen compounds for the cultivation of some microorganisms but more commonly is supplied as proteins, peptones, or amino acids. Phosphates and metals—such as magnesium and iron—are also necessary components of microbiological media. Phosphates may also serve as buffers to maintain the pH of the medium within the growth tolerance limits of the microorganism being cultivated. Various additional growth factors may also be included in the media.

Agars

Agar is the most common solidifying agent used in microbiological media. Agar is a polysaccharide extract from marine algae. It melts at 84°C and solidifies at 38°C. Agar concentrations of 15.0g/L typically are used to form solid media. Lower concentrations of 7.5–10.0g/L are used to produce soft agars or semisolid media. Below are some agars used as solidifying agents in various media.

Agar Bacteriological (Agar No. 1)
  An agar with low calcium and magnesium. Available from Oxoid Unipath.

Agar, Bacto
  A purified agar with reduced pigmented compounds, salts, and extraneous matter. Available from Difco Laboratories.

Agar, BiTek™
  Agar prepared as a special technical grade. Available from Difco Laboratories.

Agar, Flake
  A technical grade agar. Available from Difco Laboratories.

Agar, Grade A
  A select grade agar containing minerals. Available from BBL Microbiology Systems.

Agar, Granulated
  A high grade granulated agar that has been filtered, decolorized, and purified. Available from BBL Microbiology Systems.

Agarose
  A low sulfate neutral gelling fraction of agar that is a complex galactose polysaccharide of near neutral charge.

Agar, Purified
  A very high grade agar that has been filtered, decolorized, and purified by washing and extraction of refined agars. It has reduced mineral content. Available from BBL Microbiology Systems.

Agar Technical (Agar No. 3)
  A technical grade agar. Available from Difco Laboratories and Oxoid Unipath.

Ionagar
  A purified agar. Available from Oxoid Unipath.

Noble Agar
  An agar that has been extensively washed and is essentially free of impurities. Available from Difco Laboratories.

Purified Agar
  An agar that has been extensively washed and extracted with water and organic solvent. Available from Difco Laboratories and Oxoid Unipath.

Peptones

Many complex media, that is, media in which not all the specific chemical components are known, contain peptones as the source of nitrogen. Peptones are hydrolyzed proteins formed by enzymatic or acidic digestion. Casein most often is used as the protein substrate for forming peptones, but other substances, such as soybean meal, also are commonly employed. Below is a list of some of the peptones that are used as ingredients in various media.

Acidase™ Peptone
  A hydrochloric acid hydrolysate of casein. It has a nitrogen content of 8% and is deficient in cystine and tryptophan. Available from BBL Microbiology Systems.

Bacto Casitone
  A pancreatic digest of casein. Available from Difco Laboratories.

Bacto Peptamin
  A peptic digest of animal tissues. Available from Difco Laboratories.

Bacto Peptone
  An enzymatic digest of animal tissues. It has a high concentration of low molecular weight peptones and amino acids. Available from Difco Laboratories.

Bacto Proteose Peptone
  An enzymatic digest of animal tissues. It has a high concentration of high molecular weight peptones. Available from Difco Laboratories.

Bacto Soytone
  A enzymatic hydrolysate of soybean meal. Available from Difco Laboratories.

Bacto Tryptone
  A pancreatic digest of casein. Available from Difco Laboratories.

Bacto Tryptose

An enzymatic hydrolysate containing numerous peptides including those of higher molecular weights. Available from Difco Laboratories.

Biosate™ Peptone

A hydrolysate of plant and animal proteins. Available from BBL Microbiology Systems.

Casein Hydrolysate

A hydrolysate of casein prepared with hydrochloric acid digestion under pressure and neutralized with sodium hydroxide. It contains total nitrogen of 7.6% and NaCl of 28.3%. Available from Oxoid Unipath.

Gelatone

A pancreatic digest of gelatin. Available from Difco Laboratories.

Gelysate™ Peptone

A pancreatic digest of gelatin deficient in cystine and tryptophan and which has a low carbohydrate content. Available from Oxoid Unipath.

Lactoalbumin Hydrolysate

A pancreatic digest of lactoalbumin, a milk whey protein. It has high levels of amino acids. It contains total nitrogen of 11.9% and NaCl of 1.4%. Available from Difco Laboratories and Oxoid Unipath.

Liver Digest Neutralized

A papaic digest of liver that contains total nitrogen of 11.0% and NaCl of 1.6%. Available from Oxoid Unipath.

Mycological Peptone

A peptone that contains total nitrogen of 9.5% and NaCl of 1.1%. Available from Oxoid Unipath.

Myosate™ Peptone

A pancreatic digest of heart muscle. Available from BBL Microbiology Systems.

Neopeptone

An enzymatic digest of protein. Available from Difco Laboratories.

Peptone Bacteriological Neutralized

A mixed pancreatic and papaic digest of animal tissues. It contains total nitrogen of 14.0% and NaCl of 1.6%. Available from Difco Laboratories and Oxoid Unipath.

Peptone P

A peptic digest of fresh meat that has a high sulfur content and contains total nitrogen of 11.12% and NaCl of 9.3%. Available from Difco Laboratories and Oxoid Unipath.

Peptonized Milk

A pancreatic digest of high grade skim milk powder. It has a high carbohydrate and calcium concentration. It contains total nitrogen of 5.3% and NaCl of 1.6%. Available from Oxoid Unipath.

Phytone™ Peptone

A papaic digest of soybean meal. It has a high vitamin and a high carbohydrate content. Available from BBL Microbiology Systems.

Polypeptone™ Peptone

A mixture of peptones composed of equal parts of pancreatic digest of casein and peptic digest of animal tissue. Available from BBL Microbiology Systems.

Proteose Peptone

A specialized peptone prepared from a mixture of peptones that contains a wide variety of high molecular weight peptides. It contains total nitrogen of 12.7% and NaCl of 8.0%. Available from Difco Laboratories and Oxoid Unipath.

Proteose Peptone No. 2

An enzymatic digest of animal tissues with a high concentration of high molecular weight peptones. Available from Difco Laboratories.

Proteose Peptone No. 3

An enzymatic digest of animal tissues. It has a high concentration of high molecular weight peptones. Available from Difco Laboratories.

Soya Peptone

A papaic digest of soybean meal with a high carbohydrate concentration. It contains total nitrogen of 8.7% and NaCl of 0.4%. Available from Oxoid Unipath.

Soytone

A papaic digest of soybean meal. Available from Difco Laboratories and Oxoid Unipath.

Special Peptone

A mixture of peptones, including meat, plant and yeast digests. It contains a wide variety of peptides, nucleotides, and minerals. It contains total nitrogen of 11.7% and NaCl of 3.5%. Available from Oxoid Unipath.

Thiotone™ E Peptone

An enzymatic digest of animal tissue. Available from BBL Microbiology Systems.

Trypticase™ Peptone

A pancreatic digest of casein. It has a very low carbohydrate content and a relatively high tryptophan content. Available from BBL Microbiology Systems.

Tryptone

A pancreatic digest of casein. It contains total nitrogen of 12.7% and NaCl of 0.4%. Available from Oxoid Unipath.

Tryptone T

A pancreatic digest of casein with lower levels of calcium, magnesium, and iron than tryptone. It contains total nitrogen of 11.7% and NaCl of 4.9%. Available from Difco Laboratories and Oxoid Unipath.

Tryptose

An enzymatic hydrolysate containing high molecular weight peptides. It contains total nitrogen of 12.2% and NaCl of 5.7%. Available from Difco Laboratories and Oxoid Unipath.

Meat and Plant Extracts

Meat and plant infusions are aqueous extracts that are commonly used as sources of nutrients for the cultivation of microorganisms. Such infusions contain amino acids and low molecular weight peptides, carbohydrates, vitamins, minerals, and trace metals. Extracts of animal tissues contain relatively high concentrations of water soluble protein components and glycogen. Extracts of plant tissues contain relatively high concentrations of carbohydrates.

With regard to infusions, many media list as an ingredient infusion from beef heart or another animal tissue. This ingredient is prepared by boiling a given amount of the animal tissue, for example 500.0g, and then using the liquid or, more commonly, drying the broth and using the solids from the infusion. The actual weight of the dry solids extracted from the hot water used to create the infusion varies and so the ingredient typically is simply listed as 500.0g beef heart infusion, although the actual weight of solids recovered from the infusion and used in the medium is far less. Brain heart infusion is prepared from calf brains and beef heart.

Below is a list of some of the meat and plant extracts that are used as ingredients in various media.

Bacto Beef

A desiccated powder of lean beef. Available from Difco Laboratories.

Bacto Beef Extract

An extract of beef (paste). Available from Difco Laboratories.

Bacto Beef Extract Desiccated

An extract of desiccated beef. Available from Difco Laboratories.

Bacto Beef Heart for Infusion

A desiccated powder of beef heart. Available from Difco Laboratories.

Bacto Liver

A desiccated powder of beef liver. Available from Difco Laboratories.

Lab-Lemco

A meat extract powder. Available from Oxoid Unipath.

Liver Desiccated

Dehydrated ox livers. Available from Oxoid Unipath.

Malt Extract

A water soluble extract from germinated grain dried by low temperature evaporation. It has a high carbohydrate content. It contains total nitrogen of 1.1% and NaCl of 0.1%.

Growth Factors

Many microorganisms have specific growth factor requirements that must be included in media for their successful cultivation. Vitamins, amino acids, fatty acids, trace metals, and blood components often must be added to media. Most often mixtures of growth factors are used in microbiological media. Acid hydrolysates of casein commonly are used as sources of amino acids. Extracts of yeast cells also are employed as sources of amino acids and vitamins for the cultivation of microorganisms. Many media, particularly those employed in the clinical laboratory, contain blood or blood components that serve as essential nutrients for fastidious microorganisms. X factor (heme) and V factor (nicotinamide adenine dinucleotide) often are supplied by adding hemoglobin (BBL and Difco Laboratories), IsoVitaleX (BBL Microbiology Systems) and or Supplement VX (Difco Laboratories).

Below is a list of some of the growth factors that are used as ingredients in microbiological media.

Bacto Casamino Acids

A mixture of amino acids formed by acid hydrolysis of casein. Available from Difco Laboratories.

Bacto Vitamin Free Casamino Acids

A mixture of amino acids formed by acid hydrolysis of casein that is free of vitamins. Available from Difco Laboratories.

Bovine Albumin

Bovine albumin fraction V 0.2% in 0.85% saline solution. Available from BBL Microbiology Systems.

Bovine Blood, Citrated

Calf blood washed and treated with sodium citrate as an anticoagulant. Available from BBL Microbiology Systems.

Bovine Blood, Defibrinated

Calf blood treated to denature fibrinogen

without causing cell lysis. Available from BBL Microbiology Systems.

Campylobacter Growth Supplement
Sodium pyruvate, sodium metabisulfite, and $FeSO_4$.

Castenholtz Salts
Agar, $NaNO_3$, Na_2HPO_4, KNO_3, nitrilotriacetic acid, $MgSO_4 \cdot 7H_2O$, $CaSO_4 \cdot 2H_2O$, NaCl, $FeCl_3$, $MnSO_4$, H_3BO_3, $ZnSO_4$, $CoCl_2 \cdot 6H_2O$, Na_2MoO_4, $CuSO_4$, and H_2SO_4.

CVA Enrichment
Glucose, L-cysteine·HCl·H_2O, vitamin B_{12}, L-glutamine, L-cystine·2HCl, adenine, nicotinamide adenine dinucleotide, cocarboxylase, guanine·HCl, $Fe(NO_3)_3$, *p*-aminobenzoic acid, and thiamine·HCl.

Cysteine Sulfide Reducing Agent
L-Cysteine·HCl·H_2O and $Na_2S \cdot 9H_2O$.

Dubos Medium Albumin
Albumin fraction V, glucose, and saline solution. Available from Difco Laboratories.

Dubos Oleic Albumin Complex
Alkalinized oleic acid, albumin fraction V, and saline solution. Available from Difco Laboratories.

Egg Yolk Emulsion
Chicken egg yolks and whole chicken egg. Available from Difco Laboratories and Oxoid Unipath.

Egg Yolk Emulsion, 50%
Chicken egg yolks, whole chicken egg, and saline solution. Available from Difco Laboratories.

EY Tellurite Enrichment
Egg yolk suspension with potassium tellurite. Available from Difco Laboratories and Oxoid Unipath.

Fresh Yeast Extract Solution
Live, pressed, starch-free, hydrolyzed Baker's yeast.

Fildes Enrichment
A peptic digest of sheep or horse blood that is a rich source of growth factors including hemin and nicotinamide adenine dinucleotide. Available from BBL Microbiology Systems, Difco Laboratories and Oxoid Unipath.

Hemin Solution
Hemin and NaOH.

Hemoglobin
Dried bovine hemoglobin. Used to provide hemin required by many fastidious microorganisms. Available from BBL Microbiology Systems and Difco Laboratories.

Hemoglobin Solution 2%
Provides hemin required by many fastidious microorganisms. Available from BBL Microbiology Systems and Difco Laboratories.

Hoagland Trace Element Solution, Modified
H_3BO_3, $MnCl_2 \cdot 4H_2O$, $AlCl_3$, $CoCl_2$, $CuCl_2$, KI, $NiCl_2$, $ZnCl_2$, $BaCl_2$, Na_2MoO_4, $SeCl_4$, $SnCl_2 \cdot 2H_2O$, $NaVO_3 \cdot H_2O$, KBr, and LiCl.

Horse Blood, Citrated
Horse blood washed and treated with sodium citrate used as an anticoagulant. Available from BBL Microbiology Systems.

Horse Blood, Defibrinated
Horse blood treated to denature fibrinogen without causing cell lysis. Available from BBL Microbiology Systems and Oxoid Unipath.

Horse Blood, Hemolysed
Horse blood treated to lyse cells. Available from Oxoid Unipath.

Horse Blood, Oxalated
Horse blood treated with potassium oxalate as an anticoagulant. Available from Oxoid Unipath.

Horse Serum
Horse blood is allowed to clot at 2°C–8°C so that the serum separates; the serum is filter sterilized. Serum usually is inactivated by heating to 56°C for 30 minutes to eliminate lipases that would cause degradation of lipids and inactivation of complement. Available from Difco Laboratories and Oxoid Unipath.

IsoVitaleX® Enrichment
Glucose, L-cysteine·HCl, L-glutamine, L-cystine, adenine, nicotinamide adenine dinucleotide, vitamin B_{12}, thiamine pyrophosphate, guanine·HCl, $Fe(NO_3)_3 \cdot 6H_2O$, *p*-aminobenzoic acid, and thiamine·HCl. Available from BBL Microbiology Systems.

Legionella Agar Enrichment
L-Cysteine and ferric pyrophosphate. Available from Difco Laboratories.

Legionella BCYE Growth Supplement
ACES buffer/KOH, ferric pyrophosphate, L-cysteine-HCl, and α-ketoglutarate. For the enrichment of *Legionella* species. Available from Oxoid Unipath.

Leptospira Enrichment
Lyophilized pooled rabbit serum containing hemoglobin that provides long chain fatty

acids and B vitamins for growth of *Leptospira* species. Available from BBL Microbiology Systems and Difco Laboratories.

Middlebrook ADC Enrichment

NaCl, bovine albumin fraction V, glucose and catalase. The albumin binds free fatty acids that may be toxic to mycobacteria. Available from BBL Microbiology Systems and Difco Laboratories.

Middlebrook OADC Enrichment

NaCl, bovine albumin, glucose, oleic acid, and catalase. The albumin binds free fatty acids that may be toxic to mycobacteria; the enrichment provides oleic acid used by *Mycobacterium tuberculosis* for growth. Available from BBL Microbiology Systems and Difco Laboratories.

Mycoplasma Enrichment without Penicillin

Horse serum, fresh autolysate of yeast—yeast extract, and thallium acetate. Provides cholesterol and nucleic acids for growth of *Mycoplasma* species. The thallium selectively inhibits other microorganisms. Available from BBL Microbiology Systems.

Mycoplasma Supplement

Yeast extract and horse serum. Available from Difco Laboratories.

Nitsch's Trace Elements

$MnSO_4$, H_3BO_3, $ZnSO_4$, Na_2MoO_4, $CuSO_4$, $CoCl_2 \cdot 6H_2O$, and H_2SO_4.

Oleic Albumin Complex

NaCl, bovine albumin fraction V, and oleic acid. The albumin binds free fatty acids that may be toxic to mycobacteria and the enrichment provides oleic acid that is used by *Mycobacterium tuberculosis* for growth. Available from BBL Microbiology Systems.

PPLO Serum Fraction

Serum fraction A. Available from Difco Laboratories.

Rabbit Blood, Citrated

Rabbit blood washed and treated with sodium citrate as an anticoagulant. Available from BBL Microbiology Systems.

Rabbit Blood, Defibrinated

Rabbit blood treated to denature fibrinogen without causing cell lysis. Available from BBL Microbiology Systems.

RPF Supplement

Fibrinogen, rabbit plasma, trypsin inhibitor, and potassium tellurite. For the selection and nutrient supplementation of *Staphylococcus aureus*. Available from Oxoid Unipath.

Sheep Blood, Citrated

Sheep blood washed and treated with sodium citrate as an anticoagulant. Available from BBL Microbiology Systems.

Sheep Blood, Defibrinated

Sheep blood treated to denature fibrinogen without causing cell lysis. Available from Oxoid Unipath.

SLA Trace Elements

$FeCl_2 \cdot 4H_2O$, H_3BO_3, $CoCl_2 \cdot 6H_2O$, $ZnCl_2$, $MnCl_2 \cdot 4H_2O$, $NiCl_2 \cdot 6H_2O$, $CuCl_2 \cdot 2H_2O$, $Na_2MoO_4 \cdot 2H_2O$, and $Na_2SeO_3 \cdot 5H_2O$.

Soil Extract

African Violet soil and Na_2CO_3.

Supplement A

Yeast concentrate with Crystal Violet. Available from Difco Laboratories.

Supplement B

Yeast concentrate, glutamine, coenzyme, cocarboxylase, hematin and growth factors. Available from Difco Laboratories.

Supplement C

Yeast concentrate. Available from Difco Laboratories.

Supplement VX

Essential growth factors V and X. Available from Difco Laboratories.

VA Vitamin Solution

Nicotinamide, thiamine·HCl, *p*-aminobenzoic acid, biotin, calcium pantothenate, pyridoxine·2HCl, and cyanocobalamin.

Vitamin K_1 Solution

Vitamin K_1 and ethanol.

Vitox Supplement

Glucose, L-cysteine·HCl, L-glutamine, L-cystine, adenine sulfate, nicotinamide adenine dinucleotide, cocarboxylase, guanine·HCl, $Fe(NO_3)_3 \cdot 6H_2O$, *p*-aminobenzoic acid, vitamin B_{12}, and thiamine·HCl. Available from Oxoid Unipath.

Yeast Autolysate Growth Supplement

Yeast autolysate fractions, glucose, and $NaHCO_3$. Available from Oxoid Unipath.

Yeast Extract

A water soluble extract of autolyzed yeast cells. Available from BBL Microbiology Systems, Difco Laboratories, and Oxoid Unipath.

Yeastolate

A water soluble fraction of autolyzed yeast cells rich in vitamin B complex. Available from Difco Laboratories.

Selective Components

Many media contain selective components that inhibit the growth of nontarget microorganisms and favor the growth of specific organisms. Selective media are especially useful in the isolation of specific microorganisms from mixed populations. For the study of microorganisms in nature, compounds are included in many media as sole sources of carbon or nitrogen so that only a few types of microorganisms can grow. Selective toxic compounds are also frequently used to select for the cultivation of particular microbial species. The isolation of a pathogen from a stool specimen, for example, where there is a high abundance of nonpathogenic normal microbiota, requires selective media. Often antimicrobics or other selectively toxic compounds are incorporated into media to suppress the growth of the background microbiota while permitting the cultivation of the target organism of interest. Bile salts, selenite, tetrathionate, tellurite, azide, phenylethanol, sodium lauryl sulfate, high sodium chloride concentrations, and various dyes—such as eosin, Crystal Violet, and Methylene Blue—are used as selective toxic chemicals. Antimicrobial agents used to suppress specific types of microorganisms include ampicillin, chloramphenicol, colistin, cycloheximide, gentamicin, kanamycin, nali-dixic acid, sulfadiazine, and vancomycin. Various combinations of antimicrobics are effective in suppressing classes of microorganisms, such as enteric bacteria. Below are some of the selective agents, principally antimicrobic mixtures used for the selective isolation of pathogens, that are commonly used as selective agents in microbiological media.

Ampicillin Selective Supplement
: Ampicillin. Used in media for the selection of *Aeromonas hydrophila.* Available from Oxoid Unipath.

Anaerobe Selective Supplement GN
: Hemin, menadione, sodium succinate, nalidixic acid, and vancomycin. For the selection of Gram-negative anaerobes. Available from Oxoid Unipath.

Anaerobe Selective Supplement NS
: Hemin, menadione, sodium pyruvate, and nalidixic acid. For the selection of nonsporulating anaerobes. Available from Oxoid Unipath.

Bacillus cereus Selective Supplement
: Polymyxin B. For the selection of *Bacillus cereus.* Available from Oxoid Unipath.

Bordetella Selective Supplement
: Cephalexin. For the selection of *Bordetella* species. Available from Oxoid Unipath.

Brucella Selective Supplement
: Polymyxin B, bacitracin, cycloheximide, nalidixic acid, nystatin, and vancomycin. For the selection of *Brucella* species. Available from Oxoid Unipath.

Campylobacter Selective Supplement Blaser-Wang
: Vancomycin, polymyxin B, trimethoprim, amphotericin B, cephalothin. For the selection of *Campylobacter* species. Available from Oxoid Unipath.

Campylobacter Selective Supplement Butzler
: Bacitracin, cycloheximide, colistin sulfate, sodium cephazolin, and novobiocin. For the selection of *Campylobacter* species. Available from Oxoid Unipath.

Campylobacter Selective Supplement Preston
: Polymyxin B, rifampicin, trimethoprim, and cycloheximide. For the selection of *Campylobacter* species. Available from Oxoid Unipath.

Campylobacter Selective Supplement Skirrow
: Vancomycin, trimethoprim, and polymyxin B. For the selection of *Campylobacter* species. Available from Oxoid Unipath.

CCDA Selective Supplement
: Cefoperazone and amphotericin B. For the selection of *Campylobacter* species. Available from Oxoid Unipath.

Cefoperazone Selective Supplement
: Cefoperazone. For the selection of *Campylobacter* species. Available from Oxoid Unipath.

CFC Selective Supplement
: Cetrimide, fucidin, and cephaloridine. For the selection of pseudomonads. Available from Oxoid Unipath.

Chapman Tellurite Solution
: Potassium tellurite 1% solution. Available from Difco Laboratories.

Chloramphenicol Selective Supplement
: Chloramphenicol. For the selection of yeasts and filamentous fungi. Available from Oxoid Unipath.

Clostridium difficile Selective Supplement
: D-Cycloserine and cefoxitin. For the selection of *Clostridium difficile.* Available from Difco Laboratories and Oxoid Unipath.

CN Inhibitor
: Cesulodin and novobiocin. It inhibits enteric Gram-negative microorganisms. Available from BBL Microbiology Systems.

CNV Antimicrobic
: Colistin sulfate, nystatin, and vancomycin. Available from Difco Laboratories.

CNVT Antimicrobic
Colistin sulfate, nystatin, vancomycin, and trimethoprim lactate. Available from Difco Laboratories.

Colbeck's Egg Broth
Egg emulsion and saline solution. Formerly available from Difco Laboratories—replaced with egg emulsion.

Fraser Supplement
Ferric ammonium sulfate, nalidixic acid, and Acriflavin hydrochloride. For the selection of *Listeria* species. Available from Oxoid Unipath.

Gardnerella vaginalis Selective Supplement
Gentamicin sulfate, nalidixic acid, and amphotericin B. For the selection of *Gardnerella vaginalis*. Available from Oxoid Unipath.

GC Selective Supplement
Yeast autolysate, glucose, Na_2HCO_3, vancomycin, colistin methane sulfonate, nystatin, and trimethoprim. For the selection of *Neisseria* species. Available from Oxoid Unipath.

Helicobacter pylori Selective Supplement Dent
Vancomycin, trimethoprim, cefulodin and amphotericin B. For the selection of *Helicobacter pylori*.

Kanamycin Sulfate Selective Supplement
Kanamycin sulfate. For the selection of enterococci. Available from Oxoid Unipath.

LCAT Selective Supplement
Lincomycin, colistin sulfate, amphotericin B, and trimethoprim. For the selection of *Neisseria* species. Available from Oxoid Unipath.

Legionella BMPA Selective Supplement
Cefamandole, polymyxin B, and anisomycin. For the selection of *Legionella* species. Available from Oxoid Unipath.

Legionella GVPC Selective Supplement
Glycine, vancomycin hydrochloride, polymixin B sulfate, and cycloheximide. For the selection of *Legionella* species. Available from Oxoid Unipath.

Legionella MWY Selective Supplement
Glycine, polymyxin B, anisomycin, vancomycin, Bromthymol B, and Bromcresol Purple. For the selection of *Legionella* species. Available from Oxoid Unipath.

Listeria Primary Selective Enrichment Supplement
Nalidixic acid and acriflavin. For the selection of *Listeria* species. Available from Oxoid Unipath.

Listeria Selective Enrichment Supplement
Nalidixic acid, cycloheximide and acriflavin. For the selection of *Listeria* species. Available from Oxoid Unipath.

Listeria Selective Supplement MOX
Colistin and moxalactam. For the selection of *Listeria monocytogenes*. Available from Oxoid Unipath.

Listeria Selective Supplement Oxford
Cycloheximide, colistin sulfate, acriflavin, cefotetan, and fosfomycin. For the selection of *Listeria* species. Available from Oxoid Unipath.

Modified Oxford Antimicrobic Supplement
Moxalactam and colistin sulfate. Available from Difco Laboratories.

MSRV Selective Supplement
Novobiocin. For the selection of *Salmonella*. Available from Oxoid Unipath.

Mycoplasma Supplement G
Horse serum, yeast extract, thallous acetate, and penicillin. For the selection of *Mycoplasma* species. Available from Oxoid Unipath.

Mycoplasma Supplement P
Horse serum, yeast extract, thallous acetate, glucose, Phenol Red, Methylene Blue, penicillin, and *Mycoplasma* broth base. For the selection of *Mycoplasma* species. Available from Oxoid Unipath.

Mycoplasma Supplement S
Yeast extract, horse serum, thallium acetate, and penicillin. Available from Difco Laboratories.

Oxford Antimicrobic Supplement
Cycloheximide, colistin sulfate, acriflavin, cefotetan, and fosfomycin. Available from Difco Laboratories.

Oxgall
Dehydrated fresh bile. For the selection of bile tolerant bacteria. Available from Difco Laboratories.

Oxytetracycline GYE Supplement
Oxytetracycline in a buffer. For the selection of yeasts and filamentous fungi. Available from Oxoid Unipath.

PALCAM Selective Supplement
Polymyxin B, acriflavin hydrochloride, and ceftazidime. For the selection of *Listeria monocytogenes*. Available from Oxoid Unipath.

Perfringens OPSP Selective Supplement A
Sodium sulfadiazine. For the selection of

Clostridium perfringens. Available from Oxoid Unipath.

Perfringens SFP Selective Supplement A
Kanamycin sulfate and polymyxin B. For the selection of *Clostridium perfringens*. Available from Oxoid Unipath.

Perfringens TSC Selective Supplement A
D-Cycloserine. For the selection of *Clostridium perfringens*. Available from Oxoid Unipath.

Sodium Desoxycholate
Sodium salt of desoxycholic acid. Available from Difco Laboratories.

Sodium Taurocholate
Sodium salt of conjugated bile acid—75% sodium taurocholate and 25% bile salts. For the selection of bile tolerant bacteria. Available from Difco Laboratories.

STAA Selective Supplement
Streptomycin sulfate, cycloheximide, and thallous acetate. For the selection of *Brochothrix thermosphacta*. Available from Oxoid Unipath.

Staph/Strep Selective Supplement
Nalidixic acid and colistin sulfate. For the selection of *Staphylococcus* species and *Streptococcus* species. Available from Oxoid Unipath.

Streptococcus Selective Supplement COA
Colistin sulfate and oxolinic acid. For the selection of *Streptococcus* species. Available from Oxoid Unipath.

Sulfamandelate Supplement
Sodium sulfacetamide and sodium mandelate. For the selection of *Salmonella* species. Available from Oxoid Unipath.

Tellurite Solution
A solution containing potassium tellurite. Inhibits Gram-negative and most Gram-positive microorganisms. It is used for the isolation of *Corynebacterium* species, *Streptococcus* species, *Listeria* species, and *Candida albicans*. Available from BBL Microbiology Systems.

Tinsdale Supplement
Serum, potassium tellurite and sodium thiosulfate. For the selection of *Corynebacterium diphtheriae*. Available from Oxoid Unipath.

V C A Inhibitor
Vancomycin, colistin, anisomycin, and trimethoprim. Inhibits most Gram-negative and Gram-positive bacteria and yeasts. It is used for the isolation of *Neisseria* species. Available from BBL Microbiology Systems.

V C A T Inhibitor
Vancomycin, colistin, anisomycin, and trimethoprim lactate. Inhibits most Gram-negative and Gram-positive bacteria and yeasts. It is used for the isolation of *Neisseria* species. Available from BBL Microbiology Systems and Oxoid Unipath.

V C N Inhibitor
Colistin, vancomycin, and nystatin. Inhibits most Gram-negative and Gram-positive bacteria and yeasts. It is used for the isolation of *Neisseria* species. Available from BBL Microbiology Systems and Oxoid Unipath.

V C N T Inhibitor
Colistin, vancomycin, nystatin, and trimethoprim lactate. Inhibits most Gram-negative and Gram-positive bacteria and yeasts. It is used for the isolation of *Neisseria* species. Available from BBL Microbiology Systems and Oxoid Unipath.

Yersinia Selective Supplement
Cefsulodin, irgasan, and novobiocin. For the selection of *Yersinia enterocolitica*. Available from Oxoid Unipath.

pH Buffers

Maintaining the pH of media usually is accomplished by the inclusion of suitable buffers. Since microorganisms grow optimally only within certain limits of a pH range, the pH generally is maintained within a few tenths of a pH unit. Phosphate buffers commonly are used. The pH is established by using varying volumes of equimolar concentrations of Na_2HPO_4 and NaH_2PO_4.

pH	Na_2HPO_4 (mL)	NaH_2PO_4 (mL)
5.4	3.0	97.0
5.6	5.0	95.0
5.8	7.8	92.2
6.0	12.0	88.0
6.2	18.5	81.5
6.4	26.5	73.5
6.6	37.5	62.5
6.8	50.0	50.0
7.0	61.1	38.9
7.2	71.5	28.5
7.4	80.4	19.6
7.6	86.8	13.2
7.8	91.4	8.6
8.0	94.5	5.5

Differential Components

The differentiation of many microorganisms is based upon the production of acid or alkali. Some media include indicators, particularly of pH that permit the visual detection of changes in pH resulting from such metabolic reactions. Below is a list of some commonly used pH indicators and their color reactions.

pH Indicator	pH Range	Acid Color	Alkaline Color
m-Cresol Purple	0.5–2.5	Red	Yellow
Thymol Blue	1.2–2.8	Red	Yellow
Bromphenol Blue	3.0–4.6	Yellow	Blue
Bromcresol Green	3.8–5.4	Yellow	Blue
Chlorcresol Green	4.0–5.6	Yellow	Blue
Methyl Red	4.2–6.3	Red	Yellow
Chlorphenol Red	5.0–6.6	Yellow	Red
Bromcresol Purple	5.2–6.8	Yellow	Purple
Bromthymol Blue	6.0–7.6	Yellow	Blue
Phenol Red	6.8–8.4	Yellow	Red
Cresol Red	7.2–8.8	Yellow	Red
m-Cresol Purple	7.4–9.0	Yellow	Purple
Thymol Blue	8.0–9.6	Yellow	Blue
Cresolphthalein	8.2–9.8	Colorless	Red
Phenolphthalein	8.3–10.0	Colorless	Red

Manufacturers of Commercially Prepared Media

Many of the media described in the *Handbook of Media for Clinical Microbiology* are commercially available.

Company	Location
ACCUMEDIA	Columbia, MD
Baltimore Biological Laboratories (BBL)	Cockeysville, MD
Carr-Scarborough Media (CSM)	Atlanta, GA
Clinical Standards Laboratories, Inc.	Rancho Dominguez, CA
DIFCO Laboratories	Detroit, MI
Di-Med	St. Paul, MN
Edge Biologicals	Memphis, TN
Gibson Laboratories	Lexington, KY
Micro Biologics	St. Cloud, MN
Oxoid, LTD	Columbia, MD and Basingstoke, England
Portland Media Laboratories (PML)	Portland, OR
REMEL	Lenexa, KS
Smith River Biologicals	Ferrum, VA

Sources of Media

The *Handbook of Media for Clinical Microbiology* includes the media produced by major suppliers of dehydrated media—including Difco, BBL, and Oxoid Unipath.

BBL Products

The media available from Beckton Dickinson Microbiology Systems, BBL Division, PO Box 243, 250 Schilling Circle, Cockeysville MD 21030 are listed below.

BBL Product Number	Medium Name
21377	Amies Modified Transport Medium with Charcoal
97643	Amies Transport Medium without Charcoal
10926	Anaerobic Agar
97165	Anaerobic CNA Agar
97127	Andrade's Broth
10953	Antibiotic Medium 5
10965	Antibiotic Medium 8
10969	Antibiotic Medium 9
10973	Antibiotic Medium 10
10977	Antibiotic Medium 11
10986	Antibiotic Medium 13
10982	Antibiotic Medium 19
11819	Arylsulfatase Agar
97881	BCYE Differential Agar
11027	BiGGY Agar
21838	Bile Esculin Agar
97350	Bile Esculin Agar with Kanamycin
11030	Bismuth Sulfite Agar
11037	Blood Agar Base
11049	Blood Agar with Low pH
11050	Bordet Gengou Agar
11057	Brain Heart CC Agar
11059	Brain Heart Infusion
11065	Brain Heart Infusion Agar
11073	Brilliant Green Agar
11086	*Brucella* Agar
97402	*Brucella* Blood Agar with Hemin and Vitamin K1
12367	Buffered Peptone Water
21727	*Campylobacter* Agar with 5 Antimicrobics and 10% Sheep Blood
21747	*Campylobacter* Thioglycollate Medium with 5 Antimicrobics
11102	Cary and Blair Transport Medium
21733	CDC Anaerobe Blood Agar

BBL Product Number	Medium Name
21735	CDC Anaerobe Blood Agar with Kanamycin and Vancomycin
21733	CDC Anaerobe Blood Agar with Phenylethyl Alcohol
21846	CDC Anaerobe Laked Blood Agar with Kanamycin and Vancomycin
11533	Cetrimide Agar, USP
21169	Chocolate II Agar with Hemoglobin and IsoVitaleX®
12309	CIN Agar
12218	CLED Agar
21852	*Clostridium difficile* Agar
11114	*Clostridium* Selective Agar
11116	Coagulase Mannitol Agar
11124	Columbia Agar
12161	Columbia Broth
12104	Columbia CNA Agar
99933	Columbia CNA Agar, Modified with Sheep Blood
11127	Cooked Meat Medium
95982	Cooked Meat Medium with Glucose, Hemin and Vitamin K
11094	CTA Agar
11096	CTA Medium
11144	DCLS Agar
11150	Deoxycholate Agar
11154	Deoxycholate Citrate Agar
11160	Deoxycholate Lactose Agar
12330	Dermatophyte Test Medium Base
11165	Dextrose Agar
11167	Dextrose Broth
11178	DNase Test Agar
11180	Dubos Broth
11187	EC Broth
97873	Egg Yolk Agar, Modified
11199	Endo Agar
11203	Endo Agar, LES
11119	Endo Broth
12204	Enterococcosel™ Agar
12207	Enterococcosel™ Broth
11213	*Enterococcus* Agar
98118	Esculin Iron Agar
11226	Ethyl Violet Azide Broth
11230	Eugon Agar
11235	Eugon Broth
12338	FC Agar
11364	FC Broth
11245	FDA Broth
11258	Flo Agar

BBL Product Number	Medium Name
11260	Fluid Thioglycollate Medium, without Glucose or Eh Indicator
98158	GBNA Medium
97715	GC-Lect™ Agar
11279	GN Broth, Hajna
21779	Group A Selective Strep Agar with Sheep Blood
97884	HBT Bilayer Medium
12210	Hektoen Enteric Agar
11298	Indole Nitrite Medium
11303	Infusion Broth
11310	Inhibitory Mold Agar
11317	Kligler Iron Agar
11333	Lactose Broth
21858	Lecithin Lactose Agar
97920	*Legionella* Selective Agar
11341	Letheen Agar
11343	Litmus Milk
11355	Loeffler Medium
11359	Lowenstein–Jensen Medium
21896	Lowenstein–Jensen Medium with NaCl
12336	LPM Agar
11362	Lysine Iron Agar
11387	MacConkey Agar
11393	MacConkey Agar without Crystal Violet
11398	Malonate Broth, Ewing Modified
11407	Mannitol Salt Agar
21557	Martin–Lewis Agar
11429	Møller Decarboxylase Broth
21665	Møller KCN Broth Base
21517	Motility Indole Ornithine Medium
11436	Motility Test Medium
11383	MRVP Broth
11443	Mueller Hinton Broth
11438	Mueller Hinton II Agar
21411	Mycobactosel™ Agar
21413	Mycobactosel™ L–J Medium
11445	Mycophil™ Agar
11450	Mycophil™ Agar with Low pH
11452	Mycophil™ Broth
11456	*Mycoplasma* Agar Base
11461	Mycosel™ Agar
21791	Neomycin Blood Agar
21839	Nitrate Broth
11472	Nutrient Agar
11476	Nutrient Agar, 1.5%

BBL Product Number	Medium Name
11478	Nutrient Broth
11481	Nutrient Gelatin
11484	Oxidation–Fermentation Medium
20976	Petragnani Medium
11538	Phenethyl Alcohol Agar
11502	Phenol Red Agar
11506	Phenol Red Broth
11514	Phenol Red Glucose Broth
11519	Phenol Red Lactose Broth
11527	Phenol Red Mannitol Broth
11533	Phenol Red Sucrose Broth
11537	Phenylalanine Agar
11546	Phytone™ Yeast Extract Agar
11488	PKU Test Agar
11549	Potato Dextrose Agar
96349	*Pseudomonas* Isolation Agar
11558	Purple Broth
21890	Rapid Fermentation Medium
98123	Regan–Lowe Charcoal Agar
97802	SABHI Agar
11584	Sabouraud Glucose Agar
11589	Sabouraud Glucose Agar, Emmons
97815	Sabouraud Maltose Agar
11596	*Salmonella–Shigella* Agar
21820	Salt Broth, Modified
12189	Schaedler Broth
11606	Selenite Cystine Broth
11607	Selenite F Broth
11612	Sellers Agar
21183	Serum Tellurite Agar
21868	Seven H11 Agar
11576	SF Broth
11578	SIM Medium
11619	Simmons' Citrate Agar
21618	Sodium Hippurate Broth
11664	Streptosel™ Agar
11666	Streptosel™ Broth
21809	SXT Blood Agar
11685	TCBS Agar
11698	Tellurite Glycine Agar
11702	Tergitol 7 Agar
11704	Tergitol 7 Broth
11706	Tetrathionate Broth
21184	Thayer–Martin Selective Agar
11712	Thiogel ® Medium
11718	Thioglycollate Medium without Glucose
11720	Thioglycollate Medium without Indicator–135C
97885	TOC Agar
11735	Todd–Hewitt Broth
21417	Transgrow Medium with Trimethoprim
21361	Transgrow Medium without Trimethoprim
11743	Transport Medium
11747	Trichosel™ Broth, Modified
11749	Triple Sugar Iron Agar
11043	Trypticase Soy Agar
97457	Trypticase Soy Agar with Sheep Blood and Gentamicin
11768	Trypticase Soy Broth
11777	Trypticase Soy Broth with 0.1% Agar
11696	Trypticase Tellurite Agar Base
11920	Tryptophan 1% Solution
12365	Tryptose Broth
12342	Tryptose Phosphate Broth, Modified
11690	TSN Agar
11795	Urea Agar
11797	Urease Test Broth
12348	UVM Modified *Listeria* Enrichment Broth
21874	V Agar
11801	Veal Infusion Agar
11803	Veal Infusion Broth
11812	Vogel and Johnson Agar
11836	XL Agar Base
11837	XLD Agar
97886	Yeast Extract Phosphate Agar
97299	Yeast Fermentation Broth
12101	Yeast Nitrogen Base

Difco Products

The media available from Difco Laboratories, Detroit MI 28401 are listed below.

Difco Product Number	Medium Name
0744	Amies Transport Medium with Charcoal
0832	Amies Transport Medium without Charcoal
0536	Anaerobic Agar
4040	Anaerobic LKV Blood Agar
0277	Antibiotic Medium 5
0660	Antibiotic Medium 6
0667	Antibiotic Medium 8
0462	Antibiotic Medium 9
0463	Antibiotic Medium 10
0593	Antibiotic Medium 11

Difco Product Number	Medium Name
0669	Antibiotic Medium 12
0043	Antibiotic Medium 19
0635	BiGGY Agar
0879	Bile Esculin Agar
0073	Bismuth Sulfite Agar
0073	Bismuth Sulfite Broth
0045	Blood Agar Base
0696	Blood Agar No. 2
0048	Bordet–Gengou Agar
0483	Brain Heart CC Agar
0037	Brain Heart Infusion
0418	Brain Heart Infusion Agar
0236	Brewer Thioglycollate Medium
0237	Brewer Thioglycollate Medium, Modified
0285	Brilliant Green Agar
0964	*Brucella* Agar
1810	Buffered Peptone Water
3279	*Campylobacter* Agar, Blaser's
3280	*Campylobacter* Agar, Skirrows
0835	*Candida* BCG Agar Base
0854	Cetrimide Agar, USP
0513	Chlamydospore Agar
4140	Chocolate Agar
1002	Chocolate Agar, Enriched
0695	Christensen Agar
0971	CLED Agar
0501	Coagulase Agar Base
0792	Columbia Blood Agar
0944	Columbia Broth
0867	Columbia CNA Agar
0703	Cooke Rose Bengal Agar
0267	Cooked Meat Medium
0077	Crystal Violet Agar
0047	Cystine Heart Agar
0338	Czapek Dox Broth
0759	DCLS Agar
0890	Decarboxylase Base, Møller
0872	Decarboxylase Medium Base, Falkow
0273	Deoxycholate Agar
0274	Deoxycholate Citrate Agar
0420	Deoxycholate Lactose Agar
0067	Dextrose Agar
0063	Dextrose Broth
0068	Dextrose Proteose No. 3 Agar
0632	DNase Test Agar
0220	DNase Test Agar with Methyl Green
1006	Dorset Egg Medium
0942	DTM Agar
0385	Dubos Broth
0373	Dubos Oleic Agar
0017	Eijkman Lactose Medium
0076	EMB Agar
0511	EMB Agar Base
0006	Endo Agar
0736	Endo Agar, LES
0749	Endo Broth
1828	Enteric Fermentation Base
0746	*Enterococcus* Agar
0606	Ethyl Violet Azide Broth
0589	Eugon Agar
0590	Eugon Broth
0677	FC Agar
0883	FC Broth
0841	Fermentation Broth
0349	Fildes Enrichment Agar
0987	Fletcher Medium
0256	Fluid Thioglycollate Medium
0697	Fluid Thioglycollate Medium with Beef Extract
0607	Fluid Thioglycollate Medium with K Agar
0289	GC Agar
0486	GN Broth, Hajna
0451	H Broth
0044	Heart Infusion Agar
0038	Heart Infusion Broth
0853	Hektoen Enteric Agar
0647	KCN Broth
0985	K–L Virulence Agar
0086	Kligler Iron Agar
0910	Kupferberg Trichomonas Base
0911	Kupferberg Trichomonas Broth
0004	Lactose Broth
1061	Lash Serum Medium
1830	*Legionella* Agar Base
0794	*Leptospira* Medium, EMJH
0680	Letheen Agar
0681	Letheen Broth
0005	Levine EMB Agar
0294	Littman Oxgall Agar
0052	Liver Infusion Agar
0269	Liver Infusion Broth
0070	Loeffler Blood Serum Medium
0444	Lowenstein–Jensen Medium
0215	Lysine Decarboxylase Broth
0849	Lysine Iron Agar

Difco Product Number	Medium Name
0075	MacConkey Agar
0470	MacConkey Agar without Crystal Violet
0331	MacConkey Agar without Salt
0395	Malonate Broth
0569	Malonate Broth, Ewing Modified
0306	Mannitol Salt Agar
0627	Middlebrook 7H10 Agar with Middlebrook OADC Enrichment
0627	Middlebrook 7H10 Agar with Middlebrook OADC Enrichment and Hemin
0627	Middlebrook 7H10 Agar with Streptomycin
0838	Middlebrook 7H11 Agar with Middlebrook ADC Enrichment
0838	Middlebrook 7H11 Agar with Middlebrook OADC Enrichment
0838	Middlebrook 7H11 Agar with Middlebrook OADC Enrichment and Triton WR 1339
0714	Middlebrook ADC Enrichment
0801	Middlebrook OADC Enrichment with Triton WR 1339
0722	Middlebrook OADC Enrichment
1804	MIL Medium
0298	Mitis–Salivarius Agar
0450	Motility Sulfide Medium
0105	Motility Test Medium
0016	MRVP Broth
0252	Mueller Hinton Agar
0757	Mueller Hinton Broth
0264	Mueller Tellurite Medium
0689	Mycobiotic Agar
0405	Mycological Agar
0305	Mycological Agar with Low pH
0268	Nitrate Broth
0001	Nutrient Agar
0069	Nutrient Agar, 1.5%
0003	Nutrient Broth
0011	Nutrient Gelatin
0688	Oxidative Fermentative Medium
0141	Pagano–Levin Agar
1010	Petragnani Medium
0098	Phenol Red Agar
0100	Phenol Red Lactose Agar
0103	Phenol Red Mannitol Agar
0090	Phenol Red Tartrate Agar
0745	Phenylalanine Agar
0806	Phenylalanine Malonate Broth
0504	Phenylethanol Agar

Difco Product Number	Medium Name
0898	Pike Streptococcal Broth
0980	PKU Test Agar
0013	Potato Dextrose Agar
0412	PPLO Agar
0410	PPLO Broth with Crystal Violet
0554	PPLO Broth without Crystal Violet
0065	Proteose No. 3 Agar
0448	*Pseudomonas* Agar F
0449	*Pseudomonas* Agar P
0927	*Pseudomonas* Isolation Agar
0228	Purple Agar
0227	Purple Broth
0082	Purple Lactose Agar
0797	SABHI Agar
0797	SABHI Blood Agar
0109	Sabouraud Glucose Agar
0382	Sabouraud Glucose Broth
0110	Sabouraud Maltose Agar
0429	Sabouraud Maltose Broth
0074	*Salmonella–Shigella* Agar
0661	SBG Enrichment Broth
0715	SBG Sulfa Enrichment
0408	Schaedler Agar
0534	Schaedler Broth
0275	Selenite Broth
0687	Selenite Cystine Broth
0895	Sellers Agar
0315	SF Broth
0271	SIM Medium
0091	Simmons' Citrate Agar
0079	Sorbitol MacConkey Agar
0054	Stock Culture Agar
0988	Stuart Medium Base
0650	TCBS Agar
0455	Tergitol 7 Agar
0800	Tergitol 7 Agar H
0912	Tergitol 7 Broth
0580	Tetrathionate Broth
1247	Thayer–Martin Medium
0530	Thioglycollate Gelatin Medium
0363	Thioglycollate Medium without Glucose
0432	Thioglycollate Medium without Glucose and Indicator
0430	Thioglycollate Medium without Indicator
0434	Thiol Broth
0307	Thiol Medium
0786	Tinsdale Agar
5790	Tissue Culture Amino Acids, HeLa 100X

Difco Product Number	Medium Name
5785	Tissue Culture Dulbecco Solution
5351	Tissue Culture Earle Solution
5508	Tissue Culture Hanks Solution
5477	Tissue Culture Medium 199
5068	Tissue Culture Medium Eagle with Earle Balanced Salt Solution
5071	Tissue Culture Medium Eagle with Hanks Balanced Salt Solution
5651	Tissue Culture Medium Eagle, HeLa
5825	Tissue Culture Medium Ham F10
5926	Tissue Culture Medium NCTC 109
5087	Tissue Culture Medium RPMI #1640
5675	Tissue Culture Minimal Medium Eagle with Earle Balanced Salts Solution
5839	Tissue Culture Minimal Medium Eagle Spinner Modified
5556	Tissue Culture Tyrode Solution
5833	Tissue Culture Vitamins Minimal Eagle, 100X
0492	Todd–Hewitt Broth
1410	Transgrow Medium
0877	*Trichophyton* Agar 1
0874	*Trichophyton* Agar 2
0965	*Trichophyton* Agar 3
0197	*Trichophyton* Agar 4
0411	*Trichophyton Agar 5*
0524	*Trichophyton* Agar 6
0955	*Trichophyton* Agar 7
0265	Triple Sugar Iron Agar
1829	Tryptic Digest Broth
0367	Tryptic Nitrate Medium
0064	Tryptose Agar
0633	Tryptose Agar with Thiamine
0062	Tryptose Broth
0060	Tryptose Phosphate Broth
0283	Urea Agar
0272	Urea R Broth
0343	Veal Infusion Agar
0344	Veal Infusion Broth
0562	Vogel and Johnson Agar
1805	Wilkins–Chalgren Agar
0555	XL Agar Base
0788	XLD Agar
0393	Yeast Morphology Agar
0392	Yeast Nitrogen Base
1817	*Yersinia* Selective Agar Base

Oxoid Unipath Products

The media available from Unipath Ltd., Oxoid Unipath Division, 9200 Rumsey Rd., Columbia, MD 21045 and Unipath Ltd., Oxoid Unipath Division, Wade Rd., Basingstoke, Hampshire, RG24 OPW, England are listed below.

Oxoid Unipath Product Number	Medium Name
CM425	Amies Modified Transport Medium with Charcoal
CM589	BiGGY Agar
CM888	Bile Esculin Agar
CM201	Bismuth Sulfite Agar
CM55	Blood Agar Base
CM854	Blood Agar Base Sheep
CM271	Blood Agar No. 2
CM655	BMPA-α Medium
CM267	*Bordetella pertussis* Selective Medium with Bordet–Gengou Agar Base
CM119	*Bordetella pertussis* Selective Medium with Charcoal Agar Base
CM267	Bordet Gengou Agar
CM225	Brain Heart Infusion
CM375	Brain Heart Infusion Agar
CM263	Brilliant Green Agar
CM329	Brilliant Green Agar, Modified
CM691	*Brucella* Agar
CM908	*Campylobacter* Selective Medium Karmali
CM519	Cary and Blair Transport Medium
CM333	Cholera Medium TCBS
CM301	CLED Agar
CM423	CLED Agar with Andrade Indicator
CM601	*Clostridium difficile* Agar
CM331	Columbia Blood Agar
CM81	Cooked Meat Medium
CM97	Czapek Dox Agar, Modified
CM393	DCLS Agar
CM163	Deoxycholate Agar
CM35	Deoxycholate Citrate Agar
CM227	Deoxycholate Citrate Agar, Hynes
CM539	Dermasel Agar Base
CM175	Dextrose Broth
CM321	DNase Agar
CM69	EMB Agar, Modified
CM37	Endo Agar
CM479	Endo Agar, LES
CM869	Ethyl Violet Azide Broth

Oxoid Unipath Product Number	Medium Name
CM331	*Gardnerella vaginalis* Selective Medium
CM755	GBS Agar Base, Islam
CM841	GBS Medium, Rapid
CM898	*Haemophilus* Test Medium
CM419	Hektoen Enteric Agar
CM83	Hoyle Medium Base
CM471	Iso–Sensitest Agar
CM473	Iso–Sensitest Broth
CM33	Kligler Iron Agar
CM137	Lactose Broth
CM45	Litmus Milk
CM77	Liver Broth
PM1	Lowenstein–Jensen Medium
CM308	Lysine Decarboxylase Broth, Taylor Modification
CM381	Lysine Iron Agar
CM7	MacConkey Agar
CM115	MacConkey Agar No. 3
CM7b	MacConkey Agar without Salt
CM85	Mannitol Salt Agar
CM43	MRVP Medium
CM337	Mueller–Hinton Agar
CM405	Mueller–Hinton Broth
CM3	Nutrient Agar
CM1	Nutrient Broth
CM67	Nutrient Broth No. 2
CM135a	Nutrient Gelatin
CM856	Oxford Agar
CM883	Oxidative Fermentative Medium
CM139	Potato Dextrose Agar
CM559	*Pseudomonas* CN Agar
PM2A	Pyruvic Acid Egg Medium
CM41	Sabouraud Glucose Agar
CM41a	Sabouraud Maltose Agar
CM99	*Salmonella–Shigella* Agar
CM533	*Salmonella–Shigella* Agar, Modified
CM94	Salt Meat Broth
CM437	Schaedler Agar
CM497	Schaedler Broth
R39	Selenite Broth
CM399	Selenite Broth Base, Mannitol
CM699	Selenite Cystine Broth
CM409	Sensitest Agar
CM435	SIM Medium
CM155	Simmons' Citrate Agar
CM813	Sorbitol MacConkey Agar
CM331	*Streptococcus* Selective Medium
CM111	Stuart Transport Medium, Modified
CM793	Tergitol 7 Agar
CM29	Tetrathionate Broth
CM671	Tetrathionate Broth, USA
CM391	Thioglycollate Broth USP, Alternative
CM415	Thioglycollate Medium without Indicator
CM173	Thioglycollate Medium, USP
CM487	Tinsdale Agar
CM189	Todd–Hewitt Broth
CM161	*Trichomonas* Medium
R27	*Trichomonas* Medium No. 2
CM277	Triple Sugar Iron Agar
TCM129	Tryptone Soya Broth
CM87	Tryptone Water Broth
CM233	Tryptose Blood Agar
CM283	Tryptose Phosphate Broth
CM53	Urea Agar Base
CM71	Urea Broth Base
CM641	Vogel and Johnson Agar
CM619	Wilkins–Chalgren Anaerobe Agar
CM619	Wilkins–Chalgren Anaerobe Agar with G–N Supplement
CM643	Wilkins–Chalgren Anaerobe Agar with N–S Supplement
CM643	Wilkins–Chalgren Anaerobe Broth
CM469	XLD Agar
CM19	Yeast Extract Agar
CM653	*Yersinia* Selective Agar Base

Preparation of Media

The ingredients in a medium are usually dissolved and the medium is then sterilized. When agar is used as a solidifying agent the medium must be heated gently, usually to boiling, to dissolve the agar. In some cases where interactions of components, such as metals, would cause precipitates, solutions must be prepared and occasionally sterilized separately before mixing the various solutions to prepare the complete medium. The pH often is adjusted prior to sterilization, but in some cases sterile acid or base is used to adjust the pH of the medium following sterilization. Many media are sterilized by exposure to elevated temperatures. The most common method is to autoclave the medium. Different sterilization procedures are employed when heat-labile compounds are included in the formulation of the medium.

Autoclaving

Autoclaving uses exposure to steam, generally under pressure, to kill microorganisms. Exposure for 15 minutes to steam at 115 psi—121°C is most commonly used. Such exposure kills vegetative bacterial cells and bacterial endospores. However, some substances do not tolerate such exposures and lower temperatures and different exposure times are sometimes employed. Media containing carbohydrates often are sterilized at 116°C–118°C in order to prevent the decomposition of the carbohydrate and the formation of toxic compounds that would inhibit microbial growth. Below is a list of pressure–temperature relationships.

Pressure—psi	Temperature—°C
0	100
1	101.9
2	103.6
3	105.3
4	106.9
5	108.4
6	109.8
7	111.3
8	112.6
9	113.9
10	115.2
11	116.4
12	117.6
13	118.8
14	119.9
15	121.0
16	122.0
17	123.0
18	124.0
19	125.0
20	126.0
21	126.9
22	127.8
23	128.7
24	129.6
25	130.4

Tyndallization

Exposure to steam at 100°C for 30 minutes will kill vegetative bacterial cells but not endospores. Such exposure can be achieved using flowing steam in an Arnold sterilizer. By allowing the medium to cool and incubate under conditions where endospore germination will occur and by repeating the 100°C–30 minute exposure on three successive days the medium can be sterilized because all the endospores will have germinated and the heat exposure will have killed all the vegetative cells. This process of repetitive exposure to 100°C is called tyndallization, after its discoverer John Tyndall.

Inspissation

Inspissation is a heat exposure method that is employed with high protein materials, such as egg-containing media, that cannot withstand the high temperatures used in autoclaving. This process causes coagulation of the protein without greatly altering its chemical properties. Several different protocols can be followed for inspissation. Using an Arnold sterilizer or a specialized inspissator, the medium is exposed to 75°C–80°C for 2 hours on each of three successive days. Inspissation using an autoclave employs exposure to 85°C–90°C for 10 minutes achieved by having a mixture of air and steam in the chamber, followed by 15 minutes exposure during which the temperature is raised to 121°C using only steam under pressure in the chamber; the temperature then is slowly lowered to less than 60°C.

Filtration

Filtration is commonly used to sterilize media containing heat-labile compounds. Liquid media are passed through sintered glass or membranes, typically made of cellulose acetate or nitrocellulose, with small pore sizes. A membrane with a pore size of 0.2mm will trap bacterial cells and, therefore, sometimes is called a bacteriological filter. By preventing the passage of microorganisms, filtration renders fluids free of bacteria and eukaryotic microorganisms, that is, free of living organisms, and hence sterile. Many carbohydrate solutions, antibiotic solutions, and vitamin solutions are filter sterilized and added to media that have been cooled to temperatures below 50°C.

Caution about Hazardous Components

Some media contain components that are toxic or carcinogenic. Appropriate safety precautions must be taken when using media with such components. Basic fuchsin and acid fuchsin are carcinogens and caution must be used in handling media with these compounds to avoid dangerous exposure that could lead to the development of malignancies. Thallium salts, sodium azide, sodium biselenite, and cyanide are among the toxic components found in some media. These compounds are poisonous and steps must be taken to avoid ingestion, inhalation, and skin contact. Azides also react with many metals, especially copper, to form explosive metal azides. The disposal of azides must avoid contact with copper or achieve sufficient dilution to avoid the formation of such hazardous explosive compounds. Media with sulfur-containing compounds may result in the formation of hydrogen sulfide which is a toxic gas. Care must be used to ensure proper ventilation. Media with human blood or human blood components must be handled with great caution to avoid exposure to human immunodeficiency virus and other pathogens that contaminate some blood supplies. Proper handling and disposal procedures must be followed with blood-containing as well as other media that are used to cultivate microorganisms.

Diagnostic Microbiology: Isolation and Identification of Pathogens

The definitive diagnosis of an infectious disease most commonly requires the isolation and identification of the pathogenic microorganism or the identification of antigens or antibodies specifically associated with a given microbial pathogen. Specific culture methods are used for the isolation and identification of the wide variety of pathogenic microorganisms. Traditional methods for the identification of pathogens depend on observations of microscopic properties and metabolic changes that occur as a result of the growth of the pathogen. Many differential media and selective media use color changes to show microbial growth. These growth-dependent methods provide rapid and reasonably accurate means of identifying the pathogen and diagnosing many infectious diseases.

A wide range of biochemical, serological, and gene probe procedures are available for the definitive identification of microbial isolates of clinical significance. Accuracy, reliability, and speed are important factors governing the selection of clinical identification protocols. The selection of the specific procedures to be employed for the identification of pathogenic isolates is guided by the presumptive identification of the organism at the family or genus level, based on the observation of colonial morphology and other growth characteristics on the primary isolation medium, and on the microscopic observation of stained specimens.

Identification of pathogenic filamentous fungi and protozoa is generally based on the morphological characteristics of the organism, the growth appearance, and a limited number of biochemical tests. Various other morphological and biochemical characteristics are used for identifying other pathogens. Conventional identification schemes for bacteria and yeasts rely on the determination of a variety of biochemical features exhibited by growing isolates. In general, fewer than 20 tests are required to identify clinical bacterial isolates at the species level. The purpose is to distinguish the isolates present in the specimen, using the minimal number of tests to define distinct taxa accurately.

Isolation and Culture Procedures

A variety of procedures are employed for the collection, isolation, and identification of pathogenic microorganisms from different tissues. Various procedures are used for the isolation of different types of microorganisms. Clinical procedures are designed to screen and facilitate the recovery of those etiologic agents of disease that predominate in particular clinical syndromes. When the manifestation of a disease suggests that the disease may be caused by a rare pathogen and/or routine screening fails to detect a probable causative microorganism, additional specialized isolation procedures may be required.

As the clinical microbiology laboratory slowly enters the new era of "molecular diagnostics", the dependency on traditional culture methods involving the production and preparation of growth supporting media will continue to play an integral role in our efforts to detect and characterize both known and "new" microbial pathogens. Furthermore, the major goal of the clinical microbiology laboratory, as stated by Dr. Raymond Bartlett over twenty years ago, has not and will not change, "to provide information of maximal and epidemiological usefulness as rapidly as is consistent with acceptable accuracy and minimal cost". The latter two criteria serve as the driving forces that account for the slow introduction and acceptance by the clinical microbiology community of methods that have been developed for either the noncultural detection of cultural confirmation of pathogenic microorganisms. Many procedures, including the more recent nucleic acid probes and gene amplification methods, remain cost prohibitive for most laboratories, lack acceptable sensitivity and/or specificity, and are not practical or adaptable for use in the service-oriented, clinical laboratory. Until such problems are resolved, microbiologists will continue to rely on the availability and utilization of a variety of media for the isolation, characterization, and identification of primary and opportunistic microbial pathogens. The selection of a specific procedure(s) to be employed for the definitive identification of a pathogen at the family or genus level is influenced by the information gleaned from observing colonial morphology, pigment production, and other characteristics that are observed following growth of the microorganism on either selective, differential, or general purpose media.

These concepts also apply to fungi and viruses. The limited availability of practical and cost-effective detection methods requires that these microorganisms be recovered in culture before confirmation of identity can be accomplished using key morphological and biochemical characteristics, the type of cytopathic effect (CPE), or the application of culture confirmation methods such as serological, nucleic acid probes, or amplification technologies. Culture media will continue to be devised and used for the cultivation of microorganisms despite the inevitable impact that noncultural methods and molecular diagnostics systems will have on the rapid detection of microbial pathogens and ultimately affect, in a beneficial manner, patient outcome.

Media for the Isolation and Identification of Microorganisms from Clinical Specimens

The *Handbook of Media for Clinical Microbiology* includes both classic and modern media used for the identification, cultivation, and maintenance of diverse bacteria described in *The Manual of Clinical Microbiology* for medically important microorganisms. Below are some of the primary media used in clinical microbiology laboratories for the isolation of pathogens.

Medium	Use
Anaerobic Blood Agar	Cultivation of a wide variety of anaerobic pathogens
Anaerobic Kanamycin Vancomycin Laked Blood Agar	Cultivation of *Bacteroides* species
Azide Blood Agar	Cultivation of *Streptococcus pyogenes* which forms small colonies and demonstrates beta hemolysis
Bacteroides Bile Esculin Agar	Cultivation of *Bacteroides* species
Bile Esculin Agar	Cultivation of group D streptococci including *Enterococcus faecalis*
Bismuth Sulfite Agar	Cultivation of *Salmonella* including *S. typhi* from faecal specimens
Blood Agar	Cultivation of many pathogens
Blood Agar with Colistin and Nalixic Acid	Selective cultivation of Gram-positive bacteria
Brain Heart Infusion Medium	Cultivation of a wide variety of aerobic and anaerobic human pathogens
Brilliant Green Agar	Selective cultivation of *Salmonella* species
Campylobacter Agar	Selective cultivation of *Campylobacter* species
Chocolate Agar	Cultivation of a wide variety of human pathogens including *Haemophilus* and *Neisseria* species
Chopped Meat Glucose Medium	Cultivation of a wide variety of anaerobes
Deoxycholate Agar	Selective cultivation of Gram-negative enteric bacteria
Deoxycholate Citrate Agar	Selective cultivation of *Salmonella* and some *Shigella* species
Enterococcus Agar	Selective cultivation of enterococci
Eosin Methylene Blue Agar	Selective cultivation of Gram-negative enteric bacteria
GN Broth	Enrichment for *Salmonella* and *Shigella* species
Hektoen Enteric Agar	Selective cultivation of enteric bacteria including *Shigella* species
MacConkey Agar	Selective cultivation of Gram-negative enteric bacteria
Mannitol Salt Agar	Selective cultivation of *Staphylococcus* species
Martin–Lewis Agar	Selective cultivation of *Neisseria* species
New York City Agar	Selective cultivation of *Neisseria* and *Mycoplasma* species
Phenylethyl Alcohol Medium	Selective cultivation of *Staphylococcus* and *Streptococcus* species.
Selenite F Broth	Enrichment for *Salmonella* species
Tetrathionate Broth	Enrichment for *Salmonella* and some *Shigella* species
Thayer Martin Agar	Selective cultivation of *Neisseria* species
Thioglycollate Broth	Cultivation of a wide variety of anaerobes
Xylose Lysine Deoxycholate Agar	Selective cultivation of *Salmonella* and some *Shigella* species

Some Media and Procedures Used for the Diagnosis of Various Diseases

Anatomical Site	Specimen	Culture Media	Probable Organism	Syndrome
Upper respiratory tract: throat and nasopharyngeal cultures	Sterile dacron swabs	Blood agar	*Streptococcus pyogenes*	Pharyngitis, rheumatic fever
		Chocolate agar	*Haemophilus influenzae*	Epiglottitis
			Neisseria gonorrhoeae	Phayingitis
		Bordet-Gengou	*Bordetella pertussis*	Whooping cough
		Tellurite serum agar	*Corynebacterium diphtheriae*	Diphtheria
Lower respiratory tract	Sputum, transtracheal aspirate, bronchoalveolar lavage, trans–bronchial biopsy, bronchoscopy, bronchial brushing, lung biopsy, transthoracic biopsy	Blood agar	*Streptococcus pneumoniae, Staphylococcus aureus*	Pneumonia
		Chocolate agar	*Streptococcus pyogenes*	
		MacConkey's agar	*Klebsiella pneumoniae, Haemophilus influenzae*	
		Stained smears	*Histoplasma capsulatum*	
		Sabouraud's agar	*Coccidioides immitis, Candida albicans*	
		Lowenstein-Jensen	*Mycobacterium tuberculosis*	Tuberculosis
Central nervous system	Lumbar puncture for cerebro–spinal fluid	Liquid enrichment media	*Streptococcus pneumoniae*	Meningitis
		Blood agar	*Neisseria meningitidis*	
		Chocolate agar	*Haemophilus influenzae*	
Circulatory system blood	Renal puncture	Radiolabeled glucose medium	Various	Septicemia
		Roll-tube streak anaerobic culture		
Urinary tract	Midstream catch of voided urine	Blood agar	*Escherichia coli*	Urinary tract infections
		Cysteine lactose electrolyte-deficient agar	*Klebsiella, Proteus, Pseudomonas, Salmonella*	
		MacConkey's agar, EMB agar	*Serratia, E. coli, and other Gram-negative rods*	

Some Media and Procedures Used for the Diagnosis of Various Diseases

Anatomical Site	Specimen	Culture Media	Probable Organism	Syndrome
Genital tract	Urethral exudate (males) Swabs from cervix, vagina and anal canal (females)	Thayer-Martin medium and chocolate agar	*Neisseria gonorrhoeae*	Gonorrhea
Intestinal tract	Stool samples	Hektoen enteric media, xylose-lysine-desoxycholate media, brilliant green, EMB, Endo and MacConkey's agar	*Salmonella-Shigella*	
Eyes and ears	Fluids	Blood, chocolate, MacConkey's agar, Gram stain		
Skin	Swabs, aspirates, or washings from lesions	Aerobic and anaerobic culture techniques	*Clostridium tetani, C. perfringens*	

References

Below is a list of references that can be consulted for further information about media used for the isolation, cultivation, and differentiation of microorganisms.

Color Atlas and Textbook of Diagnostic Microbiology. 1992. E. W. Koneman, S. D. Allen, W. M. Janda, P. C. Schreckenberger, and W. C. Winn, Jr., eds. J. B. Lippincott Co., Philadelphia PA.

Diagnostic Microbiology. 1990. S. M. Finegold and W. J. Martin. C. V. Mosby Co., St. Louis.

Difco Manual: Dehydrated Culture Media and Reagents for Microbiology. Difco Laboratories, Detroit MI.

Food and Drug Administration Bacteriological Analytical Manual. 1992. AOAC International, Arlington VA.

Manual of BBl Products and Laboratory Procedures. 1988. D. A. Power and P. J. McCuen, eds. Beckton Dickinson and Company, Cockeysville, MD.

Manual of Clinical Microbiology. 1991. A. Ballows, W. J. Hausler, K. L. Herman, H. D. Isenberg, and H. J. Shadomy, eds. American Society for Microbiology, Washington DC.

Media for Isolation–Cultivation–Identification–Maintenance of Medical Bacteria. 1985. J. F. McFaddin. Williams and Wilkins, Baltimore MD.

The Oxoid Manual. 1990. E. Y. Bridson, ed. Unipath Ltd. Basingstoke, Hampshire England.

Acetamide Agar

Composition per liter:

Agar ... 15.0g
Acetamide ... 10.0g
NaCl ... 5.0g
K_2HPO_4 ... 1.0g
$NH_4H_2PO_4$ ... 1.0g
$MgSO_4 \cdot 7H_2O$ ... 0.2g
Bromthymol Blue ... 0.08g

pH 6.9 ± 0.2 at 25°C

Preparation of Medium: Add components to distilled/deionized water and bring volume to 1.0L. Mix thoroughly. Gently heat and bring to boiling. Adjust pH. Distribute into tubes or flasks. Autoclave for 15 min at 15 psi pressure–121°C. Cool tubes in a slanted position to produce a long slant.

Use: For the differentiation of nonfermentative Gram-negative bacteria, especially *Pseudomonas aeruginosa*. Can be used as a confirmatory test for water analysis. Bacteria that deamidate acetamide turn the medium blue.

Acetamide Agar

Composition per liter:

Agar ... 15.0g
Acetamide ... 10.0g
NaCl ... 5.0g
K_2HPO_4 ... 1.39g
KH_2PO_4 ... 0.73g
$MgSO_4 \cdot 7H_2O$ ... 0.5g
Phenol Red ... 0.012g

pH 6.9 ± 0.2 at 25°C

Source: This medium is available as a premixed powder from BBL Microbiology Systems.

Preparation of Medium: Add components to distilled/deionized water and bring volume to 1.0L. Mix thoroughly. Gently heat and bring to boiling. Adjust pH. Distribute into tubes or flasks. Autoclave for 15 min at 15 psi pressure–121°C. Cool tubes in a slanted position to produce a long slant.

Use: For the differentiation of nonfermentative Gram-negative bacteria, especially *Pseudomonas aeruginosa*. Can be used as a confirmatory test for water analysis. Bacteria that deamidate acetamide turn the medium blue.

Acetamide Broth

Composition per liter:

Acetamide ... 10.0g
NaCl ... 5.0g
K_2HPO_4 ... 1.39g
KH_2PO_4 ... 0.73g
$MgSO_4 \cdot 7H_2O$ ... 0.5g
Phenol Red ... 0.012g

pH 6.9 ± 0.2 at 25°C

Preparation of Medium: Add components to distilled/deionized water and bring volume to 1.0L. Mix thoroughly. Adjust pH. Autoclave for 15 min at 15 psi pressure–121°C.

Use: For the differentiation of nonfermentative Gram-negative bacteria, especially *Pseudomonas aeruginosa*. Can be used as a confirmatory test for water analysis. Bacteria that deamidate acetamide turn the broth purplish red.

Aeromonas hydrophila Medium

Composition per liter:

Inositol ... 10.0g
Pancreatic digest of casein ... 10.0g
L-Ornithine·HCl ... 5.0g
Proteose peptone ... 5.0g
Agar ... 3.0g
Yeast extract ... 3.0g
Mannitol ... 1.0g
Ferric ammonium citrate ... 0.5g
$Na_2S_2O_3 \cdot 5H_2O$ ... 0.4g
Bromcresol Purple ... 0.02g

pH 6.7 ± 0.2 at 25°C

Preparation of Medium: Add components to distilled/deionized water and bring volume to 1.0L. Mix thoroughly. Gently heat until dissolved. Adjust pH to 6.7. Distribute into tubes in 5.0mL volumes. Autoclave for 12 min at 15 psi pressure–121°C.

Use: For the isolation and cultivation of *Aeromonas hydrophila.*

Amies Modified Transport Medium with Charcoal

Composition per liter:

Charcoal ... 10.0g
Agar ... 4.0g
NaCl ... 3.0g
Na_2HPO_4 ... 1.15g
Sodium thioglycollate ... 1.0g
KCl ... 0.2g
$CaCl_2 \cdot 2H_2O$ ... 0.1g
$MgCl_2 \cdot 6H_2O$ ... 0.1g
KH_2PO_4 ... 0.2g

pH 7.2 ± 0.2 at 25°C

Source: This medium is available as a premixed powder from Difco Laboratories.

Preparation of Medium: Add components to distilled/deionized water and bring volume to 1.0L. Mix thoroughly. Gently heat and bring to boiling. Distribute into flasks or tubes. Autoclave for 20 min at 15 psi pressure–121°C. While cooling, turn tubes to uniformly suspend charcoal.

Use: For the transport of swab specimens to prolong the survival of microorganisms, especially *Neisseria gonorrhoeae*, between collection and culturing. Addition of charcoal to this medium neutralizes metabolic products which may be toxic to *Neisseria gonorrhoeae*. The presence of charcoal may interfere with the Gram stain.

Amies Modified Transport Medium with Charcoal

Composition per liter:

Charcoal	10.0g
NaCl	8.0g
Agar	3.6g
Na_2HPO_4	1.15g
Sodium thioglycollate	1.0g
KCl	0.2g
$CaCl_2 \cdot 2H_2O$	0.1g
$MgCl_2 \cdot 6H_2O$	0.1g
KH_2PO_4	0.2g

pH 7.2 ± 0.2 at 25°C

Source: This medium is available as a premixed powder from BBL Microbiology Systems and Oxoid Unipath.

Preparation of Medium: Add components to distilled/deionized water and bring volume to 1.0L. Mix thoroughly. Gently heat and bring to boiling. Distribute into flasks or tubes. Autoclave for 20 min at 15 psi pressure–121°C. While cooling, turn tubes to uniformly suspend charcoal.

Use: For the transport of swab specimens to prolong the survival of microorganisms, especially *Neisseria gonorrhoeae*, between collection and culturing. Addition of charcoal to this medium neutralizes metabolic products which may be toxic to *Neisseria gonorrhoeae*. The presence of charcoal may interfere with the Gram stain.

Amies Transport Medium without Charcoal

Composition per liter:

Agar	4.0g
NaCl	3.0g
Na_2HPO_4	1.15g
Sodium thioglycollate	1.0g
KCl	0.2g
$CaCl_2 \cdot 2H_2O$	0.1g
$MgCl_2 \cdot 6H_2O$	0.1g
KH_2PO_4	0.2g

pH 7.2 ± 0.2 at 25°C

Source: This medium is available as a premixed powder from Difco Laboratories.

Preparation of Medium: Add components to distilled/deionized water and bring volume to 1.0L. Mix thoroughly. Gently heat and bring to boiling. Distribute into flasks or tubes. Autoclave for 20 min at 15 psi pressure–121°C.

Use: For the transport of swab specimens to prolong the survival of microorganisms, especially *Neisseria gonorrhoeae*, between collection and culturing.

Amies Transport Medium without Charcoal

Composition per liter:

NaCl	8.0g
Agar	3.6g
Na_2HPO_4	1.15g
Sodium thioglycollate	1.0g
KCl	0.2g
$CaCl_2 \cdot 2H_2O$	0.1g
$MgCl_2 \cdot 6H_2O$	0.1g
KH_2PO_4	0.2g

pH 7.2 ± 0.2 at 25°C

Source: This medium is available as a premixed powder from BBL Microbiology Systems.

Preparation of Medium: Add components to distilled/deionized water and bring volume to 1.0L. Mix thoroughly. Gently heat and bring to boiling. Distribute into flasks or tubes. Autoclave for 20 min at 15 psi pressure–121°C.

Use: For the transport of swab specimens to prolong the survival of microorganisms, especially *Neisseria gonorrhoeae*, between collection and culturing.

Anaerobic Agar

Composition per liter:

Agar	20.0g
Pancreatic digest of casein	20.0g
Glucose	10.0g
NaCl	5.0g
Sodium thioglycollate	2.0g
Sodium formaldehyde sulfoxylate	1.0g
Methylene Blue	2.0mg

pH 7.2 ± 0.2 at 25°C

Source: This medium is available as a premixed powder from Difco Laboratories.

Preparation of Medium: Add components to distilled/deionized water and bring volume to 1.0L. Mix thoroughly. Gently heat and bring to boiling. Adjust pH to 7.2. Distribute into tubes until medium is 3 inches deep. Autoclave for 15 min at 15 psi pressure–121°C.

Use: For the cultivation of a variety of anaerobic microorganisms, especially *Clostridium* species.

Anaerobic Agar

Composition per liter:

Pancreatic digest of casein	17.5g
Agar	15.0g
Glucose	10.0g
Papaic digest of soybean meal	2.5g
NaCl	2.5g
Sodium thioglycollate	2.0g
Sodium formaldehyde sulfoxylate	1.0g
L-Cystine	0.4g
Methylene Blue	2.0mg

pH 7.2 ± 0.2 at 25°C

Source: This medium is available as a premixed powder from BBL Microbiology Systems.

Preparation of Medium: Add components to distilled/deionized water and bring volume to 1.0L. Mix thoroughly. Gently heat and bring to boiling. Autoclave for 15 minutes at 15 psi–121°C. Use with Brewer anaerobic Petri dishes or in tubes or ordinary plates and incubate in anaerobic jars.

Use: For the cultivation of a variety of anaerobic microorganisms, especially *Clostridium* species.

Anaerobic Broth

Composition per liter:

Pancreatic digest of casein	17.5g
Glucose	10.0g
NaCl	2.5g
Papaic digest of soybean meal	2.5g
Sodium thioglycollate	2.0g
Sodium formaldehyde sulfoxylate	1.0g
L-Cystine	0.4g
Methylene Blue	2.0mg

pH 7.2 ± 0.2 at 25°C

Preparation of Medium: Add components to distilled/deionized water and bring volume to 1.0L. Mix thoroughly. Gently heat and bring to boiling. Distribute into tubes or flasks. Autoclave for 15 min at 15 psi pressure–121°C.

Use: For the cultivation of a variety of anaerobic and microaerophilic microorganisms.

Anaerobic CNA Agar (Anaerobic Colistin Nalidixic Acid Agar)

Composition per liter:

Agar	13.0g
Pancreatic digest of casein	12.0g
Peptic digest of animal tissue	5.0g
Yeast extract	3.0g
Beef extract	3.0g
NaCl	5.0g
Cornstarch	1.0g
Glucose	1.0g
L-Cysteine·HCl·H_2O	0.5g
Vitamin K_1	10.0mg
Hemin	10.0mg
Colistin	10.0mg
Nalidixic acid	10.0mg
Sheep blood, defibrinated	50.0mL

Source: This medium is available as a premixed powder from BBL Microbiology Systems.

Preparation of Medium: Add components, except sheep blood, to distilled/deionized water and bring volume to 950.0mL. Mix thoroughly. Gently heat and bring to boiling. Autoclave for 15 minutes at 15 psi–121°C. Cool to 45–50°C. Aseptically add 50.0mL of sterile, defibrinated sheep blood. Mix thoroughly. Pour into sterile Petri dishes.

Use: For the selective isolation of anaerobic streptococci and other anaerobic Gram-positive bacteria.

Anaerobic Egg Yolk Agar

Composition per 1080mL:

Agar	20.0g
Proteose peptone	20.0g
NaCl	5.0g
Pancreatic digest of casein	5.0g
Yeast extract	5.0g
Egg yolk emulsion, 50%	80.0mL

pH 7.0 ± 0.2 at 25°C

Egg Yolk Emulsion, 50%:
Composition per 100mL:

Chicken egg yolks	11
Whole chicken egg	1
NaCl (0.9% solution)	50.0mL

Preparation of Egg Yolk Emulsion, 50%: Soak eggs with 1:100 dilution of saturated mercuric chloride solution for 1 min. Crack 11 eggs and separate yolks from whites. Mix egg yolks with 1 chicken egg. Measure 50.0mL of egg yolk emulsion and add to 50.0mL of 0.9% NaCl solution. Mix thoroughly. Filter sterilize. Warm to 45° to 50°C.

Preparation of Medium: Add components, except egg yolk emulsion, 50%, to distilled/deionized water and bring volume to 1.0L. Mix thoroughly. Gently heat and bring to boiling. Autoclave for 15 min at 15 psi pressure–121°C. Cool to 45° to 50°C. Aseptically add 80.0mL of sterile egg yolk emulsion, 50%. Mix thoroughly. Pour into sterile Petri dishes or distribute into sterile tubes. Allow plates to dry at 35°C for 24 hr.

Use: For the cultivation and identification of *Clostridium* species. Reactions include lecithinase, lipase, and proteolysis. Facilitates the detection of lecithinase (zone of precipitation surrounding colonies) and lipase (presence of an iridescent film in and around colonies; colonies show "pearl-like" appearance under conditions of reflected light).

Anaerobic Egg Yolk Agar

Composition per liter:

Agar	20.0g
Proteose peptone	20.0g
Pancreatic digest of casein	5.0g
NaCl	5.0g
Yeast extract	5.0g
Egg yolk emulsion, 50%	20.0mL

pH 7.0 ± 0.2 at 25°C

Egg Yolk Emulsion, 50%:
Composition per 100mL:

Chicken egg yolks	2
NaCl (0.9% solution)	10.0mL

Preparation of Egg Yolk Emulsion, 50%: Soak eggs with 1:100 dilution of saturated mercuric chloride solution for 1 min. Crack eggs and separate yolks from whites. Measure 10.0mL of egg yolk emulsion and add to 10.0mL of 0.9% NaCl solution. Mix thoroughly. Filter sterilize. Warm to 45° to 50°C.

Preparation of Medium: Add components, except egg yolk emulsion, to distilled/deionized water and bring volume to 980.0mL. Mix thoroughly. Gently heat and bring to boiling. Autoclave for 15 min at 15 psi pressure–121°C. Cool to 45° to 50°C. Aseptically add sterile egg yolk emulsion. Mix thoroughly. Pour into sterile Petri dishes. Allow plates to dry at 35°C for 24 hr.

Use: For the cultivation and identification of *Clostridium* species. Reactions include lecithinase, lipase, and proteolysis. Facilitates the detection of lecithinase (zone of precipitation surrounding colonies) and lipase (presence of an iridescent film in and around colonies; colonies show "pearl-like" appearance under conditions of reflected light).

Anaerobic LKV Blood Agar

Composition per liter:

Agar	15.0g
Pancreatic digest of casein	13.0g
Peptic digest of animal tissue	10.0g
NaCl	5.0g
Yeast extract	2.0g
Glucose	1.0g
$NaHSO_3$	0.1g
Sheep blood, laked	50.0mL
Antibiotic solution	10.0mL
Hemin	1.0mL
Vitamin K_1 solution	1.0mL

pH 7.1–7.8 at 25°C

Source: This medium is available as a premixed powder from Difco Laboratories.

Antibiotic Solution:
Composition per 10mL:

Kanamycin	0.075g
Vancomycin	7.5mg

Preparation of Antibiotic Solution: Add components to distilled/deionized water and bring volume to 10.0mL. Mix thoroughly. Filter sterilize.

Vitamin K_1 Solution:
Composition per 100mL:

Vitamin K_1	0.1g
Ethanol	99.0mL

Preparation of Vitamin K_1 Solution: Add vitamin K_1 to 99.0mL of absolute ethanol. Mix thoroughly.

Hemin Solution:
Composition per 100mL:

Hemin	0.01g
NaOH (1*N* solution)	20.0mL

Preparation of Hemin Solution: Add hemin to 20.0mL of 1*N* NaOH solution. Mix thoroughly. Bring volume to 100.0mL with distilled/deionized water.

Preparation of Medium: Add components—except sheep blood, antibiotic solution, and vitamin K_1 solution—to distilled/deionized water and bring volume to 939.0mL. Mix thoroughly. Gently heat and bring to boiling. Autoclave for 15 min at 15 psi pressure–121°C. Cool to 45° to 50°C. Aseptically add 50.0mL of sterile sheep blood, 10.0mL of sterile antibiotic solution, and 1.0mL of sterile vitamin K_1 solution. Mix thoroughly. Pour into sterile Petri dishes or distribute into sterile tubes.

Use: For the isolation and cultivation of anaerobic Gram-negative microorganisms, expecially *Bacteroides, Prevatella, Fusobacterium, and Veillonella* species from specimens containing mixed aerobic and anaerobic bacteria. Vancomycin may inhibit *Porphyromonas* species.

Andrade's Broth

Composition per liter:

Pancreatic digest of gelatin	10.0g
NaCl	5.0g
Beef extract	3.0g
Andrade's indicator	10.0mL
Carbohydrate solution	50.0mL

pH 7.4 ± 0.2 at 25°C

Source: Available as a prepared medium from BBL Microbiology Systems, in tubes containing adonitol, arabinose, cellobiose, glucose, dulcitol, fructose, galactose, inositol, lactose, maltose, mannitol, raffinose, rhamnose, salicin, sorbitol, sucrose, trehalose, or xylose.

Andrade's Indicator

Composition per 100mL:

NaOH (1*N* solution)	16.0mL
Acid Fuchsin	0.1 g

Preparation of Andrade's Indicator: Add Acid Fuchsin to NaOH solution and bring volume to 100.0mL with distilled/deionized water.

Carbohydrate Solution:

Composition per 100mL:

Carbohydrate	10.0g

Preparation of Carbohydrate Solution: Add carbohydrate to distilled/deionized water and bring volume to 100.0mL. Adonitol, arabinose, cellobiose, glucose, dulcitol, fructose, galactose, inositol, lactose, maltose, mannitol, raffinose, rhamnose, salicin, sorbitol, sucrose, trehalose, xylose, or other carbohydrates may be used. Mix thoroughly. Filter sterilize.

Preparation of Medium: Add components, except carbohydrate solution, to distilled/deionized water and bring volume to 1.0L. Mix thoroughly. Gently heat and bring to boiling. Distribute in 10.0mL volumes into test tubes containing inverted Durham tubes. Autoclave for 15 minutes at 15 psi–121°C. Cool to 25°C. Add 0.5mL of sterile carbohydrate solution to each tube.

Caution: Acid Fuchsin is a potential carcinogen and care must be taken to avoid inhalation of the powdered dye and contact with the skin.

Use: For the determination of carbohydrate fermentation reactions of microorganisms, particularly members of the Enterobacteriaceae. A Durham tube is used to collect gas produced during the fermentation reaction. Acid production is indicated by a pink color.

Antibiotic Medium 5 (Streptomycin Assay Agar with Yeast Extract)

Composition per liter:

Agar	15.0g
Pancreatic digest of gelatin	6.0g
Yeast extract	3.0g
Beef extract	1.5g

pH 7.9 ± 0.1 at 25°C

Source: This medium is available as a premixed powder from BBL Microbiology Systems, Difco Laboratories and Oxoid Unipath.

Preparation of Medium: Add components to distilled/deionized water and bring volume to 1.0L. Mix thoroughly. Gently heat and bring to boiling. Distribute into tubes or flasks. Autoclave for 15 minutes at 15 psi–121°C. Pour into sterile Petri dishes.

Use: For antibiotic assay testing. For the streptomycin assay using the cylinder plate technique and *Bacillus subtilis* as test organism.

Antibiotic Medium 6

Composition per liter:

Pancreatic digest of casein	17.0g
NaCl	5.0g
Papaic digest of soybean meal	3.0g
Glucose	2.5g
K_2HPO_4	2.5g
$MnSO_4 \cdot H_2O$	0.03g

pH 7.0 ± 0.1 at 25°C

Source: This medium is available as a premixed powder from Difco Laboratories.

Preparation of Medium: Add components to distilled/deionized water and bring volume to 1.0L. Mix thoroughly. Gently heat and bring to boiling. Distribute into tubes or flasks. Autoclave for 15 minutes at 15 psi–121°C. Pour into sterile Petri dishes.

Use: For antibiotic assay testing.

Antibiotic Medium 7

Composition per liter:

Agar	15.0g
Pancreatic digest of gelatin	6.0g
Yeast extract	3.0g
Beef extract	1.5g

pH 7.0 ± 0.1 at 25°C

Preparation of Medium: Add components to distilled/deionized water and bring volume to 1.0L. Mix thoroughly. Gently heat and bring to boiling.Adjust pH to 7.0. Distribute into tubes or flasks. Autoclave for 15 minutes at 15 psi–121°C. Pour into sterile Petri dishes.

Use: For use as a base layer in antibiotic assay testing. Especially useful for the plate assay of bacitracin and penicillin G.

Antibiotic Medium 8 (Base Agar with Low pH)

Composition per liter:

Agar	15.0g
Pancreatic digest of gelatin	6.0g
Yeast extract	3.0g
Beef extract	1.5g

pH 5.9 ± 0.1 at 25°C

Source: This medium is available as a premixed powder from Difco Laboratories, and BBL Microbiology Systems.

Preparation of Medium: Add components to distilled/deionized water and bring volume to 1.0L. Mix thoroughly. Gently heat and bring to boiling. Distribute into tubes or flasks. Autoclave for 15 minutes at 15 psi–121°C. Pour into sterile Petri dishes.

Use: For antibiotic assay testing. For use as the base agar and the seed agar in the plate assay of tetracycline. For use as the seed agar in the plate assay of vancomycin, mitomycin and mithramycin.

Antibiotic Medium 9 (Polymyxin Base Agar)

Composition per liter:

Agar	20.0g
Pancreatic digest of casein	17.0g
NaCl	5.0g
Papaic digest of soybean meal	3.0g
K_2HPO_4	2.5g
Glucose	2.5g

pH 7.2 ± 0.1 at 25°C

Source: This medium is available as a premixed powder from BBL Microbiology Systems and Difco Laboratories.

Preparation of Medium: Add components to distilled/deionized water and bring volume to 1.0L. Mix thoroughly. Gently heat and bring to boiling. Distribute into tubes or flasks. Autoclave for 15 minutes at 15 psi–121°C. Pour into sterile Petri dishes.

Use: For antibiotic assay testing. For base agar for the plate assay of carbenicillin, colistimethate, and polymyxin B.

Antibiotic Medium 10 (Polymyxin Seed Agar)

Composition per liter:

Pancreatic digest of casein	17.0g
Agar	12.0g
Polysorbate 80	10.0g
NaCl	5.0g
Papaic digest of soybean meal	3.0g
K_2HPO_4	2.5g
Glucose	2.5g

pH 7.3 ± 0.2 at 25°C

Source: This medium is available as a premixed powder from BBL Microbiology Systems and Difco Laboratories.

Preparation of Medium: Add components to distilled/deionized water and bring volume to 1.0L. Mix thoroughly. Gently heat and bring to boiling. Distribute into tubes or flasks. Autoclave for 15 minutes at 15 psi–121°C. Pour into sterile Petri dishes.

Use: For antibiotic assay testing. For seed agar for the plate assay of carbenicillin, colistimethate, and polymyxin B.

Antibiotic Medium 11 (Neomycin Assay Agar)

Composition per liter:

Agar	15.0g
Pancreatic digest of gelatin	6.0g
Pancreatic digest of casein	4.0g
Yeast extract	3.0g
Beef extract	1.5g
Glucose	1.0g

pH 8.0 ± 0.1 at 25°C

Source: This medium is available as a premixed powder from BBL Microbiology Systems, Difco Laboratories and Oxoid Unipath.

Preparation of Medium: Add components to distilled/deionized water and bring volume to 1.0L. Mix thoroughly. Gently heat and bring to boiling. Distribute into tubes or flasks. Autoclave for 15 minutes at 15 psi–121°C. Pour into sterile Petri dishes.

Use: For antibiotic assay testing. For base agar and seed agar for the plate assay to test the effectiveness of neomycin sulfate, amoxicillin, ampicillin, clindamycin, cyclacillin, erythromycin, gentamycin, neomycin, oleandomycin, and sisomycin.

Antibiotic Medium 12

Composition per liter:

Agar	25.0g
Peptone	10.0g

Glucose 10.0g
NaCl 10.0g
Yeast extract 5.0g
Beef extract 2.5g

pH 6.0 ± 0.1 at 25°C

Source: This medium is available as a premixed powder from Difco Laboratories.

Preparation of Medium: Add components to distilled/deionized water and bring volume to 1.0L. Mix thoroughly. Gently heat and bring to boiling. Distribute into tubes or flasks. Autoclave for 15 minutes at 15 psi–121°C. Pour into sterile Petri dishes.

Use: For antibiotic assay effectiveness testing.

Antibiotic Medium 13 (Sabouraud Liquid Broth, Modified) (Fluid Saboraud Medium)

Composition per liter:

Glucose 20.0g
Pancreatic digest of casein 5.0g
Peptic digest of animal tissue 5.0g

pH 5.7 ± 0.1 at 25°C

Source: This medium is available as a premixed powder from BBL Microbiology Systems and Difco Laboratories.

Preparation of Medium: Add components to distilled/deionized water and bring volume to 1.0L. Mix thoroughly. Gently heat and bring to boiling. Distribute into tubes or flasks. Autoclave for 15 minutes at 15 psi–121°C. Pour into sterile Petri dishes.

Use: For testing the effectivness of antibiotics on yeast and molds.

Antibiotic Medium 19 (Nystatin Assay Agar)

Composition per liter:

Agar 23.5g
Glucose 10.0g
NaCl 10.0g
Pancreatic digest of gelatin 9.4g
Yeast extract 4.7g
Beef extract 2.4g

pH 6.1 ± 0.2 at 25°C

Source: This medium is available as a premixed powder from BBL Microbiology Systems and also from Difco Laboratories.

Preparation of Medium: Add components to distilled/deionized water and bring volume to 1.0L. Mix thoroughly. Gently heat and bring to boiling. Distribute into tubes or flasks. Autoclave for 15 minutes at 15 psi–121°C. Pour into sterile Petri dishes.

Use: For assaying the mycostatic activity of pharmaceutical preparations. For seed agar for the plate assay to test the effectiveness of nystatin, amphotericin B, and natamycin.

Antibiotic Medium 20

Composition per liter:

Glucose 11.0g
Pancreatic digest of casein 10.0g
Yeast extract 6.5g
Pancreatic digest of gelatin 5.0g
K_2HPO_4 ....3.68g
NaCl 3.5g
Beef extract 1.5g
KH_2PO_4 ....1.32g

pH 6.6 ± 0.2 at 25°C

Preparation of Medium: Add components to distilled/deionized water and bring volume to 1.0L. Mix thoroughly. Gently heat and bring to boiling. Distribute into tubes or flasks. Autoclave for 15 minutes at 15 psi–121°C. Pour into sterile Petri dishes.

Use: For assaying the mycostatic activity of pharmaceutical preparations.

Antibiotic Medium 21

Composition per liter:

Glucose 11.0g
Pancreatic digest of gelatin 5.0g
K_2HPO_4 ....3.68g
NaCl 3.5g
Yeast extract 1.5g
Beef extract 1.5g
KH_2PO_4 ....1.32g

pH 6.6 ± 0.2 at 25°C

Preparation of Medium: Add components to distilled/deionized water and bring volume to 1.0L. Mix thoroughly. Gently heat and bring to boiling. Distribute into tubes or flasks. Autoclave for 15 minutes at 15 psi–121°C. Pour into sterile Petri dishes.

Use: For assaying the mycostatic activity of pharmaceutical preparations.

Antibiotic Sulfonamide Sensitivity Test Agar (ASS Agar)

Composition per liter:

Agar 12.0g
Proteose peptone 10.0g

Beef extract 10.0g
NaCl 3.0g
Glucose 2.0g
Na_2HPO_4 ....2.0g
Sodium acetate 1.0g
Adenine 0.01g
Guanine 0.01g
Uracil 0.01g
Xanthine 0.01g
pH 7.4 ± 0.2 at 25°C

Preparation of Medium: Add components to distilled/deionized water and bring volume to 1.0L. Mix thoroughly. Gently heat and bring to boiling. Distribute into tubes or flasks. Autoclave for 15 minutes at 15 psi–121°C. Pour into sterile Petri dishes or leave in tubes.

Use: For testing the antimicrobial effectiveness of antibiotics and sulfonamides. Also, for detecting the presence of antimicrobial substances in milk, urine and other fluids.

Arylsulfatase Agar (Wayne Sulfatase Agar)

Composition per liter:
Agar 15.0g
Na_2HPO_4 ....2.5g
L-Asparagine 1.0g
KH_2PO_4 ....1.0g
K_2HPO_4 ....1.0g
Trisodium phenolphthalein sulfate 0.65g
Pancreatic digest of casein 0.5g
Ferric ammonium citrate 0.05g
$MgSO_4·7H_2O$ 0.01g
$CaCl_2·2H_2O$ 0.5mg
$ZnSO_4·7H_2O$ 0.1mg
$CuSO_4$ 0.1mg
Glycerol 10.0mL
pH 7.0 ± 0.2 at 25°C

Source: This medium is available as a premixed powder from BBL Microbiology Systems.

Preparation of Medium: Add glycerol to approximately 800.0mL of distilled/deionized water. Mix thoroughly. Add remaining components and bring volume to 1.0L with distilled/deionized water. Mix thoroughly. Gently heat and bring to boiling. Distribute into tubes. Autoclave for 15 minutes at 15 psi–121°C. Cool tubes in an upright position.

Use: For the biochemical differentiation of species of *Mycobacterium*. Inoculate tubes with *Mycobacterium* cultures and incubate aerobically at 35°C for 3–14 days. Add 0.5–1.0mL of 2*N* Na_2CO_3 to each tube and observe color change within 30 minutes. Development of a pink color is indicative of *M. fortuitum-chelonae* complex. *M. tuberculosis* gives a negative reaction.

Aspergillus Differential Medium

Composition per liter:
Agar 15.0g
Pancreatic digest of casein 15.0g
Yeast extract 10.0g
Ferric citrate 0.5g

Preparation of Medium: Add components to distilled/deionized water and bring volume to 1.0L. Mix thoroughly. Gently heat and bring to boiling. Distribute into tubes in 7.0mL volumes. Autoclave for 15 min at 15 psi pressure–121°C. Allow tubes to cool in a slanted position.

Use: For the cultivation and differentiation of *Aspergillus flavus*. *A. flavus* appears as bright orange colonies.

Bacteroides Bile Esculin Agar (BBE Agar)

Composition per liter:
Oxgall 20.0g
Pancreatic digest of casein 15.0g
Agar 15.0g
Papaic digest of soybean meal 5.0g
NaCl 5.0g
Esculin 1.0g
Ferric ammonium citrate 0.5g
Gentamicin solution 2.5mL
Hemin solution 2.5mL
Vitamin K_1 solution 1.0mL
pH 7.0 ± 0.2 at 25°C

Source: This medium is available as a premixed powder from BBL Microbiology Systems.

Gentamicin Solution:
Composition per 10mL:
Gentamicin 0.4mg

Preparation of Gentamicin Solution: Add gentamicin to 10.0mL of distilled/deionized water. Mix thoroughly. Filter sterilize.

Hemin Solution:
Composition per 100mL:
Hemin 0.5g
NaOH (1*N* solution) 10.0mL

Preparation of Hemin Solution: Add components to 100.0mL of distilled/deionized water. Mix thoroughly. Autoclave for 15 min at 15 psi pressure–121°C. Cool to 45° to 50°C.

Vitamin K_1 Solution:
Composition per 100mL:

Vitamin K_1	1.0g
Ethanol	99.0mL

Preparation of Vitamin K_1 Solution: Add vitamin K_1 to 99.0mL of absolute ethanol. Mix thoroughly. Filter sterilize.

Preparation of Medium: Add components, except hemin solution, gentamicin solution and vitamin K_1 solution to distilled/deionized water and bring volume to 994.0mL. Mix thoroughly. Gently heat and bring to boiling. Autoclave for 15 min at 15 psi pressure–121°C. Cool to 45–50°C. Aseptically add 2.5mL sterile hemin solution, 2.5mL sterilie gentamicin solution and 1.0mL sterile vitamin K_1 solution.

Use: For the selection and presumptive identification of the *Bacteriodes fragilis* group. Also, for the differentiation of *Bacteroides* species based on hydrolysis of esculin and presence of catalase. After incubation for 48 hours, bacteria of the *B. fragilis* group appear as gray, circular, raised colonies larger than 1mm. Esculin hydrolysis is indicated by the presence of a blackened zone around the colonies.

Balamuth Medium

Composition per 200mL:

Dehydrated egg yolk	36.0g
Dried liver concentrate	1.0g
Rice starch	0.2g
Potassium phosphate buffer, pH 7.5	125.0mL
NaCl solution	125.0mL

pH 7.3 ± 0.2 at 25°C

NaCl Solution
Composition per 200mL:

NaCl	1.6g

Preparation of NaCl Solution: Add NaCl to distilled/deionized water and bring volume to 200.0mL. Mix thoroughly.

Potassium Phosphate Buffer, 0.067M
Composition per 200mL:

K_2HPO_4 (1M solution)	8.6mL
KH_2PO_4 (1M solution)	4.66mL

Preparation of Potassium Phosphate Buffer: Combine the K_2HPO_4 and KH_2PO_4 solutions. Bring volume to 200.0mL with distilled/deionized water. Adjust pH to 7.5.

Preparation of Medium: Add dehydrated egg yolk to 36.0mL of distilled/deionized water. Add 125.0mL of 0.8% NaCl. Mix thoroughly in a blender. Heat in a covered, double boiler until infusion reaches 80°C and maintain at this temperature for 20 min. Add 20.0mL of distilled/deionized H_2O. Filter through a layer of cheesecloth. To 90–100.0mL of filtrate add 0.8% NaCl solution to bring volume to 125.0mL. Autoclave for 20 min at 15 psi pressure–121°C. Cool to 4°C. Filter. To filtrate add an equal volume of 0.067M potassium phosphate buffer, pH 7.5. Add1.0g of dried liver concentrate. Mix thoroughly. Distribute into tubes or flasks in 10.0mL volumes. Autoclave for 20 min at 15 psi pressure–121°C. Prior to inoculation, add 0.01g of rice starch to each tube.

Use: For the cultivation and maintenance of *Entamoeba histolytica* and other intestinal protozoa.

Basal Mineral Medium

Composition per liter:

NH_4Cl	0.80g
K_2HPO_4	0.70g
$MgSO_4 \cdot 7H_2O$	0.01g
Disodium EDTA	9.2mg
$FeSO_4 \cdot 7H_2O$	7.0mg
$CaSO4 \cdot 2H_2O$	2.0mg
H_3BO_3	0.1mg
$ZnSO_4 \cdot 7H_2O$	0.1mg
$MnSO_4 \cdot 4H_2O$	0.02mg
$Co(NO_3)_2$	0.01mg
$NaMoO_4 \cdot 2H_2O$	0.01mg
$CuSO_4 \cdot 5H_2O$	0.5μg

Preparation of Medium: Add components to distilled/deionized water and bring volume to 1.0L. Mix thoroughly. Filter sterilize.

Use: For the cultivation of *Beggiatoa* species.

BCYEα with Alb (Buffered Charcoal Yeast Extract Agar with Albumin)

Composition per liter:

Agar	15.0g
Yeast extract	10.0g
ACES buffer (2-[(2-Amino-2-oxoethyl)-amino]-ethane sulfonic acid)	10.0g
Charcoal, activated	2.0g
α-Ketogluatrate	1.0g
Bovine serum albumin solution	10.0mL
Cysteine·HCl·H_2O solution	10.0mL
$Fe_4(P_2O_7)_3 \cdot 9H_2O$ solution	10.0mL

pH 6.9 ± 0.2 at 25°C

Bovine Serum Albumin Solution:
Composition per 10mL:

Bovine serum albumin	0.1g

Preparation of Bovine Serum Albumin Solution: Add bovine serum albumin to distilled/deionized water and bring volume to 10.0mL. Mix thoroughly. Filter sterilize.

Cysteine·HCl·H_2O Solution:
Composition per 10mL:

L-Cysteine·HCl·H_2O 0.4g

Preparation of Cysteine·HCl·H_2O Solution: Add cysteine·HCl·H_2O to distilled/deionized water and bring volume to 10.0mL. Mix thoroughly. Filter sterilize.

$Fe_4(P_2O_7)_3$·$9H_2O$ Solution:
Composition per 10mL:

$Fe_4(P_2O_7)_3$·$9H_2O$ 0.25g

Preparation of $Fe_4(P_2O_7)_3$·$9H_2O$ Solution: Add $Fe_4(P_2O_7)_3$·$9H_2O$ to distilled/deionized water and bring volume to 10.0mL. Mix thoroughly. Filter sterilize.

Preparation of Medium: Add components—except cysteine·HCl·H_2O solution, $Fe_4(P_2O_7)_3$·$9H_2O$ solution, and bovine serum albumin solution—to distilled/deionized water and bring volume to 970.0mL. Mix thoroughly. Adjust medium to pH 6.9 with 1*N* KOH. Heat gently and bring to boil for 1 minute. Autoclave for 15 min at 15 psi pressure–121°C. Cool to 50–55°C. Aseptically add the cysteine·HCl·H_2O solution, $Fe_4(P_2O_7)_3$·$9H_2O$ solution and 10.0mL of sterile bovine serum albumin solution. Mix thoroughly. Pour into sterile Petri dishes with constant agitation to keep charcoal in suspension.

Use: For the isolation, cultivation, and maintenance of *Legionella pneumophila* and other *Legionella* species from environmental and clinical specimens.

BCYEα without L-Cysteine (Buffered Charcoal Yeast Extract Agar without L-Cysteine)

Composition per liter:

Agar 15.0g
Yeast extract 10.0g
ACES buffer (2-[(2-Amino-2-oxoethyl)-amino]-ethane sulfonic acid) 10.0g
Charcoal, activated 2.0g
α-Ketogluatrate 1.0g
$Fe_4(P_2O_7)_3$·$9H_2O$ solution 10.0mL
pH 6.9 ± 0.2 at 25°C

$Fe_4(P_2O_7)_3$·$9H_2O$ Solution:
Composition per 10mL:

$Fe_4(P_2O_7)_3$·$9H_2O$ 0.25g

Preparation of $Fe_4(P_2O_7)_3$·$9H_2O$ Solution: Add $Fe_4(P_2O_7)_3$·$9H_2O$ to distilled/deionized water and bring volume to 10.0mL. Mix thoroughly. Filter sterilize.

Preparation of Medium: Add components, except $Fe_4(P_2O_7)_3$·$9H_2O$ solution, to distilled/deionized water and bring volume to 990.0mL. Mix thoroughly. Adjust medium to pH 6.9 with 1*N* KOH. Heat gently and bring to boil for 1 minute. Autoclave for 15 min at 15 psi pressure–121°C. Cool to 50–55°C. Aseptically add 10.0mL of sterile $Fe_4(P_2O_7)_3$·$9H_2O$ solution. Mix thoroughly. Pour into sterile Petri dishes with constant agitation to keep charcoal in suspension.

Use: For the isolation, cultivation, and maintenance of *Legionella pneumophila* and other *Legionella* species from environmental and clinical specimens. Many other bacteria also grow on this medium so that further characterization is necessary in order to confirm *Legionella* species.

BCYE Differential Agar (Buffered Charcoal Yeast Extract Differential Agar)

Composition per liter:

Agar 15.0g
Yeast extract 10.0g
ACES buffer (2-[(2-Amino-2-oxoethyl)-amino]-ethane sulfonic acid) 10.0g
Charcoal, activated 2.0g
α-Ketogluatrate 1.0g
L-Cysteine·HCl·H_2O 0.4g
$Fe_4(P_2O_7)_3$·$9H_2O$ 0.25g
Bromcresol Purple 0.01g
Bromthymol Blue 0.01g
pH 6.9 ± 0.2 at 25°C

Source: This medium is available as a premixed powder from BBL Microbiology Systems.

Preparation of Medium: Add components, except cysteine, to distilled/deionized water and bring volume to 1.0L. Mix thoroughly. Adjust medium to pH 6.9 with 1*N* KOH. Heat gently and bring to boil for 1 minute. Autoclave for 15 min at 15 psi pressure–121°C. Cool to 50–55°C. Add 4.0mL of a 10% solution of L-cysteine·HCl·H_2O which has been filter-sterilized. Mix thoroughly. Pour into sterile Petri dishes with constant agitation to keep charcoal in suspension.

Use: For the isolation, cultivation, and maintenance of *Legionella pneumophila* and other *Legionella* species from environmental and clinical specimens. For the presumptive differential identification of *Legionella* species based on colony color and morphology. *Legionella pneumophila* appear as light blue/green colonies. *Legionella micdadei* appear as blue/gray or dark blue colonies.

BCYE Medium, Biphasic Blood Culture (Buffered Charcoal Yeast Extract Medium, Diphasic Blood Culture)

Composition per liter:

Agar phase 1.0L
Broth phase 1.0L

pH 6.9 ± 0.2 at 25°C

Agar Phase:

Composition per liter:

Agar 20.0g
ACES buffer (2-[(2-Amino-2-oxoethyl)-amino]-ethane sulfonic acid) 10.0g
Yeast extract 10.0g
Charcoal, activated, acid washed 4.0g
KOH 2.80g
α-Ketoglutarate 1.0g
Cysteine·HCl·H_2O solution 10.0mL
$Fe_4(P_2O_7)_3 \cdot 9H_2O$ solution 10.0mL

Cysteine·HCl·H_2O Solution:

Composition per 10mL:

L-Cysteine·HCl·H_2O 0.4g

Preparation of Cysteine·HCl·H_2O Solution: Add cysteine·HCl·H_2O to distilled/deionized water and bring volume to 10.0mL. Mix thoroughly. Filter sterilize.

$Fe_4(P_2O_7)_3 \cdot 9H_2O$ Solution:

Composition per 10mL:

$Fe_4(P_2O_7)_3 \cdot 9H_2O$ 0.25g

Preparation of $Fe_4(P_2O_7)_3 \cdot 9H_2O$ Solution: Add $Fe_4(P_2O_7)_3 \cdot 9H_2O$ to distilled/deionized water and bring volume to 10.0mL. Mix thoroughly. Filter sterilize.

Preparation of Agar Phase: Add components, except cysteine·HCl·H_2O solution and $Fe_4(P_2O_7)_3$ solution, to distilled/deionized water and bring volume to 980.0mL. Mix thoroughly. Adjust medium to pH 6.9 with 1*N* KOH. Heat gently and bring to boiling for 1 minute. Autoclave for 15 min at 15 psi pressure–121°C. Cool to 50–55°C. Aseptically add the cysteine·HCl·H_2O solution and $Fe_4(P_2O_7)_3 \cdot 9H_2O$ solution. Mix thoroughly.

Broth Phase:

Composition per liter:

ACES buffer (2-[(2-Amino-2-oxoethyl)-amino]-ethane sulfonic acid) 10.0g
Yeast extract 10.0g
Charcoal, activated, acid washed 4.0g
KOH 2.40g
α-Ketoglutarate 1.0g
Sodium polyaneolsulfonate 0.30g
Cysteine·HCl·H_2O solution 10.0mL
$Fe_4(P_2O_7)_3 \cdot 9H_2O$ solution 10.0mL

Cysteine·HCl·H_2O Solution:

Composition per 10mL:

L-Cysteine·HCl·H_2O 0.4g

Preparation of Cysteine·HCl·H_2O Solution: Add cysteine·HCl·H_2O to distilled/deionized water and bring volume to 10.0mL. Mix thoroughly. Filter sterilize.

$Fe_4(P_2O_7)_3 \cdot 9H_2O$ Solution:

Composition per 10mL:

$Fe_4(P_2O_7)_3 \cdot 9H_2O$ 0.25g

Preparation of $Fe_4(P_2O_7)_3 \cdot 9H_2O$ Solution: Add $Fe_4(P_2O_7)_3 \cdot 9H_2O$ to distilled/deionized water and bring volume to 10.0mL. Mix thoroughly. Filter sterilize.

Preparation of Broth Phase: Add components, except cysteine·HCl·H_2O solution and $Fe_4(P_2O_7)_3$ solution, to distilled/deionized water and bring volume to 980.0mL. Mix thoroughly. Adjust medium to pH 6.9 with 1*N* KOH. Heat gently and bring to boiling for 1 minute. Autoclave for 15 min at 15 psi pressure–121°C. Cool to 50–55°C. Aseptically add the cysteine·HCl·H_2O solution and $Fe_4(P_2O_7)_3 \cdot 9H_2O$ solution. Mix thoroughly.

Preparation of Medium: Aseptically distribute cooled sterile agar phase into sterile blood culture bottles in 100.0mL volumes. Allow bottles to cool in a slanted position. Aseptically add 50.0mL of sterile broth phase to each blood culture bottle.

Use: For the isolation and cultivation of *Legionella pneumophila* and other *Legionella* species from blood samples.

Beef Extract Broth (ATCC Medium 225)

Composition per liter:

Beef extract 10.0g
Peptone 10.0g
NaCl 5.0g

pH 7.2 ± 0.2 at 25°C

Preparation of Medium: Add components to distilled/deionized water and bring volume to 1.0L. Mix thoroughly. Heat gently and bring to boiling. Distribute into tubes or flasks. Autoclave for 15 min at 15 psi pressure–121°C.

Use: For the cultivation of a wide variety of microorganisms including *Alcaligenes* species, *Pseudomonas aeruginosa* and *Bacillus sphaericus*. Also, for the detection of germ tubes produced by *Candida albicans*.

Beef Infusion Agar

Composition per liter:

Ground defatted beef....................................453.6g
Agar..20.0g
Peptone...10.0g
NaCl..5.0g

pH 7.6 ± 0.2 at 25°C

Preparation of Medium: Add ground beef to 1.0L of distilled/deionized water. Let stand overnight at 4°C. Gently heat and bring to 80° to 90°C for 60 min. Let stand for 2 hr. Filter through muslin. To filtrate add peptone and salt. Mix thoroughly. Adjust pH to 7.6 with 4% NaOH. Filter through Whatman #1 filter paper. Bring volume of filtrate to 1.0L. Add agar. Gently heat and bring to boiling. Distribute into tubes or flasks. Autoclave for 15 min at 15 psi pressure–121°C. Pour into sterile Petri dishes or leave in tubes.

Use: For the cultivation of a variety of microorganisms.

Beef Infusion Broth

Composition per liter:

Ground defatted beef....................................453.6g
Peptone...10.0g
NaCl..5.0g

pH 7.6 ± 0.2 at 25°C

Preparation of Medium: Add ground beef to 1.0L of distilled/deionized water. Let stand overnight at 4°C. Gently heat and bring to 80° to 90°C for 60 min. Let stand for 2 hr. Filter through muslin. To filtrate add peptone and salt. Mix thoroughly. Adjust pH to 7.6 with 4% NaOH. Filter through Whatman #1 filter paper. Bring volume of filtrate to 1.0L. Add agar. Gently heat and bring to boiling. Distribute into tubes or flasks. Autoclave for 15 min at 15 psi pressure–121°C.

Use: For the cultivation of a variety of microorganisms.

BiGGY Agar (Bismuth Sulfite Glucose Glycerin Yeast Extract Agar) (Nickerson Medium)

Composition per liter:

Agar..16.0g
Glucose...10.0g
Glycine...10.0g
Bismuth ammonium citrate................................5.0g
Na_2SO_3..3.0g
Yeast extract..1.0g

pH 6.8 ± 0.2 at 25°

Source: This medium is available as a premixed powder from Difco Laboratories, Oxoid Unipath and BBL Microbiology Systems.

Preparation of Medium: Add components to distilled/deionized water and bring volume to 1.0L. Mix thoroughly and heat with frequent agitation until boiling. Distribute into tubes or flasks. Do not autoclave. Cool to approximately 45° to 50°C. If desired, add 2mg/L neomycin sulfate. Swirl to disperse the insoluble material and pour into sterile Petri dishes.

Use: For the detection, isolation and presumptive identification of *Candida* species. Addition of neomycin helps inhibit bacterial species. *C. albicans* appears as brown to black colonies with no pigment diffusion and no sheen. *C. tropicalis* appears as dark brown colonies with black centers, black pigment diffusion and a sheen. *C. krusei* appears as shiny, wrinkled, brown to black colonies with yellow pigment diffusion. *C. pseudotropicalis* appears as flat, shiny red to brown colonies with no pigment diffusion. *C. parakrusei* appears as flat, shiny, wrinkled dark reddish-brown colonies with light reddish-brown peripheries and a yellow fringe. *C. stellatoidea* appears as flat dark brown colonies with a light fringe.

Bile Esculin Agar

Composition per liter:

Oxgall..20.0g
Agar..15.0g
Pancreatic digest of gelatin..............................5.0g
Beef extract..3.0g
Esculin...1.0g
Ferric citrate...0.5g
Horse serum...50.0mL

pH 6.8 ± 0.2 at 25°

Source: This medium is available as a premixed powder from Oxoid Unipath and BBL Microbiology Systems.

Preparation of Medium: Add components, except horse serum, to distilled/deionized water and bring volume to 950.0L. Mix thoroughly and heat with frequent agitation until boiling. Autoclave for 15 min at 15 psi pressure–121°C. Cool to 45–50°C. Aseptically add 50.0mL of filter sterilized horse serum. Distribute into sterile Petri dishes or test tubes. Cool tubes in a slanted position.

Use: For differentiation between group D streptococci and non-group D streptococci. Also, to differentiate members of the Enterobacteriaceae, particularly *Klebsiella, Enterobacter* and *Serratia* from other enteric bacteria. Also, to differentiate *Listeria monocytogenes*. Bile tolerance and esculin hydrolysis (seen as a dark brown to black complex) are presumptive for enterococci (Group D streptococci).

Bile Esculin Agar

Composition per liter:

Esculin....................1.0g
Bile esculin agar base....................1.0L
pH 6.6 ± 0.2 at 25°C

Bile Esculin Agar Base:
Composition per liter:

Oxgall....................40.0g
Agar....................15.0g
Peptone....................5.0g
Beef extract....................3.0g
Ferric citrate....................0.5g

Source: This medium is available as a premixed powder from Difco Laboratories.

Preparation of Bile Esculin Agar Base: Add components to distilled/deionized water and bring volume to 1.0L. Mix thoroughly.

Preparation of Medium: Add desired amount of esculin—typically 1.0g—to bile esculin agar base. Mix thoroughly and heat with frequent agitation until boiling. Autoclave for 15 min at 15 psi pressure–121°C. Cool to 45–50°C. Distribute into sterile Petri dishes or test tubes. Cool tubes in a slanted position.

Use: For the isolation and presumptive identification of enterococci (Group D streptococci).

Bile Esculin Agar with Kanamycin

Composition per liter:

Oxgall....................20.0g
Agar....................15.0g
Beef extract....................3.0g
Esculin....................1.0g
Ferric citrate....................0.5g
Hemin....................10.0mg
Vitamin K_1....................10.0mg
Horse serum....................50.0mL
Kanamycin solution....................10.0mL
pH 7.1 ± 0.2 at 25°

Source: This medium is available as a premixed powder from BBL Microbiology Systems.

Kanamycin Solution:
Composition per 10mL:

Kanamycin....................1.0g

Preparation of Kanamycin Solution: Add Kanamycin to distilled/deionized water and bring volume to 10.0mL. Mix thoroughly. Filter sterilize.

Preparation of Medium: Add components to distilled/deionized water and bring volume to 1.0L. Mix thoroughly and heat with frequent agitation until boiling. Autoclave for 15 min at 15 psi pressure–121°C. Cool to 45–50°C. Aseptically add 50.0mL of 5% filter sterilized horse serum and 10.0mL sterile kanamycin solution. Distribute into test tubes or flasks. Cool tubes in a slanted position.

Use: For the selective isolation and/or presumptive identification of bacteria of the *Bacteroides fragilis* group from specimens containing mixed flora. Examine colonies with a long-wavelength UV light. Pigmented colonies of the *Bacteroides* group will fluoresce red-orange. Growth on this medium with blackening of the medium is presumptive for *B. fragilis*.

Birdseed Agar (*Guizotia abyssinica* Creatinine Agar) (Niger seed Agar) (Staib Agar)

Composition per liter:

Agar....................15.0g
Glucose....................15.0g
Creatinine....................5.0g
KH_2PO_4....................3.0g
Biphenyl....................1.0g
Chloramphenicol....................0.5g
Guizotia abyssinica seed
(niger seed) extract....................1000.0mL
pH 6.7 ± 0.2 at 25°

Preparation of Medium: Prepare seed extract by grinding 50.0g of *Guizotia abyssinica* seed in 1.0L of distilled/deionized water. Boil for 30 min. Filter through cheesecloth and filter paper. Add remaining components to seed filtrate. Mix thoroughly and heat with frequent agitation until boiling. Distribute into flasks or tubes. Autoclave for 25 min at 15 psi pressure–110°C.

Use: For selective isolation and differentiation between *Cryptococcus neoformans* from other *Cryptococcus* species and other yeasts.

Bismuth Sulfite Agar

Composition per liter:

Agar....................20.0g
$Bi_2(SO_3)_3$....................8.0g
Pancreatic digest of casein....................5.0g
Peptic digest of animal tissue....................5.0g
Beef extract....................5.0g
Glucose....................5.0g
Na_2HPO_4....................4.0g
$FeSO_4 \cdot 7H_2O$....................0.3g

pH 7.5 ± 0.2 at 25°C

Source: This medium is available as a premixed powder from Difco Laboratories, Oxoid Unipath and BBL Microbiology Systems.

Preparation of Medium: Add components to distilled/deionized water and bring volume to 1.0L. Mix thoroughly and heat with frequent agitation until boiling. Boil for 1 min. Do not autoclave. Cool to 45° to 50°C. Pour into sterile Petri dishes while gently shaking flask to disperse precipitate. Use plates the same day as prepared.

Use: For the selective isolation and identification of *Salmonella* species and other enteric bacilli. *Salmonella* species appear as flat, black, "rabbit-eye" colonies surrounded by a zone of black with a metallic sheen.

Blood Agar

Composition per liter:

Agar....................15.0g
Pancreatic digest of casein....................15.0g
Papaic digest of soybean meal....................5.0g
NaCl....................5.0g
Sheep blood, defibrinated....................50.0mL

pH 7.6 ± 0.2 at 25°C

Preparation of Medium: Add components, except sheep blood, to distilled/deionized water and bring volume to 950.0mL. Mix thoroughly. Gently heat and bring to boiling. Autoclave for 15 min at 15 psi pressure–121°C. Cool to 45° to 50°C. Aseptically add 50.0mL of sterile sheep blood. Mix thoroughly. Pour into sterile Petri dishes in 20.0mL volumes.

Use: General purpose medium for the cultivation of fastidious and nonfastidious microorganisms.

Blood Agar Base (ATCC Medium 368)

Composition per liter:

Beef heart, infusion from....................500.0g
Agar....................15.0g
Tryptose....................10.0g
NaCl....................5.0g

pH 6.8 ± 0.2 at 25°C

Source: This medium is available as a premixed powder from Difco Laboratories.

Preparation of Medium: Add components to distilled/deionized water and bring volume to 1.0L. Mix thoroughly. Heat with frequent agitation and boil for 1 min to completely dissolve. Autoclave for 15 min at 15 psi–121°C. Cool the basal medium to 45–50°C. Aseptically add sterile, defibrinated blood to a final concentration of 5%. Mix thoroughly and pour into sterile Petri dishes.

Use: For the isolation, cultivation and detection of hemolytic activity of staphylococci, streptococci and other fastidious microorganisms. General purpose medium for the cultivation of fastidious and nonfastidious microorganisms.

Blood Agar Base (Infusion Agar)

Composition per liter:

Agar....................15.0g
Pancreatic digest of casein....................13.0g
NaCl....................5.0g
Yeast extract....................5.0g
Heart muscle, solids from infusion....................2.0g
Sheep blood, defibrinated....................50.0mL

pH 7.3 ± 0.2 at 25°C

Source: This medium is available as a premixed powder from BBL Microbiology Systems.

Preparation of Medium: Add components, except sheep blood, to distilled/deionized water and bring volume to 950.0mL. Mix thoroughly. Heat with frequent agitation and boil for 1 min to completely dissolve. Autoclave for 15 min at 15 psi–121°C. Cool to 45–50°C. Aseptically add 50.0mL of sterile, defibrinated sheep blood. Mix thoroughly and pour into sterile Petri dishes.

Use: For the isolation, cultivation and detection of hemolytic activity of streptococci and other fastidious microorganisms. General purpose medium for the cultivation of fastidious and nonfastidious microorganisms.

Blood Agar Base

Composition per liter:

Agar....................15.0g
Beef extract....................10.0g
NaCl....................5.0g

Peptone..10.0g
Sheep blood, defibrinated.............................50.0mL

pH 7.3 ± 0.2 at 25°C

Source: This medium is available as a premixed powder from Oxoid Unipath.

Preparation of Medium: Add components, except sheep blood, to distilled/deionized water and bring volume to 950.0mL. Mix thoroughly. Heat with frequent agitation and boil for 1 min to completely dissolve. Autoclave for 15 min at 15 psi–121°C. Cool to 45–50°C. Aseptically add 50.0mL of sterile, defibrinated sheep blood. Mix thoroughly and pour into sterile Petri dishes.

Use: For the isolation, cultivation and detection of hemolytic activity of streptococci and other fastidious microorganisms. General purpose medium for the cultivation of fastidious and nonfastidious microbes.

Blood Agar Base Sheep

Composition per liter:

Pancreatic digest of casein...............................14.0g
Agar..12.5g
NaCl...5.0g
Peptone..4.5g
Yeast extract...4.5g
Sheep blood, defibrinated.............................70.0mL

ph 7.3 ± 0.2 at 25°C

Source: This medium is available as a premixed powder from Oxoid Unipath.

Preparation: Add components to distilled/deionized water and bring volume to 1.0L. Mix thoroughly. Autoclave for 15 min at 15 psi–121°C. Cool the basal medium to 45–50°C. Aseptically add 70.0mL sterile, defibrinated sheep blood. Pour into sterile Petri dishes.

Use: For giving improved hemolytic reactions with sheep blood.

Blood Agar, Diphasic

Composition per 800mL:

Lean beef, desiccated.......................................25.0g
Agar...10.0g
Neopeptone...10.0g
NaCl...2.5g
Locke solution...200.0mL
Rabbit blood, defibrinated..........................100.0mL

pH 7.2–7.4 at 25°C

Locke Solution:

Composition per liter:

NaCl...8.0g
Glucose...2.5g
KH_2PO_4...0.3g
KCl...0.2g
$CaCl_2 \cdot 2H_2O$...0.2g

Preparation of Locke Solution: Add components to distilled/deionized water and bring volume to 1.0L. Mix thoroughly. Filter sterilize.

Preparation of Medium: Add beef to 500.0mL of distilled/deionized water. Let stand for 60 min. Gently heat and bring to 80°C for 5 min. Filter through Whatman #1 filter paper. To filtrate add remaining components, except Locke solution and rabbit blood. Mix thoroughly. Adjust pH to 7.2–7.4 with NaOH. Autoclave for 20 min at 15 psi pressure–121°C. Cool to 45° to 50°C. Aseptically add sterile rabbit blood. Mix thoroughly. Aseptically distribute into sterile tubes in 5.0mL volumes. Allow tubes to cool in a slanted position. Immediately prior to inoculation, overlay agar in each tube with 2.0mL of sterile Locke solution.

Use: For the cultivation of *Trypanosoma* species and *Leishmania* species.

Blood Agar No. 2

Composition per liter:

Proteose peptone..15.0g
Agar..12.0g
NaCl...5.0g
Yeast extract...5.0g
Liver digest...2.5g

pH 7.4 ± 0.2 at 25°C

Source: This medium is available as a premixed powder from Difco Laboratories and Oxoid Unipath.

Preparation of Medium: Add components to distilled/deionized water and bring volume to 1.0L. Mix thoroughly. Heat with frequent agitation and boil for 1 min to completely dissolve. Autoclave for 15 min at 15 psi–121°C. Cool the basal medium to 45–50°C. Aseptically add sterile, defibrinated blood to a final concentration of 7%. Pour into sterile Petri dishes.

Use: For the isolation, cultivation and detection of hemolytic activity of streptococci, pneumococci and other particularly fastidious microorganisms.

Blood Agar with Low pH

Composition per liter:

Beef heart, solids from infusion.....................500.0g
Agar..15.0g
Tryptose...10.0g
NaCl...5.0g
Sheep blood, defibrinated.............................50.0mL

pH 6. 8 ± 0.2 at 25°C

Source: This medium is available as a premixed powder from BBL Microbiology Systems.

Preparation of Medium: Add components, except sheep blood, to distilled/deionized water and bring volume to 950.0mL. Mix thoroughly. Heat with frequent agitation and boil for 1 min to completely dissolve. Autoclave for 15 min at 15 psi–121°C. Cool to 45–50°C. Aseptically add 50.0mL of sterile, defibrinated sheep blood. Mix thoroughly and pour into sterile Petri dishes.

Use: For the isolation and growth of a wide variety of microorganisms and for the detection of the hemolytic reactions of streptococci and other fastidious microorganisms. The slightly acid acid pH of this medium enhances distinct hemolytic reactions.

Blood Glucose Cystine Agar (BGC Agar)

Composition per 100mL:

Nutrient agar	85.0mL
Glucose cystine solution	10.0mL
Human blood, fresh	5.0mL

pH 6.8 ± 0.2 at 25°C

Nutrient Agar:

Composition per liter:

Agar	15.0g
Pancreatic digest of gelatin	5.0g
Beef extract	3.0g

Source: Nutrient agar is available as a premixed powder from BBL Microbiology Systems and Difco Laboratories.

Preparation of Nutrient Agar: Add components to distilled/deionized water and bring volume to 1.0L. Mix thoroughly. Gently heat while stirring and bring to boiling. Distribute into tubes or flasks. Autoclave for 15 min at 15 psi pressure–121°C. Cool to 45° to 50°C.

Glucose Cystine Solution:

Composition per 50mL:

Glucose	12.5g
Cystine·HCl	0.5g

Preparation of Glucose Cystine Solution: Add components to distilled/deionized water and bring volume to 50.0mL. Mix thoroughly. Filter sterilize.

Preparation of Medium: To 85.0mL of cooled, sterile agar solution, aseptically add 10.0mL of sterile glucose cystine solution and 5.0mL of human blood. Mix thoroughly. Pour into sterile Petri dishes or distribute into sterile tubes.

Use: For the cultivation of *Francisella tularensis*.

Blood Glucose Cystine Agar (BGC Agar)

Composition per 1050mL:

Tryptose blood agar base	33.0g
Cystine·HCl	1.0g
Sheep blood	50.0mL
Penicillin solution	10.0mL

pH 6.8 ± 0.2 at 25°C

Source: Tryptose blood agar base is available as a premixed powder from Difco Laboratories.

Penicillin Solution:

Composition per 10mL:

Penicillin G	100,000U

Preparation of PenicillinSolution: Add penicillin to distilled/deionized water and bring volume to 10.0mL. Mix thoroughly. Filter sterilize.

Preparation of Medium: Add tryptose blood agar base and cystine·HCl to distilled/deionized water and bring volume to 990.0mL. Mix thoroughly. Gently heat while stirring and bring to boiling.Autoclave for 15 min at 15 psi pressure–121°C. Cool to 45° to 50°C. Aseptically add 50.0mL of sterile, defibrinated sheep blood and10.0mL of penicillin solution. Mix thoroughly. Pour into sterile Petri dishes or distribute into sterile tubes.

Use: For the maintenance of the pathogenic phase (yeast phase) of *Blastomyces dermatitidis, Histoplasma capsulatum, Paracoccidioides brasiliensis,* and *Sprothrix schenkii*. Cultures must be transferred every three weeks to fresh media and maintained at 35 to 37°C.

BMPA–α Medium (Semiselective Medium for *Legionella pneumophila*)

Composition per liter:

Agar	15.0g
Yeast extract	10.0g
ACES buffer (2-[(2-Amino-2-oxoethyl)-amino]-ethane sulfonic acid)	10.0g
Charcoal, activated	2.0g
α-Ketogluatrate	1.0g
$Fe_4(P_2O_7)_3$·$9H_2O$	0.25g
Antibiotic inhibitor	10.0mL
L-Cysteine·HCl·H_2O solution	10.0mL

pH 6.9 ± 0.2 at 25°C

Antibiotic Inhibitor:

Composition per 10mL:

Anisomycin	0.08g
Cefamandole	4.0mg
Polymyxin B	80,000U

Preparation of Antibiotic Inhibitor: Add components to distilled/deionized water and bring volume to 10.0mL. Mix thoroughly. Filter sterilize.

L-Cysteine·HCl·H_2O Solution:
Composition per 10mL:

L-Cysteine·HCl·H_2O 0.4g

Preparation of L-Cysteine·HCl·H_2O Solution: Add L-Cysteine·HCl·H_2O to distilled/deionized water and bring volume to 10.0mL. Mix thoroughly. Filter sterilize.

Preparation of Medium: Add components, except cysteine and antibiotic inhibitor, to distilled/deionized water and bring volume to 980.0mL. Mix thoroughly. Adjust medium to pH 6.9 with 1*N* KOH. Heat gently and bring to boiling for 1 min. Autoclave for 15 min at 15 psi pressure–121°C. Cool to 50° to 55°C. Add 10.0mL of the sterile L-cysteine·HCl·H_2O solution and 10.0mL of the sterile antibiotic solution. Mix thoroughly. Pour into sterile Petri dishes with constant agitation to keep charcoal in suspension.

Use: For the selective isolation and cultivation of *Legionella pneumophila* and other *Legionella* species.

BMPA–α Medium (Edelstein BMPA–α Medium)

Composition per liter:

Agar 13.0g
Yeast extract 10.0g
ACES buffer (2-[(2-Amino-2-oxoethyl)-amino]-ethane sulfonic acid) 2.0g
Charcoal, activated 2.0g
α-Ketogluatrate 0.2g
$Fe_4(P_2O_7)_3 \cdot 9H_2O$ 0.05g
Antibiotic inhibitor 10.0mL
L-Cysteine·HCl·H_2O solution 10.0mL
pH 6.9 ± 0.2 at 25°C

Source: This medium is available as premixed vials from Oxoid Unipath.

Antibiotic Inhibitor:
Composition per 10mL:

Anisomycin 0.08g
Cefamandole 4.0mg
Polymyxin B 80,000U

Preparation of Antibiotic Inhibitor: Add components to distilled/deionized water and bring volume to 10.0mL. Mix thoroughly. Filter sterilize.

L-Cysteine·HCl·H_2O Solution:
Composition per 10mL:

L-Cysteine·HCl·H_2O 0.08g

Preparation of L-Cysteine·HCl·H_2O Solution: Add L-Cysteine·HCl·H_2O to distilled/deionized water and bring volume to 10.0mL. Mix thoroughly. Filter sterilize.

Preparation of Medium: Add components, except cysteine and antibiotic inhibitor, to distilled/deionized water and bring volume to 980.0mL. Mix thoroughly. Adjust medium to pH 6.9 with 1*N* KOH. Heat gently and bring to boiling for 1 min. Autoclave for 15 min at 15 psi pressure–121°C. Cool to 50° to 55°C. Add 10.0mL of the sterile L-cysteine·HCl·H_2O solution and 10.0mL of the sterile antibiotic solution. Mix thoroughly. Pour into sterile Petri dishes with constant agitation to keep charcoal in suspension.

Use: For the selective isolation and cultivation of *Legionella pneumophila* and other *Legionella* species.

Bordetella pertussis Selective Medium with Bordet-Gengou Agar Base

Composition 1210mL:

Bordet-Gengou agar base 1.0L
Defibrinated horse blood 200.0mL
Cephalexin solution 10.0mL
pH 6.7± 0.2 at 25°C

Source: This medium is available as a premixed powder from Oxoid Unipath.

Bordet-Gengou Agar Base:
Composition per liter:

Agar 20.0g
NaCl 5.5g
Pancreatic digest of casein 5.0g
Peptic digest of animal tissue 5.0g

Preparation of Bordet-Gengou Agar Base: Add components of Bordet-Gengou Agar Base to 1.0L of 1% glycerol solution. Autoclave for 15 min at 15 psi pressure–121°C. Cool to 50°C.

Cephalexin Solution:
Composition per 10mL:

Cephalexin 0.04g

Preparation of Cephalexin Solution: Add cephalexin to distilled/deionized water and bring volume to 10.0mL. Mix thoroughly. Filter sterilize.

Preparation of Medium: Aseptically add 10.0mL sterile cephalexin solution and 200.0mL of defibrinated horse blood to 1.0L Bordet-Gengou Agar Base. Mix thoroughly and pour into sterile Petri dishes.

Use: For selective isolation and presumptive identification of *Bordetella pertussis* and *Bordetella parapertussis*. *B. pertussis* appears as small, nearly transparent, "bisected pearl-like" colonies.

Bordetella pertussis Selective Medium with Charcoal Agar Base

Composition 1110mL:

Charcoal agar base 1.0L
Horse blood, defibrinated 100.0mL
Cephalexin solution 10.0mL

pH 6.7± 0.2 at 25°C

Source: This medium is available as a premixed powder from Oxoid Unipath.

Charcoal Agar Base:

Composition per liter:

Agar 12.0g
Beef extract 10.0g
Starch 10.0g
NaCl 5.0g
Pancreatic digest of casein 5.0g
Peptic digest of animal tissue 5.0g
Charcoal 4.0g
Nicotinic acid 1.0mg

Preparation of Charcoal Agar Base: Add components of Charcoal Agar Base to distilled/deionized water and bring volume to 1.0L. Autoclave for 15 min at 15 psi pressure–121°C. Cool to 50°C.

Cephalexin Solution:

Composition per 10mL:

Cephalexin 0.04g

Preparation of Cephalexin Solution: Add cephalexin to distilled/deionized water and bring volume to 10.0mL. Mix thoroughly. Filter sterilize.

Preparation of Medium: Aseptically add 10.0mL sterile cephalexin solution and 100.0mL defibrinated horse blood to charcoal agar base. Mix thoroughly and pour into sterile Petri dishes.

Use: For selective isolation and presumptive identification of *Bordetella pertussis* and *Bordetella parapertussis*. *B. pertussis* appears as small, pale, shiny colonies.

Bordet–Gengou Agar

Composition per liter:

Agar 20.0g
Glycerol 10.0g
NaCl 5.5g
Pancreatic digest of casein 5.0g
Peptic digest of animal tissue 5.0g
Potato, solids from infusion 4.5g
Rabbit blood 200.0mL

pH 6.7± 0.2 at 25°C

Source: This medium is available as a premixed powder from Difco Laboratories, Oxoid Unipath and BBL Microbiology Systems.

Preparation of Medium: Add 10.0g glycerol to 980.0mL distilled/deionized water. Add other components, except rabbit blood, to the glycerol solution. Mix thoroughly. Heat with occasional agitation of the medium. Boil for 1 min. Autoclave for 15 min at 15 psi pressure–121°C. Cool medium to 50°C. Aseptically add 200.0mL rabbit blood (prewarmed to 35°C) or rabbit blood to a concentration of 15%–30%. 150.0–200.0mL of sterile, defibrinated horse blood may be used in place of rabbit blood. Mix thoroughly and pour plates or prepare slants.

Use: For the detection and isolation of *Bordetella pertussis* and *Bordetella parapertussis* from clinical specimens. The medium is rendered selective by the addition of methicillin. *Bordetella pertussis* appears as small (<1mm), smooth, pearl-like colonies surrounded by a narrow zone of hemolysis. *Bordetella parapertussis* appears as brown, non-shiny colonies with a green-black coloration on the reverse side. *Bordetella bronchiseptica* appears as brown, non-shiny, moderately sized colonies with a roughly pitted surface.

Bordet–Gengou Medium (ATCC Medium 35)

Composition per liter:

Agar 20.0g
Glycerol 10.0g
Proteose peptone 10.0g
NaCl 5.5g
Pancreatic digest of casein 5.0g
Peptic digest of animal tissue 5.0g
Potato, solids from infusion 4.5g
Rabbit blood 150.0mL

pH 6.7± 0.2 at 25°C

Source: This medium is available as a premixed powder from Difco Laboratories, Oxoid Unipath and BBL Microbiology Systems.

Preparation of Medium: Add 10.0g glycerol to 980.0mL distilled/deionized water. Add other components, except rabbit blood, to the glycerol solution. Mix thoroughly. Heat with occasional agitation of the medium. Boil for 1 min. Autoclave for 15 min at 15 psi pressure–121°C. Cool medium to 50°C. Aseptically add 150.0mL rabbit blood (prewarmed to 35°C) Mix thoroughly. Pour into sterile Petri dishes or distribute into sterile tubes. Allow tubes to cool in a slanted position.

Use: For the detection and isolation of *Bordetella pertussis* and *Bordetella parapertussis* from clinical specimens. The medium is rendered selective by the addition of methicillin. *Bordetella pertussis* appears as small (<1mm), smooth, pearl-like colonies surrounded by a narrow zone of hemolysis. *Bordetella parapertussis* appears as brown, non-shiny colonies with a green-black coloration on the reverse side. *Bordetella bronchiseptica* appears as brown, non-shiny, moderately sized, colonies with a roughly pitted surface.

Borrelia Medium

Composition per 370mL:

Solution 4 .... 240.0mL
Solution 1 .... 80.0mL
Solution 2 .... 34.0mL
Rabbit serum, sterile 10.0mL
Solution 3 .... 4.0mL
Solution 5 .... 0.7mL

Solution 1:

Composition per liter:

$Na_2HPO_4 \cdot 7H_2O$ 26.52g
Glucose 12.75g
Proteose peptone No.2 .... 5.95g
Pancreatic digest of casein 2.55g
NaCl 1.20g
Sodium pyruvate 1.06g
$NaH_2PO_4 \cdot H_2O$ 1.03g
KCl 0.85g
$MgCl_2 \cdot 6H_2O$ 0.68g
N-acetylglucosamine 0.53g
Sodium citrate·$2H_2O$ 0.47g

Preparation of Solution 1: Add components to distilled/deionized water and bring volume to 1.0L. Mix thoroughly. Store at -20° C.

Solution 2:

Composition per 100mL:

Bovine albumin fraction V 10.0g

Preparation of Solution 2: Add bovine albumin to distilled/deionized water and bring volume to 100.0mL. Mix thoroughly. Adjust pH to 7.8 with NaOH. Store at –20°C.

Solution 3:

Composition per 100mL:

$NaHCO_3$ 4.5g

Preparation of Solution 3: Add $NaHCO_3$ to distilled/deionized water and bring volume to 100.0mL. Mix thoroughly. Prepare solution freshly.

Solution 4:

Composition per 100mL:

Gelatin 7.0g

Preparation of Solution 4: Add gelatin to distilled/deionized water and bring volume to 100.0mL. Mix thoroughly. Autoclave for 15 min at 10 psi pressure–115°C. Store at 4°C.

Solution 5:

Composition per 100mL:

Phenol Red 0.5g

Preparation of Solution 5: Add Phenol Red to distilled/deionized water and bring volume to 100.0mL. Mix thoroughly. Store at 4°C.

Preparation of Medium: Combine 80.0mL of solution 1, 34.0mL of solution 2, 4.0mL of solution 3, 0.7mL of solution 5 and 1.3mL of distilled/deionized water. Mix thoroughly. Filter sterilize under pressure. Aseptically distribute into sterile borosilicate screw-capped tubes in 6.0mL volumes. Melt solution 4 by immersing tube in warm water. Add 2.0mL of solution 4 to each screw-capped tube. Add 0.5mL of sterile rabbit serum to each screw-capped tube.

Use: For the cultivation of *Borrelia hermsii*, *B. turicatae*, and *B. parkeri*.

Bovine Albumin Tween™ 80 Medium, Ellinghausen and McCullough, Modified (Albumin Fatty Acid Broth, *Leptospira* Medium)

Composition per liter:

Basal medium 900.0mL
Albumin fatty acid supplement 100.0mL

Basal Medium:

Composition per liter:

Na_2HPO_4, anhydrous 1.0g
NaCl 1.0g
KH_2PO_4, anhydrous 0.3g
NH_4Cl (25% solution) 1.0mL
Glycerol (10% solution) 1.0mL
Sodium pyruvate (10% solution) 1.0mL
Thiamine·HCl (0.5% solution) 1.0mL
pH 7.4 ± 0.2 at 25°C

Preparation of Basal Medium: Add components to distilled/deionized water and bring volume to 1.0L. Mix thoroughly. Adjust pH to 7.4. Gently heat and bring to boiling. Autoclave for 15 min at 15 psi pressure–121°C. Cool to 25°C.

Albumin Fatty Acid Supplement:

Composition per 200mL:

Bovine albumin fraction V 20.0g
Polysorbate (Tween™) 80
(10% solution) 25.0mL

$FeSO_4 \cdot 7H_2O$ (0.5% solution) 20.0mL
$CaCl_2 \cdot 2H_2O$ (1.5% solution) 2.0mL
$MgCl_2 \cdot 2H_2O$ (1.5% solution) 2.0mL
Vitamin B_{12} (0.2% solution) 2.0mL
$ZnSO_4 \cdot 7H_2O$ (0.4% solution) 2.0mL
$CuSO_4 \cdot 5H_2O$ (0.3% solution) 0.2mL

Preparation of Albumin Fatty Acid Supplement: Add bovine albumin to 100.0mL of distilled/deionized water. Mix thoroughly. Add remaining components while stirring. Adjust pH to 7.4. Bring volume to 200.0mL with distilled/deionized water. Filter sterilize. Store at –20°C.

Preparation of Medium: Aseptically combine 100.0mL of sterile albumin fatty acid supplement and 900.0mL of sterile basal medium. Mix thoroughly. Aseptically distribute into sterile tubes or flasks.

Use: For the cultivation of *Leptospira* species.

Bovine Albumin Tween™ 80 Semisolid Medium, Ellinghausen and McCullough, Modified (Albumin Fatty Acid Semisolid Medium, Modified)

Composition per liter:
Basal medium 900.0mL
Albumin fatty acid supplement 100.0mL

Basal Medium:
Composition per liter:
Agar 2.2g
Na_2HPO_4, anhydrous 1.0g
NaCl 1.0g
KH_2PO_4, anhydrous 0.3g
NH_4Cl (25% solution) 1.0mL
Glycerol (10% solution) 1.0mL
Sodium pyruvate (10% solution) 1.0mL
Thiamine·HCl (0.5% solution) 1.0mL
pH 7.4 ± 0.2 at 25°C

Preparation of Basal Medium: Add components to distilled/deionized water and bring volume to 1.0L. Mix thoroughly. Adjust pH to 7.4. Gently heat and bring to boiling. Autoclave for 15 min at 15 psi pressure–121°C. Cool to 25°C.

Albumin Fatty Acid Supplement:
Composition per 200mL:
Bovine albumin fraction V 20.0g
Polysorbate (Tween™) 80 (10% solution) 25.0mL
$FeSO_4 \cdot 7H_2O$ (0.5% solution) 20.0mL
$CaCl_2 \cdot 2H_2O$ (1.5% solution) 2.0mL
$MgCl_2 \cdot 2H_2O$ (1.5% solution) 2.0mL
Vitamin B_{12} (0.2% solution) 2.0mL
$ZnSO_4 \cdot 7H_2O$ (0.4% solution) 2.0mL
$CuSO_4 \cdot 5H_2O$ (0.3% solution) 0.2mL

Preparation of Albumin Fatty Acid Supplement: Add bovine albumin to 100.0mL of distilled/deionized water. Mix thoroughly. Add remaining components while stirring. Adjust pH to 7.4. Bring volume to 200.0mL with distilled/deionized water. Filter sterilize. Store at –20°C.

Preparation of Medium: Aseptically combine 100.0mL of sterile albumin fatty acid supplement and 900.0mL of sterile basal medium. Mix thoroughly. Aseptically distribute into sterile tubes or flasks.

Use: For the cultivation of *Leptospira* species.

Bovine Serum Albumin Tween™ 80 Agar (BSA Tween™ 80 Agar)

Composition per liter:
Basal medium 900.0mL
Albumin supplement 100.0mL

Basal Medium:
Composition per liter:
Agar 11.0g
Na_2HPO_4 .... 1.0g
NaCl 1.0g
KH_2PO_4 .... 0.3g
Glycerol (10% solution) 1.0mL
NH_4Cl (25% solution) 1.0mL
Sodium pyruvate (10% solution) 1.0mL
Thiamine (0.5% solution 1.0mL

Preparation of Basal Medium: Add components to distilled/deionized water and bring volume to 1.0L. Mix thoroughly. Adjust pH to 7.4. Autoclave for 15 min at 15 psi pressure–121°C. Cool to 25°C.

Albumin Supplement:
Composition per 100mL:
Bovine albumin 10.0g
Tween™ 80 (10% solution) 12.5mL
$FeSO_4$ (0.5% solution) 10.0mL
$MgCl_2$-$CaCl_2$ solution 1.0mL
Cyanocobalamin (0.02% solution) 1.0mL
$ZnSO_4$ (0.4% solution) 1.0mL

Preparation of Albumin Supplement: Add components to distilled/deionized water and bring volume to 100.0mL. Mix thoroughly. Adjust pH to 7.4. Filter sterilize.

$MgCl_2$–$CaCl_2$ Solution:
Composition per 100mL:
$CaCl_2 \cdot 2H_2O$ 1.5g
$MgCl_2 \cdot 6H_2O$ 1.5g

Preparation of $MgCl_2$–$CaCl_2$ Solution: Add components to distilled/deionized water and bring volume to 100.0mL. Mix thoroughly.

Preparation of Medium: To 900.0mL of cooled, sterile basal medium, aseptically add 100.0mL of sterile albumin supplement. Mix thoroughly. Aseptically distribute into sterile tubes or flasks.

Use: For the cultivation and maintenance of *Leptospira* species.

Bovine Serum Albumin Tween™ 80 Broth (BSA Tween™ 80 Broth)

Composition per liter:

Basal medium	900.0mL
Albumin supplement	100.0mL

pH 7.4 ± 0.2 at 25°C

Basal Medium:

Composition per liter:

Na_2HPO_4	1.0g
NaCl	1.0g
KH_2PO_4	0.3g
Glycerol (10% solution)	1.0mL
NH_4Cl (25% solution)	1.0mL
Sodium pyruvate (10% solution)	1.0mL
Thiamine (0.5% solution	1.0mL

Preparation of Basal Medium: Add components to distilled/deionized water and bring volume to 1.0L. Mix thoroughly. Adjust pH to 7.4. Autoclave for 15 min at 15 psi pressure–121°C. Cool to 25°C.

Albumin Supplement:

Composition per 100mL:

Bovine albumin	10.0g
Tween™ 80 (10% solution)	12.5mL
$FeSO_4$ (0.5% solution)	10.0mL
$MgCl_2$-$CaCl_2$ solution	1.0mL
Cyanocobalamin (0.02% solution)	1.0mL
$ZnSO_4$ (0.4% solution)	1.0mL

Preparation of Albumin Supplement: Add components to distilled/deionized water and bring volume to 100.0mL. Mix thoroughly. Adjust pH to 7.4. Filter sterilize.

$MgCl_2$–$CaCl_2$ Solution:

Composition per 100mL:

$CaCl_2·2H_2O$	1.5g
$MgCl_2·6H_2O$	1.5g

Preparation of $MgCl_2$–$CaCl_2$ Solution: Add components to distilled/deionized water and bring volume to 100.0mL. Mix thoroughly.

Preparation of Medium: To 900.0mL of cooled, sterile basal medium, aseptically add 100.0mL of sterile albumin supplement. Mix thoroughly. Aseptically distribute into sterile tubes or flasks.

Use: For the isolation and cultivation of *Leptospira* species.

Bovine Serum Albumin Tween™ 80 Soft Agar (BSA Tween™ 80 Soft Agar) (Semisolid BSA Tween™ 80 Medium)

Composition per liter:

Basal medium	900.0mL
Albumin supplement	100.0mL

Basal Medium:

Composition per liter:

Agar	2.0g
Na_2HPO_4	1.0g
NaCl	1.0g
KH_2PO_4	0.3g
Glycerol (10% solution)	1.0mL
NH_4Cl (25% solution)	1.0mL
Sodium pyruvate (10% solution)	1.0mL
Thiamine (0.5% solution	1.0mL

Preparation of Basal Medium: Add components to distilled/deionized water and bring volume to 1.0L. Mix thoroughly. Adjust pH to 7.4. Autoclave for 15 min at 15 psi pressure–121°C. Cool to 25°C.

Albumin Supplement:

Composition per 100mL:

Bovine albumin	10.0g
Tween™ 80 (10% solution)	12.5mL
$FeSO_4$ (0.5% solution)	10.0mL
$CaCl_2$-$MgCl_2$ solution	1.0mL
Cyanocobalamin (0.02% solution)	1.0mL
$ZnSO_4$ (0.4% solution)	1.0mL

Preparation of Albumin Supplement: Add components to distilled/deionized water and bring volume to 100.0mL. Mix thoroughly. Adjust pH to 7.4. Filter sterilize.

$MgCl_2$–$CaCl_2$ Solution:

Composition per 100mL:

$CaCl_2·2H_2O$	1.5g
$MgCl_2·6H_2O$	1.5g

Preparation of $MgCl_2$–$CaCl_2$ Solution: Add components to distilled/deionized water and bring volume to 100.0mL. Mix thoroughly.

Preparation of Medium: To 900.0mL of cooled, sterile basal medium aseptically add 100.0mL of sterile albumin supplement. Mix thoroughly. Aseptically distribute into sterile tubes or flasks.

Use: For the cultivation of *Leptospira* species.

Brain Heart CC Agar (Brain Heart Cycloheximide Chloramphenicol Agar)

Composition per liter:

Pancreatic digest of casein	16.0g
Agar	13.5g
Brain heart, solids from infusion	8.0g
Peptic digest of animal tissue	5.0g
NaCl	5.0g
Na_2HPO_4	2.5g
Glucose	2.0g
Cycloheximide	0.5g
Chloramphenicol	0.05g

pH 7.4 ± 0.2 at 25°C

Source: This medium is available as a premixed powder from Difco Laboratories and BBL Microbiology Systems.

Preparation of Medium: Add components to distilled/deionized water and bring volume to 1.0L. Mix thoroughly. Distribute into tubes or flasks while shaking to distribute precipitate. Autoclave for 15 min at 15 psi–118°C.

Use: For the selective isolation of fastidious pathogenic fungi such as *Histoplasma capsulatum* and *Blastomyces dermatiditis* from specimens heavily contaminated with bacteria and other fungi. It may also be used as a base supplemented with sheep blood and gentamicin for enrichment and additional selectivity.

Brain Heart Infusion (BHI)

Composition per liter:

Pancreatic digest of gelatin	14.5g
Brain heart, solids from infusion	6.0g
Peptic digest of animal tissue	6.0g
NaCl	5.0g
Glucose	3.0g
Na_2HPO_4	2.5g

pH 7.4 ± 0.2 at 25°C

Source: This medium is available as a premixed powder from Difco Laboratories, Oxoid Unipath and BBL Microbiology Systems.

Preparation of Medium: Add components to distilled/deionized water and bring volume to 1.0L. Mix thoroughly. Distribute into tubes or flasks. Autoclave for 15 min at 15 psi–121°C.

Use: For the cultivation of fastidious and nonfastidious microorganisms, including aerobic and anaerobic bacteria, from a variety of clinical specimens and in epidemiological investigations. It is particularly useful for culturing streptococci, pneumococci and meningococci. It is also used for the preparation of inocula for use in antimicrobial susceptibility tests and as a base for blood culture.

Brain Heart Infusion Agar (BHI Agar)

Composition per liter:

Beef heart infusion	250.0g
Calf brain infusion	200.0g
Proteose peptone	10.0g
NaCl	5.0g
$Na_2HPO_4 \cdot 12H_2O$	2.5g
Glucose	2.0g

pH 7.4 ± 0.2 at 25°C

Preparation of Medium: Add components to distilled/deionized water and bring volume to 1.0L. Mix thoroughly. Gently heat and bring to boiling. Distribute into tubes or flasks. Autoclave for 15 min at 15 psi pressure–121°C. Pour into sterile Petri dishes or leave in tubes.

Use: For the cultivation of a variety of fastidious and nonfastidious, aerobic and anaerobic microorganisms.

Brain Heart Infusion Agar

Composition per liter:

Pancreatic digest of casein	16.0g
Agar	13.5g
Brain heart, solids from infusion	8.0g
Peptic digest of animal tissue	5.0g
NaCl	5.0g
Glucose	2.0g
Na_2HPO_4	2.5g

pH 7.4 ± 0.2 at 25°C

Source: This medium is available as a premixed powder from Difco Laboratories, Oxoid Unipath and BBL Microbiology Systems.

Preparation of Medium: Add components to distilled/deionized water and bring volume to 1.0L. Mix thoroughly. Distribute into tubes or flasks while shaking to distribute precipitate. Autoclave for 15 min at 15 psi–121°C.

Use: For the cultivation of a wide variety of fastidious microorganisms, including bacteria, yeast and molds.

With the addition of 10% sheep blood, it is used for the isolation and cultivation of many fungal species, including systemic fungi, from clinical specimens and in epidemiological investigations. The addition of gentamicin and chloramphenicol with 10% sheep blood produces a selective medium used for the isolation of pathogenic fungi from specimens heavily contaminated with bacteria and saprophtic fungi. It is recommended for the isolation of *Histoplasma capsulatum* and other pathogenic fungi including *Coccidioides immitis*.

Brain Heart Infusion Agar with Yeast Extract

Composition per liter:

Yeast extract	20.0g
Pancreatic digest of casein	16.0g
Agar	13.5g
Brain heart, solids from infusion	8.0g
Peptic digest of animal tissue	5.0g
NaCl	5.0g
Glucose	2.0g
Na_2HPO_4	2.5g

pH 7.4 ± 0.2 at 25°C

Source: This medium is available as a premixed powder from Difco Laboratories, Oxoid Unipath and BBL Microbiology Systems.

Preparation of Medium: Add components to distilled/deionized water and bring volume to 1.0L. Mix thoroughly. Distribute into tubes or flasks while shaking to distribute precipitate. Autoclave for 15 min at 15 psi–121°C.

Use: For the cultivation of *Mycoplasma equirhinis*.

Brain Heart Infusion Broth (BHI Broth)

Composition per liter:

Beef heart infusion	250.0g
Calf brain infusion	200.0g
Proteose peptone	10.0g
NaCl	5.0g
$Na_2HPO_4 \cdot 12H_2O$	2.5g
Glucose	2.0g

pH 7.4 ± 0.2 at 25°C

Preparation of Medium: Add components to distilled/deionized water and bring volume to 1.0L. Mix thoroughly. Distribute into tubes or flasks. Autoclave for 15 min at 15 psi pressure–121°C.

Use: For the cultivation of a variety of fastidious and nonfastidious, aerobic and anaerobic microorganisms.

Brain Heart Infusion Broth

Composition per liter:

Calf brain infusion	200.0g
Beef heart infusion	250.0g
Proteose p eptone	10.0g
Glucose	2.0g
NaCl	5.0g
Na_2HPO_4	2.5g

pH 7.4 ± 0.2 at 25°C

Preparation of Medium: Add components to distilled/deionized water and bring volume to 1.0L. Mix thoroughly. Distribute into tubes or flasks while shaking to distribute precipitate. Autoclave for 15 min at 15 psi–121°C.

Use: For the cultivation of a wide variety of microorganisms, including bacteria, yeast and molds, especially fastidious species.

Brain Heart Infusion with 0.7% Agar

Composition per liter:

Beef heart infusion	250.0g
Calf brain infusion	200.0g
Proteose peptone	10.0g
Agar	7.0g
NaCl	5.0g
$Na_2HPO_4 \cdot 12H_2O$	2.5g
Glucose	2.0g

pH 5.3 ± 0.2 at 25°C

Preparation of Medium: Add components to distilled/deionized water and bring volume to 1.0L. Mix thoroughly. Gently heat and bring to boiling. Adjust pH to 5.3 with 1*N* HCl. Distribute into tubes in 25.0mL volumes. Autoclave for 10 min at 15 psi pressure–121°C.

Use: For the cultivation of *Staphylococcus* species for the production of enterotoxin.

Brain Heart Infusion Agar with Chloramphenicol

Composition per liter:

Pancreatic digest of casein	16.0g
Agar	13.5g
Brain/heart, solids from infusion	8.0g
Peptic digest of animal tissue	5.0g
NaCl	5.0g
Glucose	2.0g
Na_2HPO_4	2.5g
Sheep blood, defibrinated	50.0mL
Chloramphenicol solution	10.0mL

pH 7.4 ± 0.2 at 25°C

Chloramphenicol Solution:
Composition per 10mL:

Chloramphenicol	0.05g

Preparation of Chloramphenicol Solution: Add chloramphenicol to distilled/deionized water and bring volume to 10.0mL. Mix thoroughly. Filter sterilize.

Preparation of Medium: Add components, except chloramphenicol solution and sheep blood, to distilled/deionized water and bring volume to 940.0mL. Mix thoroughly. Gently heat and bring to boiling. Autoclave for 15 min at 15 psi pressure–121°C. Cool to 45° to 50°C. Aseptically add sterile chloramphenicol solution and sheep blood. Mix thoroughly. Pour into sterile Petri dishes or distribute into sterile tubes.

Use: For the isolation and cultivation of a wide variety of fungal species, especially systemic fungi, from clinical specimens and in epidemiological investigations. Also used for the selective isolation of pathogenic fungi from specimens heavily contaminated with bacteria and saprophtic fungi. For maintenance of fungal species on slant cultures.

Brain Heart Infusion Agar with Penicillin and Streptomycin

Composition per liter:

Pancreatic digest of casein	16.0g
Agar	13.5g
Brain/heart, solids from infusion	8.0g
Peptic digest of animal tissue	5.0g
NaCl	5.0g
Glucose	2.0g
Na_2HPO_4	2.5g
Streptomycin	40.0mg
Penicillin	20,000U
Sheep blood, defibrinated	50.0mL

pH 7.4 ± 0.2 at 25°C

Preparation of Medium: Add components, except sheep blood, to distilled/deionized water and bring volume to 950.0mL. Mix thoroughly and while stirring bring to a boil for 1 min to completely dissolve. Autoclave for 15 min at 15 psi pressure–121°C. Cool to 50°C. Aseptically add 50.0mL defibrinated sheep blood. Mix thoroughly. Pour into sterile Petri dishes while agitating gently to distribute the precipitate through the medium.

Use: For the isolation and cultivation of a wide variety of fungal species, especially systemic fungi, from clinical specimens and in epidemiological investigations. Also used for the selective isolation of pathogenic fungi from specimens heavily contaminated with bacteria and saprophtic fungi. For maintenance of fungal species on slant cultures.

Brain Heart Infusion Agar with 10% Sheep Blood, Gentamicin and Chloramphenicol

Composition per liter:

Pancreatic digest of casein	16.0g
Agar	13.5g
Brain/heart, solids from infusion	8.0g
Peptic digest of animal tissue	5.0g
NaCl	5.0g
Glucose	2.0g
Na_2HPO_4	2.5g
Sheep blood, defibrinated	100.0mL
Antibiotic solution	10.0mL

pH 7.4 ± 0.2 at 25°C

Antibiotic Solution:
Composition per 10mL:

Chloramphenicol	0.05g
Gentamicin	0.05g

Preparation of Antibiotic Solution: Add components to distilled/deionized water and bring volume to 10.0mL. Mix thoroughly. Filter sterilize.

Preparation of Medium: Add components, except antibiotic solution and sheep blood, to distilled/deionized water and bring volume to 940.0mL. Mix thoroughly. Gently heat and bring to boiling. Autoclave for 15 min at 15 psi pressure–121°C. Cool to 45° to 50°C. Aseptically add sterile antibiotic solution and sheep blood. Mix thoroughly. Pour into sterile Petri dishes or distribute into sterile tubes.

Use: For the isolation and cultivation of a wide variety of fungal species, especially systemic fungi, from clinical specimens and in epidemiological investigations. Also used for the selective isolation of pathogenic fungi from specimens heavily contaminated with bacteria and saprophytic fungi. For maintenance of fungal species on slant cultures.

Brain Heart Infusion with Agar, Yeast Extract, Sucrose, Inactivated Horse Serum and Penicillin

Composition per liter:

Agar	12.0g
Yeast extract	5.0g
Sucrose	100.0g
Pancreatic digest of gelatin	14.5g
Brain heart, solids from infusion	6.0g

Peptic digest of animal tissue 6.0g
NaCl 5.0g
Glucose 3.0g
Na_2HPO_4 .... 2.5g
Horse serum, inactivated 100.0mL
Penicillin solution 10.0mL
pH 7.4 ± 0.2 at 25°C

Penicillin Solution:
Composition per 10mL:
Penicillin 1,000,000U

Preparation of Penicillin Solution: Add penicillin to distilled/deionized water and bring volume to 10.0mL. Mix thoroughly. Filter sterilize.

Preparation of Medium: Add components, except penicillin solution and inactivated horse serum, to distilled/deionized water and bring volume to 890.0mL. Mix thoroughly. Gently heat and bring to boiling. Autoclave for 15 min at 15 psi pressure–121°C. Cool to 45° to 50°C. Aseptically add sterile penicillin solution and horse serum. Mix thoroughly. Pour into sterile Petri dishes or distribute into sterile tubes.

Use: For the cultivation of fastidious fungi.

Brain Heart Infusion with Agar, Yeast Extract, NaCl, Inactivated Horse Serum and Penicillin

Composition per liter:
Agar 12.0g
Yeast extract 5.0g
NaCl 20.0g
Pancreatic digest of gelatin 14.5g
Brain heart, solids from infusion 6.0g
Peptic digest of animal tissue 6.0g
NaCl 5.0g
Glucose 3.0g
Na_2HPO_4 .... 2.5g
Horse serum, inactivated 100.0mL
Penicillin solution 10.0mL
pH 7.4 ± 0.2 at 25°C

Penicillin Solution:
Composition per 10mL:
Penicillin 1,000,000U

Preparation of Penicillin Solution: Add penicillin to distilled/deionized water and bring volume to 10.0mL. Mix thoroughly. Filter sterilize.

Preparation of Medium: Add components, except penicillin solution and inactivated horse serum, to distilled/deionized water and bring volume to 890.0mL. Mix thoroughly. Gently heat and bring to boiling. Autoclave for 15 min at 15 psi pressure–121°C. Cool to 45° to 50°C. Aseptically add sterile penicillin solution and horse serum. Mix thoroughly. Pour into sterile Petri dishes or distribute into sterile tubes.

Use: For the cultivation of fastidious fungi.

Brain Heart Infusion with 3% Sodium Chloride

Composition per liter:
NaCl 30.0g
Pancreatic digest of gelatin 14.5g
Brain/heart, solids from infusion 6.0g
Peptic digest of animal tissue 6.0g
Glucose 3.0g
Na_2HPO_4 .... 2.5g
pH 7.4 ± 0.2 at 25°C

Preparation of Medium: Add components to distilled/deionized water and bring volume to 1.0L. Mix thoroughly. Distribute into tubes or flasks. Autoclave for 15 min at 15 psi–121°C.

Use: For the cultivation of *Vibrio parahaemolyticus*.

Brewer Thioglycollate Medium

Composition per liter:
Beef, infusion from 500.0g
Proteose peptone 10.0g
NaCl 5.0g
Glucose 5.0g
K_2HPO_4 .... 2.0g
Sodium thioglycollate 0.5g
Agar 0.5g
Methylene Blue 2.0mg
pH 7.2 ± 0.2 at 25°C

Source: This medium is available as a premixed powder from Difco Laboratories.

Preparation of Medium: Add components to distilled/deionized water and bring volume to 1.0L. Mix thoroughly. Gently heat to boiling. Distribute into tubes or flasks. Autoclave for 15 min at 15 psi–121°C.

Use: For the cultivation and maintenance of anaerobic and microaerophilic microorganisms. Also used for testing the sterility of biological products and materials.

Brewer Thioglycollate Medium, Modified

Composition per liter:
Tryptic digest of casein 17.0g

Glucose 10.0g
NaCl 5.0g
Enzymatic hydrolysate of soybean meal.............3.0g
K_2HPO_4 ....2.0g
Sodium thioglycollate 1.0g
Agar...0.5g
Methylene Blue 2.0mg

pH 7.2 ± 0.2 at 25°C

Source: This medium is available as a premixed powder from Difco Laboratories.

Preparation of Medium: Add components to distilled/deionized water and bring volume to 1.0L. Mix thoroughly. Gently heat to boiling. Distribute into tubes or flasks. Autoclave for 15 min at 15 psi–121°C.

Use: For the cultivation and maintenance of anaerobic and microaerophilic microorganisms. Also used for testing the sterility of biological products and materials.

Brilliant Green Agar

Composition per liter:

Agar...20.0g
Lactose 10.0g
Sucrose..10.0g
Peptic digest of animal tissue.............................5.0g
Pancreatic digest of casein 5.0g
NaCl 5.0g
Phenol Red 0.08g
Brilliant Green 0.0125g

pH 6.9 ± 0.2 at 25°C

Source: This medium is available as a premixed powder from Difco Laboratories, Oxoid Unipath and BBL Microbiology Systems.

Preparation of Medium: Add components to distilled/deionized water and bring volume to1.0L. Mix thoroughly. Gently heat and bring to boiling. Distribute into tubes or flasks. Autoclave for 15 min at 15 psi pressure–121°C. Pour into sterile Petri dishes.

Use: For the selective isolation of *Salmonella* other than *S. typhi* from feces and other specimens, and food and dairy products. *Salmonella* other than *S. typhi* appear as red/pink/white colonies surrounded by a zone of red in the agar indicating nonlactose/sucrose fermentation. *Proteus* or *Pseudomonas* species may appear as small red colonies. Lactose- or sucrose-fermenting bacteria appear as yellow-green colonies surrounded by a zone of yellow-green in the agar.

Bromcresol Purple Broth

Composition per liter:

Peptone..10.0g
NaCl 5.0g
Beef extract 3.0g
Bromcresol Purple 0.04g
Carbohydrate solution 10.0mL

pH 7.0 ± 0.2 at 25°C

Carbohydrate Solution:
Composition per 10mL:

Carbohydrate...5.0g

Preparation of Carbohydrate Solution: Add carbohydrate to distilled/deionized water and bring volume to 10.0mL. Mix thoroughly. Filter sterilize.

Preparation of Medium: Add components to distilled/deionized water and bring volume to 1.0L. Mix thoroughly. Gently heat and bring to boiling. Distribute into test tubes that contain an inverted Durham tube. Autoclave for 10 min at 15 psi pressure–121°C.

Use: For the differentiation of a variety of microorganisms based on their fermentation of specific carbohydrates. Bacteria that ferment the specific carbohydrate turn the medium yellow. When bacteria produce gas, the gas is trapped in the Durham tube.

Bromcresol Purple Dextrose Broth (BCP Broth)

Composition per liter:

Glucose 10.0g
Peptone..5.0g
Beef extract 3.0g
Bromcresol Purple solution.............................2.0mL

pH 7.0 ± 0.2 at 25°C

Bromcresol Purple Solution:
Composition per 10mL:

Bromcresol Purple 0.16g
Ethanol (95% solution) 10.0mL

Preparation of Bromcresol Purple Solution: Add Bromcresol Purple to 10.0mL of ethanol. Mix thoroughly.

Preparation of Medium: Add components to distilled/deionized water and bring volume to 1.0L. Mix thoroughly. Distribute into tubes in 12–15mL volumes. Autoclave for 15 min at 15 psi pressure–121°C.

Use: For the cultivation and differentiation of bacteria based on their ability to ferment glucose. Bacteria that ferment glucose turn the medium yellow.

Bromocresol Purple Milk Solids Glucose Agar (BCP MS G Agar)

Composition per 2L:

Skim milk powder..80.0g
Glucose 40.0g
Agar...30.0g
Bromcresol Purple solution........................2.0mL

pH 6.6 ± 0.2 at 25°C

Bromcresol Purple Solution:
Composition per 10mL:

Bromcresol Purple 0.16g
Ethanol (95% solution) 10.0mL

Preparation of Bromcresol Purple Solution: Add Bromcresol Purple to 10.0mL of ethanol. Mix thoroughly.

Preparation of Medium: Add skim milk powder and Bromcresol Purple solution to distilled/deionized water and bring volume to 1.0L. Mix thoroughly. Autoclave for 8 min at 11 psi pressure–116°C. Cool to 45° to 50°C. In a separate flask add glucose to distilled/deionized water and bring volume to 200.0mL. Mix thoroughly. Autoclave for 8 min at 11 psi pressure–116°C. Cool to 45° to 50°C. In a third flask add agar to distilled/deionized water and bring volume to 800.0mL. Mix thoroughly. Gently heat and bring to boiling. Autoclave for 15 min at 15 psi pressure–121°C. Cool to 45° to 50°C. Aseptically combine the three sterile solutions. Mix thoroughly. Aseptically adjust the pH to 6.6 with sterile 1*N* HCl. Aseptically distribute into sterile tubes. Allow tubes to cool in a slanted position.

Use: For the cultivation and differentiation of *Trychophyton mentagrophytes, Trychophyton rubrum,* and *Microsporum persicolor.*

Bromcresol Purple Milk Yeast Extract with CCG

Composition per liter:

Milk solution...1.0L
Agar solution...900.0mL
Yeast extract solution 40.0mL
Chloramphenicol solution..........................10.0mL
Cycloheximide solution 10.0mL
Gentamicin solution 0.8mL

Milk Solution:
Composition per liter:

Skim milk powder..80.0g
Bromcresol Purple solution........................2.0mL

Preparation of Milk Solution: Add components to distilled/deionized water and bring volume to 1.0L. Mix thoroughly. Autoclave for 8 min at 11 psi pressure–116°C. Cool to 45° to 50°C.

Bromcresol Purple Solution:
Composition per 10mL:

Bromcresol Purple 0.16g
Ethanol (95% solution) 10.0mL

Preparation of Bromcresol Purple Solution: Add Bromcresol Purple to 10.0mL of ethanol. Mix thoroughly.

Agar Solution:
Composition per 900mL:

Agar...30.0g

Preparation of Agar Solution: Add agar to distilled/deionized water and bring volume to 900.0mL. Mix thoroughly. Gently heat and bring to boiling. Autoclave for 15 min at 15 psi pressure–121°C. Cool to 45° to 50°C.

Yeast Extract Solution:
Composition per 100mL:

Yeast extract 40.0g

Preparation of Yeast Extract Solution: Add yeast extract to distilled/deionized water and bring volume to 100.0mL. Mix thoroughly. Filter sterilize.

Chloramphenicol Solution:
Composition per 10mL:

Chloramphenicol...0.1g

Preparation of Chloramphenicol Solution: Add chloramphenicol to distilled/deionized water and bring volume to 10.0mL. Mix thoroughly. Filter sterilize.

Cycloheximide Solution:
Composition per 10mL:

Cycloheximide 0.2g

Preparation of Cycloheximide Solution: Add cycloheximide to distilled/deionized water and bring volume to 10.0mL. Mix thoroughly. Filter sterilize.

Gentamicin Solution:
Composition per 10mL:

Gentamicin 0.5g

Preparation of Gentamicin Solution: Add gentamicin to distilled/deionized water and bring volume to 10.0mL. Mix thoroughly. Filter sterilize.

Preparation of Medium: Aseptically combine the sterile milk solution and sterile agar solution. Aseptically add 40.0mL of sterile yeast extract solution, 10.0mL of sterile chloramphenicol solution, 10.0mL of sterile cycloheximide solution and 0.8mL of sterile gentamicin solution. Mix thoroughly. Aseptically distribute into sterile tubes. Allow tubes to cool in a slanted position.

Use: For the isolation, cultivation and differentiation of *Trichophyton verrucosum* and *Trichophyton schoenleinii.*

Bromthymol Blue Agar

Composition per liter:

Agar 11.0g
Peptone 10.0g
NaCl 5.0g
Yeast extract 5.0g
Lactose (33% solution) 27.0mL
Bromthymol Blue (1% solution) 10.0mL
Sodium thiosulfate (50% solution) 2.0mL
Glucose (33% solution) 1.2mL
Maranil solution (5% solution) 1.0mL
pH 7.7–7.8 at 25°C

Preparation of Medium: Add agar, peptone, NaCl, and yeast extract to distilled/deionized water and bring volume to 1.0L. Mix thoroughly. Adjust pH to 8.0. Autoclave for 20 min at 15 psi pressure–121°C. Cool to 45° to 50°C. Filter sterilize separately the lactose solution, Bromthymol Blue solution, sodium thiosulfate solution, glucose solution and maranil solution. To the cooled, sterile agar solution aseptically add 27.0mL of sterile lactose solution, 10.0mL of sterile Bromthymol Blue solution, 2.0mL of sterile sodium thiosulfate solution, 1.2mL of sterile glucose solution and1.0mL of sterile maranil solution. Mix thoroughly. Adjust pH to 7.7–7.8. Pour into sterile Petri dishes or distribute into sterile tubes.

Use: For the selective isolation and cultivation of members of the Enterobacteriaceae.

Bromthymol Blue Broth

Composition per 101.45mL:

Pancreatic digest of casein 0.7g
NaCl 0.5g
Beef extract 0.3g
Yeast extract 0.3g
Beef heart, solids from infusion 0.2g
Yeast extract 0.1g
Horse serum 10.0mL
Bromthymol Blue solution 1.0mL
Ampicillin solution 1.0mL
Urea solution 0.25mL
Nystatin solution 0.1mL
Tripeptide solution 0.1mL
pH 6.0 ± 0.2 at 25°C

Bromothymol Blue Solution:
Composition per 50mL:

Bromothymol Blue 0.2g
NaOH (0.01*N* solution) 32.0mL

Preparation of Bromothymol Blue Solution: Add Bromthymol Blue to NaOH solution. Mix thoroughly. Bring volume to 50.0mL with distilled/deionized water. Autoclave for 15 min at 15 psi pressure–121°C. Store at 25°C.

Ampicillin Solution:
Composition per 10mL:

Ampicillin 1.0g

Preparation of Ampicillin Solution: Add ampicillin to distilled/deionized water and bring volume to 10.0mL. Mix thoroughly. Filter sterilize.

Urea Solution:
Composition per 100mL:

Urea 10.0g

Preparation of Urea Solution: Add urea to distilled/deionized water and bring volume to 100.0mL. Filter sterilize. Store at –20°C.

Nystatin Solution:
Composition per 1mL:

Nystatin 50,000U

Preparation of Nystatin Solution: Add nystatin to distilled/deionized water and bring volume to 1.0mL. Filter sterilize.

Tripeptide Solution:
Composition per 10mL:

Glycyl-L-histidyl-L-lysine acetate 0.2mg

Preparation of Tripeptide Solution: Add glycyl-L-histidyl-L-lysine acetate to distilled/deionized water and bring volume to 10.0mL. Mix thoroughly. Filter sterilize. Store at –20°C.

Preparation of Medium: Add components—except horse serum, ampicillin solution, urea solution, nystatin solution and tripeptide solution—to distilled/deionized water and bring volume to 90.0mL. Mix thoroughly. Gently heat and bring to boiling. Autoclave for 15 min at 15 psi pressure–121°C. Cool to 45° to 50°C. Aseptically add 10.0mL of sterile horse serum, 1.0mL of sterile ampicillin solution, 0.25mL of sterile urea solution, 0.1mL of sterile nystatin solution and 0.1mL of sterile tripeptide solution. Mix thoroughly. Pour into sterile Petri dishes.

Use: For the cultivation of *Ureaplasma* species from clinical specimens.

Brucella Agar

Composition per liter:

Agar 15.0g
Pancreatic digest of casein 10.0g
Peptic digest of animal tissue 10.0g
NaCl 5.0g

Yeast extract 2.0g
Glucose 1.0g
$NaHSO_3$ 0.1g
Horse blood, defibrinated 100.0mL
pH 7.0 ± 0.2 at 25°C

Source: This medium is available as a premixed powder from Difco Laboratories and BBL Microbiology Systems and Oxoid Unipath.

Preparation of Medium: Add components to distilled/deionized water and bring volume to 900.0mL. Mix thoroughly. Heat gently with frequent mixing. Boil for 1 min. Autoclave for 15 min at 15 psi pressure–121°C. Cool to 45° to 50°C. Add 100.0mL sterile defibrinated horse blood. Mix gently and pour into sterile Petri dishes.

Use: For the cultivation and maintenance of *Brucella* species. Also used for the isolation and cultivation of nonfastidious and fastidious microorganisms from a variety of clinical specimens and in epidemiological investigations.

Brucella Agar Base *Campylobacter* Medium

Composition per 1100mL:

Cycloheximide (actidione) 0.05g
Sodium cephazolin 0.015g
Novobiocin 5.0mg
Bacitracin 25,000U
Colistin sulfate 10,000U
Brucella agar base 1.0L
Horse blood, defibrinated 100.0mL

Brucella Agar Base
Composition per liter:

Agar 15.0g
Pancreatic digest of casein 10.0g
Peptic digest of animal tissue 10.0g
NaCl 5.0g
Yeast extract 2.0g
Glucose 1.0g
$NaHSO_3$ 0.1g

Preparation of Brucella Agar Base: Add components to distilled/deionized water and bring volume to 1.0L. Mix thoroughly.

Optional Supplement:
Composition per 10mL:

Sodium pyruvate 0.25g
$NaHSO_3$ 0.25g
$FeSO_4 \cdot 7H_2O$ 0.25g

Preparation of Optional Supplement: Add components to distilled/deionized water and bring volume to 10.0mL. Filter sterilize.

Preparation of Medium: Add components to 1.0L of prepared Brucella Agar Base. Mix thoroughly. Autoclave for 15 min at 15 psi pressure–121°C. Cool to 45° to 50°C. Add 100.0mL sterile, defibrinated horse blood. Addition of 10.0mL of optional supplement will improve growth. Mix thoroughly. Pour into sterile Petri dishes.

Use: For the selective isolation and cultivation of *Campylobacter jejuni* from fecal specimens or rectal swabs.

Brucella Agar with Vitamin K_1

Composition per liter:

Agar 17.5g
Pancreatic digest of casein 10.0g
Peptic digest of animal tissue 10.0g
NaCl 5.0g
Yeast extract 2.0g
Glucose 1.0g
$NaHSO_3$ 0.1g
Sheep blood, defibrinated 50.0mL
Vitamin K_1 solution 1.0mL
pH 7.0 ± 0.2 at 25°C

Vitamin K_1 Solution:
Composition per 20mL:

Vitamin K_1 .... 0.2g
Ethanol, absolute 20.0mL

Preparation of Vitamin K_1 Solution: Add vitamin K_1 to 20.0mL of ethanol. Mix thoroughly. Filter sterilize.

Preparation of Medium: Add components, except sheep blood and vitamin K_1 solution, to distilled/deionized water and bring volume to 949.0mL. Mix thoroughly. Gently heat and bring to boiling. Autoclave for 15 min at 15 psi pressure–121°C. Cool to 45° to 50°C. Aseptically add 50.0mL of sterile sheep blood and 1.0mL of sterile vitamin K_1 solution. Mix thoroughly. Pour into sterile Petri dishes or distribute into sterile tubes.

Use: For the cultivation of *Brucella* species.

Brucella Albimi Broth with 0.16% Agar

Composition per liter:

Pancreatic digest of casein 10.0g
Peptic digest of animal tissue 10.0g
NaCl 5.0g
Yeast extract 2.0g
Agar 1.6g
Glucose 1.0g

$NaHSO_3$ 0.1g
Horse blood, defibrinated 100.0mL
pH 7.0 ± 0.2 at 25°C

Preparation of Medium: Add components, except horse blood, to distilled/deionized water and bring volume to 900.0mL. Mix thoroughly. Heat gently with frequent mixing. Boil for 1 min. Autoclave for 15 min at 15 psi pressure–121°C. Cool to 45° to 50°C. Aseptically add 100.0mL of sterile defibrinated horse blood. Mix thoroughly. Aseptically distribute into sterile tubes or flasks.

Use: For the cultivation and maintenance of *Campylobacter* species.

Brucella Anaerobic Blood Agar

Composition per liter:
Vitamin K_1 .... 0.01g
Anaerobic agar base 1000.0mL
Sheep blood, sterile, defibrinated 50.0mL

Anaerobic Agar Base
Composition per liter:
Pancreatic digest of casein 17.5g
Agar 15.0g
Glucose 10.0g
Papaic digest of soybean meal 2.5g
NaCl 2.5g
Sodium thioglycollate 2.0g
Sodium formaldehyde sulfoxylate 1.0g
L-Cystine·HCl·H_2O 0.4g
Methylene Blue 0.002g
pH 7.0 ± 0.2 at 25°C

Preparation of Anaerobic Agar Base: Add components to distilled/deionized water and bring volume to 1.0L. Mix thoroughly. Autoclave for 15 min at 15 psi pressure–121°C. Cool to 45° to 50°C.

Preparation of Medium: To 950.0mL of cooled, sterile anaerobic agar base, aseptically add 10.0mg of vitamin K_1 and 50.0mL sterile, defibrinated sheep blood.

Use: For the isolation of anaerobes.

Brucella Blood Agar with Hemin and Vitamin K_1

Composition per liter:
Agar 15.0g
Pancreatic digest of casein 10.0g
Peptic digest of animal tissue 10.0g
NaCl 5.0g
Yeast extract 2.0g
Glucose 1.0g
$NaHSO_3$ 0.1g
Vitamin K_1 .... 1.0mL
Hemin 1.0mL
Sheep blood, defibrinated 50.0mL

Source: Available as a prepared medium from BBL Microbiology Systems.

Vitamin K_1 Solution:
Composition per 100mL:
Vitamin K_1 .... 1.0g
Ethanol 99.0mL

Preparation of Vitamin K_1 Solution: Add vitamin K_1 to 99.0mL of absolute ethanol. Mix thoroughly. Filter sterilize.

Hemin Solution:
Composition per 100mL:
Hemin 1.0g
NaOH (1*N* solution) 20.0mL

Preparation of Hemin Solution: Add hemin to 20.0mL of 1*N* NaOH solution. Mix thoroughly. Bring volume to 100.0mL with distilled/deionized water.

Preparation of Medium: Add components, except vitamin K_1 solution and sheep blood, to distilled/deionized water and bring volume to 949.0mL. Mix thoroughly. Gently heat and bring to boiling. Autoclave for 15 min at 15 psi pressure–121°C. Cool to 45° to 50°C. Aseptically add 1.0mL sterile vitamin K_1 solution and 50.0mL of sterile defibrinated sheep blood. Mix gently and pour into sterile Petri dishes.

Use: For the isolation and cultivation of anaerobic microorganisms from clinical specimens and in epidemiological investigations. After growth on agar plates colonies should be examined under a dissecting microscope under long-wave UV light. Members of the pigmented *Bacteroides* group appear as red/orange fluorescent colonies.

Brucella Blood Culture Broth

Composition per liter:
Sucrose 100.0g
Hemin 0.5g
Sodium polyanetholsulfonate (SPS) 0.25g
Brucella broth base 1000.0mL
Vitamin K_1 solution 1.0mL
pH 7.0 ± 0.2 at 25°C

***Brucella* Broth Base:**
Composition per liter:
Pancreatic digest of casein 10.0g
Peptic digest of animal tissue 10.0g
NaCl 5.0g

Yeast extract 2.0g
Glucose 1.0g
$NaHSO_3$ 0.1g

Preparation of *Brucella* Broth Base: Add components to distilled/deionized water and bring volume to 1.0L. Mix thoroughly.

Vitamin K_1 Solution:
Composition per 100mL:
Vitamin K_1 .... 1.09g

Preparation of Vitamin K_1 Solution: Add vitamin K_1 to 99.0mL of absolute ethanol. Store in the dark at 4°C.

Preparation of Medium: Add components, except Vitamin K_1 solution, to prepared *Brucella* Broth Base. Autoclave for 15 min at 15 psi pressure–121°C. Cool to 45° to 50°C. Aseptically add 1.0mL of vitamin K_1 solution. Distribute into sterile tubes or flasks.

Use: For the isolation and cultivation of microorganisms from blood. Especially useful for the cultivation of anaerobes.

Brucella Selective Medium

Composition per liter:
Beef heart, infusion from 500.0g
Agar 15.0g
Tryptose 10.0g
NaCl 5.0g
Glucose 2.5g
Gelatin 1.0g
Sheep blood 100.0mL
Antibiotic solution 10.0mL
pH 7.4 ± 0.2 at 25°C

Antibiotic Solution:
Composition per 10mL:
Cycloheximide 1.0g
Bacitracin 250,000U
Circulin 250,000U
Polymyxin B 100,000U

Preparation of Antibiotic Solution: Add components to distilled/deionized water and bring volume to 10.0mL. Mix thoroughly. Filter sterilize.

Preparation of Medium: Add components, except sheep blood and antibiotic solution, to distilled/deionized water and bring volume to 890.0mL. Mix thoroughly. Gently heat and bring to boiling. Autoclave for 15 min at 15 psi pressure–121°C. Cool to 45° to 50°C. Aseptically add 100.0mL of sterile sheep blood and 10.0mL of sterile antibiotic solution. Mix thoroughly. Pour into sterile Petri dishes or distribute into sterile tubes.

Use: For the selective isolation and cultivation of *Brucella* species.

Buffered Charcoal Yeast Extract Differential Agar (DIFF/BCYE)

Composition per 1014mL:
Agar 17.0g
ACES (2-[(2-Amino-2-oxoethyl)-
amino]-ethane sulfonic acid) buffer 10.0g
Yeast extract 10.0g
Charcoal, activated 1.5g
$Fe_4(P_2O_7)_3 \cdot 9H_2O$ 0.25g
Bromcresol Purple 0.01g
Bromthymol Blue 0.01g
Antibiotic solution 10.0mL
Cysteine·HCl·H_2O solution 4.0mL
pH 6.9 ± 0.2 at 25°C

Antibiotic Solution:
Composition per 10mL:
Vancomycin 1.0mg
Polymyxin B 50,000U

Preparation of Antibiotic Solution: Add components to distilled/deionized water and bring volume to 10.0mL. Mix thoroughly. Filter sterilize.

Cysteine·HCl·H_2O Solution:
Composition per 10mL:
L-Cysteine·HCl·H_2O 1.0g

Preparation of Cysteine·HCl·H_2O Solution: Add L-Cysteine·HCl·H_2O to distilled/deionized water and bring volume to 10.0mL. Mix thoroughly. Filter sterilize.

Preparation of Medium: Add components, except cysteine solution and antibiotic solution, to distilled/deionized water and bring volume to 1.0L. Mix thoroughly. Adjust medium to pH 6.9 with 1*N* KOH. Heat gently and bring to boil for 1 minute. Autoclave for 15 min at 15 psi pressure–121°C. Cool to 50–55°C. Add 4.0mL of sterile L-cysteine·HCl·H_2O solution and 10.0mL of sterile antibiotic solution. Mix thoroughly. Pour into sterile Petri dishes with constant agitation to keep charcoal in suspension.

Use: For the isolation, cultivation, and maintenance of *Legionella pneumophila* and other *Legionella* species from environmental and clinical specimens. Used for the selective recovery of *Legionella pneumophila* while reducing contaminating microorganisms from environmental water samples.

Buffered Peptone Water

Composition per liter:

Pancreatic digest of gelatin	10.0g
NaCl	5.0g
Na_2HPO_4	3.5g
KH_2PO_4	1.5g

pH 7.2 ± 0.2 at 25°C

Source: This medium is available as a premixed powder from BBL Laboratories, Difco Laboratories, and Oxoid Unipath.

Preparation of Medium: Add components to distilled/deionized water and bring volume to 1.0L. Mix thoroughly. Distribute into tubes or flasks. Autoclave for 15 min at 15 psi pressure–121°C.

Use: Use as a pre-enrichment medium for the isolation of *Salmonella*, especially injured microorganisms, from various food sources.

Caffeic Acid Ferric Citrate Test Medium (CAFC Test Medium) (Caffeic Acid Agar)

Composition per liter:

Agar	20.0g
$(NH_4)_2SO_4$	5.0g
Glucose	5.0g
Yeast extract	2.0g
K_2HPO_4	0.8g
$MgSO_4 \cdot 3H_2O$	0.7g
Caffeic acid·$1/2H_2O$	0.18g
Chloramphenicol	0.05g
Ferric citrate solution	4.0mL

pH 6.5 ± 0.2 at 25°C

Ferric Citrate Solution:

Composition per 20mL:

Agar	100.0mg

Preparation of Ferric Citrate Solution: Add ferric citrate to 20.0mL of distilled/deionized water. Mix thoroughly.

Preparation of Medium: Add components, except chloramphenicol, to distilled/deionized water and bring volume to 1.0L. Mix thoroughly. Heat to boiling. Autoclave for 15 min at 15 psi pressure–121°C. Cool to 45° to 50°C. Aseptically add 0.05g of chloramphenicol. Mix thoroughly. Pour into sterile Petri dishes.

Use: For the isolation and presumptive identification of *Cryptococcus neoformans*. *C. neoformans* appears as dark brown colonies. All other *Cryptococcus* species appear as light brown or nonpigmented colonies.

CAL Agar (Cellobiose Arginine Lysine Agar) (Yersinia Isolation Agar)

Composition per liter:

Agar	20.0g
L-Arginine·HCl	6.5g
L-Lysine·HCl	6.5g
NaCl	5.0g
Cellobiose	3.5g
Yeast extract	3.0g
Sodium deoxycholate	1.5g
Neutral Red	0.03g

Preparation of Medium: Add components to distilled/deionized water and bring volume to 1.0L. Mix thoroughly. Heat to boiling. Do not autoclave. Pour into sterile Petri dishes.

Use: For isolation and characterization of *Yersinia enterocolitica* from fecal specimens and enumeration of *Y. enterocolitica* from water and other liquid specimens.

CAL Broth (Cellobiose Arginine Lysine Broth)

Composition per liter:

L-Arginine·HCl	6.5g
L-Lysine·HCl	6.5g
NaCl	5.0g
Cellobiose	3.5g
Yeast extract	3.0g
Sodium deoxycholate	1.5g
Neutral Red	0.03g

Preparation of Medium: Add components to distilled/deionized water and bring volume to 1.0L. Mix thoroughly. Heat to boiling. Do not autoclave. Distribute into sterile tubes in 6.0–8.0mL volumes.

Use: For isolation and characterization of *Yersinia enterocolitica* from fecal specimens and enumeration of *Y. enterocolitica* from water and other liquid specimens.

Campy THIO Medium

Composition per liter:

Pancreatic digest of casein	20.0g
Agar	15.0g
NaCl	2.5g
K_2HPO_4	1.5g
Sodium thioglycollate	0.6g

L-Cystine 0.4g
Na_2SO_3 .... 0.2g
Antibiotic supplement 10.0mL

Antibiotic Supplement:
Composition per 10mL:

Cephalothin 15.0mg
Vancomycin 10.0mg
Trimethoprim 5.0mg
Amphotericin B 2.0mg
Polymyxin B 2,500U

Preparation of Antibiotic Supplement: Add components to 10.0mL of distilled/deionized water. Filter sterilize.

Preparation of Medium: Add components, except antibiotic solution, to distilled/deionized water and bring volume to 990.0mL. Mix thoroughly. Gently heat and bring to boiling. Autoclave for 15 min at 15 psi pressure–121°C. Cool to 45° to 50°C. Aseptically add 10.0mL of sterile antibiotic solution. Mix thoroughly. Aseptcially distribute into sterile screw-capped tubes in 3.0mL volumes for 1.5-cm swabs or 5.0mL volumes for 3.0-cm swabs.

Use: For the maintenance—as a holding or transport medium—of *Campylobacter* species isolated from clinical specimens on swabs.

Campylobacter Agar with 5 Antimicrobics and 10% Sheep Blood

Composition per liter:

Agar 15.0g
Pancreatic digest of casein 10.0g
Peptic digest of animal tissue 10.0g
NaCl 5.0g
Yeast extract 2.0g
Glucose 1.0g
$NaHSO_3$ 0.1g
Sheep blood, defibrinated 100.0mL
Antibiotic supplement 10.0mL

pH 7.2 ± 0.2 at 25°C

Antibiotic Supplement:
Composition per 10mL:

Cephalothin 0.015g
Vancomycin 0.01g
Trimethoprim 5.0mg
Amphotericin B 2.0mg
Polymyxin B 2,500U

Preparation of Antibiotic Supplement: Add components to 10.0mL of distilled/deionized water. Filter sterilize.

Source: This medium is available as a prepared medium from BBL Microbiology Systems.

Preparation of Medium: Add components, except sheep blood and antibiotic solution, to distilled/deionized water and bring volume to 890.0mL. Mix thoroughly. Gently heat and bring to boiling. Autoclave for 15 min at 15 psi pressure–121°C. Cool to 45° to 50°C. Aseptically add 100.0mL of sterile sheep blood and 10.0mL of sterile antibiotic solution. Mix thoroughly. Pour into sterile Petri dishes or distribute into sterile tubes.

Use: For the primary selective isolation and cultivation of *Campylobacter jejuni* from human fecal specimens.

Campylobacter Agar, Blaser's (Blaser's *Campylobacter* Agar)

Composition per liter:

Campylobacter agar base 990.0mL
Supplement B 10.0mL

pH 7.4 ± 0.2 at 25°C

Campylobacter Agar Base:
Composition per liter:

Proteose peptone 15.0g
Agar 12.0g
NaCl 5.0g
Yeast extract 5.0g
Liver digest 2.5g

Source: *Campylobacter* agar base and *Campylobacter* antimicrobic supplement B are available as a premixed powder from Difco Laboratories.

Preparation of *Campylobacter* Agar Base: Add components to distilled/deionized water and bring volume to 990.0mL. Mix thoroughly. Gently heat and bring to boiling. Autoclave for 15 min at 15 psi pressure–121°C. Cool to 45° to 50°C.

Supplement B:
Composition per 10mL:

Cephalothin 15mg
Vancomycin 10mg
Trimethoprim 5mg
Amphotericin B 2mg
Polymyxin B 2,500units

Preparation of Supplement B: Add components to 10.0mL of distilled/deionized water. Filter sterilize.

Preparation of Medium: Prepare 990.0mL of *Campylobacter* Agar Base. Autoclave and cool to 45° to 50°C. Aseptically add 10.0mL of sterile Supplement B. Mix thoroughly. Pour into sterile Petri dishes.

Use: For the selective isolation of *Campylobacter jejuni* from fecal specimens, food and environmental specimens.

Campylobacter Agar, Skirrow's (Skirrow's *Campylobacter* Agar)

Composition per liter:

Campylobacter agar base 990.0mL
Supplement S 10.0mL
pH 7.4 ± 0.2 at 25°C

***Campylobacter* Agar Base:**
Composition per liter:

Proteose peptone 15.0g
Agar 12.0g
NaCl 5.0g
Yeast extract 5.0g
Liver digest 2.5g

Source: *Campylobacter* agar base and *Campylobacter* antimicrobic supplement S are available as a premixed powder from Difco Laboratories.

Preparation of *Campylobacter* Agar Base: Add components to distilled/deionized water and bring volume to 990.0mL. Mix thoroughly. Gently heat and bring to boiling. Autoclave for 15 min at 15 psi pressure–121°C. Cool to 45° to 50°C.

Supplement S:
Composition per 10mL:

Vancomycin 10mg
Trimethoprim 5mg
Polymyxin B 2,500units

Preparation of Supplement S: Add components to 10.0mL of distilled/deionized water. Filter sterilize.

Preparation of Medium: Prepare 990.0mL of *Campylobacter* Agar Base. Autoclave and cool to 45° to 50°C. Aseptically add 10.0mL of sterile Supplement S. Mix thoroughly. Pour into sterile Petri dishes.

Use: For the selective isolation of *Campylobacter jejuni* from fecal specimens, food and environmental specimens.

Campylobacter Blood-Free Selective Agar

Composition per liter:

Agar 12.0g
Beef extract 10.0g
Peptone 10.0g
Charcoal 4.0g
Casein hydrolysate 3.0g
Sodium deoxycholate 1.0g
$Fe_2SO_4 \cdot H_2O$ 0.25g
Sodium pyruvate 0.25g
Cefoperazone solution 10.0mL
pH 7.4 ± 0.2 at 25°C

Cefoperazone Solution:
Composition per 10mL:

Sodium cefoperazone 0.032g

Preparation of Cefoperazone Solution: Add sodium cefoperazone to distilled/deionized water and bring volume to 10.0mL. Mix thoroughly. Filter sterilize.

Preparation of Medium: Add components, except cefoperazone solution, to distilled/deionized water and bring volume to 990.0mL. Mix throughly. Heat with frequent agitation and boil for 1 min to completely dissolve. Autoclave for 15 min at 15 psi–121°C. Cool to 50–55°C. Add 10.0mL of sterile cefoperazone solution. Addition of 10mg/L amphotericin B improves the selectivity of the medium. Mix thoroughly. Pour into sterile Petri dishes.

Use: For the selective isolation of *Campylobacter* species, especially *C. jejuni, C. coli* and *C. laridis.*

Campylobacter Charcoal Differential Agar (CCDA) (Preston Blood–Free Medium)

Composition per liter:

Agar 12.0g
Beef extract 10.0g
Peptone 10.0g
NaCl 5.0g
Charcoal 4.0g
Casein hydrolysate 3.0g
Sodium deoxycholate 1.0g
$FeSO_4$ 0.25g
Sodium pyruvate 0.25g
Cefoperazone solution 10.0mL
pH 7.5 ± 0.2 at 25°C

Cefoperazone Solution:
Composition per 10mL:

Sodium cefoperazone 0.032g

Preparation of Cefoperazone Solution: Add sodium cefoperazone to distilled/deionized water and bring volume to 10.0mL. Mix thoroughly. Filter sterilize.

Preparation of Medium: Add components, except cefoperazone solution, to distilled/deionized water and bring volume to 990.0mL. Mix thoroughly. Gently heat and bring to boiling. Autoclave for 15 min at 15 psi pressure–121°C. Cool to 45° to 50°C. Aseptically add 10.0mL of sterile cefoperazone solution. Mix thoroughly. Pour into sterile Petri dishes or distribute into sterile tubes.

Use: For the cultivation of *Campylobacter* species.

Campylobacter Selective Medium, Blaser-Wang (Blaser-Wang *Campylobacter* Medium) (Blaser's Agar) (Campy BAP Medium)

Composition per liter:

Brucella Agar Base	890.0mL
Sheep blood	100.0mL
Antibiotic supplement	10.0mL

***Brucella* Agar Base:**
Composition per 890mL:

Agar	15.0g
Glucose	10.0g
Pancreatic digest of casein	10.0g
NaCl	5.0g
Peptic digest of animal tissue	5.0g

pH 7.5 ± 0.2 at 25°C

Preparation of *Brucella* Agar Base: Add components to distilled/deionized water and bring volume to 890.0mL. Mix thoroughly. Gently heat and bring to boiling. Autoclave for 15 min at 15 psi pressure–121°C. Cool to 45° to 50°C.

Antibiotic Supplement:
Composition per 10mL:

Cephalothin	15.0mg
Vancomycin	10.0mg
Trimethoprim	5.0mg
Amphotericin B	2.0mg
Polymyxin B	2,500U

Preparation of Antibiotic Supplement: Add components to 10.0mL of distilled/deionized water. Filter sterilize.

Preparation of Medium: Prepare 890.0mL of *Brucella* agar base. Sterilize as directed. Cool to 50–55°C and add 100.0mL sheep blood or 50.0–70.0mL laked horse blood. Laked blood is prepared by freezing whole blood overnight and thawing to room temperature. Aseptically add 10.0mL of sterile antibiotic supplement. Mix thoroughly. Pour into sterile Petri dishes.

Use: For the selective isolation of *Campylobacter* species.

Campylobacter Selective Medium, Blaser-Wang (Blaser-Wang *Campylobacter* Medium)

Composition per liter:

Columbia agar base	890.0mL
Sheep blood	100.0mL
Antibiotic supplement	10.0mL

pH 7.3 ± 0.2 at 25°C

Columbia Agar Base:
Composition per liter:

Special peptone	25.0g
Agar	10.0g
NaCl	5.0g
Starch	1.0g

Preparation of Columbia Agar Base: Add components to distilled/deionized water and bring volume to 890.0mL. Mix thoroughly. Gently heat and bring to boiling. Autoclave for 15 min at 15 psi pressure–121°C. Cool to 45° to 50°C.

Antibiotic Supplement:
Composition per 10mL:

Cephalothin	15.0mg
Vancomycin	10.0mg
Trimethoprim	5.0mg
Amphotericin B	2.0mg
Polymyxin B	2,500U

Preparation of Antibiotic Supplement: Add components to 10.0mL of distilled/deionized water. Filter sterilize.

Preparation of Medium: To 890.0mL of cooled, sterile Columbia agar base, aseptically add 100.0mL sheep blood or 50.0–70.0mL laked horse blood. Laked blood is prepared by freezing whole blood overnight and thawing to room temperature. Aseptically add 10.0mL of sterile antibiotic supplement. Mix thoroughly. Pour into sterile Petri dishes.

Use: For the selective isolation of *Campylobacter* species.

Campylobacter Selective Medium, Butzler's (Butzler's *Campylobacter* Medium)

Composition per liter:

Brucella Agar Base	940.0mL
Sheep or horse blood, defibrinated	50.0mL
Antibiotic supplement	10.0mL

pH 7.5 ± 0.2 at 25°C

***Brucella* Agar Base:**
Composition per liter:

Agar	15.0g
Glucose	10.0g
Pancreatic digest of casein	10.0g
NaCl	5.0g
Peptic digest of animal tissue	5.0g

Preparation of *Brucella* Agar Base: Add components to distilled/deionized water and bring volume to 940.0mL. Mix thoroughly. Gently heat and bring to boiling. Autoclave for 15 min at 15 psi pressure–121°C. Cool to 45° to 50°C.

Antibiotic Supplement:
Composition per 10mL:

Cycloheximide	50.0mg
Cephazolin	15.0mg
Novobiocin	5.0mg
Bacitracin	25,000U
Colistin sulfate	10,000U

Preparation of Antibiotic Supplement: Add components to 10.0mL of distilled/deionized water. Filter sterilize.

Preparation of Medium: To 940.0mL of cooled sterile *Brucella* agar base, aseptically add 50.0mL of defibrinated sheep or horse blood and 10.0mL of sterile antibiotic supplement. Mix thoroughly. For enhanced growth, medium may also be supplemented with 0.25g $Fe_2SO_4 \cdot H_2O$, 0.25g sodium metabisulfite and 0.25g sodium pyruvate. Pour into sterile Petri dishes.

Use: For the selective isolation of *Campylobacter* species.

Campylobacter Selective Medium, Butzler's (Butzler's *Campylobacter* Medium)

Composition per liter:

Columbia agar base	940.0mL
Blood, horse or sheep	50.0mL
Antibiotic supplement	10.0mL

pH 7.3 ± 0.2 at 25°C

Columbia Agar Base:
Composition per liter:

Peptone	25.0g
Agar	10.0g
NaCl	5.0g
Starch	1.0g

Preparation of Columbia Agar Base: Add components to distilled/deionized water and bring volume to 940.0mL. Mix thoroughly. Gently heat and bring to boiling. Autoclave for 15 min at 15 psi pressure–121°C. Cool to 45° to 50°C.

Antibiotic Supplement:
Composition per 10mL:

Cycloheximide	50.0mg
Cephazolin	15.0mg
Novobiocin	5.0mg
Bacitracin	25,000U
Colistin sulfate	10,000U

Preparation of Antibiotic Supplement: Add components to 10.0mL of distilled/deionized water. Filter sterilize.

Preparation of Medium: To 940.0mL of cooled, sterile Columbia agar base, aseptically add 50.0mL of defibrinated sheep or horse blood and 10.0mL of sterile antibiotic supplement. Mix thoroughly. The medium may also be supplemented with 0.25g $Fe_2SO_4 \cdot H_2O$, 0.25g sodium metabisulfite and 0.25g sodium pyruvate. Pour into sterile Petri dishes.

Use: For the selective isolation of *Campylobacter* species.

Campylobacter Selective Medium, Karmali's (Karmali's *Campylobacter* Medium)

Composition per liter:

Activated charcoal	4.0g
Columbia agar base	990.0mL
Antibiotic supplement	10.0mL

pH 7.4 ± 0.2 at 25°C

Source: This medium is available as a premixed powder from Oxoid.

Columbia Agar Base:
Composition per 990mL:

Peptone	25.0g
Agar	10.0g
NaCl	5.0g
Starch	1.0g

Preparation of Columbia Agar Base: Add components to distilled/deionized water and bring volume to 990.0mL. Mix thoroughly. Gently heat and bring to boiling. Autoclave for 15 min at 15 psi pressure–121°C. Cool to 45° to 50°C.

Antibiotic Supplement:
Composition per 10mL:

Sodium pyruvate	0.05g
Cycloheximide	0.05g
Cefoperazone	0.016g
Hemin	0.016g
Vancomycin	0.01g

Preparation of Antibiotic Supplement: Add components to 10.0mL of distilled/deionized water. Filter sterilize.

Preparation of Medium: Prepare 990.0mL of Columbia agar base. Sterilize as directed. Cool to 50–55°C. Add defibrinated sheep or horse blood to a final concentration of 5–7%. Add 10.0mL of sterile antibiotic supplement. Mix thoroughly. For enhanced growth, medium may also be supplemented with 0.25g $Fe_2SO_4 \cdot H_2O$, 0.25g sodium metabisulfite and 0.25g sodium pyruvate. Pour into sterile Petri dishes. Swirl while pouring to keep charcoal in suspension.

Use: For the selective isolation of *Campylobacter* species.

Campylobacter Selective Medium, Preston's (Preston's *Campylobacter* Medium)

Composition per liter:

Campylobacter agar base	940.0mL
Horse blood, lysed	50.0mL
Antibiotic supplement	10.0mL

pH 7.5 ± 0.2 at 25°C

***Campylobacter* Agar Base:**

Composition per liter:

Agar	12.0g
Beef extract	10.0g
Peptone	10.0g
NaCl	5.0g

Preparation of *Campylobacter* Agar Base: Add components to distilled/deionized water and bring volume to 940.0mL. Mix thoroughly. Gently heat and bring to boiling. Autoclave for 15 min at 15 psi pressure–121°C. Cool to 45° to 50°C.

Antibiotic Supplement:

Composition per 10mL:

Cycloheximide	0.1g
Rifampicin	0.01g
Trimethoprim lactate	0.01g
Polmyxin B	5,000U

Preparation of Antibiotic Supplement: Add components to 10.0mL of 50:50 acetone:distilled/deionized water. Filter sterilize.

Preparation of Medium: To 940.0mL of cooled, sterile *Campylobacter* agar base, aseptically add 50.0mL of lysed horse blood and 10.0mL of sterile antibiotic supplement. Mix thoroughly. Pour into sterile Petri dishes.

Use: For the selective isolation of *Campylobacter* species.

Campylobacter Thioglycollate Medium with 5 Antimicrobics

Composition per liter:

Pancreatic digest of casein	17.0g
Glucose	6.0g
Papaic digest of soybean meal	3.0g
NaCl	2.5g
Agar	1.6g
Sodium thioglycollate	0.5g
Na_2SO_3	0.1g
Antibiotic supplement solution	10.0mL

pH 7.0 ± 0.2 at 25°C

Source: This medium is available as a premixed powder from BBL Microbiology Systems.

Antibiotic Supplement Solution:

Composition per 10mL:

Cephalothin	0.015g
Vancomycin	0.01g
Trimethoprim	5.0mg
Amphotericin B	2.0mg
Polymyxin B	2,500U

Preparation of Antibiotic Supplement Solution: Add componenets to 10.0mL of distilled/deionized water. Mix thoroughly. Filter sterilize.

Preparation of Medium: Add components, except cephalothin, vancomycin, trimethoprim, amphotericin, and polymyxin B, to distilled deionized water and bring volume to 990.0mL. Mix thoroughly. Gently heat and bring to boiling. Autoclave for 15 min at 15 psi pressure–121°C. Cool to 45° to 50°C. Add 10.0mL of sterile antibiotic supplement. Mix thoroughly. Pour into sterile Petri dishes.

Use: For maintenenace–as a holding medium or transport medium– of fecal specimens or swabs suspected of containing *Campylobacter jejuni* or other *Campylobacter* species when immediate inoculation of *Campylobacter* growth medium is unavailable.

Candida BCG Agar Base (*Candida* Bromcresol Green Agar Base)

Composition per liter:

Glucose	40.0g
Agar	15.0g
Peptone	10.0g
Yeast extract	1.0g
Bromcresol Green	0.02g
Neomycin solution	10.0mL

pH 6.1 ± 0.1 at 25°C

Source: This medium is available as a premixed powder from Difco Laboratories.

Neomycin Solution:
Composition per 10mL:

Neomycin 0.5g

Preparation of Neomycin Solution: Add neomycin to distilled/deionized water and bring volume to 10.0mL. Mix thoroughly. Filter sterilize.

Preparation of Medium: Add components, except neomycin solution, to distilled/deionized water and bring volume to 1.0L. Mix thoroughly and heat gently until boiling. Autoclave for 15 min at 15 psi pressure–121°C. Cool to 50–55°C. Aseptically add 10.0mL sterile neomycin solution. Mix thoroughly. Pour into sterile Petri dishes or leave in tubes.

Use: For the selective isolation and identification of *Candida* species. It is a highly differential medium that is used for demonstrating morphological and biochemical reactions characterizing different *Candida* species. *C. albicans* appears as blunt conical colonies with smooth edges and yellow to blue-green color. *C. stellatoidea* appears as convex colonies with smooth edges and yellow to green color. *C. tropicalis* appears as convex colonies with wavy edges and yellow-green to green color with a dark blue-green base. *C. pseudotropicalis* appears as convex, shiny colonies with smooth edges and green color with a light green edge. *C. krusei* appears as low conical colonies with spreading edges and blue-green color. *C. stellatoidea* appears as convex colonies with smooth edges and yellow to green color.

Carbohydrate Fermentation Broth

Composition per liter:

Peptone 10.0g
NaCl 5.0g
Meat extract 3.0g
Carbohydrate solution 50.0mL
Andrade's indicator 10.0mL

pH 7.1 ± 0.2 at 25°C

Andrade's Indicator:
Composition per 100mL:

Acid Fuchsin .... 0.1 g
NaOH (1*N* solution) 16.0mL

Preparation of Andrade's Indicator: Add components to distilled/deionized water and bring volume to 100.0mL. Mix thoroughly.

Carbohydrate Solution:
Composition per 100mL:

Carbohydrate 10.0g

Preparation of Carbohydrate Solution: Add carbohydrate to distilled/deionized water and bring volume to 100.0mL. Adonitol, arabinose, cellobiose, glucose, dulcitol, fructose, galactose, inositol, lactose, maltose, mannitol, raffinose, rhamnose, salicin, sorbitol, sucrose, trehalose, xylose, or other carbohydrates may be used. Mix thoroughly. Filter sterilize.

Caution: Acid Fuchsin is a potential carcinogen and care must be taken to avoid inhalation of the powdered dye and contact with the skin.

Preparation of Medium: Add components, except carbohydrate solution, to distilled/deionized water and bring volume to 1.0L. Mix thoroughly. Gently heat and bring to boiling. Distribute in 10.0mL volumes into test tubes containing inverted Durham tubes. Autoclave for 15 minutes at 15 psi–121°C. Cool to 25°C. Add 0.5mL of sterile carbohydrate solution to each tube.

Use: For the determination of carbohydrate fermentation reactions of microorganisms, particularly members of the Enterobacteriaceae. A Durham tube is used to collect gas produced during the fermentation reaction. Acid production is indicated by a pink reaction.

Cardiobacterium hominis Medium

Composition per liter:

Glucose 5.0g
Leucine 0.43g
Threonine 0.28g
Glutamic acid 0.2g
Valine 0.19g
Glycine 0.18g
Arginine 0.16g
Histidine 0.13g
Proline 0.1g
Tyrosine 0.04g
Buffered salts solution 100.0mL
Vitamin solution 10.0mL

pH 7.0 ± 0.2 at 25°C

Buffered Salts Solution:
Composition per liter:

Na_2PHO_4 .... 284.0.g
KH_2PO_4 .... 272.0.g
NaCl 5.0g
$FeSO_4 \cdot 7H_2O$ 4.0g
$MgSO_4 \cdot 7H_2O$ 4.0g
$ZnSO_4 \cdot 7H_2O$ 0.4g
$MnSO_4 \cdot H_2O$ 0.3g
$CuSO_4 \cdot 5H_2O$ 0.05g

Preparation of Buffered Salts Solution: Add components to distilled/deionized water and bring volume to 1.0L. Mix thoroughly.

Vitamin Solution:
Composition per liter:

Pyridoxine·HCl	2.0mg
Calcium pantothenate	1.0mg
Nicotinamide	1.0mg
Thiamine·HCl	1.0mg
Biotin	0.1mg

Preparation of Vitamin Solution: Add components to distilled/deionized water and bring volume to 1.0L. Mix thoroughly.

Preparation of Medium: Add components to distilled/deionized water and bring volume to 1.0L. Mix thoroughly. Adjust pH to 7.0. Filter sterilize.

Use: For the isolation and cultivation of *Cardiobacterium hominis*.

Cary and Blair Transport Medium

Composition per liter:

Agar	5.0g
NaCl	5.0g
Sodium thioglycollate	1.5g
Na_2HPO_4	1.1g
$CaCl_2$ solution	9.0mL

pH 8.0 ± 0.5 at 25°C

Source: This medium is available as a premixed powder from BBL Microbiology Systems, Difco Laboratories and Oxoid Unipath.

$CaCl_2$ Solution:
Composition per 10mL:

$CaCl_2$	0.1g

Preparation of $CaCl_2$ Solution: Add $CaCl_2$ to distilled/deionized water and bring volume to 10.0mL. Mix thoroughly. Filter sterilize.

Preparation of Medium: Add components to distilled/deionized water and bring volume to 1.0L. Mix thoroughly and heat gently until boiling. Cool to 50°C. Add 9.0mL of a 1% $CaCl_2$ solution. Adjust the pH to 8.4. Distribute into screw-capped tubes in 7.0mL volumes. Sterilize under flowing steam for 15 min. After sterilization, tighten screw-caps.

Use: For the maintenance—as a holding medium or transport medium—of clinical specimens during collection or shipment.

Cary and Blair Transport Medium, Modified

Composition per liter:

Agar	5.0g
NaCl	5.0g
Sodium thioglycollate	1.5g
L-Cysteine·HCl·H_2O	0.5g
$CaCl_2 \cdot 2H_2O$	0.1g
Na_2HPO_4	0.1g
$NaHSO_3$	0.1g
Resazurin solution	4.0mL

pH 8.4 ± 0.2 at 25°C

Resazurin Solution:
Composition per 380mL:

Resazurin	0.05g
Ethanol (95% solution)	200.0mL

Preparation of Resazurin Solution: Add resazurin to 200.0mL of ethanol. Mix thoroughly. Bring volume to 380.0mL with distilled/deionized water.

Preparation of Medium: Add components, except L-Cysteine·HCl·H_2O, to distilled/deionized water and bring volume to 1.0L. Mix thoroughly. Gas the solution with 100% CO_2 for 10–15 min. Add the L-Cysteine·HCl·H_2O. Mix thoroughly. Adjust pH to 8.4. Anaerobically distribute into tubes under 100% N_2. Cap tubes with butyl rubber stoppers. Autoclave for 15 min at 0 psi pressure–100°C on three consecutive days.

Use: For the maintenance—as a holding medium—of clinical specimens during collection or shipment.

Casein Agar

Composition per liter:

Agar	10.0g
Skim milk	50.0mL

Preparation of Medium: Add components to distilled/deionized water and bring volume to 1.0L. Mix thoroughly. Gently heat and bring to boiling. Distribute into tubes or flasks. Autoclave for 15 min at 15 psi pressure–121°C. Pour into sterile Petri dishes or leave in tubes.

Use: For the cultivation and differentiation of aerobic actinomycetes based on casein utilization. Bacteria that utilize casein, such as *Streptomyces* and *Actinomadura* species, appear as colonies surrounded by a clear zone. *Nocardia asteroides, N. caviae,* and *Mycobacterium fortuitum* do not utilize casein.

CCVC Medium (Cephalothin Cycloheximide Vancomycin Colistin Medium

Composition per liter:

BCYE-alpha base	990.0mL
Antibiotic supplement	10.0mL

pH 6.9 ± 0.2 at 25°C

Source: This medium is available as a premixed powder from BBL Microbiology Systems.

BCYE—Alpha Base:

Composition per liter:

Agar 15.0g
Yeast extract 10.0g
ACES buffer (2-[(2-Amino-2-oxoethyl)-amino]-ethane sulfonic acid) 10.0g
Charcoal, activated 2.0g
α-Ketogluatrate 1.0g
$Fe_4(P_2O_7)_3 \cdot 9H_2O$ 0.25g
Cysteine·HCl·H_2O solution 10.0mL

Cysteine·HCl·H_2O Solution:

Composition per 10mL:

L-Cysteine·HCl·H_2O 0.4g

Preparation of Cysteine·HCl·H_2O Solution: Add cysteine·HCl·H_2O to distilled/deionized water and bring volume to 10.0mL. Mix thoroughly. Filter sterilize.

Preparation of BCYE—Alpha Base: Add components, except cysteine solution, to distilled/deionized water and bring volume to 990.0mL. Mix thoroughly. Adjust medium to pH 6.9 with 1*N* KOH. Heat gently and bring to boiling for 1 min. Autoclave for 15 min at 15 psi pressure–121°C. Cool to 50–55°C. Add 4.0mL of L-cysteine·HCl·H_2O solution. Mix thoroughly.

Antibiotic Supplement Solution:

Composition per 10.0mL:

Cycloheximide 80.0mg
Colistin 16.0mg
Cephalothin 4.0mg
Vancomycin 0.5mg

Preparation of Antibiotic Supplement Solution: Add components to 10.0mL of distilled/deionized water. Filter sterilize.

Preparation of Medium: To cooled BCYE-alpha base add 10.0mL sterile antibiotic supplement. Mix thoroughly. Adjust pH to 6.9 with sterile 1*N* KOH. Pour into sterile Petri dishes with constant agitation to keep charcoal in suspension.

Use: For the selective isolation and cultivation of *Legionella* species from environmental samples.

CDC Anaerobe Blood Agar

Composition per liter:

Agar 20.0g
Pancreatic digest of casein 15.0g
Papaic digest of soybean meal 5.0g
NaCl 5.0g
Yeast extract 5.0g
L-Cystine 0.4g
Sheep blood, defibrinated 50.0mL
Vitamin K_1 solution 1.0mL
Hemin solution 0.5mL

pH 7.5 ± 0.2 at 25°C

Source: This medium is available as a prepared medium from BBL Microbiology Systems.

Vitamin K_1 Solution:

Composition per 100mL:

Vitamin K_1 .... 1.0g
Ethanol 99.0mL

Preparation of Vitamin K_1 Solution: Add vitamin K_1 to 99.0mL of absolute ethanol. Mix thoroughly. Filter sterilize.

Hemin Solution:

Composition per 100mL:

Hemin 1.0g
NaOH (1*N* solution) 20.0mL

Preparation of Hemin Solution: Add hemin to 20.0mL of 1*N* NaOH solution. Mix thoroughly. Bring volume to 100.0mL with distilled/deionized water.

Preparation of Medium: Add components, except Vitamin K_1 and sheep blood, to distilled/deionized water and bring volume to 949.0mL. Mix thoroughly. Heat gently and bring to boiling for 1 min. Autoclave for 15 min at 15 psi pressure–121°C. Cool to 50–55°C. Aseptically add 1.0mL vitamin K_1 solution and 50.0mL sterile, defibrinated sheep blood. Mix thoroughly. Pour into sterile Petri dishes.

Use: For the isolation and cultivation of fastidious and slow-growing, obligate anaerobic bacteria from a variety of clinical specimens and in epidemiological investigations. For the isolation and cultivation of *Actinomyces israelii, Bacteroides melaninogenicus, Bacteroides thetaiotaomicron, Clostridium haemolyticum,* and *Fusobacterium necrophorum.*

CDC Anaerobe Blood Agar with Kanamycin and Vancomycin

Composition per liter:

Agar 20.0g
Pancreatic digest of casein 15.0g
NaCl 5.0g
Papaic digest of soybean meal 5.0g
Yeast extract 5.0g
L-Cystine 0.4g
Sheep blood, defibrinated 50.0mL
Antibiotic solution 10.0mL

Vitamin K_1 solution 1.0mL
Hemin solution 0.5mL

pH 7.5 ± 0.2 at 25°C

Source: This medium is available as a prepared medium from BBL Microbiology Systems.

Antibiotic Solution:
Composition per 10mL:

Kanamycin 0.1g
Vancomycin 7.5mg

Preparation of Antibiotic Solution: Add components to distilled/deionized water and bring volume to 10.0mL. Mix thoroughly. Filter sterilize.

Vitamin K_1 Solution:
Composition per 100mL:

Vitamin K_1 .... 1.0g
Ethanol 99.0mL

Preparation of Vitamin K_1 Solution: Add vitamin K_1 to 99.0mL of absolute ethanol. Mix thoroughly. Filter sterilize.

Hemin Solution:
Composition per 100mL:

Hemin 1.0g
NaOH (1*N* solution) 20.0mL

Preparation of Hemin Solution: Add hemin to 20.0mL of 1*N* NaOH solution. Mix thoroughly. Bring volume to 100.0mL with distilled/deionized water.

Preparation of Medium: Add components, except vitamin K_1 solution and sheep blood, to distilled/deionized water and bring volume to 949.0mL. Mix thoroughly. Heat gently and bring to boiling for 1 min. Autoclave for 15 min at 15 psi pressure–121°C. Cool to 50–55°C. Aseptically add 1.0mL sterile vitamin K_1 solution and 50.0mL sterile, defibrinated sheep blood. Mix thoroughly. Pour into sterile Petri dishes.

Use: For the selective isolation of fastidious and slow-growing, obligate anaerobic Gram-negative bacteria, especially *Bacteroides* species, from a variety of clinical specimens and in epidemiological investigations.

CDC Anaerobe Blood Agar with Phenylethyl Alcohol (CDC Anaerobe Blood Agar with PEA)

Composition per liter:

Agar 20.0g
Pancreatic digest of casein 15.0g
NaCl 5.0g
Papaic digest of soybean meal 5.0g
Yeast extract 5.0g
L-Cystine 0.4g
Sheep blood, defibrinated 50.0mL
Vitamin K_1 solution 10.0mL
Hemin solution 0.5mL

pH 7.5 ± 0.2 at 25°C

Source: This medium is available as a prepared medium from BBL Microbiology Systems.

Vitamin K_1 Solution:
Composition per 100mL:

Vitamin K_1 .... 0.1g
Phenylethyl alcohol 25.0g
Ethanol 74.0mL

Preparation of Vitamin K_1 Solution: Add components to 74.0mL of absolute ethanol. Mix thoroughly. Filter sterilize.

Hemin Solution:
Composition per 100mL:

Hemin 1.0g
NaOH (1*N* solution) 20.0mL

Preparation of Hemin Solution: Add hemin to 20.0mL of 1*N* NaOH solution. Mix thoroughly. Bring volume to 100.0mL with distilled/deionized water.

Preparation of Medium: Add components, except vitamin K_1 solution and sheep blood, to distilled/deionized water and bring volume to 940.0mL. Mix thoroughly. Heat gently and bring to boiling for 1 min. Autoclave for 15 min at 15 psi pressure–121°C. Cool to 50–55°C. Aseptically add 1.0mL vitamin K_1 solution add 50.0mL sterile, defibrinated sheep blood. Mix thoroughly. Pour into sterile Petri dishes.

Use: For the selective isolation of fastidious and slow-growing obligate anaerobic bacteria from a variety of clinical specimens and in epidemiological investigations.

CDC Anaerobe Laked Blood Agar with Kanamycin And Vancomycin (CDC Anaerobe Laked Blood Agar with KV)

Composition per liter:

Agar 20.0g
Pancreatic digest of casein 15.0g
Papaic digest of soybean meal 5.0g
Yeast extract 5.0g
L-Cystine 0.4g
Sheep blood, defibrinated, laked 50.0mL

Antibiotic solution 10.0mL
Vitamin K_1 solution 1.0mL
Hemin solution.. 0.5mL

pH 7.5 ± 0.2 at 25°C

Source: This medium is available as a prepared medium from BBL Microbiology Systems.

Antibiotic Solution:

Composition per 10mL:

Kanamycin 0.1g
Vancomycin.. 7.5mg

Preparation of Antibiotic Solution: Addcomponents to distilled/deionized water and bring volume to 10.0mL. Mix thoroughly. Filter sterilize.

Vitamin K_1 Solution:

Composition per 100mL:

Vitamin K_1 ....1.0g
Ethanol 99.0mL

Preparation of Vitamin K_1 Solution: Add vitamin K_1 to 99.0mL of absolute ethanol. Mix thoroughly. Filter sterilize.

Hemin Solution:

Composition per 100mL:

Hemin...1.0g
NaOH (1*N* solution).................................... 20.0mL

Preparation of Hemin Solution: Add hemin to 20.0mL of 1*N* NaOH solution. Mix thoroughly. Bring volume to 100.0mL with distilled/deionized water.

Preparation of Medium: Add components, except antibiotic solution, vitamin K_1 and laked sheep blood, to distilled/deionized water and bring volume to 939.0mL. Mix thoroughly. Heat gently and bring to boiling for 1 min. Autoclave for 15 min at 15 psi pressure–121°C. Cool to 50–55°C. Aseptically add the 1.0mL sterile vitamin K_1 solution and 10.0mL sterile antibiotic solution. Mix thoroughly. Aseptically add 50.0mL sterile, defibrinated, laked sheep blood. Laked blood is prepared by freezing whole blood overnight and thawing to room temperature. Mix thoroughly. Pour into sterile Petri dishes.

Use: For the selective isolation of fastidious and slow-growing, obligate anaerobic bacteria from a variety of clinical specimens and in epidemiological investigations.

Cereal Agar

Composition per liter:

Cereal, precooked mixed 100.0g
Agar...15.0g

Preparation of Medium: Add components to distilled/deionized water and bring volume to 1.0L. Mix thoroughly. Gently heat and bring to boiling. Autoclave for 15 min at 15 psi pressure–121°C. Pour into sterile Petri dishes or distribute into sterile tubes. Allow tubes to cool in a slanted position.

Use: For the cultivation and sporulation of fungi.

Cetrimide Agar, Non-USP

Composition per liter:

Beef heart, solids from infusion......................500.0g
Agar...15.0g
Tryptose 10.0g
NaCl 5.0g
Cetrimide...0.9g

pH 7.2 ± 0.2 at 25°C

Preparation of Medium: Add components to distilled/deionized water and bring volume to 1.0L. Mix thoroughly. Gently heat and bring to boiling. Distribute into tubes or flasks. Autoclave for 15 min at 13 psi pressure–118°C. Pour into sterile Petri dishes or leave in tubes.

Use: For the selective isolation, cultivation and identification of *Pseudomonas aeruginosa* and other Gram-negative nonfermentative bacteria.

Cetrimide Agar, USP (Pseudosel® Agar)

Composition per liter:

Pancreatic digest of gelatin 20.0g
Agar...13.6g
K_2SO_4...10.0g
$MgCl_2$...1.4g
Cetrimide...0.3g
Glycerol.. 10.0mL

pH 7.2 ± 0.2 at 25°C

Source: This medium is available as a premixed powder from Difco Laboratories and BBL Microbiology Systems.

Preparation of Medium: Add components to distilled/deionized water and bring volume to 1.0L. Mix thoroughly. Gently heat and bring to boiling. Distribute into tubes or flasks. Autoclave for 15 min at 13 psi pressure–118°C. Pour into sterile Petri dishes or leave in tubes.

Use: For the selective isolation, cultivation and identification of *Pseudomonas aeruginosa* and other Gram-negative nonfermentative bacteria.

Charcoal Agar with Horse Blood and Cephalexin

Composition per liter:

Agar....12.0g
Beef extract....10.0g
Peptone....10.0g
Starch....10.0g
NaCl....5.0g
Charcoal....4.0g
Nicotinic acid....1.0mg
Horse blood, defibrinated....100.0mL
Cephalexin solution....10.0mL

pH 7.4 ± 0.2 at 25°C

Cephalexin Solution:

Composition per 10mL:

Cephalexin....0.04g

Preparation of Cephalexin Solution: Add cephalexin to distilled/deionized water and bring volume to 10.0mL. Mix thoroughly. Filter sterilize.

Preparation of Medium: Add components, except cephalexin solution and horse blood, to distilled/deionized water and bring volume to 890.0L. Mix thoroughly. Gently heat and bring to boiling with frequent stirring. Autoclave for 15 min at 15 psi pressure–121°C. Cool to 45° to 50°C. Aseptically add 100.0mL of sterile, defibrinated horse blood and 10.0mL sterile cephalexin solution. Pour into sterile Petri dishes or distribute into tubes. Shake flask while dispensing to keep charcoal in suspension.

Use: For the cultivation and isolation of *Bordetella pertussis.*

Chlamydia Growth Medium

Composition per 500mL:

Eagle minimum essential medium with Earle salts, 10X....50.0mL
Fetal calf serum....50.0mL
L-Glutamine solution....5.0mL

pH 7.4 ± 0.2 at 25°C

Eagle Minimum Essential Medium with Earle Salts, 10X:

Composition per liter:

NaCl....6.8g
Glucose....1.0g
KCl....0.4g
$CaCl_2 \cdot 2H_2O$....0.2g
$MgCl_2 \cdot 6H_2O$....0.2g
NaH_2PO_4....0.15g
L-Arginine....0.10g
L-Lysine....0.06g
L-Isoleucine....0.05g
L-Leucine....0.05g
L-Threonine....0.05g
L-Valine....0.05g
L-Tyrosine....0.04g
L-Phenylalanine....0.03g
L-Histidine....0.03g
L-Cystine....0.02g
L-Methionine....0.02g
L-Tryptophan....0.01g
i-Inositol....2.0mg
Calcium pantothenate....1.0mg
Choline chloride....1.0mg
Folic acid....1.0mg
Nicotinamide....1.0mg
Pyridoxal....1.0mg
Thiamine·HCl....1.0mg
Riboflavin....0.1mg

Preparation of Eagle Minimum Essential Medium With Earle Salts, 10X: Add components to distilled/deionized water and bring volume to 1.0L. Mix thoroughly. Adjust pH to 7.4 with 7.5% Na_2CO_3 solution. Filter sterilize.

Glutamine Solution:

Composition per 100mL:

L-Glutamine....2.92g
NaCl (0.85% solution)....100.0mL

Preparation of Glutamine Solution: Add the glutamine to the 0.85% NaCl solution. Mix thoroughly. Filter sterilize.

Preparation of Medium: Aseptically combine 50.0mL of sterile Eagle minimum essential medium with Earle salts, 10X, 50.0mL of fetal calf serum and 5.0mL of sterile glutamine solution. Bring volume to 500.0mL with sterile distilled/deionized water. Mix thoroughly. Aseptically distribute into sterile tubes or flasks.

Use: For the cultivation of *Chlamydia* species.

Chlamydia Isolation Medium

Composition per 500mL:

Eagle minimum essential medium with Earle salts, 10X....50.0mL
Fetal calf serum....50.0mL
Selective supplement....10.0mL
L-Glutamine solution....5.0mL

pH 7.4 ± 0.2 at 25°C

Eagle Minimum Essential Medium with Earle Salts, 10X:

Composition per liter:

NaCl....6.8g
Glucose....1.0g
KCl....0.4g

$CaCl_2 \cdot 2H_2O$ 0.2g
$MgCl_2 \cdot 6H_2O$ 0.2g
NaH_2PO_4 .... 0.15g
L-Arginine 0.10g
L-Lysine 0.06g
L-Isoleucine 0.05g
L-Leucine 0.05g
L-Threonine 0.05g
L-Valine 0.05g
L-Tyrosine 0.04g
L-Phenylalanine 0.03g
L-Histidine 0.03g
L-Cystine 0.02g
L-Methionine 0.02g
L-Tryptophan 0.01g
i-Inositol 2.0mg
Calcium pantothenate 1.0mg
Choline chloride 1.0mg
Folic acid 1.0mg
Nicotinamide 1.0mg
Pyridoxal 1.0mg
Thiamine·HCl 1.0mg
Riboflavin 0.1mg

Preparation of Eagle Minimum Essential Medium With Earle Salts, 10X: Add components to distilled/deionized water and bring volume to 1.0L. Mix thoroughly. Adjust pH to 7.4 with 7.5% Na_2CO_3 solution. Filter sterilize.

Selective Supplement:

Composition per 10mL:

Glucose 0.594g
Vancomycin 0.050g
Gentamicin 0.010g
Amphotericin B 2.0mg
Cycloheximide 2.0mg

Preparation of Selective Supplement: Add components to distilled/deionized water and bring volume to 10.0mL. Mix thoroughly. Filter sterilize.

Glutamine Solution:

Composition per 100mL:

L-Glutamine 2.92g
NaCl (0.85% solution) 100.0mL

Preparation of Glutamine Solution: Add the glutamine to the 0.85% NaCl solution. Mix thoroughly. Filter sterilize.

Preparation of Medium: Aseptically combine 50.0mL of sterile Eagle minimum essential medium with Earle salts, 10X, 50.0mL of fetal calf serum, 10.0mL of selective supplement and 5.0mL of sterile glutamine solution. Bring volume to 500.0mL with sterile distilled/deionized water. Mix thoroughly. Aseptically distribute into sterile tubes or flasks.

Use: For the isolation and cultivation of *Chlamydia* species.

Chlamydospore Agar

Composition per liter:

Purified polysaccharide 20.0g
Agar 15.0g
KH_2PO_4 .... 1.0g
$(NH_4)_2SO_4$ 1.0g
Trypan Blue 0.1g
Biotin 5.0μg

pH 5.1 ± 0.2 at 25°C

Source: This medium is available as a premixed powder from Difco Laboratories.

Preparation of Medium: Add components to distilled/deionized water and bring volume to 1.0L. Mix thoroughly. Gently heat and bring to boiling. Distribute into tubes or flasks. Autoclave for 15 min at 15 psi pressure–121°C. Pour into sterile Petri dishes or leave in tubes.

Use: For differentiating *Candida albicans* from other *Candida* species of on the basis of chlamydospore formation.

Chocolate Agar

Composition per liter:

Agar 15.0g
Pantone 10.0g
Bitone 10.0g
NaCl 5.0g
Tryptic digest of beef heart 3.0g
Cornstarch 1.0g
Sheep blood, defibrinated 100.0mL
Supplement B (Difco) 10.0mL

pH 7.3 ± 0.2 at 25°C

Supplement B:

Composition per 10mL:

Supplement B contains yeast concentrate, glutamine, coenzyme, cocarboxylase, hematin and growth factors.

Preparation of Supplement B: Add components to distilled/deionized water and bring volume to 10.0mL. Mix thoroughly. Filter sterilize.

Preparation of Medium: Add components, except hemoglobin solution and sheep blood, to distilled/deionized water and bring volume to 890.0mL. Mix thoroughly. Gently heat until boiling. Autoclave for 15 min at 15 psi pressure–121°C. Cool to 45° to 50°C. Aseptically add 100.0mL of sterile defibrinated sheep blood. Gently heat while stirring and bring to 85°C for 5–10 min. Cool to 50°C. Aseptically add 10.0mL of sterile supplement B. Mix thoroughly. Pour into sterile Petri dishes or distribute into sterile tubes.

Use: For the isolation and cultivation of a variety of fastidious microorganisms.

Chocolate Agar

Composition per liter:

Proteose peptone No. 3 .... 15.0g
Agar 10.0g
NaCl 5.0g
K_2HPO_4 .... 4.0g
Cornstarch 1.0g
KH_2PO_4 .... 1.0g
Hemoglobin solution 100.0mL
Supplement B 10.0mL

pH 7.0 ± 0.2 at 25°C

Source: Available from Difco Laboratories.

Supplement B:
Composition per 10mL:

Supplement B contains yeast concentrate, glutamine, coenzyme, cocarboxylase, hematin and growth factors.

Preparation of Supplement B: Add components to distilled/deionized water and bring volume to 10.0mL. Mix thoroughly. Filter sterilize.

Hemoglobin Solution:
Composition per 100mL:

Hemoglobin 10.0g

Preparation of Hemoglobin Solution: Add hemoglobin to distilled/deionized water and bring volume to 100.0mL. Mix thoroughly. Filter sterilize.

Preparation of Medium: Add components, except hemoglobin solution and supplement B, to distilled/deionized water and bring volume to 990.0mL. Mix thoroughly. Gently heat and bring to boiling. Autoclave for 15 min at 15 psi pressure–121°C. Cool to 45° to 50°C. Aseptically add 100.0mL of sterile hemoglobin solution. Gently heat while stirring and bring to 85°C for 5–10 min. Cool to 50°C. Aseptically add 10.0mL of sterile supplement B. Mix thoroughly. Pour into sterile Petri dishes or distribute into sterile tubes.

Use: For the isolation and cultivation of fastidious microorganisms.

Chocolate Agar, Enriched

Composition per liter:

GC medium base 740.0mL
Hemoglobin solution 250.0mL
Supplement B 10.0mL

pH 7.3 ± 0.2 at 25°C

Source: Available from Difco Laboratories.

GC Medium Base:
Composition per 740mL:

Proteose peptone No. 3 .... 15.0g
Agar 20.0g
NaCl 5.0g
K_2HPO_4 .... 4.0g
Glucose 1.5g
Cornstarch 1.0g
KH_2PO_4 .... 1.0g

pH 7.2 ± 0.2 at 25°C

Preparation of GC Medium Base: Add components to distilled/deionized water and bring volume to 740.0mL. Mix thoroughly. Gently heat until boiling. Autoclave for 15 min at 15 psi pressure–121°C. Cool to 45° to 50°C.

Hemoglobin Solution:
Composition per 250mL:

Hemoglobin 10.0g

Preparation of Hemoglobin Solution: Add hemoglobin to distilled/deionized water and bring volume to 250.0mL. Mix thoroughly. Autoclave for 15 min at 15 psi pressure–121°C. Cool to 45° to 50°C.

Supplement B:
Composition per 10mL:

Supplement B contains yeast concentrate, glutamine, coenzyme, cocarboxylase, hematin and growth factors.

Preparation of Supplement B: Add components to distilled/deionized water and bring volume to 10.0mL. Mix thoroughly. Filter sterilize.

Preparation of Medium: To 740.0mL of cooled sterile GC medium base, aseptically add 250.0mL of sterile hemoglobin solution, and 10.0mL of sterile supplement B. Mix thoroughly. Pour into sterile Petri dishes or distribute into sterile tubes.

Use: For the cultivation of fastidious microorganisms, especially *Neisseria* species.

Chocolate Agar, Enriched

Composition per liter:

GC medium base 740.0mL
Hemoglobin solution 250.0mL
Supplement VX 10.0mL

pH 7.3 ± 0.2 at 25°C

Source: Available Difco Laboratories.

GC Medium Base:
Composition per 740mL:

Proteose peptone No. 3 .... 15.0g
Agar 20.0g
NaCl 5.0g
K_2HPO_4 .... 4.0g
Glucose 1.5g

Cornstarch 1.0g
KH_2PO_4 .... 1.0g

pH 7.2 ± 0.2 at 25°C

Preparation of GC Medium Base: Add components to distilled/deionized water and bring volume to 740.0mL. Mix thoroughly. Gently heat until boiling. Autoclave for 15 min at 15 psi pressure–121°C. Cool to 45° to 50°C.

Hemoglobin Solution:

Composition per 250mL:

Hemoglobin 10.0g

Preparation of Hemoglobin Solution: Add hemoglobin to distilled/deionized water and bring volume to 250.0mL. Mix thoroughly. Autoclave for 15 min at 15 psi pressure–121°C. Cool to 45° to 50°C.

Supplement VX:

Composition per 10mL:

Supplement VX contains essential growth factors.

Preparation of Supplement VX: Add components to distilled/deionized water and bring volume to 10.0mL. Mix thoroughly. Filter sterilize.

Preparation of Medium: To 740.0mL of cooled sterile GC medium base, aseptically add 250.0mL of sterile hemoglobin solution, and 10.0mL of sterile supplement B. Mix thoroughly. Pour into sterile Petri dishes or distribute into sterile tubes.

Use: For the cultivation of fastidious microorganisms, especially *Neisseria* species.

Chocolate II Agar

Composition per liter:

Agar 12.0g
Hemoglobin 10.0g
Pancreatic digest of casein 7.5g
Selected meat peptone 7.5g
NaCl 5.0g
K_2HPO_4 .... 4.0g
Cornstarch 1.0g
KH_2PO_4 .... 1.0g

Preparation of Medium: Add components to distilled/deionized water and bring volume to 1.0L. Mix thoroughly. Gently heat to boiling. Autoclave for 15 min at 15 psi pressure–121°C. Pour into sterile Petri dishes or leave in tubes.

Use: For the isolation and cultivation of fastidious microorganisms.

Chocolate II Agar with Hemoglobin and IsoVitaleX® (GCII Agar with Hemoglobin and IsoVitaleX®)

Composition per liter:

GCII agar base 990.0mL
IsoVitaleX enrichment 10.0mL

pH 7.3 ± 0.2 at 25°C

Source: Available as a prepared medium from BBL Microbiology Systems.

GCII Agar Base:

Composition per liter:

Agar 12.0g
Hemoglobin 10.0g
Pancreatic digest of casein 7.5g
Selected meat peptone 7.5g
NaCl 5.0g
K_2HPO_4 .... 4.0g
Cornstarch 1.0g
KH_2PO_4 .... 1.0g

Preparation of GCII Agar Base: Add components to distilled/deionized water and bring volume to 1.0L. Mix thoroughly. Gently heat to boiling. Autoclave for 15 min at 15 psi pressure–121°C. Cool to 45° to 50°C.

IsoVitaleX® Enrichment:

Composition per liter:

Glucose 100.0g
L-Cysteine·HCl 25.9g
L-Glutamine 10.0g
L-Cystine 1.1g
Adenine 1.0g
Nicotinamide adenine dinucleotide 0.25g
Vitamin B_{12} 0.1g
Thiamine pyrophosphate 0.1g
Guanine·HCl 0.03g
$Fe(NO_3)_3 \cdot 6H_2O$ 0.02g
p-Aminobenzoic acid 0.013g
Thiamine·HCl 3.0mg

Preparation of IsoVitaleX®: Add components to distilled/deionized water and bring volume to 1.0L. Mix thoroughly. Filter sterilize.

Preparation of Medium: Aseptically add 10.0mL of sterile IsoVitaleX® enrichment to 990.0L of sterile, cooled GCII agar base. Mix thoroughly. Pour into sterile Petri dishes or distribute into sterile tubes.

Use: For the isolation and cultivation of fastidious microorganisms, especially *Neisseria* and *Haemophilus* species, from a variety of clinical specimens.

Chocolate Tellurite Agar (Tellurite Blood Agar)

Composition per liter:

Agar	10.0g
Casein/meat (50/50) peptone	10.0g
Hemoglobin	10.0g
NaCl	5.0g
K_2HPO_4	4.0g
Cornstarch	1.0g
KH_2PO_4	1.0g
K_2TeO_3	0.1g
Bio-X enrichment	10.0mL

Bio-X Enrichment:

Composition:

Glucose	100.0g
Cysteine·HCl	25.9g
L-Glutamate	10.0g
L-Cystine	1.1g
Adenine	1.0g
Cocarboxylase	0.1g
NAD (nicotinamide adenine dinucleotide)	250.0mg
Guanine·HCl	0.03g
$FeNO_3$	0.02g
p-Aminobenzoic acid	0.013g
Vitamin B_{12}	0.01g
Thiamine·HCl	3.0mg

pH 7.2 ± 0.2 at 25°C

Preparation of Bio-X Enrichment: Add components to distilled/deionized water and bring volume to 1.0L. Mix thoroughly. Filter sterilize.

Caution: Potassium tellurite is toxic.

Preparation of Medium: Add components, except Bio-X enrichment to distilled/deionized water and bring volume to 990.0mL. Mix thoroughly. Gently heat and bring to boiling. Autoclave for 15 min at 15 psi pressure–121°C. Cool to 45° to 50°C. Aseptically add filter-sterilized Bio-X enrichment. Mix thoroughly. Pour into sterile Petri dishes or distribute into sterile tubes.

Use: For the selective isolation and cultivation of *Corynebacterium* species. *C. diphtheriae* appears as gray-black colonies.

Cholera Medium TCBS

Composition per liter:

Sucrose	20.0g
Agar	14.0g
Peptone	10.0g
NaCl	10.0g
Sodium citrate	10.0g
$Na_2S_2O_3·5H_2O$	10.0g
Ox bile	8.0 g
Yeast extract	5.0g
Ferric citrate	1.0g
Bromothymol Blue	0.04g
Thymol Blue	0.04g

pH 8.6 ± 0.2 at 25°C

Source: This medium is available as a premixed powder from Oxoid Unipath.

Preparation of Medium: Add components to distilled/deionized water and bring volume to 1.0L. Mix thoroughly. Gently heat and bring to boiling. Do not autoclave. Pour into sterile Petri dishes. Dry agar plates before using.

Use: For the growth of *Vibrio cholerae, V. parahaemolyticus,* and other *Vibrio* species.

Chopped Meat Carbohydrate Medium

Composition per 1240mL:

Peptone	30.0g
K_2HPO_4	5.0g
Yeast extract	5.0g
Cellobiose	1.0g
Maltose	1.0g
Starch	1.0g
L-Cysteine·HCl·H_2O	0.5g
Chopped meat extract filtrate	1.0L
Chopped meat extract solids	200.0mL
Resazurin (0.025% solution)	4.0mL

pH 7.0 ± 0.2 at 25°C

Chopped Meat Extract:

Composition per liter:

Beef or horse meat	500.0g
NaOH (1*N* solution)	25.0mL

Preparation of Chopped Meat Extract: Use lean beef or horse meat. Remove fat and connective tissue. Grind. Add meat and NaOH to distilled/deionized water and bring volume to 1.0L. Gently heat and bring to boiling while stirring. Cool to 25°C. Remove fat from surface. Filter. Reserve ground meat particles and filtrate. Add distilled/deionized water to filtrate and bring volume to 1.0L.

Preparation of Medium: To 1.0L of chopped meat extract filtrate, add the remaining components, except cysteine and chopped meat solids. Mix thoroughly. Gently heat to boiling. Cool to room temperature. Add the cysteine. Adjust pH to 7.0. Distribute 1 part chopped meat solids (by volume) and 5 parts of liquid (by volume) into tubes under O_2-free 97% N_2 + 3% H_2. Cap with rubber stoppers and place tubes in a press. Autoclave for 15 min at 15 psi pressure–121°C with fast exhaust.

Use: For the cultivation of anaerobic bacteria including *Clostridium* species, *Eubacterium* species and *Gemmiger formicilis*.

Chopped Meat Medium

Composition per 1205mL:

Peptone	30.0g
K_2HPO_4	5.0g
Yeast extract	5.0g
L-Cysteine·HCl·H_2O	0.5g
Chopped meat extract filtrate	1.0L
Chopped meat extract solids	200.0mL
Resazurin (0.025% solution)	4.0mL

pH 7.0 ± 0.2 at 25°C

Chopped Meat Extract:

Composition per liter:

Beef or horse meat	500.0g
NaOH (1*N* solution)	25.0mL

Preparation of Chopped Meat Extract: Use lean beef or horse meat. Remove fat and connective tissue. Grind. Add meat and NaOH to distilled/deionized water and bring volume to 1.0L. Gently heat and bring to boiling while stirring. Cool to 25°C. Remove fat from surface. Filter. Reserve ground meat particles and filtrate. Add distilled/deionized water to filtrate and bring volume to 1.0L.

Preparation of Medium: To 1.0L of chopped meat extract filtrate, add the remaining components, except cysteine and chopped meat solids. Mix thoroughly. Gently heat to boiling. Cool to room temperature. Add the cysteine. Adjust pH to 7.0. Distribute 1 part chopped meat solids (by volume) and 5 parts of liquid (by volume) into tubes under O_2-free 97% N_2 + 3% H_2. Cap with rubber stoppers and place tubes in a press. Autoclave for 15 min at 15 psi pressure–121°C with fast exhaust.

Use: For cultivation and maintenance of a variety of anaerobic bacteria including *Bacteroides* species, *Bifidobacterium* species, *Capnocytophaga* species, *Clostridium* species, *Eubacterium* species, *Fusobacterium* species, *Peptostreptococcus* species, *Prevotella* species, *Propionibacterium* species, *Ruminococcus* species and others.

Chopped Meat Medium with 1% Glucose

Composition per 1230mL:

Peptone	30.0g
Glucose	10.0g
K_2HPO_4	5.0g
Yeast extract	5.0g
L-Cysteine·HCl·H_2O	0.5g
Chopped meat extract filtrate	1.0L
Chopped meat extract solids	200.0mL
Resazurin (0.025% solution)	4.0mL

pH 7.0 ± 0.2 at 25°C

Chopped Meat Extract:

Composition per liter:

Beef or horse meat	500.0g
NaOH (1*N* solution)	25.0mL

Preparation of Chopped Meat Extract: Use lean beef or horse meat. Remove fat and connective tissue. Grind. Add meat and NaOH to distilled/deionized water and bring volume to 1.0L. Gently heat and bring to boiling while stirring. Cool to 25°C. Remove fat from surface. Filter. Reserve ground meat particles and filtrate. Add distilled/deionized water to filtrate and bring volume to 1.0L.

Preparation of Medium: To 1.0L of chopped meat extract filtrate, add the remaining components, except L-cysteine and chopped meat solids. Mix thoroughly. Gently heat to boiling. Cool to room temperature. Add the L-cysteine. Adjust pH to 7.0. Distribute 1 part chopped meat solids (by volume) and 5 parts of liquid (by volume) into tubes under O_2-free 97% N_2 + 3% H_2. Cap with rubber stoppers and place tubes in a press. Autoclave for 15 min at 15 psi pressure–121°C with fast exhaust.

Use: For the cultivation and maintenance of anaerobic bacteria including *Bacteroides disiens, Coprococcus eutastus, Eubacterium formicigenerans, Prevotella disiens, Ruminococcus torques,* and *Streptococcus hansenii.*

Chopped Meat Medium, Modified

Composition per 1230mL:

Pancreatic digest of casein	30.0g
Peptone	30.0g
Agar	20.0g
K_2HPO_4	5.0g
Yeast extract	5.0g
L-Cysteine·HCl·H_2O	0.5g
Chopped meat extract filtrate	1.0L
Chopped meat extract solids	200.0mL
Hemin solution	10.0mL
Resazurin (0.025% solution)	4.0mL
Vitamin K_1 solution	0.2mL

pH 7.0 ± 0.2 at 25°C

Chopped Meat Extract:

Composition per liter:

Beef or horse meat	500.0g
NaOH (1*N* solution)	25.0mL

Preparation of Chopped Meat Extract: Use lean beef or horse meat. Remove fat and connective tissue. Grind. Add meat and NaOH to distilled/deionized water and bring volume to 1.0L. Gently heat and bring to boiling while stirring. Cool to 25°C. Remove fat from surface. Filter. Reserve ground meat particles and filtrate. Add distilled/deionized water to filtrate and bring volume to 1.0L.

Hemin Solution:
Composition per 100mL:

Hemin 0.05g
NaOH (1*N* solution) 1.0mL

Preparation of Hemin Solution: Add components to distilled/deionized water and bring volume to 100.0mL. Mix thoroughly.

Vitamin K_1 Solution:
Composition per 30mL:

Ethanol (95% solution) 30.0mL
Vitamin K_1 .... 0.15mL

Preparation of Vitamin K_1 Solution: Mix components. Store solution protected from light at 5°C. Discard after one month.

Preparation of Medium: To 1.0L of chopped meat extract filtrate, add the remaining components, except cysteine, hemin solution, vitamin K_1 solution and chopped meat solids. Mix thoroughly. Gently heat to boiling. Cool to room temperature. Add the cysteine, hemin solution and vitamin K_1 solution. Adjust pH to 7.0. Distribute 1 part chopped meat solids (by volume) and 5 parts of liquid (by volume) into tubes under O_2-free 97% N_2 + 3% H_2. Cap with rubber stoppers and place tubes in a press. Autoclave for 15 min at 15 psi pressure–121°C with fast exhaust.

Use: For cultivation and maintenance of a variety of anaerobic bacteria including *Actinomyces* species, *Bacteroides* species, *Clostridium* species, *Eubacterium* species, *Fusobacterium* species, *Peptostreptococcus* species, *Porphyromonas* species, *Prevotella* species, *Propionibacterium* species, *Selenomonas* species and others.

Christensen Agar

Composition per liter:

Agar 15.0g
NaCl 5.0g
Sodium citrate 3.0g
KH_2PO_4 .... 1.0g
Cysteine·HCl·H_2O 0.1g
Phenol Red 12.0mg

pH 6.9 ± 0.2 at 25°C

Preparation of Medium: Add components to distilled/deionized water and bring volume to 1.0L. Mix thoroughly. Gently heat and bring to boiling. Dispense into tubes or flasks. Autoclave for 15 min at 15 psi pressure–121°C. Pour into sterile Petri dishes or leave in tubes. Allow tubes to cool in a slanted position.

Use: For the differentiation of enteric pathogens, especially members of the Enterobacteriaceae, and coliforms based on their ability to utilize citrate as a carbon source. Bacteria that can utilize citrate turn the medium pink-red.

Christensen Agar

Composition per liter:

Agar 15.0g
NaCl 5.0g
Sodium citrate 3.0g
KH_2PO_4 .... 1.0g
Yeast extract 0.5g
Glucose 0.2g
Cysteine·HCl·H_2O 0.1g
Phenol Red 12.0mg

pH 6.9 ± 0.2 at 25°C

Source: This medium is available as a premixed powder from Difco Laboratories.

Preparation of Medium: Add components to distilled/deionized water and bring volume to 1.0L. Mix thoroughly. Gently heat and bring to boiling. Dispense into tubes or flasks. Autoclave for 15 min at 15 psi pressure–121°C. Pour into sterile Petri dishes or leave in tubes. Allow tubes to cool in a slanted position.

Use: For the differentiation of enteric pathogens, especially members of the Enterobacteriaceae, and coliforms based on their ability to utilize citrate as a carbon source. Bacteria that can utilize citrate turn the medium pink-red.

Christensen Citrate Agar, Modified (Citrate Agar)

Agar 12.0g
NaCl 5.0g
Sodium citrate 3.8g
KH_2PO_4 .... 1.0g
Yeast extract 0.5g
Glucose 0.2g
Cysteine·HCl·H_2O 0.1g
Phenol Red 0.02g

pH 6.7 ± 0.2 at 25°C

Preparation of Medium: Add components to distilled/deionized water and bring volume to 1.0L. Mix thoroughly. Gently heat and bring to boiling. Dispense into tubes or flasks. Autoclave for 15 min at 15 psi pressure–121°C. Pour into sterile Petri dishes or leave in tubes. Allow tubes to cool in a slanted position.

Use: For the differentiation of enteric pathogens, especially members of the Enterobacteriaceae, and coliforms based on their ability to utilize citrate as a carbon source. Bacteria that can utilize citrate turn the medium pink-red.

Christensen Citrate Sulfide Medium

Composition per liter:

Agar	15.0g
NaCl	5.0g
Sodium citrate·$2H_2O$	3.0g
KH_2HPO_4	1.0g
Yeast extract	0.5g
Ferric citrate	0.2g
Ammonium citrate	0.2g
Glucose	0.2g
Cysteine·HCl·H_2O	0.1g
$Na_2S_2O_3 \cdot 5H_2O$	0.08g
Phenol Red	0.012g

pH 6.7± 0.2 at 25°C

Preparation of Medium: Add components to distilled/deionized water and bring volume to 1.0L. Mix thoroughly. Gently heat and bring to boiling. Dispense into tubes or flasks. Autoclave for 15 min at 15 psi pressure–121°C. Pour into sterile Petri dishes or leave in tubes. Allow tubes to cool in a slanted position.

Use: For the differentiation of enteric pathogens, especially members of the Enterobacteriaceae, and coliforms based on their ability to utilize citrate as a carbon source and production of H_2S. Bacteria that can utilize citrate turn the medium pink-red. H_2S production appears as a blackening of the butt of the tube.

CIN Agar (*Yersinia* Selective Agar) (Cefsulodin Irgasan® Novobiocin Agar)

Composition per liter:

Mannitol	20.0g
Agar	12.0g
Pancreatic digest of gelatin	10.0g
Beef extract	5.0g
Peptic digest of animal tissue	5.0g
Sodium pyruvate	2.0g
Yeast extract	2.0g
NaCl	1.0g
Sodium deoxycholate	0.5g
Neutral Red	0.03g
Cefsulodin	0.015g
Irgasan®(triclosan)	4.0mg
Novobiocin	2.5mg
Crystal Violet	1.0mg

pH 7.4 ± 0.2 at 25°C

Source: This medium is available as a premixed powder from BBL Microbiology Systems.

Preparation of Medium: Add components, except cefsulodin and novobiocin, to distilled/deionized water and bring volume to 1.0L. Heat, mixing continuously, until boiling. Do not autoclave. Cool to 45° to 50°C. Aseptically add cefsulodin and novobiocin. Mix thoroughly. Pour into sterile Petri dishes or distribute into sterile tubes.

Use: For the selective isolation and differentiation of *Yersinia enterocolitica* from a variety of clinical specimens and in epidemiological investigations based on mannitol fermentation. *Yersinia enterocolitica* appears as "bull's eye" colonies with deep red centers surrounded by a transparent periphery.

Citrate Medium, Koser's Modified

Composition per liter:

NaCl	5.0g
Citric acid	2.0g
$(NH_4)H_2PO_4$	1.0g
K_2HPO_4	1.0g
$MgSO_4 \cdot 7H_2O$	0.2g

pH 6.8 ± 0.2 at 25°C

Preparation of Medium: Add components to distilled/deionized water and bring volume to 1.0L. Mix thoroughly. Adjust pH to 6.8. Distribute into tubes in 5.0mL volumes. Autoclave for 15 min at 15 psi pressure–121°C.

Use: For the cultivation and differentiation of bacteria based on their ability to utilize citrate.

CLED Agar (Cystine Lactose Electrolyte Deficient Agar) (Brolacin Agar)

Composition per liter:

Agar	15.0g
Lactose	10.0g
Pancreatic digest of casein	4.0g
Pancreatic digest of gelatin	4.0g
Beef extract	3.0g
L-Cystine	0.128g
Bromthymol Blue	0.02g

pH 7.3 ± 0.2 at 25°C

Source: This medium is available as a premixed powder from BBL Microbiology Systems, Difco Laboratories and Oxoid Unipath.

Preparation of Medium: Add components to distilled/deionized water and bring volume to 1.0L. Mix thoroughly. Gently heat while stirring and bring to boiling. Autoclave for 15 min at 15 psi pressure–121°C. Cool to 50° to 55°C. Pour into sterile Petri dishes or distribute into sterile tubes.

Use: For the isolation, enumeration and presumptive identification of microorganisms from urine.

CLED Agar with Andrade Indicator (Cystine Lactose Electrolyte Deficient Agar with Andrade Indicator)

Composition per liter:

Agar	15.0g
Pancreatic digest of casein	10.0g
Peptone	4.0g
Beef extract	3.0g
Cystine	0.128g
Bromthymol Blue	0.02g
Andrade indicator	10.0mL

pH 7.5 ± 0.2

Source: This medium is available as a premixed powder from Oxoid Unipath.

Caution: Acid Fuchsin is a potential carcinogen and care must be taken to avoid inhalation of the powdered dye and contamination of the skin.

Andrade's Indicator:
Composition per 100mL:

NaOH (1*N* solution)	16.0mL
Acid Fuchsin	0.1g

Preparation of Andrade's Indicator: Add Acid Fuchsin to NaOH solution and bring volume to 100.0mL with distilled/deionized water.

Preparation of Medium: Add components to distilled/deionized water and bring volume to 1.0L. Mix thoroughly. Gently heat while stirring and bring to boiling. Autoclave for 15 min at 15 psi pressure–121°C. Cool to 50° to 55°C. Pour into sterile Petri dishes or distribute into sterile tubes.

Use: For the differentiation of microorganisms based on colony characteristics. For the isolation, enumeration and presumptive identification of microorganisms from urine.

Clostridium botulinum Isolation Agar (CBI Agar)

Composition per 1033mL:

Egg yolk agar base	900.0mL
Egg yolk emulsion, 50%	100.0mL
Cycloserine solution	25.0mL
Sulfamethoxazole solution	4.0mL
Trimethoprim solution	4.0mL

pH7.4 ± 0.2 at 25°C

Egg Yolk Agar Base:
Composition per 900mL:

Pancreatic digest of casein	40.0g
Agar	20.0g
Na_2HPO_4	5.0g
Yeast extract	5.0g
Glucose	2.0g
NaCl	2.0g
$MgSO_4 \cdot 7H_2O$ solution	0.2mL

Preparation of Egg Yolk Agar Base: Add components to distilled/deionized water and bring volume to 900.0mL. Mix thoroughly. Gently heat to boiling. Autoclave for 15 min at 15 psi pressure–121°C. Cool to 45° to 50°C.

$MgSO_4 \cdot 7H_2O$ Solution:
Composition per 100mL:

$MgSO_4 \cdot 7H_2O$	5.0g

Preparation of $MgSO_4 \cdot 7H_2O$ Solution: Add $MgSO_4 \cdot 7H_2O$ to distilled/deionized water and bring volume to 100.0mL. Mix thoroughly.

Cycloserine Solution:
Composition per 100mL:

Cycloserine	1.0g

Preparation of Cycloserine Solution: Add cycloserine to distilled/deionized water and bring volume to 100.0mL. Mix thoroughly. Filter sterilize.

Sulfamethoxazole Solution:
Composition per 100mL:

Sulfamethoxazole	1.9g

Preparation of Sulfamethoxazole Solution: Add sulfamethoxazole to distilled/deionized water and bring volume to 50.0mL. Add sufficient 10% NaOH to dissolve. Bring volume to 100.0mL with distilled/deionized water. Mix thoroughly. Filter sterilize.

Trimethoprim Solution:
Composition per 100mL:

Trimethoprim	0.1g

Preparation of Trimethoprim Solution: Add trimethoprim to distilled/deionized water and bring volume to 50.0mL. Gently heat to 55°C. Add sufficient 0.05*N* HCl to dissolve. Bring volume to 100.0mL with distilled/deionized water. Mix thoroughly. Filter sterilize.

Egg Yolk Emulsion, 50%:
Composition per 100mL:

Chicken egg yolks ... 11
Whole chicken egg ... 1
NaCl (0.9% solution) ... 50.0mL

Preparation of Egg Yolk Emulsion, 50%: Soak eggs with 1:100 dilution of saturated mercuric chloride solution for 1 min. Crack eggs and separate yolks from whites. Mix egg yolks with 1 chicken egg. Measure 50.0mL of egg yolk emulsion and add to 50.0mL of 0.9% NaCl solution. Mix thoroughly. Filter sterilize. Warm to 45° to 50°C.

Preparation of Medium: Aseptically add warmed, sterile egg yolk emulsion, 50%, and sterile cycloserine solution, sterile sulfamethoxazole solution, and sterile trimethoprim solution to cooled, sterile egg yolk agar base. Mix thoroughly. Pour into sterile Petri dishes.

Use: For isolation, cultivation and differentiation based on lipase activity of *Clostridium botulinum*, types A, B, and F from fecal specimens associated with food-borne and infant botulism. *C. botulinum* types A, B, and F appear as raised colonies surrounded by an opaque zone. Other *Clostridium* species and *C. botulinum* type G appear as pinpoint colonies with no opaque zone.

Clostridium difficile Agar

Composition per liter:

Clostridum difficile agar base ... 920.0mL
Clostridium difficile selective supplement ... 10.0mL
Horse blood, defibrinated ... 70.0mL
pH 7.4 ± 0.2 at 25°C

Source: This medium is available as a premixed powder from Oxoid Unipath.

***Clostridum difficile* Agar Base:**
Composition per 920mL:

Proteose peptone ... 40.0g
Agar ... 15.0g
Fructose ... 6.0g
Na_2HPO_4 ... 5.0g
NaCl ... 2.0g
KH_2PO_4 ... 1.0g
$MgSO_4 \cdot 7H_2O$ ... 0.1g

Preparation of *Clostridum difficile* Agar Base: Add components to distilled/deionized water and bring volume to 920.0mL. Mix thoroughly. Gently heat to boiling. Autoclave for 15 min at 15 psi pressure–121°C. Cool to 45° to 50°C.

***Clostridium difficile* Selective Supplement:**
Composition per 10mL:

D-Cycloserine ... 500.0mg
Cefoxitin ... 16.0mg

Preparation of *Clostridium difficile* Selective Supplement: Add components to distilled/deionized water and bring volume to 10.0mL. Mix thoroughly. Filter sterilize.

Preparation of Medium: Add 10.0mL of sterile *Clostridium difficile* selective supplement and 70.0mL of sterile, defibrinated horse blood to 920.0mL of cooled, sterile *Clostridum difficile* Agar Base. Mix thoroughly. Pour into sterile Petri dishes or distribute into sterile tubes.

Use: For the selective isolation and cultivation of *Clostridium difficile* from clinical specimens and in epidemiological investigations. *C. difficile* appears as large, yellowish, rhizoidal colonies with an odor similar to horse manure.

Clostridium difficile Agar (Cycloserine Cefoxitin Fructose Agar) (CCFA)

Composition per liter:

Peptic digest of animal tissue ... 32.0g
Agar ... 20.0g
Fructose ... 6.0g
Na_2HPO_4 ... 5.0g
NaCl ... 2.0g
KH_2PO_4 ... 1.0g
Cycloserine ... 0.25g
$MgSO_4$ ... 0.1g
Neutral Red ... 0.03g
Cefoxitin solution ... 10.0mL
pH 7.2 ± 0.2 at 25°C

Source: This medium is available as a premixed powder from BBL Microbiology Systems.

Cefoxitin Solution:
Composition per 10mL:

Cefoxitin ... 16.0mg

Preparation of Cefoxitin Solution: Add cefoxitin to distilled/deionized water and bring volume to 10.0mL. Mix thoroughly. Filter sterilize.

Preparation of Medium: Add components to distilled/deionized water and bring volume to 990.0mL. Mix thoroughly. Gently heat to boiling. Autoclave for 15 min at 15 psi pressure–121°C. Cool to 45° to

50°C. Aseptically add 10.0mL of sterile cefoxitin solution. Mix thoroughly. Pour into sterile Petri dishes or distribute into sterile tubes.

Use: For the selective isolation and cultivation of *Clostridium difficile* from clinical specimens and in epidemiological investigations.

Clostridium novyi Blood Agar

Composition per 100mL:

Agar....2.0g
Glucose....1.0g
Neopeptone....1.0g
Proteolyzed liver....0.5g
Yeast extract....0.5g
Horse blood, defibrinated....10.0mL
Reducing solution....0.75mL
Salts solution....0.5mL

pH 7.6–7.8 at 25°C

Salts Solution:

Composition per 100mL:

$MgSO_4 \cdot 7H_2O$....4.0g
$MnSO_4 \cdot 4H_2O$....0.2g
HCl....0.05g
$FeCl_3$....0.04g

Preparation of Salts Solution: Add components to distilled/deionized water and bring volume to 100.0mL. Mix thoroughly.

Reducing Solution:

Composition per 10mL:

Cysteine·HCl·H_2O....0.12g
Dithiothreitol....0.12g
Glutamine....0.06g

Preparation of Reducing Solution: Add components to distilled/deionized water and bring volume to 10.0mL. Mix thoroughly. Adjust pH to 7.6–7.8. Filter sterilize.

Preparation of Medium: Add agar to distilled/deionized water and bring volume to 50.0mL. Mix thoroughly. Gently heat and bring to boiling. In another flask add neopeptone, yeast extract, liver extract and salts solution to distilled/deionized water and bring volume to 50.0mL. Mix thoroughly. Gently heat until dissolved. Combine the two solutions. Distibute into screw-capped bottles in 18.0mL volumes. Autoclave for 10 min at 10 psi pressure–115°C. Cool to 45° to 50°C. Medium may be stored at 4°C at this point. Immediately prior to inoculation, aseptically add 2.0mL of horse blood and 0.15mL of sterile reducing solution to each tube of melted agar at 50°C. Mix thoroughly. Pour the contents of each tube into a sterile Petri dish.

Use: For the cultivation of *Clostridium novyi.*

Clostridium Selective Agar (Clostrisel Agar)

Composition per liter:

Pancreatic digest of casein....17.0g
Agar....14.0g
Glucose....6.0g
Papaic digest of soybean meal....3.0g
NaCl....2.5g
Sodium thioglycollate....1.8g
Sodium formaldehyde sulfoxylate....1.0g
L-Cystine....0.25g
NaN_3....0.15g
Neomycin sulfate....0.15g

pH 7.0 ± 0.2 at 25°C

Source: This medium is available as a premixed powder from BBL Microbiology Systems.

Preparation of Medium: Add components to distilled/deionized water and bring volume to 1.0L. Mix thoroughly. Gently heat while stirring and bring to boiling. Distribute into tubes or flasks. Autoclave for 15 min at 15 psi pressure–118°C. Pour into sterile Petri dishes or leave in tubes.

Caution: Sodium azide is toxic. Azides also react with metals and disposal must be highly diluted.

Use: For the selective isolation of pathogenic *Clostridium* species from specimens containing mixed flora, e.g., from wounds or fecal specimens.

CN Screen Medium (*Cryptococcus neoformans* Screen Medium)

Composition per liter:

Agar....15.0g
Asparagine....1.0g
Glutamine....1.0g
Glycine....1.0g
Thiamine·HCl....1.0g
Tryptophan....1.0g
K_2HPO_4....4.0g
$MgSO_4 \cdot 7H_2O$....2.5g
Glucose....1.25g
EDTA....0.6g
Biotin....0.51g
Dihydroxyphenylalanine (Dopa)....0.2g
Phenol Red....0.2g

pH 5.5–5.6 ± 0.2 at 25°C

Preparation of Medium: Add components to distilled/deionized water and bring volume to 1.0L. Mix thoroughly. Gently heat until boiling. Distribute into tubes or flasks. Autoclave for 15 min at 15 psi pressure–121°C.

Use: For the screening of yeast isolates for the presumptive identification of *Cryptococcus neoformans*. *Cryptococcus neoformans* form black colonies.

Coagulase Agar Base

Composition per liter:

Agar	25.0g
Brain heart infusion	10.5g
Pancreatic digest of casein	10.5g
D-Mannitol	10.0g
Brain heart infusion	5.0g
NaCl	3.5g
Papaic digest of soybean meal	3.5g
Bromcresol Purple	0.02g
Rabbit plasma	100.0mL

pH 7.4 ± 0.2 at 25°C

Preparation of Medium: Add components, except rabbit plasma, to distilled/deionized water and bring volume to 1.0L. Mix thoroughly. Gently heat while stirring until boiling. Distribute into tubes or flasks. Autoclave for 15 min at 15 psi pressure–121°C. Cool to 45° to 50°C. Add rabbit plasma to a final concentration of 7–15%. Mix thoroughly. Pour into sterile Petri dishes in 18.0mL volume per plate.

Use: For the cultivation and differentiation of *Staphylococcus aureus* from other *Staphylococcus* species based on coagulase production.

Coagulase Mannitol Agar

Composition per liter:

Agar	14.5g
Pancreatic digest of casein	10.5g
D-Mannitol	10.0g
Brain heart infusion	5.0g
NaCl	3.5g
Papaic digest of soybean meal	3.5g
Bromcresol Purple	0.02g
Rabbit plasma with 0.15% EDTA	100.0mL

pH 7.3 ± 0.2 at 25°C

Source: This medium is available as a premixed powder from BBL Microbiology Systems.

Preparation of Medium: Add components, except rabbit plasma, to distilled/deionized water and bring volume to 1.0L. Mix thoroughly. Gently heat while stirring until boiling. Distribute into tubes or flasks. Autoclave for 15 min at 15 psi pressure–121°C. Cool to 45° to 50°C. Add rabbit plasma with 0.15% EDTA to a final concentration of 7–15%. Mix thoroughly. Pour into sterile Petri dishes in 18.0mL volume per plate.

Use: For the cultivation and differentiation of *Staphylococcus aureus* from other *Staphylococcus* species based on coagulase production and mannitol fermentation.

COBA (Colistin Oxolinic Acid Blood Agar)

Composition per liter:

Columbia agar base	930.0mL
Horse blood, defibrinated, sterile	50.0mL
Colistin sulfate solution	10.0mL
Oxolinic acid solution	10.0mL

pH 7.3 ± 0.2 at 25°C

Columbia Agar Base:

Composition per 930mL:

Agar	13.5g
Pancreatic digest of casein	10.0g
Peptic digest of animal tissue	10.0g
NaCl	5.0g
Beef extract	3.0g
Yeast extract	3.0g
Cornstarch	1.0g

Preparation of Columbia Agar Base: Add components to distilled/deionized water and bring volume to 930.0mL. Mix thoroughly. Gently heat until boiling. Autoclave for 15 min at 15 psi pressure–121°C. Cool to 45° to 50°C.

Colistin Sulfate Solution:

Composition per 10mL:

Colistin sulfate	10.0mg

Preparation of Colistin Sulfate Solution: Add colistin sulfate to distilled/deionized water and bring volume to 10.0mL. Mix thoroughly. Filter sterilize.

Oxolinic Acid Solution:

Composition per 10mL:

Oxolinic acid	5.0–10.0mg

Preparation of Oxolinic Acid Solution: Add Oxolinic acid to distilled/deionized water and bring volume to 10.0mL. Mix thoroughly. Filter sterilize.

Preparation of Medium: To 930.0mL of sterile, cooled Columbia Agar Base, add sterile colistin sulfate, sterile oxolinic acid and sterile, defibrinated horse blood. Mix thoroughly. Pour into sterile Petri dishes.

Use: For the isolation and cultivation of streptococci in pure culture from mixed flora in clinical specimens.

Columbia Agar

Composition per liter:

Columbia agar base	950.0mL
Sheep blood	50.0mL

pH 7.3 ± 0.2 at 25°C

Source: This medium is available as a premixed powder from BBL Microbiology Systems.

Columbia Agar Base:

Composition per liter:

Agar	13.5g
Pancreatic digest of casein	12.0g
NaCl	5.0g
Peptic digest of animal tissue	5.0g
Beef extract	3.0g
Yeast extract	3.0g
Cornstarch	1.0g

Preparation of Columbia Agar Base: Add components to distilled/deionized water and bring volume to 1.0L. Mix thoroughly. Gently heat until boiling. Autoclave for 15 min at 15 psi pressure–121°C. Cool to 45° to 50°C.

Preparation of Medium: To 950.0mL of cooled, sterile Columbia Agar Base, aseptically add 50.0mL sterile, defibrinated sheep blood. Mix thoroughly. Pour into sterile Petri dishes or Distribute into sterile tubes.

Use: For the isolation and cultivation of nonfastidious and fastidious microorganisms from a variety of clinical specimens and in epidemiological investigations.

Columbia Blood Agar

Composition per liter:

Columbia blood agar base	950.0mL
Sheep blood	50.0mL

pH 7.3 ± 0.2 at 25°C

Columbia Blood Agar Base:

Composition per liter:

Agar	15.0g
Pantone	10.0g
Bitone	10.0g
NaCl	5.0g
Tryptic digest of beef heart	3.0g
Cornstarch	1.0g

Source: Columbia blood agar base is available as a premixed powder from Difco Laboratories.

Preparation of Columbia Blood Agar Base: Add components to distilled/deionized water and bring volume to 1.0L. Mix thoroughly. Gently heat until boiling. Autoclave for 15 min at 15 psi pressure–121°C. Cool to 45° to 50°C.

Preparation of Medium: To 950.0mL of cooled, sterile Columbia Blood Agar Base, aseptically add 50.0mL sterile, defibrinated sheep blood. Mix thoroughly. Pour into sterile Petri dishes or Distribute into sterile tubes.

Use: With the addition of blood or other enrichments, used for the isolation and cultivation of fastidious microorganisms.

Columbia Blood Agar

Composition per liter:

Columbia blood agar base	950.0mL
Sheep blood	50.0mL

pH 7.3 ± 0.2 at 25°C

Source: This medium is available as a premixed powder from Oxoid Unipath.

Columbia Blood Agar Base:

Composition per liter:

Special peptone	23.0g
Agar	10.0g
NaCl	5.0g
Starch	1.0g

Preparation of Columbia Blood Agar Base: Add components to distilled/deionized water and bring volume to 1.0L. Mix thoroughly. Gently heat until boiling. Autoclave for 15 min at 15 psi pressure–121°C. Cool to 45° to 50°C.

Preparation of Medium: To 950.0mL of cooled, sterile Columbia blood agar base, aseptically add 50.0mL sterile, defibrinated sheep blood. Mix thoroughly. Pour into sterile Petri dishes or distribute into sterile tubes.

Use: For the cultivation of a variety of fastidious microorganisms.

Columbia Broth

Composition per liter:

Bitone	10.0g
Pancreatic digest of casein	5.0g
Peptic digest of animal tissue	5.0g
NaCl	5.0g
Tryptic digest of beef heart	3.0g
Tris(hydroxymethyl)aminomethane·HCl	2.86g
Glucose	2.5g
Tris(hydroxymethyl)aminomethane	0.83g
Na_2CO_3	0.6g
L-Cysteine·HCl	0.1g
$MgSO_4$, anhydrous	0.1g
$FeSO_4$	0.02g

pH 7.5 ± 0.2 at 25°C

Source: This medium is available as a premixed powder from Difco Laboratories.

Preparation of Medium: Add components to distilled/deionized water and bring volume to 1.0L. Mix thoroughly. Gently heat until boiling. Distribute into tubes or flasks. Autoclave for 15 min at 15 psi pressure–121°C.

Use: For the cultivation and isolation of fastidious bacteria from clinical specimens or as a general purpose broth.

Columbia Broth

Composition per liter:

Pancreatic digest of casein	10.0g
Peptic digest of animal tissue	8.0g
NaCl	5.0g
Yeast extract	5.0g
Tris(hydroxymethyl) aminomethane·HCl buffer	2.86g
Glucose	2.5g
Tris(hydroxymethyl) aminomethane buffer	0.83g
Cysteine·HCl·H_2O	0.1g
$MgSO_4 \cdot 7H_2O$	0.05g
$FeSO_4$	0.012g

pH 7.4 ± 0.2 at 25°C

Source: This medium is available as a premixed powder from BBL Microbiology Systems.

Preparation of Medium: Add components to distilled/deionized water and bring volume to 1.0L. Mix thoroughly. Gently heat until boiling. Distribute into tubes or flasks. Autoclave for 15 min at 15 psi pressure–121°C.

Use: For the cultivation of a wide variety of microorganisms. Use as a general purpose medium.

Columbia CNA Agar (Columbia Colistin Nalidixic Acid Agar)

Composition per liter:

Columbia blood agar base	950.0L
Sheep blood	50.0mL

pH 7.3 ± 0.2 at 25°C

Source: This medium is available as a premixed powder from BBL Microbiology Systems and Difco Laboratories.

Columbia Blood Agar Base:

Composition per liter:

Agar	13.5g
Pancreatic digest of casein	12.0g
NaCl	5.0g
Peptic digest of animal tissue	5.0g
Beef extract	3.0g
Yeast extract	3.0g
Cornstarch	1.0g
Nalidixic acid	15.0mg
Colistin	10.0mg

Preparation of Columbia Blood Agar Base: Add components to distilled/deionized water and bring volume to 1.0L. Mix thoroughly. Gently heat until boiling. Autoclave for 15 min at 15 psi pressure–121°C. Cool to 45° to 50°C.

Preparation of Medium: To 950.0mL of cooled, sterile Columbia blood agar base, aseptically add 50.0mL of sterile, defibrinated sheep blood. Mix thoroughly. Pour into sterile Petri dishes or distribute into sterile tubes.

Use: For the selective isolation, cultivation and differentiation of Gram-positive cocci, especially Group A streptococci, from clinical specimens and in epidemiological investigations.

Columbia CNA Agar, Modified with Sheep Blood

Composition per liter:

Columbia blood agar base	950.0L
Sheep blood, defribinated	50.0mL

pH 7.3 ± 0.2 at 25°C

Source: This medium is available as a premixed powder from BBL Microbiology Systems.

Columbia Blood Agar Base:

Composition per liter:

Agar	13.5g
Pancreatic digest of casein	12.0g
NaCl	5.0g
Peptic digest of animal tissue	5.0g
Beef extract	3.0g
Yeast extract	3.0g
Cornstarch	1.0g
Nalidixic acid	5.0mg
Colistin	10.0mg

Preparation of Columbia Blood Agar Base: Add components to distilled/deionized water and bring volume to 1.0L. Mix thoroughly. Gently heat until boiling. Autoclave for 15 min at 15 psi pressure–121°C. Cool to 45° to 50°C.

Preparation of Medium: To 950.0L of cooled, sterile Columbia blood agar base, aseptically add 50.0mL of sterile, defibrinated sheep blood. Mix thoroughly. Pour into sterile Petri dishes or distribute into sterile tubes.

Use: For the selective isolation, cultivation and differentiation of Gram-positive cocci, especially Group A streptococci, from clinical specimens and in epidemiological investigations.

Converse Liquid Medium, Levine Modification

Composition per liter:

Ionagar No. 2 or Noble agar 10.0g
Glucose 4.0g
Ammonium acetate 1.23g
K_2HPO_4 ....0.52g
Tamol 0.5g
$MgSO_4 \cdot 7H_2O$ 0.4g
KH_2PO_4 ....0.4g
NaCl 0.014g
Na_2CO_3 ....0.012g
$CaCl_2 \cdot 2H_2O$ 0.002g
$ZnSO_4 \cdot 7H_2O$ 0.002g

Preparation of Medium: Add components to distilled/deionized water and bring volume to 1.0L. Mix thoroughly. Gently heat and bring to boiling. Autoclave for 15 min at 15 psi pressure–121°C. Pour into sterile Petri dishes in 15.0mL volumes.

Use: For the cultivation and induction of spherules of *Coccidioides immitis.*

Cooke Rose Bengal Agar

Composition per liter:

Agar 20.0g
Glucose 10.0g
Enzymatic hydrolysate of soybean meal 5.0g
KH_2PO_4 ....1.0g
$MgSO_4 \cdot 7H_2O$ 0.5g
Rose Bengal 35.0mg

pH 6.0 ± 0.2 at 25°C

Source: This medium is available as a premixed powder from Difco Laboratories.

Preparation of Medium: Add components to distilled/deionized water and bring volume to 1.0L. Mix thoroughly. Gently heat until boiling. Distribute into tubes or flasks. Autoclave for 15 min at 15 psi pressure–121°C. Pour into sterile Petri dishes or leave in tubes.

Use: For the isolation of fungi.

Cooked Meat Medium

Composition per liter:

Beef heart 454.0g
Proteose peptone 20.0g
NaCl 5.0g
Glucose 2.0g

pH 7.2 ± 0.2 at 25°C

Source: This medium is available as a premixed powder from Difco Laboratories.

Preparation of Medium: Finely chop beef heart. Add approximately 1.5g of heart particles to test tubes. Add remaining components to distilled/deionized water and bring volume to 1.0L. Mix thoroughly. Distribute into tubes in 10.0mL volumes. Autoclave for 15 min at 15 psi pressure–121°C. Slowly cool tubes to prevent expulsion of meat particles.

Use: For cultivation and maintenance of anaerobic microorganisms.

Cooked Meat Medium

Composition per liter:

Heart muscle 454.0g
Beef extract 10.0g
Peptone..10.0g
NaCl 5.0g
Glucose 2.0g

pH 7.2 ± 0.2 at 25°C

Source: This medium is available as a premixed powder from Oxoid Unipath.

Preparation of Medium: Finely chop beef heart. Add approximately 1.5g of heart particles to test tubes. Add remaining components to distilled/deionized water and bring volume to 1.0L. Mix thoroughly. Distribute into tubes in 10.0mL volumes. Autoclave for 15 min at 15 psi pressure–121°C. Slowly cool tubes to prevent expulsion of meat particles.

Use: For the cultivation and maintenance of aerobic and anaerobic microorganisms. Used for the cultivation of anaerobes, especially pathogenic clostridia.

Cooked Meat Medium

Composition per liter:

Heart tissue granules 98.0g
Peptic digest of animal tissue 20.0g
NaCl 5.0g
Glucose 2.0g

pH 7.2 ± 0.2 at 25°C

Source: This medium is available as a premixed powder from BBL Microbiology Systems.

Preparation of Medium: Add approximately 1.0g of heart tissue granules to test tubes. Add remaining components to distilled/deionized water and bring volume to 1.0L. Mix thoroughly. Distribute into tubes in 10.0mL volumes. Autoclave for 15 min at 15 psi pressure–121°C. Slowly cool tubes to prevent expulsion of meat particles.

Use: For the cultivation of anaerobes, especially pathogenic clostridia.

Cooked Meat Medium, Modified

Composition per liter:

Cooked meat medium 66.0g
Solution A 1.0L

pH 6.8 ± 0.2 at 25°C

Solution A:

Composition per liter:

Pancreatic digest of casein 10.0g
Glucose 2.0g
Soluble starch..1.0g
Sodium thioglycollate 1.0g
Neutral Red (1% aqueous) 5.0mL

Preparation of Solution A: Add components to distilled/deionized water and bring volume to 1.0L. Mix thoroughly. Gently heat until dissolved.

Preparation of Medium: Add 1.0g of cooked meat medium to each of 66 test tubes. Add 15.0mL of solution A to each test tube. Allow meat particles to rehydrate. Autoclave for 15 min at 15 psi pressure–121°C.

Use: For the cultivation of a variety of anaerobic microorganisms.

Cooked Meat Medium with Glucose, Hemin and Vitamin K

Composition per liter:

Heart tissue granules 98.0g
Peptic digest of animal tissue..........................20.0g
NaCl 5.0g
Glucose 5.0g
Yeast extract..5.0g
Hemin.. 5.0mg
Vitamin K... 1.0mg

pH 7.2 ± 0.2 at 25°C

Source: This medium is available as a premixed powder from BBL Microbiology Systems.

Preparation of Medium: Add approximately 1.0g of heart tissue granules to test tubes. Add remaining components to distilled/deionized water and bring volume to 1.0L. Mix thoroughly. Distribute into tubes in 10.0mL volumes. Autoclave for 15 min at 15 psi pressure–121°C. Slowly cool tubes to prevent expulsion of meat particles.

Use: For the cultivation of anaerobes, especially pathogenic clostridia.

Corynebacterium Liquid Enrichment Medium

Composition per 2000mL:

Fosfomycin 0.15g
Glucose 6-phosphate..0.03g
Solution A 985.0mL
Bovine serum 100.0mL
Egg yolk suspension 10eggs
Nystatin solution 1.15mL
L-Cystine (1% solution) 1.0mL

pH 7.4 ± 0.2 at 25°C

Solution A:

Composition per liter:

Meat extract 9.0g
Proteose peptone no. 3 ....9.0g
NaCl 2.7g
Glucose 1.8g
$Na_2HPO_4 \cdot 12H_2O$...1.8g
K_2TeO_3 (2% solution).................................75.0mL
L-Cystine (1% solution) 10.0mL

Preparation of Solution A: Add components to distilled/deionized water and bring volume to 985.0mL. Mix thoroughly. Filter sterilize.

Egg Yolk Emulsion:

Composition:

Chicken egg yolks..9
Whole chicken egg...1

Preparation of Egg Yolk Emulsion: Soak eggs with 1:100 dilution of saturated mercuric chloride solution for 1 min. Crack eggs and separate yolks from whites. Mix egg yolks with 1 chicken egg. Filter sterilize.

Nystatin Solution:

Composition per 10mL:

Nystatin..10,000U

Preparation of Nystatin Solution: Add nystatin to distilled/deionized water and bring volume to 10.0mL. Mix thoroughly. Filter sterilize.

Cystine Solution:

Composition per 10mL:

L-Cystine 0.1g

Preparation of Cystine Solution: Add cystine to distilled/deionized water and bring volume to 10.0mL. Mix thoroughly. Filter sterilize.

Caution: Potassium tellurite is toxic.

Preparation of Medium: To 985.0mL of sterile solution A, aseptically add the remaining components. Mix thoroughly. Aseptically distribute into sterile tubes in 2.0–3.0mL volumes.

Use: For the isolation and cultivation of *Corynebacterium diphtheriae.*

Cotton Seed Agar

Composition per liter:

Pharmamedia..20.0g
Glucose 20.0g
Agar...20.0g

pH 6.0± 0.2 at 25°C

Source: This medium is available from Traders Protein Division of Traders Oil Mill Co., Fort Worth, Texas.

Preparation of Medium: Add components to distilled/deionized water and bring volume to 1.0L. Mix thoroughly. Gently heat until boiling. Distribute into tubes or flasks. Autoclave for 15 min at 15 psi pressure–121°C.

Use: For the conversion of mycelium to the yeast form for *Blastomyces dermatitidis* only.

Crystal Violet Agar

Composition per liter:

Agar 15.0g
Lactose 10.0g
Proteose peptone 5.0g
Beef extract 3.0g
Crystal Violet 3.3mg

pH 6.8 ± 0.1 at 25°C

Preparation of Medium: Add components to distilled/deionized water and bring volume to 1.0L. Mix thoroughly. Gently heat until boiling. Distribute into tubes or flasks. Autoclave for 15 min at 15 psi pressure–121°C. Pour into sterile Petri dishes or leave in tubes.

Use: For the differentiation of pathogenic from nonpathogenic staphylococci. Hemolytic and coagulase-producing strains of *Staphylococcus aureus* appear as purple or yellow colonies. Nonhemolytic and noncoagulating strains of *Staphylococcus* species appear as white colonies.

Crystal Violet Azide Esculin Agar

Composition per liter:

Agar...15.0g
Glucose 5.0g
NaCl 5.0g
Proteose peptone 5.0g
Pancreatic digest of casein 5.0g
Meat extract 3.0g
Esculin...1.0g
NaN_3..1.0g
Crystal Violet 0.1g
Bovine blood, citrated 100.0mL

pH 7.5 ± 0.2 at 25°C

Preparation of Medium: Add components, except citrated bovine blood, to distilled/deionized water and bring volume to 900.0mL. Mix thoroughly. Gently heat and bring to boiling. Autoclave for 15 min at 15 psi pressure–121°C. Cool to 45° to 50°C. Aseptically add sterile citrated bovine blood. Mix thoroughly. Pour into sterile Petri dishes or distribute into sterile tubes.

Caution: Sodium azide is toxic. Azides also react with metals and disposal must be highly diluted.

Use: For the cultivation of *Erysipelothrix rhusiopathiae.*

Crystal Violet Esculin Agar

Composition per liter:

Agar...15.0g
Glucose 5.0g
NaCl 5.0g
Proteose peptone 5.0g
Pancreatic digest of casein 5.0g
Meat extract 3.0g
Esculin...1.0g
Crystal Violet 2.0mg
Blood, citrated.. 100.0mL

pH 7.5 ± 0.2 at 25°C

Preparation of Medium: Add components, except citrated blood, to distilled/deionized water and bring volume to 900.0mL. Mix thoroughly. Gently heat and bring to boiling. Autoclave for 15 min at 15 psi pressure–121°C. Cool to 45° to 50°C. Aseptically add citrated blood. Mix thoroughly. Pour into sterile Petri dishes or distribute into sterile tubes.

Use: For the cultivation of *Erysipelothrix rhusiopathiae.*

CTA Agar (Cystine Trypticase™ Agar)

Composition per liter:

Pancreatic digest of casein 20.0g
Agar...14.0g
NaCl 5.0g
L-Cystine 0.5g
Na_2SO_3..0.5g
Phenol Red 0.017g

pH 7.3 ± 0.2 at 25°C

Preparation of Medium: Add components to distilled/deionized water and bring volume to 1.0L. Mix thoroughly. Gently heat until boiling. Distribute into tubes or flasks. Autoclave for 15 min at 15 psi pressure–118°C. Pour into sterile Petri dishes or leave in tubes. Two drops of sterile rabbit serum added per tube prior to solidification enhance the recovery of *Corynebacterium diphtheriae.*

Use: For cultivation and maintenance of a variety of fastidious microorganisms. Used for carbohydrate fermentation tests in the differentiation of *Neisseria* species.

CTA Medium (Cystine Trypticase™ Agar Medium) (Cystine Tryptic Agar)

Composition per liter:

Pancreatic digest of casein	20.0g
NaCl	5.0g
Carbohydrate	5.0g
Agar	2.5g
L-Cystine	0.5g
Na_2SO_3	0.5g
Phenol Red	0.017g

pH 7.3 ± 0.2 at 25°C

Source: Available as a premixed powder from BBL Microbiology Systems and Difco Laboratories.

Preparation of Medium: Add components to distilled/deionized water and bring volume to 1.0L. Mix thoroughly. Adjust pH to 7.3. Gently heat until boiling. Distribute into tubes or flasks. Autoclave for 15 min at 15 psi pressure–118°C. Cool tubes in an upright position. Store at room temperature.

Use: For cultivation and maintenance of a variety of fastidious microorganisms. Used for the detection of bacterial motility. Used, with added specific carbohydrate, for fermentation reactions of fastidious microorganisms, especially, *Neisseria* species, pneumococci, streptococci and non-sporeforming anaerobes.

CTA Medium with Yeast Extract and Rabbit Serum (Cystine Trypticase™ Agar Medium with Yeast Extract and Rabbit Serum)

Composition per liter:

Yeast extract	50.0g
Pancreatic digest of casein	20.0g
NaCl	5.0g
Carbohydrate	5.0g
Agar	2.5g
L-Cystine	0.5g
Na_2SO_3	0.5g
Phenol Red	0.017g
Rabbit serum	250.0mL

pH 7.3 ± 0.2 at 25°C

Preparation of Medium: Add components, except rabbit serum to distilled/deionized water and bring volume to 750.0mL. Mix thoroughly. Adjust pH to 7.3. Gently heat until boiling. Autoclave for 15 min at 15 psi pressure–118°C. Cool to 50°C. Aseptically add sterile rabbit serum. Mix thoroughly. Distribute into sterile tubes. Store at room temperature—do not refrigerate.

Use: For the cultivation and maintenance of fastidious microorganisms, especially mycoplasmas and related microorganisms.

CVA Medium (Cefoperazone Vancomycin Amphotericin Medium)

Composition per liter:

Agar	15.0g
Casein peptone	10.0g
Meat peptone	10.0g
NaCl	5.0g
Yeast autolysate	2.0g
Glucose	1.0g
$NaHSO_3$	0.1g
Sheep blood, defibrinated	50.0mL
CVA antibiotic solution	10.0mL

pH 7.0 ± 0.2 at 25°C

CVA Antibiotic Solution:

Composition per 10mL:

Cefoperazone	20.0mg
Vancomycin	10.0mg
Amphotericin B	2.0mg

Preparation of CVA Antibiotic Solution: Add components to distilled/deionized water and bring volume to 10.0mL. Mix thoroughly. Filter sterilize.

Preparation of Medium: Add components, except CVA antibiotic solution and sheep blood, to distilled/deionized water and bring volume to 940.0mL. Mix thoroughly. Gently heat until boiling. Autoclave for 15 min at 15 psi pressure–121°C. Cool to 45° to 50°C. Aseptically add sterile CVA antibiotic solution and sterile, defibrinated sheep blood. Mix thoroughly. Pour into sterile Petri dishes.

Use: For the isolation and cultivation of *Campylobacter* species from clinical specimens.

Cycloserine Cefoxitin Egg Yolk Fructose Agar

Composition per liter:

Proteose peptone No. 2	40.0g
Agar	25.0g
Fructose	6.0g
Na_2HPO_4	5.0g
NaCl	2.0g
KH_2PO_4	1.0g
$MgSO_4 \cdot 7H_2O$	0.1g
Egg yolk emulsion	100.0mL
Antibiotic solution	10.0mL
Neutral Red solution	3.0mL
Hemin solution	1.0mL

Egg Yolk Emulsion:
Composition:

Chicken egg yolks	11
Whole chicken egg	1

Preparation of Egg Yolk Emulsion: Soak eggs with 1:100 dilution of saturated mercuric chloride solution for 1 min. Crack eggs. Separate yolks from whites for 11 eggs. Mix egg yolks with 1 chicken egg.

Antibiotic Solution:
Composition per 10mL:

Cycloserine	0.5g
Cefoxitin	0.016g

Preparation of Antibiotic Solution: Add components to distilled/deionized water and bring volume to 10.0mL. Mix thoroughly. Filter sterilize.

Neutral Red Solution:
Composition per 10mL:

Neutral Red	0.1g
Ethanol	10.0mL

Preparation of Neutral Red Solution: Add Neutral Red to 10.0mL of ethanol. Mix thoroughly.

Hemin Solution:
Composition per 100mL:

Hemin	0.5g
NaOH (1*N* solution)	10.0mL

Preparation of Hemin Solution: Add hemin to 10.0mL of 1*N* NaOH solution. Mix thoroughly. Bring volume to 100.0mL with distilled/deionized water.

Preparation of Medium: Add components, except egg yolk emulsion and antibiotic solution, to distilled/deionized water and bring volume to 890.0mL. Mix thoroughly. Gently heat and bring to boiling. Autoclave for 15 min at 15 psi pressure–121°C. Cool to 45° to 50°C. Aseptically add sterile egg yolk emulsion and antibiotic solution. Mix thoroughly. Pour into sterile Petri dishes.

Use: For the selective isolation and cultivation of *Clostridium difficile* from feces.

CYE Agar (Charcoal Yeast Extract Agar)

Composition per liter:

Agar	17.0g
Yeast extract	10.0g
Charcoal, activated, acid-washed	2.0g
L-Cysteine·HCl·H_2O solution	10.0mL
$Fe_4(P_2O_7)_3$ solution	10.0mL

pH 6.9 ± .05 at 50°C

L-Cysteine·HCl·H_2O Solution:
Composition per 10mL:

L-Cysteine·HCl·H_2O	0.4g

Preparation of L-Cysteine·HCl·H_2O solution: Add L-Cysteine·HCl·H_2O to distilled/deionized water and bring volume to 10.0mL. Mix thoroughly. Filter sterilize.

$Fe_4(P_2O_7)_3$ Solution:
Composition per liter:

$Fe_4(P_2O_7)_3$	0.25g

Preparation of $Fe_4(P_2O_7)_3$ Solution: Add soluble $Fe_4(P_2O_7)_3$ to distilled/deionized water and bring volume to 10.0mL. Mix thoroughly. Filter sterilize. The soluble $Fe_4(P_2O_7)_3$ must be kept dry and in the dark. Do not use if brown or yellow. Prepare solutions freshly. Do not heat over 60°C to dissolve. The mixture dissolves readily in a 50°C water bath.

Preparation of Medium: Add components, except L-cysteine·HCl·H_2O solution and $Fe_4(P_2O_7)_3$ solution to distilled/deionized water and bring volume to 980.0mL. Mix thoroughly. Gently heat to boiling. Autoclave for 15 min at 15 psi pressure–121°C. Cool to 50° C. Add 10.0mL of sterile L-cysteine·HCl·H_2O solution. Add 10.0mL of sterile $Fe_4(P_2O_7)_3$ solution. Adjust pH to 6.9 at 50°C by adding 4.0–4.5mL of 1.0 *N* KOH. This is a critical step. Mix thoroughly. Pour in 20.0mL volumes into sterile Petri dishes. Swirl medium while pouring to keep charcoal in suspension.

Use: For the cultivation and maintenance of *Legionella* species and *Tatlockia micdadei*.

CYE Agar, Buffered (Charcoal Yeast Extract Agar, Buffered)

Composition per liter:

Agar	17.0g
ACES buffer (N-2-acetamido-2-aminoethane sulfonic acid)	10.0g

Yeast extract..10.0g
Charcoal, activated, acid-washed........................2.0g
L-Cysteine·HCl·H_2O solution........................10.0mL
$Fe_4(P_2O_7)_3$ solution.......................................10.0mL

pH 6.9 ± .05 at 50°C

L-Cysteine·HCl·H_2O Solution:
Composition per 10mL:

L-Cysteine·HCl·H_2O..0.4g

Preparation of L-Cysteine·HCl·H_2O solution: Add L-Cysteine·HCl·H_2O to distilled/deionized water and bring volume to 10.0mL. Mix thoroughly. Filter sterilize.

$Fe_4(P_2O_7)_3$ solution:
Composition per liter:

$Fe_4(P_2O_7)_3$..0.25g

Preparation of $Fe_4(P_2O_7)_3$ solution: Add soluble $Fe_4(P_2O_7)_3$ to distilled/deionized water and bring volume to 10.0mL. Mix thoroughly. Filter sterilize. The soluble $Fe_4(P_2O_7)_3$ must be kept dry and in the dark. Do not use if brown or yellow. Prepare solutions freshly. Do not heat over 60°C to dissolve. The mixture dissolves readily in a 50°C water bath.

Preparation of Medium: Add components, except L-cysteine·HCl·H_2O solution and $Fe_4(P_2O_7)_3$ solution to distilled/deionized water and bring volume to 980.0mL. Mix thoroughly. Gently heat to boiling. Autoclave for 15 min at 15 psi pressure–121°C. Cool to 50° C. Add 10.0mL of sterile L-cysteine·HCl·H_2O solution. Add 10.0mL of sterile $Fe_4(P_2O_7)_3$ solution. Adjust pH to 6.9 at 50°C by adding 4.0 to 4.5mL of 1.0 *N* KOH. This is a critical step. Mix thoroughly. Pour in 20.0mL volumes into sterile Petri dishes. Swirl medium while pouring to keep charcoal in suspension.

Use: For the cultivation and maintenance of *Legionella* species and *Xylella fastidiosa.*

Cystine Heart Agar

Composition per liter:

Beef heart, solids from infusion.....................500.0g
Agar..15.0g
Glucose..10.0g
Proteose peptone...10.0g
NaCl...5.0g
L-Cystine...1.0g
Hemoglobin solution.................................100.0mL

pH 6.8 ± 0.2 at 25°C

Source: This medium is available as a premixed powder from Difco Laboratories.

Hemoglobin Solution:
Composition per 100mL:

Hemoglobin..2.0g

Preparation of Hemoglobin Solution: Add hemoglobin to cold distilled/deionized water and bring volume to 100.0mL. Mix thoroughly by shaking for 10–15 min. Autoclave for 15 min at 15 psi pressure–121°C. Cool to 50° to 60°C.

Preparation of Medium: Add components, except hemoglobin solution, to distilled/deionized water and bring volume to 900.0mL. Mix thoroughly. Gently heat until boiling. Autoclave for 15 min at 15 psi pressure–121°C. Cool to 50–60°C. Aseptically add 100.0mL of sterile cooled hemoglobin solution. Mix thoroughly. Pour into sterile Petri dishes or distribute into sterile tubes.

Use: For the cultivation and maintenance of *Francisella tularensis* and *Francisella philomiragia.* Without the hemoglobin enrichment, it supports excellent growth of Gram-negative cocci and other pathogenic microorganisms.

Cystine Heart Agar with Rabbit Blood

Composition per liter:

Beef heart, solids from infusion.....................500.0g
Agar..15.0g
Glucose..10.0g
Proteose peptone...10.0g
NaCl...5.0g
L-Cystine...1.0g
Rabbit blood, defibrinated...........................50.0mL

pH 6.8 ± 0.2 at 25°C

Source: This medium is available as a premixed powder from Difco Laboratories.

Preparation of Medium: Add components, except rabbit blood, to distilled/deionized water and bring volume to 950.0mL. Mix thoroughly. Gently heat until boiling. Autoclave for 15 min at 15 psi pressure–121°C. Cool to 50–60°C. Aseptically add 50.0mL of sterile, defibrinated rabbit blood. Mix thoroughly. Pour into sterile Petri dishes or distribute into sterile tubes.

Use: For the cultivation and maintenance of *Francisella tularensis* and *Francisella philomiragia.* Without the hemoglobin enrichment, it supports excellent growth of Gram-negative cocci and other pathogenic microorganisms.

Cystine Tellurite Blood Agar

Composition per 120mL:

Heart infusion agar.....................................100.0mL
K_2TeO_3 solution..15.0mL
Sheep blood...5.0mL
L-Cystine...5.0mg

pH 7.4 ± 0.2 at 25°C

Heart Infusion Agar:
Composition per liter:

Beef heart, infusion from	500.0g
Agar	20.0g
Tryptose	10.0g
Yeast extract	5.0g
NaCl	5.0g

Preparation of Heart Infusion Agar: Add components to distilled/deionized water and bring volume to 1.0L. Mix thoroughly. Autoclave for 15 min at 15 psi pressure–121°C. Cool to 45° to 50°C.

K_2TeO_3 Solution:
Composition per 100mL:

K_2TeO_3	0.3g

Preparation of K_2TeO_3 Solution: Add K_2TeO_3 to distilled/deionized water and bring volume to 100.0mL. Mix thoroughly. Autoclave for 15 min at 15 psi pressure–121°C.

Caution: Potassium tellurite is toxic.

Preparation of Medium: Add sterile K_2TeO_3 solution, sterile defibrinated sheep blood and sterile solid L-cystine to sterile, cooled Heart Infusion Agar. Mix thoroughly. Pour into sterile Petri dishes or distribute into sterile tubes.

Use: For the isolation, differentiation and cultivation of *Corynebacterium diphtheriae. C. diphtheriae* appear as dark gray to black colnies.

Cystine Tellurite Blood Agar

Composition per liter:

Heart infusion agar	900.0mL
K_2TeO_3 solution	75.0mL
Rabbit blood	25.0mL
L-Cystine	22.0mg

pH 7.4 ± 0.2 at 25°C

Heart Infusion Agar:
Composition per 900mL:

Beef heart, solids from infusion	500.0g
Agar	20.0g
Tryptose	10.0g
Yeast extract	5.0g
NaCl	5.0g

Preparation of Heart Infusion Agar: Add components to distilled/deionized water and bring volume to 900.0mL. Mix thoroughly. Autoclave for 15 min at 15 psi pressure–121°C. Cool to 45° to 50°C.

K_2TeO_3 Solution:
Composition per 100mL:

K_2TeO_3	0.3g

Preparation of K_2TeO_3 Solution: Add K_2TeO_3 to distilled/deionized water and bring volume to 100.0mL. Mix thoroughly. Autoclave for 15 min at 15 psi pressure–121°C.

Caution: Potassium tellurite is toxic.

Preparation of Medium: Add sterile K_2TeO_3 solution, sterile rabbit blood and sterile solid L-cystine to sterile, cooled Heart Infusion Agar. Mix thoroughly. Pour into sterile Petri dishes or distribute into sterile tubes.

Use: For the isolation, differentiation and cultivation of *Corynebacterium diphtheriae. C. diphtheriae* appear as dark gray to black colnies.

Czapek Agar (ATCC Medium 312)

Composition per liter:

Sucrose	30.0g
Agar	15.0g
$NaNO_3$	3.0g
K_2HPO_4	1.0g
KCl	0.5g
$MgSO_4 \cdot 7H_20$	0.5g
$FeSO_4 \cdot 7H_2O$	0.01g

pH 7.3 ± 0.2 at 25°C

Preparation of Medium: Add components, except sucrose, to distilled/deionized water and bring volume to 900.0mL. Mix thoroughly. Distribute into tubes or flasks. In a separate flask, add sucrose to distilled/deionized water and bring volume to 100.0mL. Mix thoroughly. Autoclave both solutions separately for 15 min at 15 psi pressure–121°C. Cool to 50°C. Combine the sterile solutions. Mix thoroughly. Pour into sterile Petri dishes or distribute into sterile tubes.

Use: For the cultivation and maintenance of *Streptomyces* species. Also, for the cultivation of Actinoplanaceae.

Czapek Agar with Peptone (ATCC Medium 522)

Composition per liter:

Sucrose	30.0g
Agar	15.0g
Peptone	5.0g
$NaNO_3$	3.0g
K_2HPO_4	1.0g
KCl	0.5g
$MgSO_4 \cdot 7H_20$	0.5g
$FeSO_4 \cdot 7H_2O$	0.01g

pH 7.3 ± 0.2 at 25°C

Preparation of Medium: Add components, except sucrose, to distilled/deionized water and bring volume to 900.0mL. Mix thoroughly. Distribute into tubes or flasks. In a separate flask, add sucrose to distilled/deionized water and bring volume to 100.0mL. Mix thoroughly. Autoclave both solutions separately for 15 min at 15 psi pressure–121°C. Cool to 50°C. Combine the sterile solutions. Mix thoroughly. Pour into sterile Petri dishes or distribute into sterile tubes.

Use: For the cultivation and maintenance of *Streptomyces* species. Also, for the cultivation of Actinoplanaceae.

Czapek Dox Agar, Modified

Composition per liter:

Sucrose	30.0g
Agar	12.0g
$NaNO_3$	2.0g
Magnesium glycerophosphate	0.5g
KCl	0.5g
K_2SO_4	0.35g
$FeSO_4$	0.01g

pH 6.8 ± 0.2 at 25°C

Source: This medium is available as a premixed powder from Oxoid Unipath.

Preparation of Medium: Add components to distilled/deionized water and bring volume to 1.0L. Mix thoroughly. Distribute into tubes or flasks. Autoclave for 15 min at 15 psi pressure–121°C. Pour into sterile Petri dishes or leave in tubes.

Use: For the cultivation and maintenance of numerous fungal species. For chlamydospore production by *Candida albicans* and identification of *Aspergillus* species.

Czapek Dox Broth

Composition per liter:

Sucrose	30.0g
$NaNO_3$	3.0g
K_2HPO_4	1.0g
$MgSO_4 \cdot 7H_2O$	0.5g
KCl	0.5g
$FeSO_4 \cdot 7H_2O$	0.01g

pH 7.3 ± 0.2 at 25°C

Source: This medium is available as a premixed powder from Difco Laboratories.

Preparation of Medium: Add components to distilled/deionized water and bring volume to 1.0L. Mix thoroughly. Distribute into tubes or flasks. Autoclave for 15 min at 15 psi pressure–121°C.

Use: For the cultivation and maintenance of a variety of fungal and bacterial species that can use nitrate as sole nitrogen source.

DCLS Agar (Deoxycholate Citrate Lactose Sucrose Agar)

Composition per liter:

Agar	12.0g
Sodium citrate·$3H_2O$	10.5g
Lactose	5.0g
$Na_2S_2O_3$	5.0g
Sucrose	5.0g
Pancreatic digest of casein	3.5g
Peptic digest of animal tissue	3.5g
Beef extract	3.0g
Sodium deoxycholate	2.5g
Neutral Red	0.03g

pH 7.2 ± 0.1 at 25°C

Source: This medium is available as a premixed powder from BBL Microbiology Systems, Difco Laboratories and Oxoid Unipath Unipath.

Preparation of Medium: Add components to distilled/deionized water and bring volume to 1.0L. Mix thoroughly. Gently heat while stirring and bring to boiling. Do not overheat. Do not autoclave. Pour into sterile Petri dishes in 20.0mL volumes.

Use: For the selective isolation of *Salmonella* species, *Shigella* species and *Vibrio* species from fecal specimens.

Decarboxylase Base, Møller

Composition per liter:

Amino acid	10.0g
Beef extract	5.0g
Peptone	5.0g
Glucose	0.5g
Bromcresol Purple	0.01g
Cresol Red	5.0mg
Pyridoxal	5.0mg
Mineral oil	200.0mL

pH 6.0 ± 0.2 at 25°C

Source: This medium is available as a premixed powder Difco Laboratories.

Preparation of Medium: Add components, except mineral oil, to distilled/deionized water and bring volume to 1.0L. For amino acid, use L-arginine, L-lysine or L-ornithine. Mix thoroughly. Distribute into screw-capped tubes in 5.0mL volumes. Autoclave medium and mineral oil separately for 15 min at 15 psi pressure–121°C. After inoculation, overlay medium with 1.0mL of sterile mineral oil per tube.

Use: For the cultivation and differentiation of bacteria based on their ability to decarboxylate the amino acid. Bacteria that decarboxylate arginine, lysine or ornithine turn the medium turbid purple.

Decarboxylase Medium Base, Falkow

Composition per liter:

Amino acid	5.0g
Peptone	5.0g
Yeast extract	3.0g
Glucose	1.0g
Bromcresol Purple	0.02g
Mineral oil	200.0mL

pH 6.8 ± 0.2 at 25°C

Source: This medium is available as a premixed powder from Difco Laboratories.

Preparation of Medium: Add components, except mineral oil, to distilled/deionized water and bring volume to 1.0L. For amino acid, use L-arginine, L-lysine or L-ornithine. Mix thoroughly. Distribute into screw-capped tubes in 5.0mL volumes. Autoclave medium and mineral oil separately for 15 min at 15 psi pressure–121°C. After inoculation, overlay medium with 1.0mL of sterile mineral oil per tube.

Use: For the cultivation and differentiation of bacteria based on their ability to decarboxylate the amino acid. Bacteria that decarboxylate arginine, lysine or ornithine turn the medium turbid purple.

Decarboxylase Medium, Ornithine Modified

Composition per liter:

L-Ornithine	10.0g
Meat peptone	5.0g
Yeast extract	3.0g
Bromcresol Purple solution	5.0mL

pH 5.5 ± 0.2 at 25°C

Bromcresol Purple Solution:
Composition per 100mL:

Bromcresol Purple	0.2g
Ethanol	50.0mL

Preparation of Bromcresol Purple Solution: Add bromcresol purple to ethanol. Mix thoroughly. Bring volume to 100.0mL with distilled/deionized water. Mix thoroughly. Filter sterilize.

Preparation of Medium: Add components to distilled/deionized water and bring volume to 1.0L. Mix thoroughly. Gently heat until dissolved. Adjust pH to 5.5 with HCl or NaOH. Distribute into screw-capped tubes. Autoclave for 15 min at 15 psi pressure–121°C.

Use: For the cultivation and differentiation of bacteria based on their ability to decarboxylate ornithine. Bacteria that decarboxylate ornithine turn the medium turbid purple.

Deoxycholate Agar

Composition per liter:

Agar	16.0g
Lactose	10.0g
NaCl	5.0g
Pancreatic digest of casein	5.0g
Peptic digest of animal tissue	5.0g
K_2HPO_4	2.0g
Ferric citrate	1.0g
Sodium citrate	1.0g
Sodium deoxycholate	1.0g
Neutral Red	0.033g

pH 7.3 ± 0.2 at 25°C

Source: This medium is available as a premixed powder from BBL Microbiology Systems.

Preparation of Medium: Add components to distilled/deionized water and bring volume to 1.0L. Mix thoroughly. Gently heat and bring to boiling. Do not autoclave. Cool to 45° to 50°C. Pour into sterile Petri dishes.

Use: For the selective isolation, cultivation, enumeration and differentiation of Gram-negative enteric microorganisms from a variety of clinical specimens and in epidemiological investigations. *Escherichia coli* appears as large, flat rose-red colonies. *Enterobacter* and *Klebsiella* species appear as large mucoid pale colonies with a pink center. *Proteus* and *Salmonella* species appear as large, colorless to tan colonies. *Shigella* species appear as colorless to pink colonies. *Pseudomonas* species appear as irregular colorless to brown colonies.

Deoxycholate Agar (Desoxycholate Agar)

Composition per liter:

Agar	15.0g
Lactose	10.0g
Peptone	10.0g
NaCl	5.0g
K_2HPO_4	2.0g
Ferric citrate	1.0g
Sodium citrate	1.0g
Sodium deoxycholate	1.0g
Neutral Red	0.03g

pH 7.3 ± 0.2 at 25°C

Source: This medium is available as a premixed powder from Oxoid Unipath and Difco Laboratories.

Preparation of Medium: Add components to distilled/deionized water and bring volume to 1.0L. Mix thoroughly. Gently heat and bring to boiling. Do not autoclave. Cool to 50°C. Pour into sterile Petri dishes.

Use: For the selective isolation, cultivation, enumeration and differentiation of Gram-negative enteric microorganisms from a variety of clinical specimens and in epidemiological investigations. *Escherichia coli* appears as large, flat rose-red colonies. *Enterobacter* and *Klebsiella* species appear as large mucoid pale colonies with a pink center. *Proteus* and *Salmonella* species appear as large, colorless to tan colonies. *Shigella* species appear as colorless to pink colonies. *Pseudomonas* species appear as irregular colorless to brown colonies.

Deoxycholate Citrate Agar

Composition per liter:

Sodium citrate 50.0g
Agar 15.0g
Lactose 10.0g
Beef extract 5.0g
Peptone 5.0g
$Na_2S_2O_3 \cdot 5H_2O$ 5.0g
Sodium deoxycholate 2.5g
Ferric citrate 1.0g
Neutral Red 0.025g

pH 7.3 ± 0.2 at 25°C

Source: This medium is available as a premixed powder from Oxoid Unipath.

Preparation of Medium: Add components to distilled/deionized water and bring volume to 1.0L. Mix thoroughly. Gently heat and bring to boiling. Do not autoclave. Cool to 45° to 50°C. Pour into sterile Petri dishes. Dry the agar surface before use.

Use: For the selective isolation and cultivation of enteric pathogens, especially *Salmonella* and *Shigella* species.

Deoxycholate Citrate Agar (Desoxycholate Citrate Agar)

Composition per liter:

Pork infusion 330.0g
Sodium citrate 20.0g
Agar 13.5g
Lactose 10.0g
Proteose peptone No. 3 .... 10.0g
Sodium deoxycholate 5.0g
Ferric ammonium citrate 2.0g
Neutral Red 0.02g

pH 7.5 ± 0.2 at 25°C

Source: This medium is available as a premixed powder from Difco Laboratories.

Preparation of Medium: Add components to distilled/deionized water and bring volume to 1.0L. Mix thoroughly. Gently heat and bring to boiling. Do not autoclave. Cool to 45° to 50°C. Pour into sterile Petri dishes. Dry the agar surface before use.

Use: For the selective isolation and cultivation of enteric pathogens, especially *Salmonella* and *Shigella* species.

Deoxycholate Citrate Agar

Composition per liter:

Sodium citrate 20.0g
Agar 17.0g
Lactose 10.0g
Meat, solids from infusion 10.0g
Peptic digest of animal tissue 10.0g
Sodium deoxycholate 5.0g
Ferric citrate 1.0g
Neutral Red 0.02g

pH 7.3 ± 0.2 at 25°C

Source: This medium is available as a premixed powder from BBL Microbiology Systems.

Preparation of Medium: Add components to distilled/deionized water and bring volume to 1.0L. Mix thoroughly. Gently heat and bring to boiling. Do not autoclave. Cool to 45° to 50°C. Pour into sterile Petri dishes. Dry the agar surface before use.

Use: For the selective isolation and cultivation of enteric pathogens, especially *Salmonella* and *Shigella* species.

Deoxycholate Citrate Agar, Hynes

Composition per liter:

Agar 12.0g
Lactose 10.0g
Sodium citrate 8.5g
$Na_2S_2O_3 \cdot 5H_2O$ 5.4g
Beef extract powder 5.0g
Peptone 5.0g
Sodium deoxycholate 5.0g
Ferric citrate 1.0g
Neutral Red 0.02g

pH 7.3 ± 0.2 at 25°C

Source: This medium is available as a premixed powder from Oxoid Unipath.

Preparation of Medium: Add components to distilled/deionized water and bring volume to 1.0L. Mix

thoroughly. Gently heat and bring to boiling. Do not autoclave. Cool to 45° to 50°C. Pour into sterile Petri dishes. Dry the agar surface before use.

Use: For the selective isolation, cultivation, and differentiation of enteric pathogens, especially *Salmonella* and *Shigella* species. Lactose-fermenting bacteria appear as pink colonies that may or may not be surrounded by a zone of precipitated deoxycholate. Nonlactose-fermenting bacteria appear as colorless colonies that are surrounded by a clear orange-yellow zone.

Deoxycholate Lactose Agar

Composition per liter:

Agar 15.0g
Lactose 10.0g
NaCl 5.0g
Pancreatic digest of casein 5.0g
Peptic digest of animal tissue 5.0g
Sodium citrate 2.0g
Sodium deoxycholate 0.5g
Neutral Red 0.033g

pH 7.1 ± 0.2 at 25°C

Source: This medium is available as a premixed powder from BBL Microbiology Systems and Difco Laboratories.

Preparation of Medium: Add components to distilled/deionized water and bring volume to 1.0L. Mix thoroughly. Gently heat and bring to boiling. Do not autoclave. Cool to 45° to 50°C. Pour into sterile Petri dishes. Dry the agar surface before use.

Use: For the selective isolation, cultivation, and differentiation of enteric pathogens, especially *Salmonella* and *Shigella* species. Lactose-fermenting bacteria appear as pink colonies that may or may not be surrounded by a zone of precipitated deoxycholate. Nonlactose-fermenting bacteria appear as colorless colonies that are surrounded by a clear orange-yellow zone. Also used for the enumeration of coliform bacteria from water, milk and dairy products.

Dermasel Agar Base

Composition per liter:

Glucose 20.0g
Agar 14.5g
Papaic digest of soybean meal 10.0g
Antibiotic inhibitor 10.0mL

pH 6.8–7.0 at 25°C

Source: This medium is available as a premixed powder from Oxoid Unipath.

Antibiotic Inhibitor:
Composition per 10mL:

Cycloheximide 0.4g
Chloramphenicol 0.05g
Acetone 10.0mL

Preparation of Antibiotic Inhibitor: Add cycloheximide and chloramphenicol to 10.0mL of acetone. Mix thoroughly.

Preparation of Medium: Add components to distilled/deionized water and bring volume to 990.0mL. Mix thoroughly. Gently heat and bring to boiling. Do not overheat. Add antibiotic inhibitor. Mix thoroughly. Autoclave for 10 min at 15 psi pressure–121°C. Pour into sterile Petri dishes.

Use: For the isolation and cultivation of dermatophytic fungi isolated from hair, nails or skin scrapings.

Dermatophyte Test Medium Base

Composition per liter:

Agar 20.0g
Glucose 10.0g
Papaic digest of soybean meal 10.0g
Cycloheximide 0.5g
Phenol Red 0.2g
Gentamycin sulfate 0.1g
Chlortetracycline 0.1g

pH 5.5 ± 0.2 at 25°C

Source: This medium is available as a premixed powder from BBL Microbiology Systems.

Preparation of Medium: Add components, except gentamycin sulfate and chlortetracycline, to distilled/deionized water and bring volume to 1.0L. Mix thoroughly. Gently heat while stirring and bring to boiling. Autoclave for 15 min at 15 psi pressure–121°C. Cool to 45° to 50°C. Aseptically add gentamycin sulfate and chlortetracycline. Mix thoroughly. Pour into sterile Petri dishes.

Use: For the selective isolation and cultivation of pathogenic fungi from cutaneous sources.

Dextrose Agar

Composition per liter:

Agar 15.0g
Glucose 10.0g
Tryptose 10.0g
NaCl 5.0g
Beef extract 3.0g

pH 7.3 ± 0.2 at 25°C

Source: This medium is available as a premixed powder from Difco Laboratories.

Preparation of Medium: Add components to distilled/deionized water and bring volume to 1.0L. Mix thoroughly. Gently heat and bring to boiling. Distribute into tubes or flasks. Autoclave for 15 min at 15 psi pressure–121°C. Pour into sterile Petri dishes or leave in tubes.

Use: For the cultivation of a wide variety of microorganisms. For use as a base for the preparation of blood agar and for general laboratory procedures.

Dextrose Broth

Composition per liter:

Tryptose 10.0g
Glucose 5.0g
NaCl 5.0g
Beef extract 3.0g
pH 7.2 ± 0.2 at 25°C

Source: This medium is available as a premixed powder from Difco Laboratories and Oxoid Unipath.

Preparation of Medium: Add components to distilled/deionized water and bring volume to 1.0L. Mix thoroughly. Distribute into tubes or flasks. Autoclave for 15 min at 15 psi pressure–121°C.

Use: For the isolation and enrichment of fastidious or damaged microorganisms.

Dextrose Broth

Composition per liter:

Pancreatic digest of casein 10.0g
Glucose 5.0g
NaCl 5.0g
pH 7.3 ± 0.2

Source: This medium is available as a premixed powder from BBL Microbiology Systems.

Preparation of Medium: Add components to distilled/deionized water and bring volume to 1.0L. Mix thoroughly. Distribute into tubes or flasks. Autoclave for 15 min at 15 psi pressure–121°C.

Use: For the cultivation and differentiation of microorganisms based on their ability to ferment glucose. If desired, a Durham tube may be added to the test tubes to determine gas production.

Dextrose Proteose No. 3 Agar

Composition per liter:

Proteose peptone No. 3 .... 20.0g
Agar 13.0g
NaCl 5.0g
Glucose 2.0g
Tellurite blood solution 50.0mL
pH 7.4 ± 0.2 at 25°C

Tellurite Blood Solution:
Composition per 60mL:

Sheep blood, defibrinated 50.0mL
Chapman tellurite solution 10.0mL

Preparation of Tellurite Blood Solution: Aseptically combine 10.0mL of Chapman tellurite solution with 50.0mL of sterile, defibrinated sheep blood. Mix thoroughly.

Chapman Tellurite Solution:
Composition per 100mL:

K_2TeO_3 .... 1.0g

Preparation of Chapman Tellurite Solution: Add K_2TeO_3 to distilled/deionized water and bring volume to 100.0mL. Mix thoroughly. Filter sterilize.

Caution: Potassium tellurite is toxic.

Preparation of Medium: Add components, except tellurite blood solution, to distilled/deionized water and bring volume to 940.0mL. Mix thoroughly. Gently heat and bring to boiling. Autoclave for 15 min at 15 psi pressure–121°C. Cool to 75° to 80°C. Aseptically add 50.0mL of sterile tellurite blood solution. Mix thoroughly. Maintain at 75° to 80°C for 10 to 15 min or until the agar becomes chocolatized. Cool slowly to 50°C. Pour into sterile Petri dishes or distribute into sterile tubes.

Use: For propagating pure cultures of *Neisseria gonorrhoeae* and other fastidious microorganisms.

Diamonds Medium, Modified

Composition per liter:

Pancreatic digest of casein 20.0g
Yeast extract 1.0g
L-Cysteine·HCl·H_2O 0.5g
Maltose 0.5g
L-Ascorbic acid 0.02g
Horse serum, inactivated 100.0mL
Antibiotic inhibitor 10.0mL
pH 6.5 ± 0.2 at 25°C

Antibiotic Inhibitor:
Composition per 10mL:

Streptomycin sulfate 0.15g
Amphotericin B 0.2mg
Penicillin G 100,000U

Preparation of Antibiotic Inhibitor: Add components to distilled/deionized water and bring volume to 10.0mL. Mix thoroughly. Filter sterilize.

Preparation of Medium: Add components, except antibiotic inhibitor and horse serum, to distilled/deionized water and bring volume to 890.0mL. Mix thoroughly. Gently heat and bring to boiling. Autoclave for 15 min at 15 psi pressure–121°C. Cool to

25°C. Aseptically add sterile antibiotic inhibitor and horse serum. Mix thoroughly. Aseptically distribute into sterile tubes in 5.0mL volumes.

Use: For the cultivation of *Trichomonas* species.

Differential Agar Medium A8 for *Ureaplasma urealyticum*

Composition per 103.1mL:

Basal agar...80.0mL
Horse serum, unheated................................20.0mL
Fresh yeast extract solution............................1.0mL
Urea solution..1.0mL
CVA enrichment...0.5mL
L-Cysteine·HCl·H_2O solution.........................0.5mL
GHL tripeptide solution..................................0.1mL
pH 5.5 ± 0.2 at 25°C

Basal Agar:

Composition per 80mL:

Tryptic soy broth 2.4g
Noble agar...1.05g
Putrescine·2HCl 0.17g
$CaCl_2 \cdot 2H_2O$...0.015g

Preparation of Basal Agar: Add components to distilled/deionized water and bring volume to 80.0mL. Mix thoroughly. Adjust pH to 5.5 with 2*N* HCl. Gently heat and bring to boiling. Autoclave for 15 min at 15 psi pressure–121°C. Cool to 50° to 55°C.

Fresh Yeast Extract Solution:

Composition per 100mL:

Baker's yeast, live, pressed, starch-free 25.0g

Preparation of Fresh Yeast Extract Solution: Add the live Baker's yeast to 100.0mL of distilled/deionized water. Autoclave for 90 min at 15 psi pressure–121°C. Allow to stand. Remove supernatant solution. Adjust pH to 6.6 to 6.8. Filter sterilize.

Urea Solution:

Composition per 30mL:

Urea...3.0g

Preparation of Urea Solution: Add urea to distilled/deionized water and bring volume to 30.0mL. Mix thoroughly. Filter sterilize.

CVA Enrichment:

Composition per liter:

Glucose 100.0g
L-Cysteine·HCl·H_2O...25.9g
L-Glutamine...10.0g
Adenine 1.0g
L-Cystine·2HCl...1.0g
Nicotinamide adenine dinucleotide..................0.25g
Cocarboxylase..0.1g
Guanine·HCl 0.03g
$Fe(NO_3)_3$ 0.02g
Vitamin B_{12} 0.01g
p-Aminobenzoic acid 0.013g
Thiamine·HCl...3.0mg

Preparation of CVA Enrichment: Add components to distilled/deionized water and bring volume to 1.0L. Mix thoroughly. Filter sterilize.

Cysteine·HCl·H_2O Solution:

Composition per 50mL:

Cysteine·HCl·H_2O...1.0g

Preparation of Cysteine·HCl·H_2O Solution: Add cysteine·HCl·H_2O to distilled/deionized water and bring volume to 50.0mL. Mix thoroughly. Filter sterilize.

GHL Tripeptide Solution:

Composition per 10mL:

GHL tripeptide 0.2mg

Preparation of GHL Tripeptide Solution: Add GHL tripeptide (glycyl-L-histidyl-L-lysine acetate) to distilled/deionized water and bring volume to 10.0mL. Mix thoroughly. Filter sterilize.

Preparation of Medium: To 80.0mL of cooled, sterile basal agar, aseptically add 20.0mL of sterile horse serum, 1.0mL of sterile fresh yeast extract solution, 1.0mL of sterile urea solution, 0.5mL of sterile CVA enrichment, 0.5mL of sterile cysteine·HCl·H_2O solution and 0.1mL of sterile GHL tripeptide solution. Mix thoroughly. Pour into sterile Petri dishes in 20.0mL volumes.

Use: For the cultivation and maintenance of *Ureaplasma urealyticum.*

DNase Agar

Composition per liter:

Tryptose 20.0g
Agar...12.0g
NaCl 5.0g
Deoxyribonucleic acid 2.0g
pH 7.3 ± 0.2 at 25°C

Source: This medium is available as a premixed powder from Oxoid Unipath.

Preparation of Medium: Add components to distilled/deionized water and bring volume to 1.0L. Mix thoroughly. Gently heat and bring to boiling. Distribute into tubes or flasks. Autoclave for 15 min at 15 psi pressure–121°C. Pour into sterile Petri dishes or leave in tubes.

Use: For the differentiation of microorganisms, especially *Staphylococcus* species and *Serratia marce-*

scens, based on their production of deoxyribonuclease.

DNase Medium

Composition per liter:

Agar	15.0g
Pancreatic digest of casein	10.0g
Peptic digest of animal tissue	10.0g
L-Arabinose	10.0g
NaCl	5.0g
Deoxyribonucleic acid	2.0g
Methyl Green	0.09g
Phenol Red	0.05g
Antibiotic solution	10.0mL

pH 7.3 ± 0.2 at 25°C

Antibiotic Solution:

Composition per 10mL:

Cephalothin	0.01g
Ampicillin	5.0mg
Colistimethate	5.0mg
Amphotericin B	2.5mg

Preparation of Antibiotic Solution: Add components to distilled/deionized water and bring volume to 10.0mL. Mix thoroughly. Filter sterilize.

Preparation of Medium: Add components, except antibiotic solution, to distilled/deionized water and bring volume to 990.0mL. Mix thoroughly. Gently heat and bring to boiling. Autoclave for 15 min at 15 psi pressure–121°C. Cool to 45° to 50°C. Aseptically add sterile components. Mix thoroughly. Pour into sterile Petri dishes or distribute into sterile tubes.

Use: For the isolation and cultivation of *Serratia marcescens*.

DNase Test Agar

Composition per liter:

Agar	15.0g
Pancreatic digest of casein	15.0g
NaCl	5.0g
Papaic digest of soybean meal	5.0g
Deoxyribonucleic acid	2.0g

pH 7.3 ± 0.2 at 25°C

Source: This medium is available as a premixed powder from BBL Microbiology Systems and Difco Laboratories.

Preparation of Medium: Add components to distilled/deionized water and bring volume to 1.0L. Mix thoroughly. Gently heat while stirring and bring to boiling. Distribute into tubes or flasks. Autoclave for 15 min at 13 psi pressure–118°C. Pour into sterile Petri dishes or leave in tubes.

Use: For the differentiation of microorganisms, especially *Staphylococcus* species and *Serratia marcescens*, based on their production of deoxyribonuclease.

DNase Test Agar with Methyl Green

Composition per liter:

Agar	15.0g
Pancreatic digest of casein	10.0g
Peptic digest of animal tissue	10.0g
NaCl	5.0g
Deoxyribonucleic acid	2.0g
Methyl Green	0.05g

pH 7.3 ± 0.2 at 25°C

Source: This medium is available as a premixed powder from Difco Laboratories.

Preparation of Medium: Add components to distilled/deionized water and bring volume to 1.0L. Mix thoroughly. Gently heat while stirring and bring to boiling. Distribute into tubes or flasks. Autoclave for 15 min at 13 psi pressure–118°C. Pour into sterile Petri dishes or leave in tubes.

Use: For the differentiation of microorganisms, especially *Staphylococcus* species and *Serratia marcescens*, based on their production of deoxyribonuclease.

DNase Test Agar with Toluidine Blue

Composition per liter:

Agar	15.0g
Pancreatic digest of casein	10.0g
Peptic digest of animal tissue	10.0g
NaCl	5.0g
Deoxyribonucleic acid	2.0g
Toluidine Blue	0.1g

pH 7.3 ± 0.2

Preparation of Medium: Add components to distilled/deionized water and bring volume to 1.0L. Mix thoroughly. Gently heat while stirring and bring to boiling. Distribute into tubes or flasks. Autoclave for 15 min at 13 psi pressure–118°C. Pour into sterile Petri dishes or leave in tubes.

Use: For the differentiation of microorganisms, especially *Staphylococcus* species and *Serratia marcescens*, based on their production of deoxyribonuclease.

DTM Agar (Dermatophyte Test Medium Agar)

Composition per liter:

Agar	20.0g
Enzymatic digest of soybean meal	10.0g
Glucose	10.0g
Cycloheximide	0.5g
Phenol Red	0.2g
Chlortetracycline	0.1g
Gentamicin	0.1g

pH 7.3 ± 0.2 at 25°C

Source: Available as a prepared medium from Difco Laboratories.

Preparation of Medium: Add components to distilled/deionized water and bring volume to 1.0L. Mix thoroughly. Gently heat and bring to boiling. Distribute into tubes or flasks. Autoclave for 15 min at 15 psi pressure–121°C. Pour into sterile Petri dishes or leave in tubes.

Use: For the isolation and cultivation of dermatophytic fungi. The observation of a deep red color in the presence of fungal growth is suggestive of a dermatophyte. Some saprophytic fungi such as *Aspergillus* species will also cause the medium to turn red.

Dubos Broth with Horse Serum

Composition per liter:

Na_2HPO_4	2.5g
L-Asparagine	2.0g
KH_2PO_4	1.0g
Pancreatic digest of casein	0.5g
Tween™ 80	0.2g
$CaCl_2 \cdot 2H_2O$	0.5mg
$CuSO_4$	0.1mg
$ZnSO_4 \cdot 7H_2O$	0.1mg
Ferric ammonium citrate	0.05g
$MgSO_4 \cdot 7H_2O$	0.01g
Horse serum	50.0mL

pH 6.5 ± 0.2 at 25°C

Preparation of Medium: Add components, except horse serum, to distilled/deionized water and bring volume to 950.0mL. Mix thoroughly. Gently heat and bring to boiling. Autoclave for 15 min at 15 psi pressure–121°C. Cool to 45° to 50°C. Aseptically add sterile horse serum. Mix thoroughly. Aseptically distribute into sterile tubes.

Use: For the cultivation and maintenance of the *Corynebacterium* species.

Dubos Oleic Agar

Composition per liter:

Agar	15.0g
Na_2HPO_4	2.5g
KH_2PO_4	1.0g
L-Asparagine	1.0g
Pancreatic digest of casein	0.5g
Ferric ammonium citrate	0.05g
$MgSO_4 \cdot 7H_2O$	0.01g
$CaCl_2 \cdot 2H_2O$	0.5mg
$CuSO_4$	0.1mg
$ZnSO_4 \cdot 7H_2O$	0.1mg
Dubos oleic albumin complex	20.0mL
Penicillin solution	10.0mL

pH 6.6 ± 0.2 at 25°C

Source: This medium is available as a premixed powder from Difco Laboratories.

Dubos Oleic Albumin Complex:

Composition per 100mL:

Bovine serum albumin, fraction V	5.0g
Oleic acid, sodium salt	0.05g
NaCl (0.85% solution)	100.0mL

Preparation of Dubos Oleic Albumin Complex: Add bovine serum albumin and oleic acid to 100.0mL of NaCl solution. Mix thoroughly. Filter sterilize.

Penicillin Solution:

Composition per 10mL:

Penicillin	10,000U

Preparation of Penicillin Solution: Add penicillin to distilled/deionized water and bring volume to 10.0mL. Mix thoroughly. Filter sterilize.

Preparation of Medium: Add components, except Dubos oleic albumin complex and penicillin solution, to distilled/deionized water and bring volume to 970.0mL. Mix thoroughly. Gently heat and bring to boiling. Autoclave for 15 min at 15 psi pressure–121°C. Cool to 45° to 50°C. Aseptically add sterile Dubos oleic albumin complex and penicillin solution. Mix thoroughly. Pour into sterile Petri dishes or distribute into sterile tubes. Allow tubes to cool in a slanted position.

Use: For the isolation of *Mycobacterium tuberculosis* and determining its sensitivity to chemotherapeutic agents.

Duncan–Strong Sporulation Medium, Modified (DS Sporulation Medium, Modified) (Sporulation Medium, Modified)

Composition per liter:

Proteose peptone	15.0g
$Na_2HPO_4 \cdot 7H_2O$	10.0g
Raffinose	4.0g
Yeast extract	4.0g
Sodium thioglycollate	1.0g

pH 7.8 ± 0.2 at 25°C

Preparation of Medium: Add components to distilled/deionized water and bring volume to 1.0L. Mix thoroughly. Gently heat and bring to boiling. Distribute into tubes or flasks. Autoclave for 15 min at 15 psi pressure–121°C. Adjust pH to 7.8 with filter-sterilized 0.66M Na_2CO_3. Pour into sterile Petri dishes or leave in tubes.

Use: For the cultivation and induction of sporulation of *Clostridium perfringens*.

Dunkelberg Carbohydrate Medium, Modified

Composition per 100mL:

Proteose peptone No. 3	1.5g
Carbohydrate	1.0g
$Na_2HPO_4 \cdot 2H_2O$	0.207g
Phenol Red	0.055g
$NaH_2PO_4 \cdot H_2O$	0.038g
Horse serum	5.0mL

pH 7.4 ± 0.2 at 25°C

Preparation of Medium: Add components, except horse serum, to distilled/deionized water and bring volue to 95.0mL. For carbohydrate use glucose, maltose or starch. Mix thoroughly. Filter sterilize. Aseptically add sterile horse serum. Mix thoroughly. Aseptically distribute into sterile tubes or flasks.

Use: For the cultivation and differentiation of *Gardnerella vaginalis* based on their ability to ferment glucose, maltose or starch.

Dunkelberg Maintenance Medium

Composition per liter:

Proteose peptone No. 3	20.0g
Soluble starch	10.0g
Agar	8.0g
Glucose	2.0g
Na_2HPO_4	1.0g
NaH_2PO_4	1.0g

pH 6.8 ± 0.2 at 25°C

Preparation of Medium: Add starch to approximately 100.0mL of cold distilled/deionized water. Mix thoroughly. Add starch solution to 400.0mL of boiling distilled/deionized water. Add remaining components. Mix thoroughly. Bring volume to 1.0L with distilled/deionized water. Distribute into screw-capped tubes. Autoclave for 12 min at 8 psi pressure–112°C.

Use: For the cultivation and maintenance of *Gardnerella vaginalis.*

Dunkelberg Semisolid Carbohydrate Fermentation Medium

Composition per liter:

Proteose peptone No. 3	20.0g
Agar	5.0g
Carbohydrate	10.0g
Bromcresol Purple solution	1.0mL

pH 7.4 ± 0.2 at 25°C

Bromcresol Purple Solution:

Composition per 10mL:

Bromcresol Purple	0.16g
Ethanol (95% solution)	10.0mL

Preparation of Bromcresol Purple Solution: Add bromcresol purple to 10.0mL of ethanol. Mix thoroughly. Filter sterilize.

Preparation of Medium: Add components to distilled/deionized water and bring volue to 1.0L. For carbohydrate use glucose, maltose or starch. Mix thoroughly. Gently heat and bring to boiling. Filter sterilize. Aseptically distribute into sterile tubes or flasks.

Use: For the cultivation and differentiation of *Gardnerella vaginalis* based on their ability to ferment glucose, maltose or starch.

Eagle Medium

Composition per 99.1mL:

Eagle MEM in Hanks BSS	87.0mL
Fetal bovine serum	10.0mL
$NaHCO_3$ (7.5% solution)	1.0mL
Penicillin-streptomycin solution	1.0mL
Amphotericin B solution	0.1mL

pH 7.2–7.4 at 25°C

Eagle MEM in Hanks BSS:
Composition per liter:

NaCl	8.0g
Glucose	1.0g
KCl	0.4g
$CaCl_2 \cdot 2H_2O$	0.14g
$MgSO_4 \cdot 7H_2O$	0.1g
KH_2PO_4	0.06g
Na_2HPO_4	0.05g
L-Isoleucine	0.026g
L-Leucine	0.026g
L-Lysine	0.026g
L-Threonine	0.024g
L-Valine	0.0235g
L-Tyrosine	0.018g
L-Arginine	0.0174g
L-Phenylalanine	0.0165g
L-Cystine	0.012g
L-Histidine	8.0mg
L-Methionine	7.5mg
Phenol Red	5.0mg
L-Tryptophan	4.0mg
Inositol	1.8mg
Biotin	1.0mg
Folic acid	1.0mg
Calcium pantothenate	1.0mg
Choline chloride	1.0mg
Nicotinamide	1.0mg
Pyridoxal·HCl	1.0mg
Thiamine·HCl	1.0mg
Riboflavin	0.1mg

Preparation of Eagle MEM in Hanks BSS: Add components to distilled/deionized water and bring volume to 1.0L. Mix thoroughly.

Penicillin-Streptomycin Solution:
Composition per 1mL:

Streptomycin	0.01g
Penicillin	10,000U

Preparation of Penicillin-Streptomycin Solution: Add components to distilled/deionized water and bring volume to 1.0mL. Mix thoroughly.

Amphotericin B Solution:
Composition per 1mL:

Amphotericin B	1.0mg

Preparation of Amphotericin B Solution: Add amphotericin B to distilled/deionized water and bring volume to 1.0mL. Mix thoroughly.

Preparation of Medium: Combine components. Mix thoroughly. Filter sterilize.

Use: For the cultivation of animal tissue culture cell lines.

Eagle Medium

Composition per liter:

Hanks balanced salt solution (10X)	100.0mL
Calf serum	50.0mL
$NaHCO_3$ (7.5% solution)	29.6mL
Tissue culture amino acids (50X)	20.0mL
Tissue culture vitamins (100X)	10.0mL
Glutamine solution	10.0mL
Phenol Red (0.5% solution)	4.0mL
Penicillin solution	1.0mL
Streptomycin solution	0.4mL

pH 7.0 ± 0.2 at 25°C

Hanks Balanced Salt Solution (10X):
Composition per 100mL:

NaCl	8.0g
Glucose	1.0g
KCl	0.4g
$NaHCO_3$	0.35g
$CaCl_2 \cdot 2H_2O$	0.14g
$MgCl_2 \cdot 6H_2O$	0.1g
$MgSO_4 \cdot 7H_2O$	0.1g
Na_2HPO_4	0.06g
KH_2PO_4	0.06g
Phenol Red	0.02g

Preparation of Hanks Balanced Salt Solution (10X): Add components to distilled/deionized water and bring volume to 100.0mL. Mix thoroughly.

Tissue Culture Amino Acids (50X):
Composition per liter:

L-Arginine	0.1g
L-Lysine	0.058g
L-Isoleucine	0.052g
L-Leucine	0.052g
L-Threonine	0.048g
L-Valine	0.046g
L-Tyrosine	0.036g
L-Phenylalanine	0.032g
L-Histidine	0.031g
L-Cystine	0.024g
L-Methionine	0.015g
L-Tryptophan	0.010g

Preparation of Tissue Culture Amino Acids, Minimal Eagle 50X: Add components to distilled/deionized water and bring volume to 1.0L. Mix thoroughly.

Tissue Culture Vitamins (100X):
Composition per liter:

Inositol	2.0mg
Calcium pantothenate	1.0mg
Choline chloride	1.0mg
Folic acid	1.0mg
Nicotinamide	1.0mg

Pyridoxal 1.0mg
Thiamine·HCl 1.0mg
Riboflavin 0.1mg

Preparation of TC Vitamins, Minimal Eagle 100X: Add components to distilled/deionized water and bring volume to 1.0L. Mix thoroughly.

Glutamine Solution:
Composition per 100mL:
L-Glutamine 2.9g

Preparation of Glutamine Solution: Add glutamine to distilled/deionized water and bring volume to 100.0mL. Mix thoroughly.

Penicillin Solution:
Composition per 1mL:
Penicillin 200,000U

Preparation of Penicillin Solution: Add penicillin to distilled/deionized water and bring volume to 1.0mL. Mix thoroughly.

Streptomycin Solution:
Composition per 1mL:
Streptomycin 0.5g

Preparation of Streptomycin Solution: Add streptomycin to distilled/deionized water and bring volume to 1.0mL. Mix thoroughly.

Preparation of Medium: Combine components. Mix thoroughly. Adjust pH to 7.0 with 1*N* NaOH. Filter sterilize.

Use: For the cultivation of animal tissue culture cell lines especially for use with rhinoviruses.

Eagle Medium

Composition per 100.1mL:
Eagle MEM in Earle BSS 94.0mL
$NaHCO_3$ (7.5% solution) 3.0mL
Fetal bovine serum, inactivated 2.0mL
Penicillin-streptomycin solution 1.0mL
Amphotericin B solution 0.1mL
pH 7.2–7.4 at 25°C

Eagle MEM in Earle BSS:
Composition per liter:
NaCl 6.8g
Glucose 1.0g
KCl 0.4g
$CaCl_2 \cdot 2H_2O$ 0.2g
$MgCl_2 \cdot 6H_2O$ 0.2g
NaH_2PO_4 .... 0.15g
L-Arginine 0.10g
L-Lysine 0.06g
L-Isoleucine 0.05g
L-Leucine 0.05g
L-Threonine 0.05g
L-Valine 0.05g
L-Tyrosine 0.04g
L-Phenylalanine 0.03g
L-Histidine 0.03g
L-Cystine 0.02g
L-Methionine 0.02g
L-Tryptophan 0.01g
i-Inositol 2.0mg
Calcium pantothenate 1.0mg
Choline chloride 1.0mg
Folic acid 1.0mg
Nicotinamide 1.0mg
Pyridoxal 1.0mg
Thiamine·HCl 1.0mg
Riboflavin 0.1mg

Preparation of Eagle MEM in Earle BSS: Add components to distilled/deionized water and bring volume to 1.0L. Mix thoroughly.

Penicillin-Streptomycin Solution:
Composition per 1mL:
Streptomycin 0.01g
Penicillin 10,000U

Preparation of Penicillin-Streptomycin Solution: Add components to distilled/deionized water and bring volume to 1.0mL. Mix thoroughly.

Amphotericin B Solution:
Composition per 1mL:
Amphotericin B 1.0mg

Preparation of Amphotericin B Solution: Add amphotericin B to distilled/deionized water and bring volume to 1.0mL. Mix thoroughly.

Preparation of Medium: Combine components. Mix thoroughly. Filter sterilize.

Use: For the cultivation of animal tissue culture cell lines.

Eagle Medium, Modified

Composition per liter:
Eagle MEM (10X) 100.0mL
Fetal bovine serum 100.0mL
Glucose solution 20.0mL
HEPES (N-2-Hydroxyethyl
piperazine-N′-2-ethanesulfonic acid)
buffer, 1M, pH 7.2 .... 20.0mL
Glutamine solution 10.0mL
$NaHCO_3$ (7.5% solution) 7.5mL
Gentamicin sulfate solution 0.2mL
pH 7.2 ± 0.2 at 25°C

Eagle MEM (10X):

Composition per 100mL:

Sterile salt solution....................................97.0mL
TC amino acids, minimal Eagle 50X.............2.0mL
TC vitamins, minimal Eagle 100X.................1.0mL

Preparation of Eagle MEM (10X): Combine components. Mix thoroughly. Filter sterilize.

Sterile Salt Solution:

Composition per 100mL:

NaCl..6.8g
Glucose...1.0g
KCl...0.4g
$CaCl_2$...0.2g
$MgCl_2$...0.2g
NaH_2PO_4..0.15g

Preparation of Sterile Salt Solution: Add components to distilled/deionized water and bring volume to 100.0mL. Mix thoroughly. Filter sterilize.

Tissue Culture Amino Acids, Minimal Eagle 50X:

Composition per liter:

L-Arginine..0.1g
L-Lysine...0.06g
L-Isoleucine..0.05g
L-Leucine..0.05g
L-Threonine..0.05g
L-Valine...0.05g
L-Tyrosine..0.04g
L-Phenylalanine.....................................0.03g
L-Histidine...0.03g
L-Cystine...0.02g
L-Methionine..0.02g
L-Tryptophan..0.01g

Preparation of Tissue Culture Amino Acids, Minimal Eagle 50X: Add components to distilled/deionized water and bring volume to 1.0L. Mix thoroughly. Adjust pH to 7.2–7.4. Filter sterilize.

TC Vitamins, Minimal Eagle 100X:

Composition per liter:

Inositol.. 2.0mg
Calcium pantothenate.......................... 1.0mg
Choline chloride.................................. 1.0mg
Folic acid.. 1.0mg
Nicotinamide....................................... 1.0mg
Pyridoxal... 1.0mg
Thiamine·HCl...................................... 1.0mg
Riboflavin.. 0.1mg

Preparation of TC Vitamins, Minimal Eagle 100X: Add components to distilled/deionized water and bring volume to 1.0L. Mix thoroughly. Filter sterilize.

Glucose Solution:

Composition per 100mL:

Glucose...27.0g

Preparation of Glucose Solution: Add glucose to distilled/deionized water and bring volume to 100.0mL. Mix thoroughly. Filter sterilize.

Glutamine Solution:

Composition per 10mL:

L-Glutamine..5.0g

Preparation of Glutamine Solution: Add glutamine to distilled/deionized water and bring volume to 10.0mL. Mix thoroughly. Filter sterilize.

Gentamicin Solution:

Composition per 1mL:

Gentamicin sulfate...............................0.05g

Preparation of Gentamicin Solution: Add gentamicin sulfate to distilled/deionized water and bring volume to 1.0mL. Mix thoroughly. Filter sterilize.

Preparation of Medium: Combine components. Mix thoroughly. Filter sterilize.

Use: For the cultivation of animal tissue culture cell lines, especially for McCoy cells.

EC Broth (*Escherichia coli* Broth) (EC Medium)

Composition per liter:

Pancreatic digest of casein..................20.0g
Lactose...5.0g
NaCl...5.0g
K_2HPO_4..4.0g
Bile salts mixture................................1.5g
KH_2PO_4..1.5g

pH 6.9 ± 0.2 at 25°C

Source: This medium is available as a premixed powder from BBL Microbiology Systems and Difco Laboratories.

Preparation of Medium: Add components to distilled/deionized water and bring volume to 1.0L. Mix thoroughly. Distribute into test tubes that contain an inverted Durham tube. Autoclave for 12 min at 15 psi pressure–121°C. Cool broth as quickly as possible.

Use: For the cultivation and differentiation of coliform bacteria at 37°C and of *Escherichia coli* at 45.5°C.

Egg Yolk Agar

Composition per liter:

Proteose peptone No. 2	40.0g
Agar	25.0g
Na_2HPO_4	5.0g
Glucose	2.0g
NaCl	2.0g
KH_2PO_4	1.0g
$MgSO_4 \cdot 7H_2O$	0.1g
Egg yolk emulsion	100.0mL
Hemin solution	1.0mL

pH 7.6 ± 0.2 at 25°C

Hemin Solution:

Composition per 100mL:

Hemin	0.5g
NaOH (1*N* solution)	20.0mL

Preparation of Hemin Solution: Add hemin to 20.0mL of 1*N* NaOH solution. Mix thoroughly. Bring volume to 100.0mL with distilled/deionized water.

Egg Yolk Emulsion:

Composition:

Chicken egg yolks	11
Whole chicken egg	1

Preparation of Egg Yolk Emulsion: Soak eggs with 1:100 dilution of saturated mercuric chloride solution for 1 min. Crack eggs and separate yolks from whites. Mix egg yolks with 1 chicken egg.

Preparation of Medium: Add components, except egg yolk emulsion, to distilled/deionized water and bring volume to 900.0mL. Mix thoroughly. Gently heat and bring to boiling. Autoclave for 15 min at 15 psi pressure–121°C. Cool to 45° to 50°C. Aseptically add sterile egg yolk emulsion. Mix thoroughly. Pour into sterile Petri dishes.

Use: For the isolation, cultivation and differentiation of *Clostridium* species and some other anaerobic bacteria.

Egg Yolk Agar, Modified

Composition per liter:

Agar	20.0g
Pancreatic digest of casein	15.0g
Vitamin K_1	10.0g
NaCl	5.0g
Papaic digest of soybean meal	5.0g
Yeast extract	5.0g
L-Cystine	0.4g
Hemin	5.0mg
Egg yolk emulsion	100.0mL

Source: Available as a prepared medium from BBL Microbiology Systems.

Egg Yolk Emulsion:

Composition:

Chicken egg yolks	11
Whole chicken egg	1

Preparation of Egg Yolk Emulsion: Soak eggs with 1:100 dilution of saturated mercuric chloride solution for 1 min. Crack eggs and separate yolks from whites. Mix egg yolks with 1 chicken egg.

Preparation of Medium: Add components, except egg yolk emulsion, to distilled/deionized water and bring volume to 900.0mL. Mix thoroughly. Gently heat and bring to boiling. Autoclave for 15 min at 15 psi pressure–121°C. Cool to 45° to 50°C. Aseptically add sterile egg yolk emulsion. Mix thoroughly. Pour into sterile Petri dishes.

Use: For the isolation, cultivation and differentiation of *Clostridium* species and some other anaerobic bacteria.

Eijkman Lactose Medium

Composition per liter:

Pancreatic digest of casein	15.0g
K_2HPO_4	10.0g
KH_2PO_4	4.0g
Lactose	3.0g
NaCl	2.5g

pH 6.8 ± 0.1 at 25°C

Preparation of Medium: Add components to distilled/deionized water and bring volume to 1.0L. Mix thoroughly. Distribute into test tubes that contain an inverted Durham tube. Autoclave for 15 min at 15 psi pressure–121°C.

Use: For the cultivation and differentiation of *Escherichia coli* from other coliform organisms based on their ability to ferment lactose and produce gas.

Eijkman Lactose Medium

Composition per liter:

Tryptose	15.0g
NaCl	5.0g
K_2HPO_4	4.0g
Lactose	3.0g
KH_2PO_4	1.5g

pH 6.8 ± 0.1 at 25°C

Preparation of Medium: Add components to distilled/deionized water and bring volume to 1.0L. Mix thoroughly. Distribute into test tubes that contain an inverted Durham tube. Autoclave for 15 min at 15 psi pressure–121°C.

Use: For the cultivation and differentiation of *Escherichia coli* from other coliform organisms based on their ability to ferment lactose and produce gas.

EMB Agar (Eosin Methylene Blue Agar)

Composition per liter:

Agar	13.5g
Pancreatic digest of casein	10.0g
Lactose	5.0g
Sucrose	5.0g
K_2HPO_4	2.0g
Eosin Y	0.4g
Methylene Blue	0.065g

pH 7.2 ± 0.2 at 25°C

Source: This medium is available as a premixed powder from BBL Microbiology Systems and Difco Laboratories.

Preparation of Medium: Add components to distilled/deionized water and bring volume to 1.0L. Mix thoroughly. Gently heat and bring to boiling. Distribute into tubes or flasks. Autoclave for 15 min at 15 psi pressure–121°C. Pour into sterile Petri dishes.

Use: For the isolation, cultivation and differentiation of Gram-negative enteric bacteria based on lactose fermentation. Bacteria that ferment lactose, especially the coliform bacterium *Escherichia coli*, appear as colonies with a green metallic sheen or blue-black to brown color. Bacteria that do not ferment lactose appear as colorless or transparent light purple colonies.

EMB Agar Base

Composition per liter:

Agar	15.0g
Peptone	10.0g
K_2HPO_4	2.0g
Eosin Y	0.4g
Methylene Blue	0.065g

pH 7.3 ± 0.2 at 25°C

Preparation of Medium: Add components to distilled/deionized water and bring volume to 1.0L. Mix thoroughly. Gently heat and bring to boiling. Distribute into tubes or flasks. Autoclave for 15 min at 15 psi pressure–121°C. Pour into sterile Petri dishes.

Use: For the isolation, cultivation and differentiation of Gram-negative enteric bacteria based on lactose fermentation. Bacteria that ferment lactose, especially the coliform bacterium *Escherichia coli*, appear as colonies with a green metallic sheen or blue-black to brown color. Bacteria that do not ferment lactose appear as colorless or transparent light purple colonies.

EMB Agar, Modified (Eosin Methylene Blue Agar, Modified)

Composition per liter:

Agar	15.0g
Lactose	10.0g
Pancreatic digest of gelatin	10.0g
K_2HPO_4	2.0g
Eosin Y	0.4g
Methylene Blue	0.065g

pH 6.8 ± 0.2 at 25°C

Source: This medium is available as a premixed powder from Oxoid.

Preparation of Medium: Add components to distilled/deionized water and bring volume to 1.0L. Mix thoroughly. Gently heat and bring to boiling. Distribute into tubes or flasks. Autoclave for 15 min at 15 psi pressure–121°C. Cool to 60°C. Shake medium to oxidize methylene blue. Pour into sterile Petri dishes. Swirl flask while pouring plates to distribute precipitate.

Use: For the isolation, cultivation and differentiation of Gram-negative enteric bacteria based on lactose fermentation. Bacteria that ferment lactose, especially the coliform bacterium *Escherichia coli*, appear as colonies with a green metallic sheen or blue-black to brown color. Bacteria that do not ferment lactose appear as colorless or transparent light purple colonies.

Endo Agar

Composition per liter:

Agar	15.0g
Lactose	10.0g
Peptic digest of animal tissue	10.0g
K_2HPO_4	3.5g
Na_2SO_3	2.5g
Basic Fuchsin	0.5g

pH 7.4 ± 0.2 at 25°C

Source: This medium is available as a premixed powder from BBL Microbiology Systems and Difco Laboratories.

Caution: Basic Fuchsin is a potential carcinogen and care must be taken to avoid inhalation of the powdered dye and contact with the skin.

Preparation of Medium: Add components to distilled/deionized water and bring volume to 1.0L. Mix thoroughly. Gently heat and bring to boiling. Autoclave for 15 min at 15 psi pressure–121°C. Cool to 45° to 50°C. Pour into sterile Petri dishes. Swirl flask while pouring plates to keep precipitate in suspension. Protect from the light.

Use: For the selective isolation, cultivation and differentiation of coliform and other enteric microorganisms based on their ability to ferment lactose. Lactose-fermenting bacteria appear as dark red colonies with a gold metallic sheen. Lactose-nonfermenting bacteria appear as colorless or translucent colonies.

Endo Agar

Composition per liter:

Agar 10.0g
Lactose 10.0g
Peptic digest of animal tissue 10.0g
K_2HPO_4 .... 3.5g
Na_2SO_3 .... 2.5g
Basic Fuchsin solution 4.0mL
pH 7.5 ± 0.2 at 25°C

Source: This medium is available as a premixed powder from Oxoid.

Caution: Basic Fuchsin is a potential carcinogen and care must be taken to avoid inhalation of the powdered dye and contact with the skin.

Basic Fuchsin Solution:
Composition per 10mL:

Basic Fuchsin 1.0g
Ethanol (95% solution) 10.0mL

Preparation of Basic Fuchsin Solution: Add Basic Fuchsin to 10.0mL of ethanol. Mix thoroughly.

Preparation of Medium: Add components to distilled/deionized water and bring volume to 1.0L. Mix thoroughly. Gently heat and bring to boiling. Autoclave for 15 min at 15 psi pressure–121°C. Cool to 45° to 50°C. Pour into sterile Petri dishes. Swirl flask while pouring plates to keep precipitate in suspension. Protect from the light.

Use: For the selective isolation, cultivation and differentiation of coliform and other enteric microorganisms based on their ability to ferment lactose. Lactose-fermenting bacteria appear as dark red colonies with a gold metallic sheen. Lactose-nonfermenting bacteria appear as colorless or translucent colonies.

Endo Agar, LES (Endo Agar, Laurance Experimental Station) (m-Endo Agar, LES) (m–LES, Endo Agar)

Composition per liter:

Agar 14.0g
Lactose 9.4g
Peptones (pancreatic digest of casein 65% and yeast extract 35%) 7.5g
NaCl 3.7g
Pancreatic digest of casein 3.7g
Peptic digest of animal tissue 3.7g
K_2HPO_4 .... 3.3g
Na_2SO_3 .... 1.6g
Yeast extract 1.2g
KH_2PO_4 .... 1.0g
Basic Fuchsin 0.8g
Sodium lauryl sulfate 0.05g
Ethanol 20.0mL
pH 7.2 ± 0.2 at 25°C

Source: This medium is available as a premixed powder from BBL Microbiology Systems and Difco Laboratories.

Caution: Basic Fuchsin is a potential carcinogen and care must be taken to avoid inhalation of the powdered dye and contact with the skin.

Preparation of Medium: Add ethanol to approximately 900.0mL of distilled/deionized water. Add remaining components. Bring volume to 1.0L with distilled/deionized water. Mix thoroughly. Gently heat and bring to boiling. Autoclave for 15 min at 15 psi pressure–121°C. Pour into sterile 60-mm Petri dishes in 4.0mL volumes. Protect from the light.

Use: For the cultivation and enumeration of coliform bacteria by the membrane filter method.

Endo Agar, LES (m-Endo Agar, LES)

Composition per liter:

Agar 10.0g
Lactose 9.4g
Tryptose 7.5g
NaCl 3.7g
Peptone 3.7g
Pancreatic digest of casein 3.7g
K_2HPO_4 .... 3.3g
Na_2SO_3 .... 1.6g
Yeast extract 1.2g
KH_2PO_4 .... 1.0g
Sodium deoxycholate 0.1g
Sodium lauryl sulfate 0.05g
Basic Fuchsin solution 8.0mL
pH 7.2 ± 0.2 at 25°C

Caution: Basic Fuchsin is a potential carcinogen and care must be taken to avoid inhalation of the powdered dye and contact with the skin.

Basic Fuchsin Solution:
Composition per 10mL:

Basic Fuchsin 1.0g
Ethanol (95% solution) 10.0mL

Preparation of Basic Fuchsin Solution: Add Basic Fuchsin to 10.0mL of ethanol. Mix thoroughly.

Preparation of Medium: Add components to distilled/deionized water and bring volume to 1.0L. Mix thoroughly. Gently heat and bring to boiling. Autoclave for 15 min at 15 psi pressure–121°C. Cool to 45° to 50°C. Pour into sterile Petri dishes. Swirl flask while pouring plates to keep precipitate in suspension. Protect from the light.

Use: For the cultivation and enumeration of coliform bacteria from water by the membrane filter method.

Endo Broth (m-Endo Broth)

Composition per liter:

Lactose	12.5g
Peptone	10.0g
NaCl	5.0g
Pancreatic digest of casein	5.0g
Peptic digest of animal tissue	5.0g
K_2HPO_4	4.375g
Na_2SO_3	2.1g
Yeast extract	1.5g
KH_2PO_4	1.375g
Basic Fuchsin	1.05g
Sodium deoxycholate	0.1g
Ethanol (95% solution)	20.0mL

pH 7.2 ± 0.1 at 25°C

Source: This medium is available as a premixed powder from BBL Microbiology Systems and Difco Laboratories.

Caution: Basic Fuchsin is a potential carcinogen and care must be taken to avoid inhalation of the powdered dye and contact with the skin.

Preparation of Medium: Add ethanol to approximately 900.0mL of distilled/deionized water. Add remaining components. Bring volume to 1.0L with distilled/deionized water. Mix thoroughly. Gently heat and bring to boiling. Rapidly cool broth below 45°C. Do not autoclave. Use 1.8–2.0mL for each filter pad. Protect from the light. Prepare broth freshly.

Use: For the cultivation and enumeration of coliform bacteria from water by the membrane filter method.

Enrichment Broth for *Aeromonas hydrophila*

Composition per liter:

NaCl	5.0g
Maltose	3.5g
Yeast extract	3.0g
Bile salts No. 3	1.0g
L-Cysteine·HCl·H_2O	0.3g
Bromthymol Blue	0.03g
Novobiocin	5.0mg

pH 7.0 ± 0.2 at 25°C

Preparation of Medium: Add components to distilled/deionized water and bring volume to 1.0L. Mix thoroughly. Distribute into tubes or flasks. Autoclave for 15 min at 15 psi pressure–121°C.

Use: For the cultivation and enrichment of *Aeromonas hydrophila.*

Entamoeba Medium (Endamoeba Medium)

Composition per liter:

Liver infusion	272.0g
Rice powder	14.2g
Agar	11.0g
Proteose peptone	5.5g
Sodium glycerophosphate	3.0g
NaCl	2.7g
Horse serum	50.0mL

pH 7.0 ± 0.2 at 25°C

Source: This medium is available as a premixed powder from Difco Laboratories.

Rice Powder:

Composition per 15g:

Rice powder	15.0g

Preparation of Rice Powder: Sterilize rice powder at 160°C for 60 min. Do not overheat or rice powder will scorch.

Preparation of Medium: Add components, except horse serum and rice powder, to distilled/deionized water and bring volume to 994.0mL. Mix thoroughly. Gently heat and bring to boiling. Distribute into tubes in 7.0mL volumes. Autoclave for 15 min at 15 psi pressure–121°C. Allow tubes to cool in a slanted position. Aseptically add enough sterile horse serumm to each tube to cover about half the slant. Aseptically add 0.1g of sterile rice powder to each tube.

Use: For the cultivation of *Entamoeba histolytica.*

Enteric Fermentation Base (Fermentation Base for *Campylobacter*)

Composition per liter:

Peptic digest of animal tissue	10.0g
NaCl	5.0g

Beef extract 3.0g
Carbohydrate solution 100.0mL
Andrade's indicator 10.0mL
pH 7.2 ± 0.1 at 25°C

Source: This medium is available as a premixed powder from Difco Laboratories.

Carbohydrate Solution:
Composition per 100mL:
Carbohydrate 10.0g

Preparation of Carbohydrate Solution: Add carbohydrate to distilled/deionized water and bring volume to 100.0mL. Mix thoroughly. Filter sterilize. Glucose, lactose, mannitol, sucrose, adonitol, arabinose, cellobiose, dulcitol, glycerol, inositol, salicin, xylose or other carbohydrates may be used. For the preparation of expensive carbohydrate solutions (adonitol, arabinose, cellobiose, dulcitol, glycerol, inositol, salicin or xylose), 5.0g of carbohydrate per 100.0mL of distilled/deionized water may be used.

Andrade's Indicator:
Composition per 100mL:
NaOH (1*N* solution) 16.0mL
Acid Fuchsin ....0.1 g

Caution: Acid Fuchsin is a potential carcinogen and care must be taken to avoid inhalation of the powdered dye and contact with the skin.

Preparation of Andrade's Indicator: Add components to distilled/deionized water and bring volume to 100.0mL. Mix thoroughly.

Preparation of Medium: Add components, except carbohydrate solution, to distilled/deionized water and bring volume to 900.0mL. Mix thoroughly. Gently heat and bring to boiling. Distribute into tubes that contain an inverted Durham tube in 9.0mL volumes. Autoclave for 15 min at 15 psi pressure–121°C. Cool to 25°C. Aseptically add 1.0mL of sterile carbohydrate solution per tube. Mix thoroughly.

Use: For the cultivation and differentiation of a variety of bacteria based on their ability to ferment different carbohydrates. Bacteria that produce acid from carbohydrate fermentation turn the medium dark pink to red. Bacteria that produce gas have a bubble trapped in the Durham tube.

Enterobacter Medium

Composition per 800mL:
Casein hydrolysate 2.0g
K_2HPO_4 ....1.4g
K_2SO_4 ....1.0g
Yeast extract 1.0g
KH_2PO_4 ....0.6g
$MgSO_4$ 0.5g
Glycerol 20.0mL

Preparation of Medium: Add components to distilled/deionized water and bring volume to 800.0mL. Mix thoroughly. Distribute into tubes or flasks. Autoclave for 15 min at 15 psi pressure–121°C.

Use: For the cultivation and maintenance of *Enterobacter* species and *Klebsiella pneumoniae.*

Enterococci Confirmatory Agar

Composition per liter:
Agar 15.0g
Glucose 5.0g
Pancreatic digest of casein 5.0g
Yeast extract 5.0g
NaN_3 ....0.4g
Methylene Blue 10.0mg
pH 8.0 ± 0.2 at 25°C

Source: This medium is available as a premixed powder from Difco Laboratories.

Caution: Sodium azide is toxic. Azides also react with metals and disposal must be highly diluted.

Preparation of Medium: Add components to distilled/deionized water and bring volume to 1.0L. Mix thoroughly. Gently heat and bring to boiling. Distribute into tubes. Autoclave for 15 min at 15 psi pressure–121°C. Allow tubes to cool in a slanted position. Add sufficient amount of Enterococci confirmatory broth to cover half the slant.

Use: For the identification of enterococci from water by the confirmatory test.

Enterococcosel™ Agar

Composition per liter:
Pancreatic digest of casein 17.0g
Agar 13.5g
Oxgall 10.0g
NaCl 5.0g
Yeast extract 5.0g
Peptic digest of animal tissue 3.0g
Esculin 1.0g
Sodium citrate 1.0g
Ferric ammonium citrate 0.5g
NaN_3 ....0.25g
pH 7.1 ± 0.2 at 25°C

Source: This medium is available as a premixed powder from BBL Microbiology Systems.

Caution: Sodium azide is toxic. Azides also react with metals and disposal must be highly diluted.

Preparation of Medium: Add components to distilled/deionized water and bring volume to 1.0L. Mix thoroughly. Gently heat while stirring and bring to boiling. Distribute into tubes or flasks. Autoclave for 15 min at 15 psi pressure–121°C. Pour into sterile Petri dishes or leave in tubes.

Use: For the rapid, selective isolation, cultivation and enumeration of fecal Group D streptococci (enterococci). For the cultivation of staphylococci and *Listeria monocytogenes.*

Enterococcosel™ Broth

Composition per liter:

Pancreatic digest of casein	17.0g
Oxgall	10.0g
NaCl	5.0g
Yeast extract	5.0g
Peptic digest of animal tissue	3.0g
Esculin	1.0g
Sodium citrate	1.0g
Ferric ammonium citrate	0.5g
NaN_3	0.25g

pH 7.1 ± 0.2 at 25°C

Source: This medium is available as a premixed powder from BBL Microbiology Systems.

Caution: Sodium azide is toxic. Azides also react with metals and disposal must be highly diluted.

Preparation of Medium: Add components to distilled/deionized water and bring volume to 1.0L. Mix thoroughly. Gently heat while stirring until dissolved. Distribute into tubes or flasks. Autoclave for 15 min at 15 psi pressure–121°C.

Use: For the cultivation and differentiation of Group D streptococci (enterococci).

Enterococcus Agar (m-*Enterococcus* Agar) (Azide Agar)

Composition per liter:

Pancreatic digest of casein	15.0g
Agar	10.0g
Papaic digest of soybean meal	5.0g
Yeast extract	5.0g
KH_2PO_4	4.0g
Glucose	2.0g
NaN_3	0.4g
Triphenyltetrazolium chloride	0.1g

pH 7.2 ± 0.2 at 25°C

Source: This medium is available as a premixed powder from BBL Microbiology Systems and Difco Laboratories.

Caution: Sodium azide is toxic. Azides also react with metals and disposal must be highly diluted.

Preparation of Medium: Add components to distilled/deionized water and bring volume to 1.0L. Mix thoroughly. Gently heat and bring to boiling. Cool to 45° to 50°C. Do not autoclave. Pour into sterile Petri dishes.

Use: For the isolation, cultivation and enumeration of entercocci in water, sewage and feces by the membrane filter method. Also used for the direct plating of specimens for detection and enumeration of fecal streptococci.

Esculin Agar

Composition per liter:

Agar	15.0g
Pancreatic digest of casein	13.0g
NaCl	5.0g
Yeast extract	5.0g
Heart muscle, solids from infusion	2.0g
Esculin	1.0g
Ferric citrate	0.5g

pH 7.3 ± 0.2 at 25°C

Preparation of Medium: Add components to distilled/deionized water and bring volume to 1.0L. Mix thoroughly. Gently heat and bring to boiling. Distribute into screw-capped tubes in 3.0mL volumes. Autoclave for 15 min at 15 psi pressure–121°C. Allow tubes to cool in a slanted position.

Use: For the cultivation and differentiation of bacteria based on their ability to hydrolyze esculin and produce H_2S. Bacteria that hydrolyze esculin appear as colonies surrounded by a reddish-brown to dark brown zone. Bacteria that produce H_2S appear as black colonies.

Esculin Broth

Composition per liter:

Beef heart, solids from infusion	500.0g
Tryptose	10.0g
NaCl	5.0g
Agar	1.0g
Esculin	1.0g

pH 7.0 ± 0.2 at 25°C

Preparation of Medium: Add components to distilled/deionized water and bring volume to 1.0L. Mix thoroughly. Gently heat and bring to boiling. Distribute into screw-capped tubes in 7.0mL volumes. Autoclave for 15 min at 15 psi pressure–121°C.

Use: For the cultivation and differentiation of bacteria based on their ability to hydrolyze esculin. Bacteria that hydrolyze esculin turn the medium brown-black to black.

Esculin Iron Agar

Composition per liter:

Agar....................15.0g
Esculin....................1.0g
Ferric ammonium citrate....................0.5g

pH 7.1 ± 0.2 at 25°C

Source: This medium is available as a premixed powder from BBL Microbiology Systems.

Preparation of Medium: Add components to distilled/deionized water and bring volume to 1.0L. Mix thoroughly. Gently heat and bring to boiling. Distribute into tubes or flasks. Autoclave for 15 min at 15 psi pressure–121°C. Pour into sterile Petri dishes.

Use: For the cultivation and identification of enterococci based on their ability to hydrolyze esculin. Used in conjunction with E agar and the membrane filter method.

Esculin Mannitol Agar

Composition per liter:

Agar....................13.5g
Polypeptone™....................10.0g
D-Mannitol....................10.0g
Pancreatic digest of casein....................5.0g
Yeast extract....................5.0g
NaCl....................5.0g
Heart peptone....................3.0g
Cornstarch....................1.0g
Esculin....................1.0g
Ferric ammonium citrate....................0.5g
Phenol Red....................0.025g
Nalidixic acid solution....................10.0mL
Colistin solution....................10.0mL

pH 7.3 ± 0.2 at 25°C

Nalidixic Acid Solution:
Composition per 10mL:

Nalidixic acid....................0.015g

Preparation of Nalidixic Acid Solution: Add nalidixic acid to distilled/deionized water and bring volume to 10.0mL. Mix thoroughly. Filter sterilize.

Colistin Solution:
Composition per 10mL:

Colistin....................0.01g

Preparation of Colistin Solution: Add colistin to distilled/deionized water and bring volume to 10.0mL. Mix thoroughly. Filter sterilize.

Preparation of Medium: Add components, except nalidixic acid solution and colistin solution, to distilled/deionized water and bring volume to 980.0mL. Mix thoroughly. Gently heat and bring to boiling. Autoclave for 15 min at 15 psi pressure–121°C. Cool to 45° to 50°C. Aseptically add sterile nalidixic acid solution and colistin solution. Mix thoroughly. Pour into sterile Petri dishes or distribute into sterile tubes.

Use: For the selective isolation, cultivation and differentiation of *Staphylococcus aureus* and Group D streptococci based on mannitol fermentation and hydrolysis of esculin. Bacteria that ferment mannitol appear as yellow colonies surrounded by a yellow zone. Bacteria that hydrolyze esculin appear as dark brown to black colonies surrounded by a dark brown to black zone.

ETGPA (Egg Tellurite Glycine Pyruvate Agar)

Composition per liter:

Agar....................17.0g
Glycine....................12.0g
Sodium pyruvate....................10.0g
Pancreatic digest of casein....................10.0g
Beef extract....................5.0g
LiCl....................5.0g
Yeast extract....................1.0g
Egg yolk emulsion....................50.0mL
K_2TeO_3 solution....................10.0mL

pH 7.0 ± 0.2 at 25°C

Source: This medium is available as a premixed powder from BBL Microbiology Systems.

Egg Yolk Emulsion:
Composition:

Chicken egg yolks....................11
Whole chicken egg....................1

Preparation of Egg Yolk Emulsion: Soak egg with 1:100 dilution of saturated mercuric chloride solution for 1 min. Crack eggs and separate yolks from whites. Mix egg yolks with 1 chicken egg.

K_2TeO_3 Solution:
Composition per 100mL:

K_2TeO_3....................1.0g

Preparation of K_2TeO_3 Solution: Add K_2TeO_3 to distilled/deionized water and bring volume to 100.0mL. Mix thoroughly. Filter sterilize.

Caution: Potassium tellurite is toxic.

Preparation of Medium: Add components to distilled/deionized water and bring volume to 940.0mL. Mix thoroughly. Gently heat and bring to boiling. Autoclave for 15 min at 15 psi pressure–121°C. Cool to 45 to 50°C. Add 10.0mL of sterile 1% tellurite solution and 50.0mL of sterile egg yolk emulsion. If desired, add sulphamethazine to a final concentration of 50mg/mL. Mix thoroughly but gently and pour into sterile Petri dishes.

Use: For the selective isolation and enumeration of coagulase-positive staphylococci from food, skin, soil, air and other materials. For the differentiation and identification of staphylococci on the basis of their ability to clear egg yolk. Addition of sulphamethazine inhibits the growth of *Proteus*. Gray-black colonies surrounded by a clear zone are diagnostic for *Staphylococcus aureus*.

Ethyl Violet Azide Broth (EVA Broth)

Composition per liter:

Pancreatic digest of casein	13.5g
Yeast extract	6.5g
Glucose	5.0g
NaCl	5.0g
K_2HPO_4	2.7g
KH_2PO_4	2.7g
NaN_3	0.4g
Ethyl Violet	0.83mg

pH 7.0 ± 0.2 at 25°C

Source: This medium is available as a premixed powder from BBL Microbiology Systems.

Preparation of Medium: Add components to distilled/deionized water and bring volume to 1.0L. Mix thoroughly. Gently heat and bring to boiling. Distribute into tubes in 10.0mL volumes. Autoclave for 15 min at 15 psi pressure–121°C.

Caution: Sodium azide is toxic. Azides also react with metals and disposal must be highly diluted.

Use: For the isolation, cultivation and enumeration of enterococci from water and other specimens. Fecal enterococci turn the medium turbid with a purple sediment on the bottom of the tube.

Ethyl Violet Azide Broth (EVA Broth)

Composition per liter:

Tryptose	20.0g
Glucose	5.0g
NaCl	5.0g
K_2HPO_4	2.7g
KH_2PO_4	2.7g
NaN_3	0.4g
Ethyl Violet	0.83mg

pH 7.0 ± 0.2 at 25°C

Source: This medium is available as a premixed powder from Difco Laboratories.

Preparation of Medium: Add components to distilled/deionized water and bring volume to 1.0L. Mix thoroughly. Gently heat and bring to boiling. Distribute into tubes in 10.0mL volumes. Autoclave for 15 min at 15 psi pressure–121°C.

Caution: Sodium azide is toxic. Azides also react with metals and disposal must be highly diluted.

Use: For the isolation, cultivation and enumeration of enterococci from water and other specimens. Fecal enterococci turn the medium turbid with a purple sediment on the bottom of the tube.

Ethyl Violet Azide Broth (EVA Broth)

Composition per liter:

Tryptose	20.0g
Glucose	5.0g
NaCl	5.0g
K_2HPO_4	2.7g
KH_2PO_4	2.7g
NaN_3	0.3g
Ethyl Violet	0.5mg

pH 6.8 ± 0.2 at 25°C

Source: This medium is available as a premixed powder from Oxoid.

Preparation of Medium: Add components to distilled/deionized water and bring volume to 1.0L. Mix thoroughly. Gently heat and bring to boiling. Distribute into tubes in 10.0mL volumes. Autoclave for 15 min at 15 psi pressure–121°C.

Caution: Sodium azide is toxic. Azides also react with metals and disposal must be highly diluted.

Use: For the isolation, cultivation and enumeration of enterococci from water and other specimens. Fecal enterococci turn the medium turbid with a purple sediment on the bottom of the tube.

Eugon Agar

Composition per liter:

Agar	15.0g
Pancreatic digest of casein	15.0g
Glucose	5.5g
Papaic digest of soybean meal	5.0g
NaCl	4.0g
L-Cystine	0.3g
Na_2SO_3	0.2g

pH 7.0 ± 2.0 at 25°C

Source: This medium is available as a premixed powder fromand Difco Laboratories.

Preparation of Medium: Add components to distilled/deionized water and bring volume to 1.0L. Mix thoroughly. Gently heat and bring to boiling. Distribute into tubes or flasks. Autoclave for 15 min at 13 psi pressure–118°C. Pour into sterile Petri dishes.

Use: For the cultivation and maintenance of a variety of fastidious microorganisms. For the cultivation and maintenance of *Bifidobacterium* species.

Eugon Blood Agar (Eugonagar™)

Composition per liter:

Agar	15.0g
Pancreatic digest of casein	15.0g
Glucose	5.5g
Papaic digest of soybean meal	5.0g
NaCl	4.0g
L-Cystine	0.3g
Na_2SO_3	0.2g
Sheep blood, defibrinated	100.0mL

pH 7.0 ± 2.0 at 25°C

Source: This medium is available as a premixed powder from BBL Microbiology Systems.

Preparation of Medium: Add components, except sheep blood, to distilled/deionized water and bring volume to 900.0mL. Mix thoroughly. Gently heat and bring to boiling. Autoclave for 15 min at 13 psi pressure–118°C. Cool to 45° to 50°C. Aseptically add sterile sheep blood. Mix thoroughly. Pour into sterile Petri dishes or distribute into sterile tubes. If desired, medium may be chocolatized by maintaining at 80° to 85°C for 20 min after the addition of sheep blood.

Use: For the cultivation and maintenance of fastidious microorganisms. For the cultivation and maintenance of *Bifidobacterium* species.

Eugon Broth (Eugonbroth™)

Composition per liter:

Pancreatic digest of casein	15.0g
Glucose	5.5g
Papaic digest of soybean meal	5.0g
NaCl	4.0g
L-Cystine	0.3g
Na_2SO_3	0.2g

pH 7.0 ± 0.2 at 25°C

Source: This medium is available as a premixed powder from BBL Microbiology Systems and Difco Laboratories.

Preparation of Medium: Add components to distilled/deionized water and bring volume to 1.0L. Mix thoroughly. Gently heat while stirring and bring to boiling. Distribute into tubes or flasks. Autoclave for 15 min at 13 psi pressure–118°C.

Use: For the cultivation and maintenance of a variety of fastidious microorganisms.

Fastidious Anaerobe Agar (FAA)

Composition per liter:

Peptone	23.0g
Agar	12.0g
NaCl	5.0g
Glucose	1.0g
L-Arginine	1.0g
Sodium pyruvate	1.0g
Soluble starch	1.0g
Cysteine·HCl·H_2O	0.5g
Sodium succinate	0.5g
$NaHCO_3$	0.4g
$Na_4P_2O_7 \cdot 10H_2O$	0.25g
Sheep blood, defibrinated	50.0mL
Hemin solution	1.0mL
Vitamin K_1	0.1mL

pH 7.2 ± 0.2 at 25°C

Vitamin K_1 Solution:

Composition per 100mL:

Vitamin K_1	1.0g
Ethanol	99.0mL

Preparation of Vitamin K_1 Solution: Add vitamin K_1 to 99.0mL of absolute ethanol. Mix thoroughly.

Hemin Solution:

Composition per 100mL:

Hemin	1.0g
NaOH (1*N* solution)	20.0mL

Preparation of Hemin Solution: Add hemin to 20.0mL of 1*N* NaOH solution. Mix thoroughly. Bring volume to 100.0mL with distilled/deionized water.

Preparation of Medium: Add components, except defibrinated sheep blood, to distilled/deionized water and bring volume to 950.0mL. Mix thoroughly. Gently heat and bring to boiling. Autoclave for 15 min at 15 psi pressure–121°C. Cool to 45° to 50°C. Aseptically add 50.0mL sterile defibrinated sheep blood. Mix thoroughly. Pour into sterile Petri dishes or distribute into sterile tubes.

Use: For the cultivation of a variety of fastidious anaerobes from clinical specimens and in epidemiological investigations.

Fastidious Anaerobe Agar, Alternative Selective (FAA Alternative Selective)

Composition per liter:

Peptone	23.0g
Agar	12.0g
NaCl	5.0g
Glucose	1.0g
L-Arginine	1.0g
Sodium pyruvate	1.0g
Soluble starch	1.0g
Cysteine·HCl·H_2O	0.5g
Sodium succinate	0.5g
$NaHCO_3$	0.4g
$Na_4P_2O_7 \cdot 10H_2O$	0.25g
Sheep blood, defibrinated	50.0mL
Hemin solution	1.0mL
Vitamin K_1	0.1mL

pH 7.2 ± 0.2 at 25°C

Vitamin K_1 Solution:
Composition per 100mL:

Vitamin K_1	1.0g
Ethanol	99.0mL

Preparation of Vitamin K_1 Solution: Add vitamin K_1 to 99.0mL of absolute ethanol. Mix thoroughly.

Hemin Solution:
Composition per 100mL:

Hemin	1.0g
NaOH (1*N* solution)	20.0mL

Preparation of Hemin Solution: Add hemin to 20.0mL of 1*N* NaOH solution. Mix thoroughly. Bring volume to 100.0mL with distilled/deionized water.

Preparation of Medium: Add components, except defibrinated sheep blood, to distilled/deionized water and bring volume to 950.0mL. Mix thoroughly. Gently heat and bring to boiling. Autoclave for 15 min at 15 psi pressure–121°C. Cool to 45° to 50°C. Aseptically add 50.0mL sterile defibrinated sheep blood. Mix thoroughly. Pour into sterile Petri dishes or distribute into sterile tubes.

Use: For the cultivation of a variety of fastidious anaerobes from clinical specimens and in epidemiological investigations.

Fastidious Anaerobe Agar, Alternative Selective with Neomycin, Vancomycin, and Josamycin (FAA Alternative Selective with Neomycin, Vancomycin, and Josamycin)

Composition per liter:

Peptone	23.0g
Agar	12.0g
NaCl	5.0g
Glucose	1.0g
L-Arginine	1.0g
Sodium pyruvate	1.0g
Soluble starch	1.0g
Cysteine·HCl·H_2O	0.5g
Sodium succinate	0.5g
$NaHCO_3$	0.4g
$Na_4P_2O_7 \cdot 10H_2O$	0.25g
Neomycin	0.1g
Sheep blood, defibrinated	50.0mL
Vancomycin solution	10.0mL
Josamycin	10.0mL
Hemin solution	1.0mL
Vitamin K_1	0.1mL

pH 7.2 ± 0.2 at 25°C

Vitamin K_1 Solution:
Composition per 100mL:

Vitamin K_1	1.0g
Ethanol	99.0mL

Preparation of Vitamin K_1 Solution: Add vitamin K_1 to 99.0mL of absolute ethanol. Mix thoroughly.

Hemin Solution:
Composition per 100mL:

Hemin	1.0g
NaOH (1*N* solution)	20.0mL

Preparation of Hemin Solution: Add hemin to 20.0mL of 1*N* NaOH solution. Mix thoroughly. Bring volume to 100.0mL with distilled/deionized water.

Vancomycin Solution:
Composition per 10mL:

Vancomycin	5.0mg

Preparation of Vancomycin Solution: Add vancomycin to distilled/deionized water and bring volume to 10.0mL. Mix thoroughly. Filter sterilize.

Josamycin Solution:
Composition per 10mL:

Josamycin	3.0mg

Preparation of Josamycin Solution: Add josamycin to distilled/deionized water and bring volume to 10.0mL. Mix thoroughly. Filter sterilize.

Preparation of Medium: Add components, except defibrinated sheep blood, vancomycin solution, and josamycin solution, to distilled/deionized water and bring volume to 930.0mL. Mix thoroughly. Gently heat and bring to boiling. Autoclave for 15 min at 15 psi pressure–121°C. Cool to 45° to 50°C. Aseptically add 50.0mL sterile defibrinated sheep blood, 10.0mL vancomycin solution, and 10.0mL josamycin solution. Mix thoroughly. Pour into sterile Petri dishes or distribute into sterile tubes.

Use: For the selective cultivation of *Fusobacterium* species from clinical specimens and in epidemiological investigations.

Fastidious Anaerobe Agar, Selective (FAA Selective)

Composition per liter:

Peptone	23.0g
Agar	12.0g
NaCl	5.0g
Glucose	1.0g
L-Arginine	1.0g
Sodium pyruvate	1.0g
Soluble starch	1.0g
Cysteine·HCl·H_2O	0.5g
Sodium succinate	0.5g
$NaHCO_3$	0.4g
$Na_4P_2O_7 \cdot 10H_2O$	0.25g
Sheep blood, defibrinated	50.0mL
Hemin solution	1.0mL
Vitamin K_1	0.1mL

pH 7.2 ± 0.2 at 25°C

Vitamin K_1 Solution:
Composition per 100mL:

Vitamin K_1	1.0g
Ethanol	99.0mL

Preparation of Vitamin K_1 Solution: Add vitamin K_1 to 99.0mL of absolute ethanol. Mix thoroughly.

Hemin Solution:
Composition per 100mL:

Hemin	1.0g
NaOH (1*N* solution)	20.0mL

Preparation of Hemin Solution: Add hemin to 20.0mL of 1*N* NaOH solution. Mix thoroughly. Bring volume to 100.0mL with distilled/deionized water.

Preparation of Medium: Add components, except defibrinated sheep blood, to distilled/deionized water and bring volume to 950.0mL. Mix thoroughly. Gently heat and bring to boiling. Autoclave for 15 min at 15 psi pressure–121°C. Cool to 45° to 50°C. Aseptically add 50.0mL sterile defibrinated sheep blood. Mix thoroughly. Pour into sterile Petri dishes or distribute into sterile tubes.

Use: For the cultivation of a variety of fastidious anaerobes from clinical specimens and in epidemiological investigations.

Fastidious Anaerobe Agar, Selective with Neomycin and Vancomycin (FAA Selective with Neomycin and Vancomycin)

Composition per liter:

Peptone	23.0g
Agar	12.0g
NaCl	5.0g
Glucose	1.0g
L-Arginine	1.0g
Sodium pyruvate	1.0g
Soluble starch	1.0g
Cysteine·HCl·H_2O	0.5g
Sodium succinate	0.5g
$NaHCO_3$	0.4g
$Na_4P_2O_7 \cdot 10H_2O$	0.25g
Neomycin	0.1g
Sheep blood, defibrinated	50.0mL
Vancomycin solution	10.0mL
Hemin solution	1.0mL
Vitamin K_1	0.1mL

pH 7.2 ± 0.2 at 25°C

Vitamin K_1 Solution:
Composition per 100mL:

Vitamin K_1	1.0g
Ethanol	99.0mL

Preparation of Vitamin K_1 Solution: Add vitamin K_1 to 99.0mL of absolute ethanol. Mix thoroughly.

Hemin Solution:
Composition per 100mL:

Hemin	1.0g
NaOH (1*N* solution)	20.0mL

Preparation of Hemin Solution: Add hemin to 20.0mL of 1*N* NaOH solution. Mix thoroughly. Bring volume to 100.0mL with distilled/deionized water.

Vancomycin Solution:
Composition per 10mL:

Vancomycin	7.5mg

Preparation of Vancomycin Solution: Add vancomycin to distilled/deionized water and bring volume to 10.0mL. Mix thoroughly. Filter sterilize.

Preparation of Medium: Add components, except defibrinated sheep blood and vancomycin solution, to distilled/deionized water and bring volume to 940.0mL. Mix thoroughly. Gently heat and bring to boiling. Autoclave for 15 min at 15 psi pressure–

121°C. Cool to 45° to 50°C. Aseptically add 50.0mL sterile defibrinated sheep blood and 10.0mL vancomycin solution. Mix thoroughly. Pour into sterile Petri dishes or distribute into sterile tubes.

Use: For the selective cultivation of *Fusobacterium* species from clinical specimens and in epidemiological investigations.

FC Agar (Fecal Coliform Agar) (m-FC Agar) (m–Fecal Coliform Agar)

Composition per liter:

Agar	15.0g
Lactose	12.5g
NaCl	5.0g
Proteose peptone No. 3	5.0g
Yeast extract	3.0g
Bile salts	1.5g
Aniline Blue	0.1g
Rosolic acid solution	10.0mL

pH 7.4 ± 0.2 at 25°C

Source: This medium is available as a premixed powder from BBL Microbiology Systems and Difco Laboratories.

Rosolic Acid Solution:
Composition per 100mL:

Rosolic acid	1.0g

Preparation of Rosolic Acid Solution: Add rosolic acid to 0.2*N* NaOH and bring volume to 100.0L. Mix thoroughly.

Preparation of Medium: Add 10.0mL rosolic acid solution to 950.0mL distilled/deionized water. Mix thoroughly. Add other components and bring volume to 1.0L with distilled/deionized water. Mix thoroughly. Gently heat and bring to boiling with frequent mixing. Do not autoclave. Pour into sterile Petri dishes or leave in tubes.

Use: For the cultivation of fecal coliform bacteria from waters and the enumeration of coliform bacteria using the membrane filtration method.

FC Agar (Fecal Coliform Agar) (m-FC Agar) (m–Fecal Coliform Agar)

Composition per liter:

Agar	15.0g
Lactose	12.5g
Tryptose	10.0g
NaCl	5.0g
Proteose peptone No. 3	5.0g
Yeast extract	3.0g
Bile salts	1.5g
Aniline Blue	0.1g
Rosolic acid solution	10.0mL

pH 7.4 ± 0.2 at 25°C

Rosolic Acid Solution:
Composition per 100mL:

Rosolic acid	1.0g

Preparation of Rosolic Acid Solution: Add rosolic acid to 0.2*N* NaOH and bring volume to 100.0L. Mix thoroughly.

Preparation of Medium: Add 10.0mL rosolic acid solution to 950.0mL distilled/deionized water. Mix thoroughly. Add other components and bring volume to 1.0L with distilled/deionized water. Mix thoroughly. Gently heat and bring to boiling with frequent mixing. Do not autoclave. Pour into sterile Petri dishes or leave in tubes.

Use: For the cultivation of fecal coliform bacteria from waters and the enumeration of coliform bacteria using the membrane filtration method.

FC Broth (Fecal Coliform Broth) (m-FC Broth) (m–Fecal Coliform Broth)

Composition per liter:

Lactose	12.5g
Tryptose	10.0g
NaCl	5.0g
Proteose peptone No. 3	5.0g
Yeast extract	3.0g
Bile salts	1.5g
Aniline Blue	0.1g
Rosolic acid solution	10.0mL

pH 7.4 ± 0.2 at 25°C

Rosolic Acid Solution:
Composition per 100mL:

Rosolic acid	1.0g

Preparation of Rosolic Acid Solution: Add rosolic acid to 0.2*N* NaOH and bring volume to 100.0L. Mix thoroughly.

Preparation of Medium: Add 10.0mL rosolic acid solution to 950.0mL distilled/deionized water. Mix thoroughly. Add other components and bring volume to 1.0L with distilled/deionized water. Mix thoroughly. Gently heat and bring to boiling with frequent mixing. Do not autoclave. Pour into sterile Petri dishes or leave in tubes.

Use: For the cultivation of fecal coliform bacteria from waters and the enumeration of coliform bacteria using the membrane filtration method.

FC Broth (Fecal Coliform Broth) (m-FC Broth) (m–Fecal Coliform Broth)

Composition per liter:

Lactose....................12.5g
NaCl....................5.0g
Proteose peptone No. 3....................5.0g
Yeast extract....................3.0g
Bile salts....................1.5g
Aniline Blue....................0.1g
Rosolic acid solution....................10.0mL

pH 7.4 ± 0.2 at 25°C

Source: This medium is available as a premixed powder from BBL Microbiology Systems and Difco Laboratories.

Rosolic Acid Solution:

Composition per 100mL:

Rosolic acid....................1.0g

Preparation of Rosolic Acid Solution: Add rosolic acid to 0.2*N* NaOH and bring volume to 100.0L. Mix thoroughly.

Preparation of Medium: Add 10.0mL rosolic acid solution to 950.0mL distilled/deionized water. Mix thoroughly. Add other components and bring volume to 1.0L with distilled/deionized water. Mix thoroughly. Gently heat and bring to boiling with frequent mixing. Do not autoclave. Pour into sterile Petri dishes or leave in tubes.

Use: For the cultivation of fecal coliform bacteria from waters and the enumeration of coliform bacteria using the membrane filtration method.

FDA Agar (ATCC Medium 182) (AATCC Bacteriostasis Agar) (American Association of Textile Chemists and Colorists Bacteriostasis Agar)

Composition per liter:

Agar....................15.0g
Peptic digest of animal tissue....................10.0g
Beef extract....................5.0g
NaCl....................5.0g

pH 6.8 ± 0.1 at 25°C

Source: This medium is available as a premixed powder from BBL Microbiology Systems.

Preparation of Medium: Add components to distilled/deionized water and bring volume to 1.0L. Mix thoroughly. Gently heat and bring to boiling. Distribute into tubes or flasks. Autoclave for 15 min at 15 psi pressure–121°C. Pour into sterile Petri dishes or leave in tubes.

Use: For testing the antibacterial activities of antiseptics and disinfectants.

FDA Broth (AATCC Bacteriostasis Broth) (American Association of Textile Chemists and Colorists Bacteriostasis Broth)

Composition per liter:

Peptic digest of animal tissue....................10.0g
Beef extract....................5.0g
NaCl....................5.0g

pH 6.8 ± 0.1 at 25°C

Source: This medium is available as a premixed powder from BBL Microbiology Systems.

Preparation of Medium: Add components to distilled/deionized water and bring volume to 1.0L. Mix thoroughly. Distribute into tubes or flasks. Autoclave for 15 min at 15 psi pressure–121°C.

Use: For testing the antibacterial activities of antiseptics and disinfectants.

Fecal Coliform Agar, Modified (m–Fecal Coliform Agar, Modified) (FCIC)

Composition per liter:

Agar....................15.0g
Inositol....................10.0g
Tryptose....................10.0g
Proteose peptone No. 3....................5.0g
NaCl....................5.0g
Yeast extract....................3.0g
Bile salts No. 3....................1.5g
Aniline Blue....................0.1g

pH 7.4 ± 0.2 at 25°C

Preparation of Medium: Add components and bring volume to 1.0L. Mix thoroughly. Gently heat and bring to boiling. Do not autoclave. Cool to 50°C. Adjust pH to 7.4. Pour into sterile Petri dishes in 20.0mL volumes. Allow surface of plates to dry before using.

Use: For the isolation, cultivation and enumeration of *Klebsiella* species using the membrane filter method.

Fermentation Broth (CHO Medium)

Composition per liter:

Pancreatic digest of casein	15.0g
Yeast extract	7.0g
NaCl	2.5g
Agar	0.75g
Sodium thioglycollate	0.5g
L-Cystine	0.25g
Ascorbic acid	0.1g
Bromthymol Blue	0.01g
Carbohydrate or starch solution	100.0mL

pH 7.0 ± 0.1 at 25°C

Source: This medium is available as a premixed powder from Difco Laboratories.

Carbohydrate Solution:

Composition per 100mL:

Carbohydrate	6.0g

Preparation of Carbohydrate Solution: Add carbohydrate to distilled/deionized water and bring volume to 10.0mL. Mix thoroughly. Filter sterilize.

Starch Solution:

Composition per 100mL:

Starch	2.5g

Preparation of Starch Solution: Add starch to distilled/deionized water and bring volume to 100.0mL. Mix thoroughly. Filter sterilize.

Preparation of Medium: Add components, except carbohydrate solution, to distilled/deionized water and bring volume to 900.0mL. Mix thoroughly. Distribute into tubes or flasks. Autoclave for 15 min at 15 psi pressure–121°C. Cool to 45° to 50°C. Aseptically add 100.0mL sterile carbohydrate solution. Mix thoroughly. Aseptically distribute into sterile tubes or flasks. Loosen caps on tubes. Place in an anaerobic chamber under an atmosphere of 85% N_2, 10% H_2, and 5% CO_2. Fasten the caps securely or maintain in an anaerobic chamber.

Use: For differentiation of anaerobic bacteria based upon carbohydrate fermentation. Bacteria that ferment carbohydrates turn the medium yellow.

Fermentation Medium

Composition per liter:

Glucose or mannitol	10.0g
Pancreatic digest of casein	10.0g
Agar	2.2g
Yeast extract	1.0
Bromcresol Purple	0.04g

pH 7.0 ± 0.2 at 25°C

Preparation of Medium: Add components to distilled/deionized water and bring volume to 1.0L. Mix thoroughly. Gently heat and bring to boiling. Distribute into tubes or flasks. Autoclave for 10 min at 15 psi pressure–121°C. Pour into sterile Petri dishes or leave in tubes.

Use: For differentiating *Staphylococcus* and *Micrococcus* species based upon the fermentation of glucose and mannitol.

F–G Agar (Feeley–Gorman Agar)

Composition per liter:

Casein, acid hydrolyzed	17.5g
Agar	17.0g
Beef extract	3.0g
Starch	1.5g
Cysteine solution	10.0mL
$Fe_4(P_2O_7)_3$ solution	10.0mL

pH 6.9 ± 0.05 at 25°C

Cysteine Solution:

Composition per 10mL:

L-Cysteine $HCl{\cdot}H_2O$	0.4g

Preparation of Cysteine Solution: Add L-Cysteine $HCl{\cdot}H_2O$ to distilled/deionized water and bring volume to 10.0mL. Mix thoroughly. Filter sterilize.

$Fe_4(P_2O_7)_3$ Solution:

Composition per 10mL:

$Fe_4(P_2O_7)_3$	0.25g

Preparation of $Fe_4(P_2O_7)_3$ Solution: Add $Fe_4(P_2O_7)_3$ to distilled/deionized water and bring volume to 10.0mL. Mix thoroughly. Filter sterilize.

Preparation of Medium: Add components, except cysteine solution and $Fe_4(P_2O_7)_3$ solution, to distilled/deionized water and bring volume to 980.0mL. Mix thoroughly. Gently heat and bring to boiling. Autoclave for 15 min at 15 psi pressure–121°C. Cool to 45° to 50°C. Aseptically add 10.0mL cysteine solution. Mix thoroughly. Aseptically add 10.0mL $Fe_4(P_2O_7)_3$ solution. Mix thoroughly. Adjust pH to 6.9. Pour into sterile Petri dishes or distribute into sterile tubes.

Use: For the isolation and cultivation of *Legionella pneumophila.*

F–G Agar with Selenium (Feeley–Gorman Agar with Selenium)

Composition per liter:

Casein, acid hydrolyzed 17.5g
Agar 17.0g
Beef extract 3.0g
Starch 1.5g
Cysteine solution 10.0mL
$Fe_4(P_2O_7)_3$ solution 10.0mL
$Na_2SeO_3 \cdot 5H_2O$ solution 10.0mL

pH 6.9 ± 0.05 at 25°C

Cysteine Solution:

Composition per 10mL:

L-Cysteine $HCl \cdot H_2O$ 0.4g

Preparation of Cysteine Solution: Add L-cysteine $HCl \cdot H_2O$ to distilled/deionized water and bring volume to 10.0mL. Mix thoroughly. Filter sterilize.

$Fe_4(P_2O_7)_3$ Solution:

Composition per 10mL:

$Fe_4(P_2O_7)_3$ 0.25g

Preparation of $Fe_4(P_2O_7)_3$ Solution: Add $Fe_4(P_2O_7)_3$ to distilled/deionized water and bring volume to 10.0mL. Mix thoroughly. Filter sterilize.

$Na_2SeO_3 \cdot 5H_2O$ Solution:

Composition per 10mL:

$Na_2SeO_3 \cdot 5H_2O$ 0.010g

Preparation of $Na_2SeO_3 \cdot 5H_2O$ Solution: Add $Na_2SeO_3 \cdot 5H_2O$ to distilled/deionized water and bring volume to 10.0mL. Mix thoroughly. Filter sterilize.

Preparation of Medium: Add components—except cysteine solution, $Fe_4(P_2O_7)_3$ solution, and $Na_2SeO_3 \cdot 5H_2O$ solution—to distilled/deionized water and bring volume to 970.0mL. Mix thoroughly. Gently heat and bring to boiling. Autoclave for 15 min at 15 psi pressure–121°C. Cool to 45° to 50°C. Aseptically add 10.0mL of sterile cysteine solution. Mix thoroughly. Aseptically add 10.0mL of sterile $Fe_4(P_2O_7)_3$ solution and 10.0mL of sterile $Na_2SeO_3 \cdot 5H_2O$ solution. Mix thoroughly. Adjust pH to 6.9. Pour into sterile Petri dishes or distribute into sterile tubes.

Use: For the isolation and cultivation of *Legionella pneumophila.*

F–G Broth (Feeley–Gorman Broth)

Composition per liter:

Casein, acid hydrolyzed 17.5g
Beef extract 3.0g
Starch 1.5g
Cysteine solution 10.0mL
$Fe_4(P_2O_7)_3$ solution 10.0mL

pH 6.9 ± 0.05 at 25°C

Cysteine Solution:

Composition per 10mL:

L-Cysteine $HCl \cdot H_2O$ 0.4g

Preparation of Cysteine Solution: Add L-cysteine $HCl \cdot H_2O$ to distilled/deionized water and bring volume to 10.0mL. Mix thoroughly. Filter sterilize.

$Fe_4(P_2O_7)_3$ Solution:

Composition per 10mL:

$Fe_4(P_2O_7)_3$ 0.25g

Preparation of $Fe_4(P_2O_7)_3$ Solution: Add $Fe_4(P_2O_7)_3$ to distilled/deionized water and bring volume to 10.0mL. Mix thoroughly. Filter sterilize.

Preparation of Medium: Add components, except cysteine solution and $Fe_4(P_2O_7)_3$ solution, to distilled/deionized water and bring volume to 980.0mL. Mix thoroughly. Gently heat and bring to boiling. Autoclave for 15 min at 15 psi pressure–121°C. Cool to 45° to 50°C. Aseptically add 10.0mL cysteine solution. Mix thoroughly. Aseptically add 10.0mL $Fe_4(P_2O_7)_3$ solution. Mix thoroughly. Adjust pH to 6.9. Aseptically distribute into sterile tubes or flasks.

Use: For the cultivation of *Legionella pneumophila.*

Fildes Enrichment Agar

Composition per liter:

Agar 15.0g
Peptone 5.0g
Beef extract 3.0g
Fildes enrichment solution 50.0mL

Fildes Enrichment Solution:

Composition 206mL:

Pepsin 1.0g
NaCl (0.85% solution) 150.0mL
Sheep blood, defibrinated 50.0mL
HCl 6.0mL

pH 7.0–7.2 at 25°C

Source: Fildes enrichment solution is available from Difco Laboratories.

Preparation of Fildes Enrichment Solution: Combine components. Mix thoroughly. Incubate at 56°C for 4 h. Bring pH to 7.0 with 20% NaOH. Adjust pH to 7.2 with HCl. Do not autoclave. Add 0.25 mL of chloroform and store at 4°C. Before use heat to 56°C to remove chloroform.

Preparation of Medium: Add components, except Fildes enrichment solution, to distilled/deionized water and bring volume to 950.0mL. Mix thoroughly. Gently heat and bring to boiling. Autoclave for 15 min at 15 psi pressure–121°C. Cool to 56°C. Aseptically add 50.0mL sterile Fildes enrichment solution. Mix thoroughly. Pour into sterile Petri dishes or distribute into sterile tubes.

Use: For the isolation and cultivation of *Haemophilus influenzae.*

Fletcher Medium

Composition per liter:

Agar	1.5g
NaCl	0.5g
Peptone	0.3g
Beef extract	0.2g
Rabbit serum	50.0mL

pH 7.9 ± 0.1 at 25°C

Source: This medium is available as a premixed powder from Difco Laboratories.

Preparation of Medium: Add components, except rabbit serum, to distilled/deionized water and bring volume to 950.0mL. Mix thoroughly. Gently heat and bring to boiling. Autoclave for 15 min at 15 psi pressure–121°C. Cool to 50° to 55°C. Aseptically add 50.0mL sterile rabbit serum. Mix thoroughly. Aseptically distribute into sterile tubes or flasks.

Use: For the isolation, cultivation and maintenance of cultures of *Leptospira* species.

Fletcher Medium with Fluorouracil (Flurouracil Leptospira Medium)

Composition per liter:

Agar	1.5g
NaCl	0.5g
Peptone	0.3g
Beef extract	0.2g
Rabbit serum	50.0mL
Fluorouracil solution	20.0mL

pH 7.9 ± 0.1 at 25°C

Fluorouracil Solution:

Composition per 100mL:

Fluorouracil	10.0g

Preparation of Fluorouracil Solution: Add fluorouracil to 50.0mL distilled/deionized water. Add 1.0mL 2*N* NaOH and bring volume to 100.0mL. Gently heat to 56°C for 2 hr. Adjust pH to 7.4–7.6 with NaOH. Mix thoroughly. Filter sterilize.

Preparation of Medium: Add components, except rabbit serum and fluorouracil solution, to distilled/deionized water and bring volume to 930.0mL. Mix thoroughly. Gently heat and bring to boiling. Autoclave for 15 min at 15 psi pressure–121°C. Cool to 50° to 55°C. Aseptically add 80.0mL sterile rabbit serum. Mix thoroughly. Aseptically distribute into sterile tubes or flasks. Immediately prior to use add 0.1mL fluorouracil solution per 5.0mL medium.

Use: For the isolation, cultivation and maintenance of cultures of *Leptospira* species.

Fletcher's Semisolid Medium

Composition per 2120mL:

Agar	1.5g
NaCl	0.5g
Peptone	0.3g
Beef extract	0.2g
Rabbit serum	240.0mL

pH 7.9 ± 0.1 at 25°C

Preparation of Medium: Add components, except rabbit serum, to distilled/deionized water and bring volume to 1880.0mL. Mix thoroughly. Gently heat and bring to boiling. Autoclave for 15 min at 15 psi pressure–121°C. Cool to 50° to 55°C. Aseptically add 240.0mL sterile rabbit serum. Mix thoroughly. Aseptically distribute into sterile tubes or flasks.

Use: For the isolation, cultivation and maintenance of cultures of *Leptospira* species.

FLN Medium (Fluorescence Lactose Nitrate Medium)

Composition per liter:

Lactose	20.0g
Agar	15.0g
Proteose peptone No. 3	10.0g
KNO_3	2.0g
K_2HPO_4	1.5g
$MgSO_4 \cdot 7H_2O$	1.5g
$NaNO_2$	0.5g
Phenol Red.	0.02g

pH 7.2 ±0.2 at 25°C

Preparation of Medium: Add components to distilled/deionized water and bring volume to 1.0L. Mix thoroughly. Gently heat and bring to boiling. Distribute into tubes or flasks. Autoclave for 15 min at 15 psi pressure–121°C. Pour into sterile Petri dishes or leave in tubes.

Use: For the differentiation of pseudomonads from other non–fermentative bacilli. Lactose fermentation is indicated by the medium turning yellow. *Pseudomonas cepacia* often produces acid from lactose. Denitrification from nitrate or nitrite is indicated by the formation of gas bubbles in the solid medium. *P. aeruginosa*, *P. mendocina*, and *P. denitrificans* are positive for denitrification. Fluorescein production is indicated by fluorescence under UV light. *P. aeruginosa* is positive for fluorescein production; *P. denitrificans* does not produce fluorescein.

Flo Agar

Composition per liter:

Agar	14.0g
Pancreatic digest of casein	10.0g
Peptic digest of animal tissue	10.0g
K_2HPO_4	1.5g
$MgSO_4 \cdot 7H_2O$	1.5g

pH 7.2 ± 0.2 at 25°C

Source: This medium is available as a premixed powder from BBL Microbiology Systems.

Preparation of Medium: Add components to distilled/deionized water and bring volume to 1.0L. Mix thoroughly. Gently heat and bring to boiling. Distribute into tubes or flasks. Autoclave for 15 min at 15 psi pressure–121°C. Pour into sterile Petri dishes or leave in tubes.

Use: For cultivation of fluorescent *Pseudomonas* species.

Fluid Thioglycollate Medium

Composition per liter:

Pancreatic digest of casein	15.0g
Glucose	5.5g
Yeast extract	5.0g
NaCl	2.5g
Agar	0.75g
L-Cystine	0.5g
Sodium thioglycollate	0.5g
Resazurin	1.0mg

pH 7.1 ± 0.2 at 25°C

Source: This medium is available as a premixed powder from Difco Laboratories.

Preparation of Medium: Add components to distilled/deionized water and bring volume to 1.0L. Mix thoroughly. Gently heat and bring to boiling. Distribute into tubes or flasks. Autoclave for 15 min at 15 psi pressure–121°C. If medium becomes oxidized before use (resazurin turns red) heat in a boiling water bath to expel absorbed O_2. Cool to 25°C.

Use: For the cultivation of anaerobic, microaerophilic and aerobic microorganisms. For use in sterility testing of a variety of specimens.

Fluid Thioglycollate Medium with Beef Extract

Composition per liter:

Pancreatic digest of casein	15.0g
Glucose	5.5g
Yeast extract	5.0g
Beef extract	5.0g
NaCl	2.5g
Agar	0.75g
L-Cystine	0.5g
Sodium thioglycollate	0.5g
Resazurin	1.0mg

pH 7.2 ± 0.2 at 25°C

Source: This medium is available as a premixed powder from Difco Laboratories.

Preparation of Medium: Add components to distilled/deionized water and bring volume to 1.0L. Mix thoroughly. Gently heat and bring to boiling. Distribute into tubes or flasks. Autoclave for 15 min at 15 psi pressure–121°C. If medium becomes oxidized before use (resazurin turns red) heat in a boiling water bath to expel absorbed O_2. Cool to 25°C.

Use: For the cultivation of anaerobic, microaerophilic and aerobic microorganisms. For use in sterility testing of a variety of specimens.

Fluid Thioglycollate Medium with Rabbit Serum

Composition per liter:

Pancreatic digest of casein	15.0g
Glucose	5.5g
Yeast extract	5.0g
NaCl	2.5g
Agar	0.75g
L-Cystine	0.5g
Sodium thioglycollate	0.5g
Resazurin	1.0mg
Rabbit serum	100.0mL

pH 7.1 ± 0.2 at 25°C

Preparation of Medium: Add components, except rabbit serum, to distilled/deionized water and bring volume to 900.0mL. Mix thoroughly. Gently heat and bring to boiling. Distribute into tubes in 9.0mL volumes. Autoclave for 15 min at 15 psi pressure–121°C. If medium becomes oxidized before use (resazurin turns red) heat in a boiling water bath to expel absorbed O_2. Cool to 25°C. Immediately prior to inoculation, aseptically and anerobically add 1.0mL of sterile rabbit serum to each tube. Mix thoroughly.

Use: For the cultivation of anaerobic, microaerophilic and aerobic microorganisms. For use in sterility testing of a variety of specimens.

Fluid Thioglycollate Medium without Glucose or E_h Indicator

Composition per liter:

Pancreatic digest of casein	20.0g
NaCl	2.5g
Agar	0.75g
L-Cystine	0.5g
Sodium thioglycollate	0.5g

pH 7.1 ± 0.2 at 25°C

Source: This medium is available as a premixed powder from BBL Microbiology Systems.

Preparation of Medium: Add components to distilled/deionized water and bring volume to 1.0L. Mix thoroughly. Gently heat and bring to boiling. Distribute into tubes or flasks. Autoclave for 15 min at 15 psi pressure–121°C. Aseptically distribute into sterile screw cap tubes or flasks.

Use: For cultivation of anaerobic bacteria. For use as a basal medium in carbohydrate fermentation tests for differentiating anaerobic bacteria. For carbohydrate fermentation tests 1.0 mL of a 10% filter sterilized carbohydrate solution is aseptically added to 9.0mL of fluid thioglycollate medium.

Fluid Thioglycollate Medium with K Agar

Composition per liter:

Pancreatic digest of casein	15.0g
Glucose	5.0g
Yeast extract	5.0g
KCl	2.5g
K agar	0.45g
L-Cystine	0.5g
Resazurin	1.0mg
Thioglycollic acid	0.3mL

pH 7.2 ± 0.2 at 25°C

Source: This medium is available as a premixed powder from Difco Laboratories.

Preparation of Medium: Add components to distilled/deionized water and bring volume to 1.0L. Mix thoroughly. Gently heat and bring to boiling. Distribute into tubes or flasks. Autoclave for 15 min at 15 psi pressure–121°C. If medium becomes oxidized before use (resazurin turns red) heat in a boiling water bath to expel absorbed O_2. Cool to 25°C.

Use: For the cultivation of anaerobic, microaerophilic and aerobic microorganisms. For use in sterility testing of a variety of specimens.

FN Medium (Fluorescence Denitrification Medium)

Composition per liter:

Agar	15.0g
Proteose peptone No. 3	10.0g
KNO_3	2.0g
K_2HPO_4	1.5g
$MgSO_4 \cdot 7H_2O$	1.5g
$NaNO_2$	0.5g

pH 7.2 ±0.2 at 25°C

Preparation of Medium: Add components to distilled/deionized water and bring volume to 1.0L. Mix thoroughly. Gently heat and bring to boiling. Distribute into tubes or flasks. Autoclave for 15 min at 15 psi pressure–121°C. Pour into sterile Petri dishes or leave in tubes.

Use: For the differentiation of pseudomonads from other non–fermentative bacilli. Denitrification from nitrate or nitrite is indicated by the formation of gas bubbles in the solid medium. *P. aeruginosa, P. mendocina,* and *P. denitrificans* are positive for denitrification. Fluorescein production is indicated by fluorescence under UV light. *P. aeruginosa* is positive for fluorescein production; *P. denitrificans* does not produce fluorescein.

FRAG Agar (Fragilis Agar)

Composition per 1025mL:

L-Cysteine·HCl·H_2O	0.5g
Basal solution	995.0mL
Glucuronic acid solution	25.0mL
Gentamicin solution	1.0mL
Hemin-vitamin K_1 solution	1.0mL
Ferric sulfate solution	1.0mL
Mineral solution	1.0mL
Phenol Red (1% solution)	1.0mL
Vitamin B_{12} solution	0.05mL

pH 7.0 ± 0.1 at 25°C

Basal Solution:

Composition per 995mL:

Oxgall	20.0g
Agar	15.4g
K_2HPO_4	2.26g
Yeast extract	2.0g

Pancreatic digest of casein 1.4g
$(NH_4)_2HPO_4$ 1.0g
K_2HPO_4 .... 0.9g
Papaic digest of soybean meal 0.12g
NaCl 0.12g

Preparation of Basal Solution: Add components to distilled/deionized water and bring volume to 995.0mL. Mix thoroughly. Gently heat and bring to boiling. Autoclave for 15 min at 15 psi pressure–121°C. Cool to 45° to 50°C.

Glucuronic Acid Solution:
Composition per 100mL:
D-Glucuronic acid 40.0g

Preparation of Glucuronic Acid Solution: Add glucuronic acid to distilled/deionized water and bring volume to 100.0mL. Mix thoroughly. Filter sterilize.

Gentamicin Solution:
Composition per 10mL:
Gentamicin 0.1mg

Preparation of Gentamicin Solution: Add gentamicin to distilled/deionized water and bring volume to 10.0mL. Mix thoroughly. Filter sterilize.

Vitamin K_1–Hemin Solution:
Composition per liter:
Vitamin K_1 solution 10.0mL
Hemin solution 10.0mL

Preparation of Vitamin K_1–Hemin Solution: Add components to distilled/deionized water and bring the volume to 1.0L. Mix thoroughly. Filter sterilize.

Vitamin K_1 Solution:
Composition per 100mL:
Vitamin K_1 .... 1.0g
Ethanol 99.0mL

Preparation of Vitamin K_1 Solution: Add vitamin K_1 to 99.0mL of absolute ethanol. Mix thoroughly.

Hemin Solution:
Composition per 10mL:
Hemin 0.5g
NaOH 0.4g

Preparation of Hemin Solution: Add hemin and NaOH to distilled/deionized water and bring volume to 10.0mL. Mix thoroughly.

Ferric Sulfate Solution:
Composition per 100mL:
$FeSO_4 \cdot 9H_2O$ 0.04g

Preparation of Ferric Sulfate Solution: Add $FeSO_4 \cdot 9H_2O$ to distilled/deionized water and bring volume to 100.0mL. Mix thoroughly. Filter sterilize.

Mineral Solution:
Composition per 100mL:
NaCl 9.0g
$CaCl_2 \cdot 2H_2O$ 0.27g
$MgCl_2 \cdot 6H_2O$ 0.2g
$CoCl_2 \cdot 6H_2O$ 0.1g
$MnCl_2 \cdot 4H_2O$ 0.1g

Preparation of Mineral Solution: Add components to distilled/deionized water and bring volume to 100.0mL. Mix thoroughly. Filter sterilize.

Vitamin B_{12} Solution:
Composition per 10mL:
Vitamin B_{12} 0.1mg

Preparation of Vitamin B_{12} Solution: Add vitamin B_{12} to distilled/deionized water and bring volume to 10.0mL. Mix thoroughly. Filter sterilize.

Preparation of Medium: To 995.0mL of cooled, sterile basal solution aseptically add, 0.5g of cysteine·HCl·H_2O, 25.0mL of sterile glucuronic acid solution, 1.0mL of sterile gentamicin solution, 1.0mL of sterile hemin-vitamin K_1 solution, 1.0mL of sterile ferric sulfate solution, 1.0mL of sterile mineral solution, 1.0mL of Phenol Red solution, and 0.05mL of sterile vitamin B_{12} solution. Mix thoroughly. Pour into sterile Petri dishes or distribute into sterile tubes.

Use: For the isolation, cultivation and differentiation of the *Bacteroides fragilis* Group (*B. fragilis, B. thetaiotaomicron, B. vulgatus, B. distasonis, B. ovatus,* and *B. uniformis*) from clinical specimens.

Francisella tularensis Isolation Medium

Composition per liter:
Agar 10.0g
Glucose 10.0g
Pancreatic digest of casein 10.0g
Peptic digest of animal tissue 10.0g
Cysteine·HCl·H_2O 5.0g
NaCl 5.0g
Sodium thioglycollate 2.0g
Glucose 1.0g
Thiamine·HCl 5.0mg
pH 7.2 ± 0.2 at 25°C

Preparation of Medium: Add components, except agar, to distilled/deionized water and bring volume to 1.0L. Mix thoroughly. Adjust pH to 7.2. Add agar. Gently heat and bring to boiling. Autoclave for 20 min at 15 psi pressure–121°C. Cool to 45° to 50°C. Pour into sterile Petri dishes.

Use: For the isolation and cultivation of *Francisella tularensis*.

Fusobacterium Medium

Composition per liter:

Agar	15.0g
Pancreatic digest of casein	15.0g
Glucose	5.0g
NaCl	5.0g
Yeast extract	5.0g
L-Cysteine	0.75g
Crystal Violet	0.01g
Bovine serum	50.0mL
Streptomycin solution	10.0mL

pH 7.2 ± 0.2 at 25°C

Streptomycin Solution:
Composition per 10mL:

Streptomycin	0.01g

Preparation of Streptomycin Solution: Add streptomycin to distilled/deionized water and bring volume to 10.0mL. Mix thoroughly. Filter sterilize.

Preparation of Medium: Add components, except bovine serum and streptomycin solution, to distilled/deionized water and bring volume to 940.0mL. Mix thoroughly. Gently heat and bring to boiling. Autoclave for 15 min at 15 psi pressure–121°C. Cool to 45° to 50°C. Aseptically add 50.0mL of sterile bovine serum and 10.0mL of sterile streptomycin solution. Mix thoroughly. Pour into sterile Petri dishes or distribute into sterile tubes.

Use: For the cultivation of *Fusobacterium* species.

Gardnerella vaginalis Selective Medium

Composition per liter:

Columbia blood agar base	940.0mL
Rabbit or horse serum	50.0mL
Antibiotic inhibitor solution	10.0mL

pH 7.2 ± 0.2 at 25°C

Source: This medium is available as a premixed powder from Oxoid Unipath.

Columbia Blood Agar Base:
Composition per liter:

Special peptone	23.0g
Agar	10.0g
NaCl	5.0g
Starch	1.0g

Source: Columbia blood agar base is available as a premixed powder from Oxoid Unipath.

Preparation of Columbia Blood Agar Base: Add components to distilled/deionized water and bring volume to 1.0L. Mix thoroughly. Gently heat until boiling. Autoclave for 15 min at 15 psi pressure–121°C. Cool to 45° to 50°C.

Antibiotic Inhibitor Solution:
Composition per 10mL:

Nalidixic acid	0.035g
Gentamicin sulFate	4.0mg
Amphotericin B	2.0mg
Ethanol	4.0mL

Preparation of Antibiotic Inhibitor Solution: Add components to distilled/deionized water and bring volume to 10.0mL. Mix thoroughly. Filter sterilize.

Preparation of Medium: To 940.0mL of cooled sterile Columbia blood agar base, aseptically add 50.0mL rabbit or horse blood serum and 10.0mL sterile antibiotic inhibitor solution. Pour into sterile Petri dishes or distribute into sterile tubes.

Use: For the selective isolation, cultivation and differentiation of *Gardnerella vaginalis* from clinical specimens, such as the vaginal discharge of patients with vaginitis. *G. vaginalis* exhibits β hemolysis on this medium.

GBNA Medium (Gum Base Nalidixic Acid Medium)

Composition per liter:

Gellan gum	8.0g
Pancreatic digest of casein	5.7g
NaCl	1.7g
Papaic digest of soybean meal	1.0g
Glucose	0.83g
K_2HPO_4	0.83g
$MgCl_2 \cdot 6H_2O$	0.33g
Nalidixic acid	0.05g

pH 7.2 ± 0.2 at 25°C

Source: This medium is available as a premixed powder from BBL Microbiology Systems.

Preparation of Medium: Add components to tap water and bring volume to 1.0L. Mix thoroughly. Gently heat and bring to boiling. Distribute into tubes or flasks. Autoclave for 15 min at 15 psi pressure–121°C. Pour into sterile Petri dishes or leave in tubes.

Use: For the isolation and cultivation of *Listeria monocytogenes* from clinical specimens and in epidemiological investigations.

GBS Agar Base, Islam (Group B Streptococci Agar) (Islam GBS Agar)

Composition per liter:

Proteose peptone	23.0g
Agar	10.0g

Na_2HPO_4 .... 5.75g
Soluble starch 5.0g
NaH_2PO_4 .... 1.5g
Horse serum, heat inactivated 50.0mL
pH 7.5 ± 0.1 at 25°C

Source: This medium is available as a premixed powder from Oxoid Unipath.

Preparation of Medium: Add components, except horse serum, to distilled/deionized water and bring volume to 950.0mL. Mix thoroughly. Gently heat and bring to boiling. Autoclave for 15 min at 15 psi pressure–121°C. Cool to 45° to 50°C. Aseptically add 50.0mL sterile inactivated horse serum. Mix thoroughly. Pour into sterile Petri dishes or distribute into sterile tubes.

Use: For the isolation and detection of Group B streptococci from clinical specimens. Group B streptococci produce orange pigmented colonies when incubated under anaerobic conditions.

GBS Medium, Rapid (Group B Streptococci Medium)

Starch 80.0g
Proteose peptone 23.0g
Na_2HPO_4 .... 5.75g
NaH_2PO_4 .... 1.5g
Horse serum, inactivated 50.0mL
Antibiotic inhibitor solution .... 10.0 mL
pH 7.5 ± 0.2 at 25°C

Source: This medium is available as a premixed powder from Oxoid Unipath.

Antibiotic Inhibitor Solution:
Composition per 10mL:
Metronidazole 10.0mg
Gentamicin 2.0mg

Preparation of Antibiotic Inhibitor Solution: Add components to distilled/deionized water and bring volume to 10.0mL. Mix thoroughly. Filter sterilize.

Preparation of Medium: Add components, except horse serum and antibiotic inhibitor solution, to distilled/deionized water and bring volume to 940.0mL. Mix thoroughly. Gently heat and bring to boiling. Autoclave for 15 min at 15 psi pressure–121°C. Cool to 45° to 50°C. Aseptically add 50.0mL of sterile heat inactivated horse serum and 10.0mL sterile antibiotic inhibitor solution. Mix thoroughly. Aseptically distribute into sterile tubes or flasks. Cool to 5°C and hold at that temperature for 12 hr prior to use.

Use: For the rapid isolation and cultivation of Group B streptococci from clinical specimens.

GC Agar (ATCC Medium 814)

Composition per 1010mL:
GC agar base, 2× 500.0mL
Hemoglobin solution 500.0mL
Supplement solution 10.0mL
pH 7.2 ± 0.2 at 25°C

GC Agar Base, 2X:
Composition 500mL:
Agar 10.0g
Pancreatic digest of casein 7.5g
Peptic digest of animal tissue 7.5g
NaCl 5.0g
K_2HPO_4 .... 4.0g
Cornstarch 1.0g
KH_2PO_4 .... 1.0g

Source: GC agar base is available as a premixed powder from BBL Microbiology Systems. This base may be replaced by GC medium base available from Difco Laboratories.

Preparation of GC Agar Base, 2X: Add components to distilled/deionized water and bring volume to 500.0mL. Mix thoroughly. Gently heat until boiling. Autoclave for 15 min at 15 psi pressure–121°C. Cool to 45° to 50°C.

Hemoglobin Solution:
Composition 500mL:
Bovine hemoglobin 10.0g

Preparation of Hemoglobin Solution: Add bovine hemoglobin to distilled/deionized water and bring volume to 500.0mL. Mix thoroughly. Autoclave for 15 min at 15 psi pressure–121°C. Cool to 45° to 50°C.

Supplement Solution:
Composition per liter:
Glucose 100.0g
L-Cysteine·HCl 25.9g
L-Glutamine 10.0g
L-Cystine 1.1g
Adenine 1.0g
Nicotinamide adenine dinucleotide 0.25g
Vitamin B_{12} 0.1g
Thiamine pyrophosphate 0.1g
Guanine·HCl 0.03g
$Fe(NO_3)_3 \cdot 6H_2O$ 0.02g
p-Aminobenzoic acid 0.013g
Thiamine·HCl 3.0mg

Source: The supplement solution IsoVitaleX® enrichment is available from BBL Microbiology Laboratories. This enrichment may be replaced by supplement VX from Difco Laboratories.

Preparation of Supplement Solution: Add components to distilled/deionized water and bring volume to 1.0L. Mix thoroughly. Filter sterilize.

Preparation of Medium: To 500.0mL of sterile GC agar base aseptically add 500.0mL of sterile hemoglobin solution at 45° to 50°C. Mix thoroughly. Aseptically add 10.0mL sterile supplement solution. Mix thoroughly. Pour into sterile Petri dishes or distribute into sterile tubes.

Use: For the isolation and cultivation of fastidious bacteria, especially *Neisseria* and *Haemophilus* species. For the cultivation and maintenance of *Moraxella catarrhalis, Campylocbacter pylori, Eikenella corrodens, Helicobacter pylori, Moraxella nonliquefaciens, Morococcus cerebrosis, Oligella ureolytica. O. urethralis, Pasteurella volantium, Proteus mirabilis,* and *Taylorella equigenitalis.*

GC Agar (GC Medium) (ATCC Medium 1351)

Composition per 1.0L:

GC agar base 950.0mL
Blood, defibrinated 50.0mL

pH 7.2 ± 0.2 at 25°C

GC Agar Base:
Composition per liter:

Agar 10.0g
Pancreatic digest of casein 7.5g
Peptic digest of animal tissue 7.5g
NaCl 5.0g
K_2HPO_4 .... 4.0g
Cornstarch 1.0g
KH_2PO_4 .... 1.0g

Source: GC agar base is available as a premixed powder from BBL Microbiology Systems. This base may be replaced by GC medium base available from Difco Laboratories.

Preparation of GC Agar Base: Add components to distilled/deionized water and bring volume to 1.0L. Mix thoroughly. Gently heat until boiling. Autoclave for 15 min at 15 psi pressure–121°C. Cool to 75° to 80°C.

Preparation of Medium: To 950.0mL of sterile GC agar base aseptically add 50.0mL sterile defibrinated blood with thorough mixing and maintain at 75° to 80°C for 15–20 min until the medium is chocolatized. Pour into sterile Petri dishes or distribute into sterile tubes.

Use: For the isolation and cultivation of fastidious bacteria, especially *Neisseria* and *Haemophilus* species. For the cultivation and maintenance of *Moraxella catarrhalis, Campylocbacter pylori, Eikenella corrodens, Helicobacter pylori, Moraxella nonliquefaciens, Morococcus cerebrosis, Oligella ureolytica. O. urethralis, Pasteurella volantium, Proteus mirabilis,* and *Taylorella equigenitalis.*

GC Agar with Ampicillin

Composition per 1020mL:

GC agar base, 2x 500.0mL
Hemoglobin solution 500.0mL
Supplement solution 10.0mL
Ampicillin solution 10.0mL

pH 7.2 ± 0.2 at 25°C

GC Agar Base, 2X:
Composition 500mL:

Agar 10.0g
Pancreatic digest of casein 7.5g
Peptic digest of animal tissue 7.5g
NaCl 5.0g
K_2HPO_4 .... 4.0g
Cornstarch 1.0g
KH_2PO_4 .... 1.0g

Source: GC agar base is available as a premixed powder from BBL Microbiology Systems. This base may be replaced by GC medium base available from Difco Laboratories.

Preparation of GC Agar Base, 2X: Add components to distilled/deionized water and bring volume to 500.0mL. Mix thoroughly. Gently heat until boiling. Autoclave for 15 min at 15 psi pressure–121°C. Cool to 45° to 50°C.

Hemoglobin Solution:
Composition 500mL:

Bovine hemoglobin 10.0g

Preparation of Hemoglobin Solution: Add bovine hemoglobin to distilled/deionized water and bring volume to 500.0mL. Mix thoroughly. Autoclave for 15 min at 15 psi pressure–121°C. Cool to 45° to 50°C.

Supplement Solution:
Composition per liter:

Glucose 100.0g
L-Cysteine·HCl 25.9g
L-Glutamine 10.0g
L-Cystine 1.1g
Adenine 1.0g
Nicotinamide adenine dinucleotide 0.25g
Vitamin B_{12} 0.1g
Thiamine pyrophosphate 0.1g
Guanine·HCl 0.03g
$Fe(NO_3)_3 \cdot 6H_2O$ 0.02g
p-Aminobenzoic acid 0.013g
Thiamine·HCl 3.0mg

Source: The supplement solution IsoVitaleX® enrichment is available from BBL Microbiology Laboratories. This enrichment may be replaced by supplement VX from Difco Laboratories.

Preparation of Supplement Solution: Add components to distilled/deionized water and bring volume to 1.0L. Mix thoroughly. Filter sterilize.

Ampicillin Solution:
Composition per 10mL:

Ampicillin	0.02g

Preparation of Ampicillin Solution: Add ampicillin to distilled/deionized water and bring volume to 10.0mL. Mix thoroughly. Filter sterilize.

Preparation of Medium: To 500.0mL of sterile GC agar base aseptically add 500.0mL of sterile hemoglobin solution at 45° to 50°C. Mix thoroughly. Aseptically add 10.0mL of sterile supplement solution and 10.0mL of sterile ampicillin solution. Mix thoroughly. Pour into sterile Petri dishes or distribute into sterile tubes.

Use: For the cultivation and maintenance of ampicillin-resistant strains of *Moraxella catarrhalis, Haemophilus influenzae,* and *H. parainfluenzae.*

GC Agar with Ampicillin and Gentamicin

Composition per 1030mL:

GC agar base, 2×	500.0mL
Hemoglobin solution	500.0mL
Supplement solution	10.0mL
Ampicillin solution	10.0mL
Gentamicin solution	10.0mL

pH 7.2 ± 0.2 at 25°C

GC Agar Base, 2X:
Composition 500mL:

Agar	10.0g
Pancreatic digest of casein	7.5g
Peptic digest of animal tissue	7.5g
NaCl	5.0g
K_2HPO_4	4.0g
Cornstarch	1.0g
KH_2PO_4	1.0g

Source: GC agar base is available as a premixed powder from BBL Microbiology Systems. This base may be replaced by GC medium base available from Difco Laboratories.

Preparation of GC Agar Base, 2X: Add components to distilled/deionized water and bring volume to 500.0mL. Mix thoroughly. Gently heat until boiling. Autoclave for 15 min at 15 psi pressure–121°C. Cool to 45° to 50°C.

Hemoglobin Solution:
Composition 500mL:

Bovine hemoglobin	10.0g

Preparation of Hemoglobin Solution: Add bovine hemoglobin to distilled/deionized water and bring volume to 500.0mL. Mix thoroughly. Autoclave for 15 min at 15 psi pressure–121°C. Cool to 45° to 50°C.

Supplement Solution:
Composition per liter:

Glucose	100.0g
L-Cysteine·HCl	25.9g
L-Glutamine	10.0g
L-Cystine	1.1g
Adenine	1.0g
Nicotinamide adenine dinucleotide	0.25g
Vitamin B_{12}	0.1g
Thiamine pyrophosphate	0.1g
Guanine·HCl	0.03g
$Fe(NO_3)_3 \cdot 6H_2O$	0.02g
p-Aminobenzoic acid	0.013g
Thiamine·HCl	3.0mg

Source: The supplement solution IsoVitaleX® enrichment is available from BBL Microbiology Laboratories. This enrichment may be replaced by supplement VX from Difco Laboratories.

Preparation of Supplement Solution: Add components to distilled/deionized water and bring volume to 1.0L. Mix thoroughly. Filter sterilize.

Ampicillin Solution:
Composition per 10mL:

Ampicillin	0.01g

Preparation of Ampicillin Solution: Add ampicillin to distilled/deionized water and bring volume to 10.0mL. Mix thoroughly. Filter sterilize.

Gentamicin Solution:
Composition per 10mL:

Gentamicin	2.0mg

Preparation of Gentamicin Solution: Add gentamicin to distilled/deionized water and bring volume to 10.0mL. Mix thoroughly. Filter sterilize.

Preparation of Medium: To 500.0mL of sterile GC agar base aseptically add 500.0mL of sterile hemoglobin solution at 45° to 50°C. Mix thoroughly. Aseptically add 10.0mL sterile supplement solution, 10.0mL sterile ampicillin solution, and 10.0mL sterile gentamicin solution. Mix thoroughly. Pour into sterile Petri dishes or distribute into sterile tubes.

Use: For the cultivation and maintenance of *Haemophilus parainfluenzae.*

GC Agar with Ampicillin and Tetracycline

Composition per 1030mL:

GC agar base, 2×....500.0mL
Hemoglobin solution....500.0mL
Supplement solution....10.0mL
Ampicillin solution....10.0mL
Tetracycline solution....10.0mL

pH 7.2 ± 0.2 at 25°C

GC Agar Base, 2X:

Composition 500mL:

Agar....10.0g
Pancreatic digest of casein....7.5g
Peptic digest of animal tissue....7.5g
NaCl....5.0g
K_2HPO_4....4.0g
Cornstarch....1.0g
KH_2PO_4....1.0g

Source: GC agar base is available as a premixed powder from BBL Microbiology Systems. This base may be replaced by GC medium base available from Difco Laboratories.

Preparation of GC Agar Base, 2X: Add components to distilled/deionized water and bring volume to 500.0mL. Mix thoroughly. Gently heat until boiling. Autoclave for 15 min at 15 psi pressure–121°C. Cool to 45° to 50°C.

Hemoglobin Solution:

Composition 500mL:

Bovine hemoglobin....10.0g

Preparation of Hemoglobin Solution: Add hemoglobin to distilled/deionized water and bring volume to 500.0mL. Mix thoroughly. Autoclave for 15 min at 15 psi pressure–121°C. Cool to 45° to 50°C.

Supplement Solution:

Composition per liter:

Glucose....100.0g
L-Cysteine·HCl....25.9g
L-Glutamine....10.0g
L-Cystine....1.1g
Adenine....1.0g
Nicotinamide adenine dinucleotide....0.25g
Vitamin B_{12}....0.1g
Thiamine pyrophosphate....0.1g
Guanine·HCl....0.03g
$Fe(NO_3)_3 \cdot 6H_2O$....0.02g
p-Aminobenzoic acid....0.013g
Thiamine·HCl....3.0mg

Source: The supplement solution IsoVitaleX® enrichment is available from BBL Microbiology Laboratories. This enrichment may be replaced by supplement VX from Difco Laboratories.

Preparation of Supplement Solution: Add components to distilled/deionized water and bring volume to 1.0L. Mix thoroughly. Filter sterilize.

Ampicillin Solution:

Composition per 10mL:

Ampicillin....0.01g

Preparation of Ampicillin Solution: Add ampicillin to distilled/deionized water and bring volume to 10.0mL. Mix thoroughly. Filter sterilize.

Tetracycline Solution:

Composition per 10mL:

Tetracycline....5.0mg

Preparation of Tetracycline Solution: Add tetracycline to distilled/deionized water and bring volume to 10.0mL. Mix thoroughly. Filter sterilize.

Preparation of Medium: To 500.0mL of sterile GC agar base aseptically add 500.0mL of sterile hemoglobin solution at 45° to 50°C. Mix thoroughly. Aseptically add 10.0mL sterile supplement solution, 10.0mL sterile ampicillin solution, and 10.0mL sterile tetracycline solution. Mix thoroughly. Pour into sterile Petri dishes or distribute into sterile tubes.

Use: For the cultivation and maintenance of *Haemophilus parainfluenzae.*

GC Agar with Chloramphenicol, Tetracycline, and Ampicillin

Composition per 1040mL:

GC agar base, 2×....500.0mL
Hemoglobin solution....500.0mL
Supplement solution....10.0mL
Ampicillin solution....10.0mL
Tetracycline solution....10.0mL
Chloramphenicol solution....10.0mL

pH 7.2 ± 0.2 at 25°C

GC Agar Base, 2X:

Composition per 500mL:

Agar....10.0g
Pancreatic digest of casein....7.5g
Peptic digest of animal tissue....7.5g
NaCl....5.0g
K_2HPO_4....4.0g
Cornstarch....1.0g
KH_2PO_4....1.0g

Source: GC agar base is available as a premixed powder from BBL Microbiology Systems. This base may be replaced by GC medium base available from Difco Laboratories.

Preparation of GC Agar Base, 2X: Add components to distilled/deionized water and bring volume to 500.0mL. Mix thoroughly. Gently heat until boiling. Autoclave for 15 min at 15 psi pressure–121°C. Cool to 45° to 50°C.

Hemoglobin Solution:
Composition per500mL:

Bovine hemoglobin 10.0g

Preparation of Hemoglobin Solution: Add bovine hemoglobin to distilled/deionized water and bring volume to 500.0mL. Mix thoroughly. Autoclave for 15 min at 15 psi pressure–121°C. Cool to 45° to 50°C.

Supplement Solution:
Composition per liter:

Glucose 100.0g
L-Cysteine·HCl 25.9g
L-Glutamine 10.0g
L-Cystine 1.1g
Adenine 1.0g
Nicotinamide adenine dinucleotide 0.25g
Vitamin B_{12} 0.1g
Thiamine pyrophosphate 0.1g
Guanine·HCl 0.03g
$Fe(NO_3)_3 \cdot 6H_2O$ 0.02g
p-Aminobenzoic acid 0.013g
Thiamine·HCl 3.0mg

Source: The supplement solution IsoVitaleX® enrichment is available from BBL Microbiology Laboratories. This enrichment may be replaced by supplement VX from Difco Laboratories.

Preparation of Supplement Solution: Add components to distilled/deionized water and bring volume to 1.0L. Mix thoroughly. Filter sterilize.

Ampicillin Solution:
Composition per 10mL:

Ampicillin 0.01g

Preparation of Ampicillin Solution: Add ampicillin to distilled/deionized water and bring volume to 10.0mL. Mix thoroughly. Filter sterilize.

Tetracycline Solution:
Composition per 10mL:

Tetracycline 5.0mg

Preparation of Tetracycline Solution: Add tetracycline to distilled/deionized water and bring volume to 10.0mL. Mix thoroughly. Filter sterilize.

Chloramphenicol Solution:
Composition per 10mL:

Chloramphenicol 5.0mg

Preparation of Chloramphenicol Solution: Add chloramphenicol to distilled/deionized water and bring volume to 10.0mL. Mix thoroughly. Filter sterilize.

Preparation of Medium: To 500.0mL of sterile GC agar base aseptically add 500.0mL of sterile hemoglobin solution at 45° to 50°C. Mix thoroughly. Aseptically add 10.0mL sterile supplement solution, 10.0mL sterile ampicillin solution, 10.0mL sterile chloramphenicol, and 10.0mL sterile tetracycline solution. Mix thoroughly. Pour into sterile Petri dishes or distribute into sterile tubes.

Use: For the cultivation and maintenance of *Haemophilus parainfluenzae.*

GC Agar with Defined Supplements

GC agar base 990.0mL
Defined supplements solution 10.0mL

pH 7.2 ± 0.2 at 25°C

GC Agar Base:
Composition per liter:

Agar 10.0g
Pancreatic digest of casein 7.5g
Peptic digest of animal tissue 7.5g
NaCl 5.0g
K_2HPO_4 .... 4.0g
Cornstarch 1.0g
KH_2PO_4 .... 1.0g

Source: GC agar base is available as a premixed powder from BBL Microbiology Systems. This base may be replaced by GC medium base available from Difco Laboratories.

Preparation of GC Agar Base: Add components to distilled/deionized water and bring volume to 1.0L. Mix thoroughly. Gently heat until boiling. Autoclave for 15 min at 15 psi pressure–121°C. Cool to 45° to 50°C.

Defined Supplements Solution:
Composition per 100mL:

Glucose 40.0g
Glutamine 1.0g
$Fe(NO_3)_3 \cdot 6H_2O$ 0.05g
Cocarboxylase 2.0mg

Preparation of Defined Supplements Solution: Add components to distilled/deionized water and bring volume to 100.0mL. Mix thoroughly. Filter sterilize.

Preparation of Medium: To 990.0mL of sterile GC agar base aseptically add 10.0mL sterile defined supplements solution. Mix thoroughly. Pour into sterile Petri dishes or distribute into sterile tubes.

Use: For the cultivation and maintenance of *Neisseria gonorrhoeae.*

GC Agar with Penicillin G

Composition per 1020mL:

GC medium base, 2× 500.0mL
Hemoglobin solution 500.0mL
Supplement solution 10.0mL
Penicillin G solution 10.0mL
pH 7.2 ± 0.2 at 25°C

GC Medium Base, 2X:
Composition 500mL:

Proteose peptone No. 3 .... 15.0g
Agar 10.0g
NaCl 5.0g
K_2HPO_4 .... 4.0g
Cornstarch 1.0g
KH_2PO_4 .... 1.0g

Source: GC medium base is available as a premixed powder from Difco Laboratories.

Preparation of GC Medium Base, 2X: Add components to distilled/deionized water and bring volume to 500.0mL. Mix thoroughly. Gently heat until boiling. Autoclave for 15 min at 15 psi pressure–121°C. Cool to 45° to 50°C.

Hemoglobin Solution:
Composition 500mL:

Bovine hemoglobin 10.0g

Preparation of Hemoglobin Solution: Add bovine hemoglobin to distilled/deionized water and bring volume to 500.0mL. Mix thoroughly. Autoclave for 15 min at 15 psi pressure–121°C. Cool to 45° to 50°C.

Supplement Solution:
Composition per liter:

Glucose 100.0g
L-Cysteine·HCl 25.9g
L-Glutamine 10.0g
L-Cystine 1.1g
Adenine 1.0g
Nicotinamide adenine dinucleotide 0.25g
Vitamin B_{12} 0.1g
Thiamine pyrophosphate 0.1g
Guanine·HCl 0.03g
$Fe(NO_3)_3 \cdot 6H_2O$ 0.02g
p-Aminobenzoic acid 0.013g
Thiamine·HCl 3.0mg

Source: The supplement solution IsoVitaleX® enrichment is available from BBL Microbiology Laboratories. This enrichment may be replaced by supplement VX from Difco Laboratories.

Preparation of Supplement Solution: Add components to distilled/deionized water and bring volume to 1.0L. Mix thoroughly. Filter sterilize.

Penicillin G Solution:
Composition per 10mL:

Penicillin G 0.05g

Preparation of Penicillin G Solution: Add penicillin G to distilled/deionized water and bring volume to 10.0mL. Mix thoroughly. Filter sterilize.

Preparation of Medium: To 500.0mL of sterile GC medium base aseptically add 500.0mL of sterile hemoglobin solution at 45° to 50°C. Mix thoroughly. Aseptically add 10.0mL of sterile supplement solutionand 10.0mL of sterile penicillin G solution. Mix thoroughly. Pour into sterile Petri dishes or distribute into sterile tubes.

Use: For the cultivation and maintenance of penicillin-resistant strains of *Neisseria gonorrhoeae*.

GC Agar with Supplement A

Composition per 1020mL:

GC medium base, 2× 500.0mL
Hemoglobin solution 500.0mL
Supplement solution 10.0mL
Supplement A, Difco 10.0mL
pH 7.2 ± 0.2 at 25°C

GC Medium Base, 2X:
Composition 500mL:

Proteose peptone No. 3 .... 15.0g
Agar 10.0g
NaCl 5.0g
K_2HPO_4 .... 4.0g
Cornstarch 1.0g
KH_2PO_4 .... 1.0g

Source: GC medium base is available as a premixed powder from Difco Laboratories.

Preparation of GC Medium Base, 2X: Add components to distilled/deionized water and bring volume to 500.0mL. Mix thoroughly. Gently heat until boiling. Autoclave for 15 min at 15 psi pressure–121°C. Cool to 45° to 50°C.

Hemoglobin Solution:
Composition 500mL:

Bovine hemoglobin 10.0g

Preparation of Hemoglobin Solution: Add bovine hemoglobin to distilled/deionized water and bring volume to 500.0mL. Mix thoroughly. Autoclave for 15 min at 15 psi pressure–121°C. Cool to 45° to 50°C.

Supplement Solution:
Composition per liter:

Glucose 100.0g
L-Cysteine·HCl 25.9g
L-Glutamine 10.0g

L-Cystine 1.1g
Adenine 1.0g
Nicotinamide adenine dinucleotide 0.25g
Vitamin B_{12} 0.1g
Thiamine pyrophosphate 0.1g
Guanine·HCl 0.03g
$Fe(NO_3)_3 \cdot 6H_2O$ 0.02g
p-Aminobenzoic acid 0.013g
Thiamine·HCl 3.0mg

Source: The supplement solution IsoVitaleX® enrichment is available from BBL Microbiology Laboratories. This enrichment may be replaced by supplement VX from Difco Laboratories.

Preparation of Supplement Solution: Add components to distilled/deionized water and bring volume to 1.0L. Mix thoroughly. Filter sterilize.

Supplement A:
Composition per 10mL:
Supplement A contains yeast concentrate with Crystal Violet.

Preparation of Supplement A: Add components to distilled/deionized water and bring volume to 10.0mL. Mix thoroughly. Filter sterilize.

Preparation of Medium: To 500.0mL of sterile GC medium base aseptically add 500.0mL of sterile hemoglobin solution at 45° to 50°C. Mix thoroughly. Aseptically add 10.0mL of sterile supplement solution and 10.0mL of sterile supplement A solution. Mix thoroughly. Pour into sterile Petri dishes or distribute into sterile tubes.

Use: For the cultivation of *Neisseria gonorrhoeae*, other *Neisseria* species and *Haemophilus* species.

GC Agar with Supplement A and with VCN Inhibitor

Composition per 1030mL:
GC medium base, 2× 500.0mL
Hemoglobin solution 500.0mL
Supplement solution 10.0mL
Supplement A, Difco 10.0mL
VCN inhibitor 10.0mL
pH 7.2 ± 0.2 at 25°C

GC Medium Base, 2X:
Composition 500mL:
Proteose peptone No. 3 .... 15.0g
Agar 10.0g
NaCl 5.0g
K_2HPO_4 .... 4.0g
Cornstarch 1.0g
KH_2PO_4 .... 1.0g

Source: GC medium base is available as a premixed powder from Difco Laboraotries.

Preparation of GC Medium Base, 2X: Add components to distilled/deionized water and bring volume to 500.0mL. Mix thoroughly. Gently heat until boiling. Autoclave for 15 min at 15 psi pressure–121°C. Cool to 45° to 50°C.

Hemoglobin Solution:
Composition 500mL:
Bovine hemoglobin 10.0g

Preparation of Hemoglobin Solution: Add bovine hemoglobin to distilled/deionized water and bring volume to 500.0mL. Mix thoroughly. Autoclave for 15 min at 15 psi pressure–121°C. Cool to 45° to 50°C.

Supplement Solution:
Composition per liter:
Glucose 100.0g
L-Cysteine·HCl 25.9g
L-Glutamine 10.0g
L-Cystine 1.1g
Adenine 1.0g
Nicotinamide adenine dinucleotide 0.25g
Vitamin B_{12} 0.1g
Thiamine pyrophosphate 0.1g
Guanine·HCl 0.03g
$Fe(NO_3)_3 \cdot 6H_2O$ 0.02g
p-Aminobenzoic acid 0.013g
Thiamine·HCl 3.0mg

Source: The supplement solution IsoVitaleX® enrichment is available from BBL Microbiology Laboratories. This enrichment may be replaced by supplement VX from Difco Laboratories.

Preparation of Supplement Solution: Add components to distilled/deionized water and bring volume to 1.0L. Mix thoroughly. Filter sterilize.

Supplement A:
Composition per 10mL:
Supplement A contains yeast concentrate with Crystal Violet.

Preparation of Supplement A: Add components to distilled/deionized water and bring volume to 10.0mL. Mix thoroughly. Filter sterilize.

VCN Inhibitor:
Composition per 10mL:
Colistin 7.5mg
Vancomycin 3.0mg
Nystatin 12,500U

Preparation of VCN Inhibitor: Add components to distilled/deionized water and bring volume to 10.0mL. Mix thoroughly. Filter sterilize.

Preparation of Medium: To 500.0mL of sterile GC medium base aseptically add 500.0mL of sterile hemoglobin solution at 45° to 50°C. Mix thoroughly. Aseptically add 10.0mL of sterile supplement solution, 10.0mL of sterile supplement A solution and 10.0mL sterile VCN Inhibitor. Mix thoroughly. Pour into sterile Petri dishes or distribute into sterile tubes.

Use: For the cultivation of *Neisseria gonorrhoeae*, other *Neisseria* species and *Haemophilus* species.

GC Agar with Supplement A and with VCTN Inhibitor

Composition per 1030mL:

GC medium base, 2×....................500.0mL
Hemoglobin solution....................500.0mL
Supplement solution....................10.0mL
Supplement A, Difco....................10.0mL
VCTN inhibitor....................10.0mL

pH 7.2 ± 0.2 at 25°C

GC Medium Base, 2X:
Composition 500mL:

Proteose peptone No. 3....................15.0g
Agar....................10.0g
NaCl....................5.0g
K_2HPO_4....................4.0g
Cornstarch....................1.0g
KH_2PO_4....................1.0g

Source: GC medium base is available as a premixed powder from Difco Laboraotries.

Preparation of GC Medium Base, 2X: Add components to distilled/deionized water and bring volume to 500.0mL. Mix thoroughly. Gently heat until boiling. Autoclave for 15 min at 15 psi pressure–121°C. Cool to 45° to 50°C.

Hemoglobin Solution:
Composition 500mL:

Bovine hemoglobin....................10.0g

Preparation of Hemoglobin Solution: Add bovine hemoglobin to distilled/deionized water and bring volume to 500.0mL. Mix thoroughly. Autoclave for 15 min at 15 psi pressure–121°C. Cool to 45° to 50°C.

Supplement Solution:
Composition per liter:

Glucose....................100.0g
L-Cysteine·HCl....................25.9g
L-Glutamine....................10.0g
L-Cystine....................1.1g
Adenine....................1.0g
Nicotinamide adenine dinucleotide....................0.25g
Vitamin B_{12}....................0.1g
Thiamine pyrophosphate....................0.1g
Guanine·HCl....................0.03g
$Fe(NO_3)_3 \cdot 6H_2O$....................0.02g
p-Aminobenzoic acid....................0.013g
Thiamine·HCl....................3.0mg

Source: The supplement solution IsoVitaleX® enrichment is available from BBL Microbiology Laboratories. This enrichment may be replaced by supplement VX from Difco Laboratories.

Preparation of Supplement Solution: Add components to distilled/deionized water and bring volume to 1.0L. Mix thoroughly. Filter sterilize.

Supplement A:
Composition per 10mL:

Supplement A contains yeast concentrate with Crystal Violet.

Preparation of Supplement A: Add components to distilled/deionized water and bring volume to 10.0mL. Mix thoroughly. Filter sterilize.

VCTN Inhibitor:
Composition per liter:

Vancomycin....................4.0mg
Colistin....................7.5mg
Trimethoprim lactate....................5.0mg
Nystatin....................12,500U

Preparation of VCTN Inhibitor: Add components to distilled/deionized water and bring volume to 10.0mL. Mix thoroughly. Filter sterilize.

Preparation of Medium: To 500.0mL of sterile GC medium base aseptically add 500.0mL of sterile hemoglobin solution at 45° to 50°C. Mix thoroughly. Aseptically add 10.0mL of sterile supplement solution, 10.0mL of sterile supplement A, and 10.0mL sterile VCTN inhibitor. Mix thoroughly. Pour into sterile Petri dishes or distribute into sterile tubes.

Use: For the cultivation of *Neisseria gonorrhoeae*, other *Neisseria* species and *Haemophilus* species.

GC Agar with Supplement B

GC agar base....................990.0mL
Supplement B....................10.0mL

pH 7.2 ± 0.2 at 25°C

GC Agar Base:
Composition per liter:

Agar....................10.0g
Pancreatic digest of casein....................7.5g
Peptic digest of animal tissue....................7.5g
NaCl....................5.0g

K_2HPO_4 .... 4.0g
Cornstarch 1.0g
KH_2PO_4 .... 1.0g

Source: GC agar base is available as a premixed powder from BBL Microbiology Systems. This base may be replaced by GC medium base available from Difco Laboratories.

Preparation of GC Agar Base: Add components to distilled/deionized water and bring volume to 1.0L. Mix thoroughly. Gently heat until boiling. Autoclave for 15 min at 15 psi pressure–121°C. Cool to 45° to 50°C.

Supplement B:
Composition per 10mL:
Supplement B contains yeast concentrate, glutamine, coenzyme, cocarboxylase, hematin and growth factors.

Preparation of Supplement B: Add components to distilled/deionized water and bring volume to 10.0mL. Mix thoroughly. Filter sterilize.

Preparation of Medium: To 990.0mL of sterile GC agar base aseptically add 10.0mL sterile defined supplement B. Mix thoroughly. Pour into sterile Petri dishes or distribute into sterile tubes.

Use: For the cultivation and maintenance of *Neisseria gonorrhoeae*.

GC Medium, New York City Formulation

Composition per liter:
GC agar base 850.0mL
Horse blood, lysed 100.0mL
Yeast autolysate supplement 30.0mL
LCAT antibiotic solution 20.0mL
pH 7.3 ± 0.2 at 25°C

GC Agar Base:
Composition per 850mL:
Special peptone 15.0g
Agar 10.0g
NaCl 5.0g
K_2HPO_4 .... 4.0g
Cornstarch 1.0g
KH_2PO_4 .... 1.0g
pH 7.2 ± 0.2 at 25°C

Preparation of GC Agar Base: Add components of GC medium base and the hemoglobin to distilled/deionized water and bring volume to 850.0mL. Mix thoroughly. Gently heat until boiling. Autoclave for 15 min at 15 psi pressure–121°C. Cool to 45° to 50°C.

Horse Blood, Lysed:
Composition per 100mL:
Saponin 0.5g
Horse blood, defibrinated 100.0mL

Preparation of Horse Blood, Lysed: Add saponin to defibrinated horse blood. Mix thoroughly. Allow blood to lyse.

Yeast Autolysate Supplement:
Composition per 30.0mL:
Yeast autolysate 10.0g
Glucose 1.0g
$NaHCO_3$ 0.15g

Preparation of Yeast Autolysate Supplement: Add components to distilled/deionized water and bring volume to 30.0mL. Mix thoroughly. Filter sterilize.

LCAT Antibiotic Solution:
Composition per 20mL:
Colistin 6.0mg
Trimethoprim lactate 5.0mg
Lincomycin 1.0mg
Amphotericin B 1.0mg

Preparation of LCAT Antibiotic Solution: Add components to distilled/deionized water and bring volume to 20.0mL. Mix thoroughly. Filter sterilize.

Preparation of Medium: To 850.0mL of cooled sterile GC agar base, aseptically add 100.0mL of sterile lysed horse blood, 30.0mL of sterile yeast autolysate supplement, and 20.0mL of LCAT antibiotic solution. Mix thoroughly. Pour into sterile Petri dishes or distribute into sterile tubes.

Use: For the selective isolation and cultivation of fastidious microorganisms, especially *Neisseria* species.

GCII Agar

Composition per liter:
GCII agar base, 2× 490.0mL
Hemoglobin solution 490.0mL
Supplement solution 10.0mL
pH 7.2 ± 0.2 at 25°C

GCII Agar Base, 2X:
Composition per liter:
Agar 10.0g
Pancreatic digest of casein 7.5g
Selected meat peptone 7.5g
NaCl 5.0g
K_2HPO_4 .... 4.0g
Cornstarch 1.0g
KH_2PO_4 .... 1.0g

Source: GCII agar base is available as a premixed powder from BBL Microbiology Systems.

Preparation of GCII Agar Base, 2X: Add components to distilled/deionized water and bring volume to 500.0mL. Mix thoroughly. Gently heat until boiling. Autoclave for 15 min at 15 psi pressure–121°C. Cool to 45° to 50°C.

Hemoglobin Solution:
Composition 500mL:
Hemoglobin..10.0g

Preparation of Hemoglobin Solution: Add hemoglobin to distilled/deionized water and bring volume to 500.0mL. Mix thoroughly. Autoclave for 15 min at 15 psi pressure–121°C. Cool to 45° to 50°C.

Supplement Solution:
Composition per liter:
Glucose 100.0g
L-Cysteine·HCl 25.9g
L-Glutamine 10.0g
L-Cystine 1.1g
Adenine 1.0g
Nicotinamide adenine dinucleotide..................0.25g
Vitamin B_{12} 0.1g
Thiamine pyrophosphate....................................0.1g
Guanine·HCl 0.03g
$Fe(NO_3)_3 \cdot 6H_2O$ 0.02g
p-Aminobenzoic acid 0.013g
Thiamine·HCl... 3.0mg

Preparation of Supplement Solution: Add components to distilled/deionized water and bring volume to 1.0L. Mix thoroughly. Filter sterilize.

Preparation of Medium: To 490.0mL of sterile GC II agar base aseptically add 490.0mL of sterile hemoglobin solution at 45° to 50°C. Mix thoroughly. Aseptically add 10.0mL sterile supplement solution. Mix thoroughly. Pour into sterile Petri dishes or distribute into sterile tubes.

Use: For the isolation and cultivation of fastidious microorganisms, especially *Neisseria* and *Haemophilus* species, from clinical specimens.

GCII Agar

Composition per liter:
GCII agar base 950.0mL
Blood, defibrinated.......................................50.0mL
pH 7.2 ± 0.2 at 25°C

GCII Agar Base with Extra Agar:
Composition per liter:
Agar..12.0g
Pancreatic digest of casein 7.5g
Selected meat peptone.......................................7.5g
NaCl 5.0g
K_2HPO_4 ....4.0g
Cornstarch 1.0g
KH_2PO_4 ....1.0g

Source: GCII agar base is available as a premixed powder from BBL Microbiology Systems.

Preparation of GCII Agar Base Extra Agar: Add components to distilled/deionized water and bring volume to 1.0L. Mix thoroughly. Gently heat until boiling. Autoclave for 15 min at 15 psi pressure–121°C. Cool to 45° to 50°C.

Preparation of Medium: To 950.0mL of sterile GC agar base aseptically add 50.0mL sterile defibrinated blood with thorough mixing and maintain at 75° to 80°C for 15–20 min until the medium is chocolatized. Pour into sterile Petri dishes or distribute into sterile tubes.

Use: For the isolation and cultivation of fastidious microorganisms, especially *Neisseria* and *Haemophilus* species, from clinical specimens.

GCA Agar with Thiamine

Composition per liter:
Glucose 25.0g
Agar...14.0g
Papaic digest of soybean meal 10.0g
NaCl 5.0g
Pancreatic digest of heart muscle.......................3.0g
$Cysteine \cdot HCl \cdot H_2O$ 1.0g
Thiamine 0.05mg
Rabbit blood, defibrinated...........................50.0mL
pH 6.8 ± 0.2 at 25°C

Preparation of Medium: Add components, except rabbit blood, to distilled/deionized water and bring volume to 950.0mL. Mix thoroughly. Gently heat and bring to boiling. Autoclave for 15 min at 15 psi pressure–121°C. Cool to 45° to 50°C. Aseptically add sterile rabbit blood. Mix thoroughly. Pour into sterile Petri dishes or distribute into sterile tubes.

Use: For the isolation and cultivation of *Francisella tularensis*.

GC–Lect™ Agar

Composition per liter:
GCII agar base, 2× 500.0mL
Hemoglobin solution...................................500.0mL
Supplement solution.....................................10.0mL
Selective agent solution 10.0mL
pH 7.2 ± 0.2 at 25°C

Source: This medium is available as a prepared medium from BBL Microbiology Systems.

GCII Agar Base, 2X with Extra Agar:
Composition per liter:
Agar..12.0g
Pancreatic digest of casein 7.5g
Selected meat peptone.......................................7.5g
NaCl 5.0g

K_2HPO_4 .... 4.0g
Cornstarch 1.0g
KH_2PO_4 .... 1.0g

Source: GCII agar base is available as a premixed powder from BBL Microbiology Systems.

Preparation of GCII Agar Base, 2X with Extra Agar: Add components to distilled/deionized water and bring volume to 500.0mL. Mix thoroughly. Gently heat until boiling. Autoclave for 15 min at 15 psi pressure–121°C. Cool to 45° to 50°C.

Hemoglobin Solution:
Composition 500mL:
Hemoglobin 10.0g

Preparation of Hemoglobin Solution: Add hemoglobin to distilled/deionized water and bring volume to 500.0mL. Mix thoroughly. Autoclave for 15 min at 15 psi pressure–121°C. Cool to 45° to 50°C.

Supplement Solution:
Composition per liter:
Glucose 100.0g
L-Cysteine·HCl 25.9g
L-Glutamine 10.0g
L-Cystine 1.1g
Adenine 1.0g
Nicotinamide adenine dinucleotide 0.25g
Vitamin B_{12} 0.1g
Thiamine pyrophosphate 0.1g
Guanine·HCl 0.03g
$Fe(NO_3)_3 \cdot 6H_2O$ 0.02g
p-Aminobenzoic acid 0.013g
Thiamine·HCl 3.0mg

Source: The supplement solution IsoVitaleX® enrichment is available from BBL Microbiology Laboratories. This enrichment may be replaced by supplement VX from Difco Laboratories.

Preparation of Supplement Solution: Add components to distilled/deionized water and bring volume to 1.0L. Mix thoroughly. Filter sterilize.

Selective Agents:
Composition per 10mL:
Selective agents 0.017g

Preparation of Selective Agents: Add components to distilled/deionized water and bring volume to 10.0mL. Mix thoroughly. Filter sterilize.

Preparation of Medium: To 500.0mL of sterile GC II agar base aseptically add 500.0mL of sterile hemoglobin solution at 45° to 50°C. Mix thoroughly. Aseptically add 10.0mL sterile supplement solution and 10.0mL selective agents solution. Mix thoroughly. Pour into sterile Petri dishes or distribute into sterile tubes.

Use: For the isolation and cultivation of *Neisseria gonorrhoeae* from clinical specimens.

Gelatin Agar (GA Medium)

Composition per liter:
Solution 1 .... 950.0mL
Solution 2 .... 50.0mL
pH 7.2 ± 0.2 at 25°C

Solution 1:
Composition per 950mL:
Gelatin 30.0g
Agar 15.0g
Pancreatic digest of casein 10.0g
NaCl 2.0g
D-Mannitol 1.0g
Glucose 1.0g
KNO_3 .... 1.0g
Sodium acetate 1.0g
Sodium formate 1.0g
Sodium succinate 1.0g
Yeast extract 1.0g
Sodium lactate (60% solution) 5.0mL

Preparation of Solution 1: Add components to distilled/deionized water and bring volume to 950.0mL. Mix thoroughly. Gently heat and bring to boiling. Autoclave for 15 min at 15 psi pressure–121°C.

Solution 2:
Composition per 50mL:
Na_2HPO_4 .... 1.0g
L-Cysteine·HCl·H_2O 0.5g
$Na_2CO_3 \cdot H_2O$ 0.5g
Sucrose 0.5g
Dithiothreitol 0.1g
Menadione solution 2.0mL

Preparation of Solution 2: Add components to distilled/deionized water and bring volume to 50.0mL. Mix thoroughly. Filter sterilize.

Menadione Solution:
Composition per 100mL:
Menadione (vitamin K_3) 0.05g
Ethanol 99.0mL

Preparation of Menadione Solution: Add menadione to 99.0mL of absolute ethanol. Mix thoroughly.

Preparation of Medium: Aseptically combine sterile solution 1 with sterile solution 2. Mix thoroughly. Pour into sterile Petri dishes.

Use: For the cultivation and differentiation of microorganisms from dental plaque based on their ability to produce gelatinase. For the differentiation of aerobic, anaerobic and facultative microorganisms of clinical significance.

Gelatin Agar

Composition per liter:

Agar	15.0g
Gelatin	15.0g
Peptone	4.0g
Yeast extract	1.0g

pH 7.2 ± 0.2 at 25°C

Preparation of Medium: Add components to distilled/deionized water and bring volume to 1.0L. Mix thoroughly. Gently heat and bring to boiling. Distribute into tubes or flasks. Autoclave for 15 min at 15 psi pressure–121°C. Pour into sterile Petri dishes or leave in tubes.

Use: For the cultivation of a variety of heterotrophic bacteria based upon their utilization of gelatin.

Gelatin Infusion Broth

Composition per liter:

Beef heart, solids from infusion	500.0g
Gelatin	40.0g
Tryptose	10.0g
NaCl	5.0g

pH 7.4 ± 0.2 at 25°C

Preparation of Medium: Add components to distilled/deionized water and bring volume to 1.0L. Mix thoroughly. Gently heat and bring to boiling. Distribute into tubes or flasks. Autoclave for 15 min at 15 psi pressure–121°C. Pour into sterile Petri dishes or leave in tubes.

Use: For the cultivation and differentiation of a variety of heterotrophic bacteria based upon their production of gelatinase. The gelatinase liquifies the medium.

Gelatin Medium

Composition per liter:

Gelatin	4.0g

pH 7.0 ± 0.2 at 25°C

Preparation of Medium: Add gelatin to distilled/deionized water and bring volume to 1.0L. Mix thoroughly. Gently heat and bring to boiling. Distribute into tubes. Autoclave for 15 min at 15 psi pressure–121°C.

Use: For the cultivation and differentiation of *Nocardia* and *Streptomyces* species based on utilization of gelatin. *Nocardia asteroides* usually exhibits no growth. *Nocardia brasiliensis* shows good growth and round compact colonies. *Streptomyces* species show varying degrees of growth.

Gelatin Medium

Composition per liter:

Gelatin	120.0g
Pancreatic digest of casein	13.0g
Sodium chloride	5.0g
Yeast extract	5.0g
Heart muscle, solids from infusion	2.0g
Sodium thioglycollate	0.5g

pH 7.0 ± 0.2 at 25°C

Preparation of Medium: Add components to distilled/deionized water and bring volume to 1.0L. Mix thoroughly. Gently heat and bring to boiling. Distribute into tubes. Autoclave for 15 min at 15 psi pressure–121°C. Pour into sterile Petri dishes or leave in tubes.

Use: For the cultivation of gelatin-utilizing *Clostridium* species.

Gelatin Metronidazole Cadmium Medium (GMC Medium)

Composition per liter:

Solution 1	950.0mL
Solution 2	50.0mL

pH 7.2 ± 0.2 at 25°C

Solution 1:

Composition per 950mL:

Gelatin	30.0g
Agar	15.0g
Pancreatic digest of casein	10.0g
NaCl	2.0g
D-Mannitol	1.0g
Glucose	1.0g
KNO_3	1.0g
Sodium acetate	1.0g
Sodium formate	1.0g
Sodium succinate	1.0g
Yeast extract	1.0g
$CdSO_4 \cdot 8H_2O$	0.020g
Metronidazole	0.010g
Sodium lactate (60% solution)	5.0mL

Preparation of Solution 1: Add components to distilled/deionized water and bring volume to 950.0mL. Mix thoroughly. Gently heat and bring to boiling. Autoclave for 15 min at 15 psi pressure–121°C.

Solution 2:

Composition per 50mL:

Na_2HPO_4	1.0g
L-Cysteine·HCl·H_2O	0.5g
$Na_2CO_3 \cdot H_2O$	0.5g
Sucrose	0.5g

Dithiothreitol..0.1g
Menadione solution......................................2.0mL

Preparation of Solution 2: Add components to distilled/deionized water and bring volume to 50.0mL. Mix thoroughly. Filter sterilize.

Menadione Solution:

Composition per 100mL:

Menadione (vitamin K_3) 0.05g
Ethanol 99.0mL

Preparation of Menadione Solution: Add menadione to 99.0mL of absolute ethanol. Mix thoroughly.

Preparation of Medium: Aseptically combine sterile solution 1 with sterile solution 2. Mix thoroughly. Pour into sterile Petri dishes.

Use: For the cultivation and differentiation of microorganisms from dental plaque based on their ability to produce gelatinase. For the differentiation of microorganisms of clinical significance.

Glucose Tetrazolium Medium

Composition per liter:

Agar...15.0g
Pancreatic digest of gelatin 6.0g
Yeast extract..3.0g
Beef extract 1.5g
1,3,5-Triphenyl tetrazolium chloride 0.05g
Glucose solution..100.0mL

pH 6.6 ± 0.1 at 25°C

Glucose Solution:

Composition per 100mL:

Glucose 10.0g

Preparation of Glucose Solution: Add glucose to distilled/deionized water and bring volume to 100.0mL. Mix thoroughly. Filter sterilize.

Preparation of Medium: Add components, except glucose solution, to distilled/deionized water and bring volume to 900.0mL. Mix thoroughly. Gently heat and bring to boiling. Autoclave for 15 min at 15 psi pressure–121°C. Cool to 45° to 50°C. Aseptically add glucose solution. Mix thoroughly. Pour into sterile Petri dishes or distribute into sterile tubes.

Use: For the cultivation and maintenance of *Streptococcus mutans.*

Glucose Yeast Broth with NaCl

Composition per liter:

NaCl 50.0g
Agar...15.0g
Yeast extract..3.0g
Glucose 1.0g

pH 7.0 ± 0.2 at 25°C

Preparation of Medium: Add components to distilled/deionized water and bring volume to 1.0L. Mix thoroughly. Gently heat and bring to boiling. Distribute into tubes or flasks. Autoclave for 15 min at 15 psi pressure–121°C. Pour into sterile Petri dishes or leave in tubes.

Use: For the cultivation and maintenance of *Pediococcus halophilus.*

Glycerol–Free Medium

Composition per liter:

L-Asparagine 4.0g
L-Glutamine..4.0g
Monosodium glutamate 4.0g
Na_2HPO_4 ....2.5g
Pancreatic digest of casein 1.0g
KH_2PO_4 ....1.0g
Triton®WR 1339 ....0.25g
Ferric ammonium citrate..................................0.05g
$CaCl_2$ 1.0mg
$CuSO_4$ 0.5mg
$ZnSO_4$ 0.5mg

Preparation of Medium: Add components to distilled/deionized water and bring volume to 1.0L. Mix thoroughly. Distribute into tubes or flasks. Autoclave for 15 min at 15 psi pressure–121°C.

Use: For the cultivation of *Mycobacterium tuberculosis* Bacille Calmette-Guèrin (BCG) for vaccine production.

Glycerol Solution

Composition per 130mL:

Gylcerol...30.0 mL

Preparation of Medium: Add glycerol to distilled/deionized water and bring volume to 130.0mL. Mix thoroughly. Autoclave for 15 min at 15 psi pressure–121°C. Aseptically dispense 1.0mL aliquots into sterile 2.0mL vials.

Use: For the preservation of yeasts, including *Candida* and *Saccharomyces* speices. Not recommended for *Cryptococcus* species. Remove one loopful of yeast cells grown on primary medium, such as Sabouraud's agar, and suspend in the glycerol solution; freeze at minus 70°C.

Glycine Cycloheximide Phenol Red Agar

Composition per liter:

Solution B 800.0mL
Solution A 200.0mL

Solution A:

Composition per 200mL:

Glycine 10.0g
$(NH_4)_2SO_4$ 5.0g
KH_2PO_4 .... 1.0g
$MgSO_4 \cdot 7H_2O$ 0.5g
NaCl 0.1g
$CaCl_2 \cdot 2H_2O$ 0.1g
DL-Methionine 0.02g
DL-Tryptophan 0.02g
L-Histidine·HCl 0.01g
Inositol 2.0mg
H_3BO_3 .... 0.5mg
$ZnSO_4 \cdot 7H_2O$ 0.4mg
$MnSO_4 \cdot 4H_2O$ 0.4mg
Thiamine·HCl 0.4mg
Pyroxidine·HCl 0.4mg
Niacin 0.4mg
Calcium pantothenate 0.4mg
p-Aminobenzoic acid 0.2mg
Riboflavin 0.2mg
$FeCl_3$ 0.2mg
$Na_2MoO_4 \cdot 4H_2O$ 0.2mg
KI 0.1mg
$CuSO_4 \cdot 5H_2O$ 0.04mg
Folic acid 2.0μg
Biotin 2.0μg
Cycloheximide solution 1.6mL

Preparation of Solution A: Add components to distilled/deionized water and bring volume to 200.0mL. Mix thoroughly. Filter sterilize.

Cycloheximide Solution:

Composition per 100mL:

Cycloheximide 0.1g

Preparation of Cycloheximide Solution: Add cylcohexamide to distilled/deionized water and bring volume to 100.0mL. Mix thoroughly. Filter sterilize.

Solution B:

Composition per 800mL:

Agar 20.0g
Phenol Red solution 30.0mL

Preparation of Solution B: Add components to distilled/deionized water and bring volume to 800.0mL. Mix thoroughly. Gently heat and bring to boiling. Autoclave for 15 min at 15 psi pressure–121°C. Cool to 50° to 55°C.

Phenol Red Solution:

Composition per 100mL:

Phenol Red 0.5g

Preparation of Phenol Red Solution: Add phenol Red to distilled/deionized water and bring volume to 100.0mL. Mix thoroughly.

Preparation of Medium: To 800.0mL sterile solution B, aseptically add add 200.0mL sterile solution A at 55°C. Mix thoroughly. Pour into sterile Petri dishes or distribute into sterile tubes.

Use: For the selective cultivation and differentiation of fungi from clinical specimens and in epidemiological investigations. *Cryptococcus neoformans* turns the medium bright red.

GN Broth, Hajna

Composition per liter:

Pancreatic digest of casein 10.0g
Peptic digest of animal tissue 10.0g
NaCl 5.0g
Sodium citrate 5.0g
K_2HPO_4 .... 4.0g
D-Mannitol 2.0g
KH_2PO_4 .... 1.5g
Glucose 1.0g
Sodium deoxycholate 0.5g

pH 7.0 ± 0.2 at 25°C

Source: This medium is available as a premixed powder from BBL Microbiology Systems and Difco Laboratories.

Preparation of Medium: Add components to distilled/deionized water and bring volume to 1.0L. Mix thoroughly. Gently heat and bring to boiling. Distribute into tubes or flasks. Autoclave for 15 min at 13 psi pressure–118°C. Pour into sterile Petri dishes or leave in tubes.

Use: For the selective cultivation of *Salmonella* and *Shigella* species.

Gonococcus Medium

Composition per 623mL:

Part II 500.0mL
Part I 123.0mL

pH 7.4 ± 0.2 at 25°C

Part I:

Composition per 123mL:

K_2HPO_4 .... 10.5g
Glucose 7.5g
NaCl 5.25g
KH_2PO_4 .... 4.5g
Sodium acetate 1.5g
L-Cysteine·HCl·H_2O 1.2g
Sodium citrate 1.13g
$NaHCO_3$ 1.0g
K_2SO_4 .... 0.9g
Na_2SO_4 .... 0.75g
$MgCl_2 \cdot 6H_2O$ 0.45g

KCl	0.3g
NH_4Cl	0.3g
L-Arginine·HCl	0.25g
L-Proline	0.25g
Oxaloacetate	0.25g
L-Glutamic acid	0.19g
L-Methionine	0.19g
L-Asparagine·H_2O	0.13g
L-Isoleucine	0.13g
L-Serine	0.13g
L-Cystine	0.05g
Calcium pantothenate	0.02g
Thiamine·HCl	0.02g
Thiamine pyrophosphate chloride	0.02g
Nicotinamide adenine dinucleotide	0.01g
$CaCl_2 \cdot 2H_2O$	5.0mg
$Fe(NO_3)_3 \cdot 9H_2O$	5.0mg
Uracil	5.0mg
Biotin	4.0mg
Hypoxanthine	2.5mg
Sodium thioglycollate	0.025mg

Preparation of Part I: Add components to distilled/deionized water and bring volume to 123.0mL. Mix thoroughly. Adjust pH to 7.2 with 6*N* NaOH. Warm to 50°C for 45 min. Filter sterilize.

Part II:
Composition per 500mL:

Agar	10.0g
Soluble starch	7.5g

Preparation of Part II: Add components to distilled/deionized water and bring volume to 500.0mL. Mix thoroughly. Gently heat and bring to boiling. Autoclave for 15 min at 15 psi pressure–121°C. Cool to 45° to 50°C.

Preparation of Medium: Aseptically combine 123.0mL of sterile part I and 500.0mL of cooled, sterile part II. Mix thoroughly. Pour into sterile Petri dishes.

Use: For the cultivation of gonococci.

Granada Medium

Composition per liter:

Starch, soluble	150.0g
Proteose peptone No. 3	38.0g
NaCl	3.0g
Trimethoprim lactate	0.015g
Sodium phosphate buffer (0.06M, pH 7.4)	900.0mL
Horse serum, coagulated	100.0mL

pH 7.4 ± 0.2 at 25°C

Preparation of Medium: Add proteose peptone No. 3 and NaCl to 200.0mL of sodium phosphate buffer and bring to boiling. Add 400.0mL of cold sodium phosphate buffer, starch and trimethoprim lactate. Mix thoroughly. Bring volume to 900.0mL with sodium phosphate buffer. Gently heat while stirring in a boiling water bath for exactly 20 min. Do not autoclave. Cool to 90° to 95°C. Add horse serum. Mix thoroughly. Cool to 60° to 65°C while stirring. Pour into sterile Petri dishes. Medium will solidify in 2–3 hr.

Use: For the early selective isolation and cultivation of Group B streptococci from clinical specimens.

Group A Selective Strep Agar with Sheep Blood

Composition per liter:

Pancreatic digest of casein	14.5g
Agar	14.0g
NaCl	5.0g
Papaic digest of soybean meal	5.0g
Sheep blood	50.0mL
Growth factor solution	10.0mL
Selective agents solution	10.0mL

pH 7.4 ± 0.2 at 25°C

Source: This medium is available as a prepared medium from BBL Microbiology Systems.

Growth Factor Solution:
Composition per 10mL:

Growth factors, BBL	1.5g

Preparation of Growth Factor Solution: Add growth factors to distilled/deionized water and bring volume to 10.0mL. Mix thoroughly. Filter sterilize.

Selective Agent Solution:
Composition per 10mL:

Selective agents	0.042g

Preparation of Selective Agent Solution: Add selective agents to distilled/deionized water and bring volume to 10.0mL. Mix thoroughly. Filter sterilize.

Preparation of Medium: Add components, except sheep blood, growth factor solution, and selective agent solution, to distilled/deionized water and bring volume to 930.0mL. Mix thoroughly. Gently heat and bring to boiling. Autoclave for 15 min at 15 psi pressure–121°C. Cool to 45° to 50°C. Aseptically add 50.0mL sheep blood, 10.0mL sterile growth factor solution, and 10.0mL sterile selective agent components. Mix thoroughly. Pour into sterile Petri dishes or distribute into sterile tubes.

Use: For the selective cultivation and primary isolation of group A streptococci, especially *Streptococcus pyogenes* from clinical specimens.

H Agar (Hominis Agar)

Composition per 98mL:

Base agar 65.0mL
Horse serum 20.0mL
Yeast dialysate 10.0mL
Penicillin solution 2.0mL
Thallium acetate solution 1.0mL

pH 7.3 ± 0.2 at 25°C

Base Agar:

Composition per liter:

Papaic digest of soybean meal 20.0g
Agarose 10.0g
NaCl 5.0g
Phenol Red (2% solution) 1.0mL

Preparation of Base Agar: Add components to distilled/deionized water and bring volume to 1.0L. Mix thoroughly. Gently heat and bring to boiling. Adjust pH to 7.3. Autoclave for 15 min at 15 psi pressure–121°C. Cool to 45° to 50°C.

Yeast Dialysate:

Composition per 10mL:

Active, dried yeast 450.0g

Preparation of Yeast Dialysate: Add active, dried yeast to distilled/deionized water and bring volume to 1250.0mL. Gently heat and bring to 40°C. Autoclave for 15 min at 15 psi pressure–121°C. Put into dialysis tubing. Dialyze against 1.0L of distilled/deionized water for 2 days at 4°C. Discard tubing and its contents. Autoclave dialysate for 15 min at 15 psi pressure–121°C. Store at –20°C.

Penicillin Solution:

Composition per 10mL:

Penicillin 100,000U

Preparation of Penicillin Solution: Add penicillin to distilled/deionized water and bring volume to 10.0mL. Mix thoroughly. Filter sterilize.

Thallium Acetate Solution:

Composition per 10mL:

Thallium acetate 0.33g

Preparation of Thallium Acetate Solution: Add thallium acetate to distilled/deionized water and bring volume to 10.0mL. Mix thoroughly. Filter sterilize.

Caution: Thallium salts are toxic.

Preparation of Medium: To 65.0mL of cooled, sterile base agar, aseptically add 10.0mL of sterile yeast dialysate, 20.0mL of horse serum, 2.0mL of sterile penicillin solution and 1.0mL of sterile thallium acetate solution. Mix thoroughly. Pour into 10mm × 35mm Petri dishes in 5.0mL volumes. Allow plates to stand overnight at 25°C to remove excess surface moisture.

Use: For isolation of *Mycoplasma pneumoniae* and *M. hominis.*

H Broth

Composition per liter:

NaCl 5.0g
Pancreatic digest of casein 5.0g
Peptone 5.0g
Beef extract 3.0g
K_2HPO_4 .... 2.5g
Glucose 1.0g

pH 7.2 ± 0.2 at 25°C

Preparation of Medium: Add components to distilled/deionized water and bring volume to 1.0L. Mix thoroughly. Distribute into tubes in 4.0mL volumes. Autoclave for 15 min at 10 psi pressure–115°C.

Use: For the preparation of the H agglutination antigen used in the differentiation and identification of *Salmonella* species types and subtypes.

H Broth (Hominis Broth)

Composition per 99mL:

Base broth 65.0mL
Horse serum 20.0mL
Yeast dialysate 10.0mL
Penicillin solution 2.0mL
Glucose solution 1.0mL
Thallium acetate solution 1.0mL

pH 7.3 ± 0.2 at 25°C

Base Broth:

Composition per liter:

Papaic digest of soybean meal 20.0g
NaCl 5.0g
Phenol Red (2% solution) 1.0mL

Preparation of Base Broth: Add components to distilled/deionized water and bring volume to 1.0L. Mix thoroughly. Gently heat and bring to boiling. Adjust pH to 7.3. Autoclave for 15 min at 15 psi pressure–121°C. Cool to 25°C.

Yeast Dialysate:

Composition per 10mL:

Active, dried yeast 450.0g

Preparation of Yeast Dialysate: Add active, dried yeast to distilled/deionized water and bring volume to 1250.0mL. Gently heat and bring to 40°C. Autoclave for 15 min at 15 psi pressure–121°C. Put into dialysis tubing. Dialyze against 1.0L of distilled/deionized water for 2 days at 4°C. Discard tubing and its contents. Autoclave dialysate for 15 min at 15 psi pressure–121°C. Store at –20°C.

Penicillin Solution:
Composition per 10mL:

Penicillin	100,000U

Preparation of Penicillin Solution: Add penicillin to distilled/deionized water and bring volume to 10.0mL. Mix thoroughly. Filter sterilize.

Glucose Solution:
Composition per 10mL:

D-Glucose	1.8g

Preparation of Glucose Solution: Add D-glucose to distilled/deionized water and bring volume to 10.0mL. Mix thoroughly. Filter sterilize.

Thallium Acetate Solution:
Composition per 10mL:

Thallium acetate	0.33g

Preparation of Thallium Acetate Solution: Add thallium acetate to distilled/deionized water and bring volume to 10.0mL. Mix thoroughly. Filter sterilize.

Caution: Thallium salts are toxic.

Preparation of Medium: To 65.0mL of cooled, sterile base broth, aseptically add 10.0mL of sterile yeast dialysate, 20.0mL of horse serum, 2.0mL of sterile penicillin solution, 1.0mL of sterile glucose solution and 1.0mL of sterile thallium acetate solution. Mix thoroughly. Aseptically distribute into sterile screw-capped tubes in 5.0mL volumes. Screw caps down tightly.

Use: For isolation and cultivation of *Mycoplasma pneumoniae.*

H Broth (Hominis Broth)

Composition per 100mL:

Base broth	65.0mL
Horse serum	20.0mL
Yeast dialysate	10.0mL
Penicillin solution	2.0mL
Arginine solution	2.0mL
Thallium acetate solution	1.0mL

pH 7.3 ± 0.2 at 25°C

Base Broth:
Composition per liter:

Papaic digest of soybean meal	20.0g
NaCl	5.0g
Phenol Red (2% solution)	1.0mL

Preparation of Base Broth: Add components to distilled/deionized water and bring volume to 1.0L. Mix thoroughly. Gently heat and bring to boiling. Adjust pH to 7.3. Autoclave for 15 min at 15 psi pressure–121°C. Cool to 25°C.

Yeast Dialysate:
Composition per 10mL:

Active, dried yeast	450.0g

Preparation of Yeast Dialysate: Add active, dried yeast to distilled/deionized water and bring volume to 1250.0mL. Gently heat and bring to 40°C. Autoclave for 15 min at 15 psi pressure–121°C. Put into dialysis tubing. Dialyze against 1.0L of distilled/deionized water for 2 days at 4°C. Discard tubing and its contents. Autoclave dialysate for 15 min at 15 psi pressure–121°C. Store at –20°C.

Penicillin Solution:
Composition per 10mL:

Penicillin	100,000U

Preparation of Penicillin Solution: Add penicillin to distilled/deionized water and bring volume to 10.0mL. Mix thoroughly. Filter sterilize.

Arginine Solution:
Composition per 10mL:

L-Arginine	1.74g

Preparation of Arginine Solution: Add L-arginine to distilled/deionized water and bring volume to 10.0mL. Mix thoroughly. Filter sterilize.

Thallium Acetate Solution:
Composition per 10mL:

Thallium acetate	0.33g

Preparation of Thallium Acetate Solution: Add thallium acetate to distilled/deionized water and bring volume to 10.0mL. Mix thoroughly. Filter sterilize.

Caution: Thallium salts are toxic.

Preparation of Medium: To 65.0mL of cooled, sterile base broth, aseptically add 10.0mL of sterile yeast dialysate, 20.0mL of horse serum, 2.0mL of sterile penicillin solution, 1.0mL of sterile glucose solution and 1.0mL of sterile thallium acetate solution. Mix thoroughly. Aseptically distribute into sterile screw-capped tubes in 5.0mL volumes. Screw caps down tightly.

Use: For isolation and cultivation of *Mycoplasma hominis.*

H Diphasic Medium

Composition per 197mL:

Base agar	65.0mL
Base broth	65.0mL
Horse serum	40.0mL
Yeast dialysate	20.0mL
Penicillin solution	4.0mL
Glucose solution	1.0mL
Thallium acetate solution	2.0mL

pH 7.3 ± 0.2 at 25°C

Base Agar:

Composition per liter:

Papaic digest of soybean meal	20.0g
Agarose	10.0g
NaCl	5.0g
Phenol Red (2% solution)	1.0mL

Preparation of Base Agar: Add components to distilled/deionized water and bring volume to 1.0L. Mix thoroughly. Gently heat and bring to boiling. Adjust pH to 7.3. Autoclave for 15 min at 15 psi pressure–121°C. Cool to 45° to 50°C.

Base Broth:

Composition per liter:

Papaic digest of soybean meal	20.0g
NaCl	5.0g
Phenol Red (2% solution)	1.0mL

Preparation of Base Broth: Add components to distilled/deionized water and bring volume to 1.0L. Mix thoroughly. Gently heat and bring to boiling. Adjust pH to 7.3. Autoclave for 15 min at 15 psi pressure–121°C. Cool to 25°C.

Yeast Dialysate:

Composition per 10mL:

Dried yeast, active	450.0g

Preparation of Yeast Dialysate: Add active, dried yeast to distilled/deionized water and bring volume to 1250.0mL. Gently heat and bring to 40°C. Autoclave for 15 min at 15 psi pressure–121°C. Put into dialysis tubing. Dialyze against 1.0L of distilled/deionized water for 2 days at 4°C. Discard tubing and its contents. Autoclave dialysate for 15 min at 15 psi pressure–121°C. Store at –20°C.

Penicillin Solution:

Composition per 10mL:

Penicillin	100,000U

Preparation of Penicillin Solution: Add penicillin to distilled/deionized water and bring volume to 10.0mL. Mix thoroughly. Filter sterilize.

Glucose Solution:

Composition per 10mL:

D-Glucose	1.8g

Preparation of Glucose Solution: Add D-glucose to distilled/deionized water and bring volume to 10.0mL. Mix thoroughly. Filter sterilize.

Thallium Acetate Solution:

Composition per 10mL:

Thallium acetate	0.33g

Preparation of Thallium Acetate Solution: Add thallium acetate to distilled/deionized water and bring volume to 10.0mL. Mix thoroughly. Filter sterilize.

Caution: Thallium salts are toxic.

Preparation of Medium: To 65.0mL of cooled, sterile base agar, aseptically add 10.0mL of sterile yeast dialysate, 20.0mL of horse serum, 2.0mL of sterile penicillin solution and 1.0mL of sterile thallium acetate solution. Mix thoroughly. Aseptically distribute into screw-capped tubes in 3.0mL volumes. Allow agar to solidify. To 65.0mL of cooled, sterile base broth, aseptically add 10.0mL of sterile yeast dialysate, 20.0mL of horse serum, 2.0mL of sterile penicillin solution, 1.0mL of sterile glucose solution and 1.0mL of sterile thallium acetate solution. Mix thoroughly. Aseptically distribute 3.0mL of broth solution on top of the 3.0mL of solidified base agar in each tube. Screw caps down tightly.

Use: For the isolation and cultivation of *Mycoplasma pneumoniae*.

Haemophilus ducreyi Medium

Composition per liter:

Columbia blood agar base	675.0mL
Rabbit blood	300.0mL
Fresh yeast extract solution	25.0mL

pH 6.5–7.0 at 25°C

Columbia Blood Agar Base:

Composition per 675mL:

Agar	15.0g
Pantone	10.0g
Bitone	10.0g
NaCl	5.0g
Tryptic digest of beef heart	3.0g
Cornstarch	1.0g

Preparation of Columbia Blood Agar Base: Add components to distilled/deionized water and bring volume to 675.0mL. Mix thoroughly. Gently heat until boiling. Autoclave for 15 min at 15 psi pressure–121°C. Cool to 45° to 50°C.

Fresh Yeast Extract Solution:

Composition per 100mL:

Baker's yeast live, pressed, starch-free	25.0g

Preparation of Fresh Yeast Extract Solution: Add the live Baker's yeast to 100.0mL of distilled/deionized water. Autoclave for 90 min at 15 psi pressure–121°C. Allow to stand. Remove supernatant solution. Adjust pH to 6.6–6.8. Filter sterilize.

Preparation of Medium: To 675.0mL of cooled, sterile Columbia blood agar base, aseptically add rabbit blood and sterile fresh yeast extract solution. Aseptically adjust pH to 6.5–7.0.

Use: For the cultivation and maintenance of *Haemophilus ducreyi*.

Haemophilus ducreyi Medium, Revised (Ducreyi Medium, Revised)

Composition per 1010mL:

Solution B	500.0mL
Solution A	400.0mL
Solution C	110.0mL

pH 7.4 ± 0.2 at 25°C

Solution A:

Composition per 400mL:

Beef heart, infusion from	500.0g
Agar	15.0g
Tryptose	10.0g
NaCl	5.0g

Preparation of Solution A: Add components to distilled/deionized water and bring volume to 400.0L. Mix thoroughly. Gently heat and bring to boiling. Autoclave for 15 min at 15 psi pressure–121°C. Cool to 50°C.

Solution B:

Composition per 500mL:

Hemoglobin	10.0g

Preparation of Solution B: Add components to distilled/deionized water and bring volume to 500.0L. Mix thoroughly. Gently heat and bring to boiling. Autoclave for 15 min at 15 psi pressure–121°C. Cool to 45° to 50°C.

Solution C:

Composition per 110mL:

Fetal bovine serum	100.0mL
Supplement solution	10.0mL

Supplement Solution:

Composition per liter:

Glucose	100.0g
L-Cysteine·HCl	25.9g
L-Glutamine	10.0g
L-Cystine	1.1g
Adenine	1.0g
Nicotinamide adenine dinucleotide	0.25g
Vitamin B_{12}	0.1g
Thiamine pyrophosphate	0.1g
Guanine·HCl	0.03g
$Fe(NO_3)_3 \cdot 6H_2O$	0.02g
p-Aminobenzoic acid	0.013g
Thiamine·HCl	3.0mg

Source: The supplement solution IsoVitaleX® enrichment is available from BBL Microbiology Laboratories. This enrichment may be replaced by supplement VX from Difco Laboratories.

Preparation of Supplement Solution: Add components to distilled/deionized water and bring volume to 1.0L. Mix thoroughly. Filter sterilize.

Preparation of Solution C: Combine components. Mix thoroughly. Filter sterilize. Warm to 45° to 50°C.

Preparation of Medium: Aseptically combine solution A, solution B and solution C. Mix thoroughly. Pour into sterile Petri dishes or distribute into sterile tubes.

Use: For the cultivation and maintenance of *Haemophilus ducreyi.*

Haemophilus influenzae Defined Medium MI

Composition per liter:

NaCl	5.8g
K_2HPO_4	3.5g
Glycerol	3.0g
KH_2PO_4	2.7g
Inosine	2.0g
L-Glutamic acid	1.3g
K_2SO_4	1.0g
Sodium lactate	0.8g
L-Aspartic acid	0.5g
Nitrilotriethanol	0.4g
L-Arginine	0.3g
L-Leucine	0.3g
L-Cystine	0.2g
$MgCl_2$	0.2g
L-Tyrosine	0.2g
L-Methionine	0.1g
L-Serine	0.1g
Uracil	0.1g
L-Lysine	0.05g
Glycine	0.03g
$CaCl_2$	0.022g
Hypoxanthine	0.02g
Polyvinyl alcohol	0.02g
Tween™ 80	0.02g
Hemin	0.01g
L-Histidine	0.01g
Calcium pantothenate	4.0mg
Ethylenediaminetetraacetate	4.0mg
Nicotinamide adenine dinucleotide	4.0mg
Thiamine	4.0mg

Preparation of Medium: Add components to distilled/deionized water and bring volume to 1.0L. Mix thoroughly. Filter sterilize.

Use: For the cultivation of *Haemophilus influenza* in a chemically defined medium.

Haemophilus influenzae Defined Medium MI–Cit

Composition per liter:

NaCl	5.8g
K_2HPO_4	3.5g
Glycerol	3.0g
KH_2PO_4	2.7g
Inosine	2.0g
L-Glutamic acid	1.3g
K_2SO_4	1.0g
Sodium lactate	0.8g
L-Aspartic acid	0.5g
Nitrilotriethanol	0.4g
L-Leucine	0.3g
L-Cystine	0.2g
$MgCl_2$	0.2g
L-Tyrosine	0.2g
Citrulline	0.15g
L-Methionine	0.1g
L-Serine	0.1g
L-Lysine	0.05g
Glycine	0.03g
$CaCl_2$	0.022g
Hypoxanthine	0.02g
Polyvinyl alcohol	0.02g
Tween™ 80	0.02g
Hemin	0.01g
L-Histidine	0.01g
Calcium pantothenate	4.0mg
Ethylenediaminetetraacetate	4.0mg
Nicotinamide adenine dinucleotide	4.0mg
Thiamine	4.0mg

Preparation of Medium: Add components to distilled/deionized water and bring volume to 1.0L. Mix thoroughly. Filter sterilize.

Use: For the cultivation of *Haemophilus influenza* in a chemically defined medium.

Haemophilus Test Medium (HTM)

Composition per liter:

Beef infusion	300.0g
Acid hydrolysate of casein	17.5g
Agar	17.0g
Starch	1.5g
Yeast extract	5.0g
HTM supplement	10.0mL

pH 7.4 ± 0.2 at 25°C

Source: This medium is available as a premixed powder from Oxoid Unipath.

HTM Supplement:

Composition per 10mL:

Nicotinamide adenine dinucleotide	0.03g
Hematin	0.03g

Preparation of HTM Supplement: Add components to distilled/deionized water and bring volume to 10.0mL. Mix thoroughly. Filter sterilize.

Preparation of Medium: Add components, except HTM supplement, to distilled/deionized water and bring volume to 990.0mL. Mix thoroughly. Gently heat and bring to boiling. Autoclave for 15 min at 15 psi pressure–121°C. Cool to 45° to 50°C. Aseptically add 10.0mL of sterile HTM supplement. Mix thoroughly. Pour into sterile Petri dishes or distribute into sterile tubes.

Use: For antimicrobial susceptibility testing of *Haemophilus influenzae.*

Ham's F–10 Medium

Composition per liter:

NaCl	7.4g
$NaHCO_3$	1.2g
Glucose	1.1g
$NaH_2PO_4{\cdot}H_2O$	0.29g
KCl	0.28g
L-Arginine·HCl	0.21g
L-Glutamine	0.15g
$MgSO_4{\cdot}7H_2O$	0.15g
Sodium pyruvate	0.11g
KH_2PO_4	0.08g
$CaCl_2{\cdot}2H_2O$	0.04g
L-Cystine·2HCl	0.04g
L-Histidine·HCl·H_2O	0.02g
L-Lysine·HCl	0.02g
L-Asparagine-H_2O	0.01g
L-Aspartic Acid	0.01g
L-Glutamic acid	0.01g
L-Leucine	0.01g
L-Proline	0.01g
L-Serine	0.01g
L-Alanine	8.9mg
Glycine	7.5mg
D-Phenylalanine	5.0mg
L-Methionine	4.5mg
Hypoxanthine	4.1mg
L-Threonine	3.6mg
L-Valine	3.5mg
L-Isoleucine	2.6mg
L-Tyrosine	1.8mg
Vitamin B_{12}	1.4mg
Folic acid	1.3mg
Phenol red	1.2mg
Thiamine·HCl	1.0mg

$FeSO_4·7H_2O$	0.8mg
Choline chloride	0.7mg
D-Calcium pantothenate	0.7mg
Thymidine	0.7mg
Niacinamide	0.6mg
L-Tryptophan	0.6mg
Isoinositol	0.5mg
Riboflavin	0.4mg
Lipoic acid	0.2mg
Pyridoxine·HCl	0.2mg
$ZnSO_4·7H_2O$	0.03mg
Biotin	0.02mg
$CuSO_4·5H_2O$	3.0μg

pH 7.0 ± 0.2 at 25°C

Preparation of Medium: Add components to distilled/deionized water and bring volume to 1.0L. Mix thoroughly. Filter sterilize.

Use: For the growth of Y-1 cell cultures used in the mouse adrenal assay for heat-labile toxin of enterotoxigenic *Escherichia coli* and *Vibrio* species.

HBT Bilayer Medium (Human Blood Tween™ Bilayer Medium)

Composition per 1062.5mL:

Agar	13.5g
Pancreatic digest of casein	12.0g
Casein/meat peptone	10.0g
NaCl	5.0g
Peptic digest of animal tissue	5.0g
Beef extract	3.0g
Yeast extract	3.0g
Cornstarch	1.0g
Human blood, anticoagulated	25.0mL
Colistin solution	10.0mL
Nalidixic acid solution	10.0mL
Amphotericin B solution	10.0mL
Polysorbate 80 (Tween™ 80) solution	7.5mL

pH 7.3 ± 0.2 at 25°C

Source: This medium is available as a premixed powder from BBL Microbiology Systems.

Colistin Solution:
Composition per liter:

Colistin	0.010g

Preparation of Colistin: Add colistin to distilled/deionized water and bring volume to 10.0mL. Mix thoroughly. Filter sterilize.

Nalidixic Acid Solution:
Composition per liter:

Nalidixic acid	0.020g

Preparation of Nalidixic Acid Solution: Add nalidixic acid to distilled/deionized water and bring volume to 10.0mL. Mix thoroughly. Filter sterilize.

Amphotericin B Solution:
Composition per liter:

Amphotericin B	3.0mg

Preparation of Amphotericin B Solution: Add Amphotericin B to distilled/deionized water and bring volume to 10.0mL. Mix thoroughly. Filter sterilize.

Tween™ 80 Solution:
Composition per 100mL:

Tween™ 80	1.0mL

Preparation of Tween™ 80 Solution: Add Tween™ 80 to distilled/deionized water and bring volume to 100.0mL. Mix thoroughly. Adjust pH to 7.3. Filter sterilize.

Preparation of Medium: Add components, except amphotericin B, Tween™ 80 and human blood, to distilled/deionized water and bring volume to 1.0L. Mix thoroughly. Gently heat and bring to boiling. Divide the medium into two 500.0mL fractions. Autoclave both flasks of media for 15 min at 15 psi pressure–121°C. Cool to 45° to 50°C. To one flask aseptically add 5.0mL of sterile colistin solution, 5.0mL of sterile nalidixic acid solution, 5.0mL of sterile amphotericin B solution, and 3.75mL of Tween™ 80 solution. Mix thoroughly. Pour into sterile Petri dishes in 7.0mL volumes. Allow agar to harden. To remaining flask aseptically add 5.0mL of sterile colistin solution, 5.0mL of sterile nalidixic acid solution, 5.0mL of sterile amphotericin B solution, 3.75mL of sterile Tween™ 80 solution, and 25.0mL of sterile human blood. Mix thoroughly. Pour into the same Petri dishes that each contain 7.0mL of the agar medium without blood. The top layer should be approximately 14.0mL per plate.

Use: For the selective isolaltion, cultivation and differentiation of *Gardnerella vaginalis* from clinical specimens.

Heart Infusion Agar

Composition per liter:

Beef heart, infusion from	500.0g
Agar	15.0g
Tryptose	10.0g
NaCl	5.0g

pH 7.4 ± 0.2 at 25°C

Source: This medium is available as a premixed powder from Difco Laboratories.

Preparation of Medium: Add components to distilled/deionized water and bring volume to 1.0L. Mix thoroughly. Gently heat and bring to boiling. Distribute into tubes or flasks. Autoclave for 15 min at 15 psi pressure–121°C. Pour into sterile Petri dishes or leave in tubes.

Use: For the isolation and cultivation of a wide variety of fastidious microorganisms. It can also be used as a base for the preparation of blood agar in determining hemolytic reactions. For the cultivation and maintenance of *Bacillus anthracis, Bacillus cereus, Bacillus mycoides, Serratia rubidaea, Staphylococcus aureus, Tsatumella ptyseos,* and *Vibrio vulnificus.*

Heart Infusion Broth

Composition per liter:

Beef heart, infusion from	500.0g
Tryptose	10.0g
NaCl	5.0g

pH 7.4 ± 0.2 at 25°C

Source: This medium is available as a premixed powder from Difco Laboratories.

Preparation of Medium: Add components to distilled/deionized water and bring volume to 1.0L. Mix thoroughly. Distribute into tubes or flasks. Autoclave for 15 min at 15 psi pressure–121°C.

Use: For the isolation and cultivation of a wide variety of fastidious microorganisms.

Heart Infusion Broth with Additives for *Streptobacillus*

Composition per liter:

Beef heart, infusion from	500.0g
Tryptose	10.0g
Peptone	10.0g
NaCl	5.0g
Glucose	0.5g
Horse serum, inactivated	200.0mL

pH 7.5 ± 0.2 at 25°C

Preparation of Medium: Add components, except horse serum, to distilled/deionized water and bring volume to 800.0mL. Mix thoroughly. Gently heat and bring to boiling. Autoclave for 15 min at 15 psi pressure–121°C. Cool to 45° to 50°C. Aseptically add sterile horse serum. Mix thoroughly. Aseptically distribute into sterile tubes or flasks.

Use: For the cultivation and maintenance of *Streptobacillus moniliformis.*

Heart Infusion Tyrosine Agar

Composition per liter:

Beef heart, infusion from	500.0g
Agar	15.0g
Tryptose	10.0g
NaCl	5.0g
L-Tyrosine	1.0g

pH 7.4 ± 0.2 at 25°C

Preparation of Medium: Add components to distilled/deionized water and bring volume to 1.0L. Mix thoroughly. Gently heat and bring to boiling. Distribute into tubes. Autoclave for 15 min at 15 psi pressure–121°C. Allow tubes to cool in a slanted position.

Use: For the cultivation and differentiation of *Bordetella parapertussis* based on browning of blood-free medium.

Hektoen Enteric Agar

Composition per liter:

Agar	13.5g
Lactose	12.0g
Peptic digest of animal tissue	12.0g
Sucrose	12.0g
Bile salts	9.0g
NaCl	5.0g
$Na_2S_2O_3$	5.0g
Yeast extract	3.0g
Salicin	2.0g
Ferric ammonium citrate	1.5g
Acid Fuchsin	0.1g
Bromthymol Blue	0.064g

pH 7.6 ± 0.2 at 25°C

Source: This medium is available as a premixed powder from BBL Microbiology Systems, Difco Laboratories and Oxoid Unipath.

Caution: Acid Fuchsin is a potential carcinogen and care must be taken to avoid inhalation of the powdered dye and contact with the skin.

Preparation of Medium: Add components to distilled/deionized water and bring volume to 1.0L. Mix thoroughly. Gently heat while stirring until components are dissolved. Do not autoclave. Pour into sterile Petri dishes. Allow agar to solidify with the Petri dish covers partially off.

Use: For the isolation and cultivation of Gram-negative enteric microorganisms from a variety of clinical specimens and in epidemiological investigations based on lactose or sucrose fermentation and H_2S production. It is used for the isolation and differentiation of *Salmonella* and *Shigella.* Bacteria that ferment lactose or sucrose appear as yellow to orange colonies. Bacteria that produce H_2S appear as colonies with black centers.

Helicobacter pylori Isolation Agar

Composition per liter:

Agar	15.0g
Bitone	10.0g
Pancreatic digest of casein	5.0g
NaCl	5.0g
Peptic digest of animal tissue	5.0g
Tryptic digest of beef heart	3.0g
Cornstarch	1.0g
Horse blood, laked	35.0mL
Antibiotic inhibitor solution	10.0mL

pH 7.3 ± 0.2 at 25°C

Antibiotic Inhibitor Solution:

Composition per 10mL:

Vancomycin	0.010g
Amphotericin B	5.0mg
Cefsulodin	5.0mg
Trimethoprim lactate	5.0mg

Preparation of Antibiotic Inhibitor Solution: Add components to distilled/deionized water and bring volume to 10.0mL. Mix thoroughly. Filter sterilize.

Preparation of Medium: Add components, except horse blood and antibiotic inhibitor solution, to distilled/deionized water and bring volume to 955.0mL. Mix thoroughly. Gently heat and bring to boiling. Autoclave for 15 min at 15 psi pressure–121°C. Cool to 45° to 50°C. Aseptically add sterile horse blood and sterile antibiotic inhibitor solution. Mix thoroughly. Pour into sterile Petri dishes or distribute into sterile tubes.

Use: For the isolation and cultivation of *Helicobacter pylori* from clinical specimens.

Hemo ID Quad Plate with Growth Factors (*Hemophilus* Identification Quadrant Plate with Growth Factors)

Composition per plate:

Quadrant I	5.0mL
Quadrant II	5.0mL
Quadrant III	5.0mL
Quadrant IV	5.0mL

Quadrant I:

Composition per 5mL:

Hemin	0.1mg
Brain heart infusion agar	5.0mL

Quadrant II:

Composition per 5mL:

Brain heart infusion agar	5.0mL
Supplement solution	0.05mL

Quadrant III:

Composition per 5mL:

Hemin	0.1mg
Brain heart infusion agar	5.0mL
Supplement solution	0.05mL

Quadrant IV:

Composition per 5mL:

Hemin	0.1mg
Brain heart infusion agar	5.0mL
Horse blood	0.25mL
Supplement solution	0.05mL

Source: The supplement solution IsoVitaleX® enrichment is available from BBL Microbiology Laboratories. This enrichment may be replaced by supplement VX from Difco Laboratories.

Use: For the differentiation and presumptive identification of *Haemophilus* species. The Hemo ID Quad Plate is a four-sectored plate each containing a different medium.

Hemorrhagic coli Agar (HC Agar)

Composition per liter:

Sorbitol	20.0g
Pancreatic digest of casein	20.0g
Agar	15.0g
NaCl	5.0g
Bile salts No. 3	1.12g
Bromcresol Purple	0.015g

pH 7.2 ± 0.2 at 25°C

Preparation of Medium: Add components to distilled/deionized water and bring volume to 1.0L. Mix thoroughly. Gently heat and bring to boiling. Distribute into tubes or flasks. Autoclave for 15 min at 15 psi pressure–121°C. Pour into sterile Petri dishes.

Use: For the isolation and cultivation of enterohemorraghic *Escherichia coli* O157:H7 from food.

Histoplasma capsulatum Agar

Composition per liter:

Agar	15.0g
Glucose	10.0g
Potato starch	2.0g
α-Ketoglutaric acid	1.0g
L-Cystine·HCl·H_2O	1.0g
Glutathione, reduced	0.5g

L-Asparagine 0.1g
L-Tryptophan 0.02g
Solution 1 .... 250.0mL
Solution 3 .... 40.0mL
Solution 2 .... 10.0mL
Solution 4 .... 10.0mL
Solution 8 .... 10.0mL
Solution 5 .... 1.0mL
Solution 6 .... 0.1mL
Solution 7 .... 0.1mL

pH 6.5 ± 0.2 at 25°C

Solution 1:

Composition per liter:

KH_2PO_4 .... 8.0g
$(NH_4)_2SO_4$ 8.0g
$MgSO_4 \cdot 7H_2O$ 0.86g
$CaCl_2$, anhydrous 0.08g

Preparation of Solution 1: Add components to distilled/deionized water and bring volume to 500.0mL. Mix thoroughly. Bring volume to 1.0L with distilled/deionized water. Store at 5°C.

Solution 2:

Composition per liter:

$FeSO_4 \cdot 7H_2O$ 5.70g
$MnCl_2 \cdot 6H_2O$ 0.80g
$NaMoO_4 \cdot 2H_2O$ 0.15g
HCl, concentrated 1.0mL

Preparation for Solution 2: Add the 1.0mL of concentrated HCl to 100.0mL of distilled water in a 1.0L volumetric flask. Dissolve each component completely in the sequence given. Bring volume to 1.0L with distilled/deionized water. Store at 5°C. Discard if red color or red precipitate appears.

Solution 3:

Composition per 100mL:

Casein, acid-hydrolyzed, vitamin-free 10.0g

Preparation for Solution 3: Add casein to distilled/deionized water and bring volume to 100.0mL. Do not use enzymatically-digested casein.

Solution 4:

Composition per liter:

Calcium pantothenate 0.2g
Inositol 0.2g
Riboflavin 0.2g
Thiamine·HCl 0.2g
Nicotinamide 0.1g
Biotin 0.01g

Preparation for Solution 4: Add components to distilled/deionized water and bring volume to 1.0L. Mix thoroughly. Store at –20°C.

Solution 5:

Composition per 100mL:

Hemin 0.2g
NH_4OH, concentrated 0.3mL

Preparation for Solution 5: Add hemin to approximately 30.0mL of distilled/deionized water. Add NH_4OH. Mix thoroughly until dissolved. Bring volume to 100.0mL with distilled/deionized water. Store at 5° C.

Solution 6:

Composition per 10mL:

DL-Thioctic acid 0.01g
Ethanol (95% solution) 10.0mL

Preparation for Solution 6: Add DL-thioctic acid to 10.0mL of ethanol. Mix thoroughly. Store at –20°C.

Solution 7:

Composition per 10mL:

Coenzyme A 0.01g
$Na_2S \cdot 5H_2O$ (0.05% solution) 0.2mL

Preparation for Solution 7: Prepare $Na_2S \cdot 5H_2O$ solution in freshly boiled distilled/deionized water. Add coenzyme A to 9.8mL of distilled/deionized water. Mix thoroughly. Add freshly prepared $Na_2S \cdot 5H_2O$ solution. Mix thoroughly. Store the solution at –20°C.

Solution 8:

Composition per 100mL:

Oleic acid 0.10g

Preparation for Solution 8: Add oleic acid to 50.0mL of distilled/deionized water. Adjust pH to 9.0 with NaOH. Gently heat until dissolved. Bring volume to 100.0mL with distilled/deionized water. Store at 5°C.

Preparation of Medium: Add components—except agar, potato starch, and solution 8—to distilled/deionized water and bring volume to 400.0mL. Mix thoroughly. Adjust pH to 6.5 with 20% KOH solution. Filter sterilize. In a separate flask add potato starch to 50.0mL of distilled/deionized water. Add the starch solution to 450.0mL of boiling distilled/deionized water. Add 10.0mL of solution 8 and the agar. Mix thoroughly. Autoclave for 15 min at 15 psi pressure–121°C. Cool to 70°C. Aseptically combine the two sterile solutions. Pour into sterile Petri dishes or distribute into sterile tubes.

Use: For cultivation and maintenance of *Histoplasma capsulatum* in the mycelial phase.

Histoplasma capsulatum Broth

Composition per liter:

Glucose 10.0g
Citric acid 10.0g

α-Ketoglutaric acid 1.0g
L-Cystine·HCl·H_2O 1.0g
Potato starch 0.5g
Glutathione, reduced 0.5g
L-Asparagine 0.1g
L-Tryptophan 0.02g
Solution 1 .... 250.0mL
Solution 3 .... 40.0mL
Solution 2 .... 10.0mL
Solution 4 .... 10.0mL
Solution 5 .... 1.0mL
Solution 8 .... 1.0mL
Solution 6 .... 0.1mL
Solution 7 .... 0.1mL
pH 6.5 ± 0.2 at 25°C

Solution 1:
Composition per liter:
KH_2PO_4 .... 8.0g
$(NH_4)_2SO_4$ 8.0g
$MgSO_4 \cdot 7H_2O$ 0.86g
$CaCl_2$, anhydrous 0.08g
$ZnSO_4 \cdot 7H_2O$ 0.05g

Preparation of Solution 1: Add components to distilled/deionized water and bring volume to 500.0mL. Mix thoroughly. Bring volume to 1.0L with distilled/deionized water. Store at 5°C.

Solution 2:
Composition per liter:
$FeSO_4 \cdot 7H_2O$ 5.70g
$MnCl_2 \cdot 6H_2O$ 0.80g
$NaMoO_4 \cdot 2H_2O$ 0.15g
HCl, concentrated 1.0mL

Preparation for Solution 2: Add 1.0mL of concentrated HCl to 100mL of distilled water in a 1.0L volumetric flask. Dissolve each component completely in the sequence given. Bring volume to 1.0L with distilled/deionized water. Store at 5°C. Discard if red color or red precipitate appears.

Solution 3:
Composition per 100mL:
Casein, acid-hydrolyzed, vitamin-free 10.0g

Preparation for Solution 3: Add casein to distilled/deionized water and bring volume to 100.0mL. Do not use enzymatically-digested casein.

Solution 4:
Composition per liter:
Calcium pantothenate 0.2g
Inositol 0.2g
Riboflavin 0.2g
Thiamine·HCl 0.2g
Nicotinamide 0.1g
Biotin 0.01g

Preparation for Solution 4: Add components to distilled/deionized water and bring volume to 1.0L. Mix thoroughly. Store at –20°C.

Solution 5:
Composition per 100mL:
Hemin 0.2g
NH_4OH, concentrated 0.3mL

Preparation for Solution 5: Add hemin to approximately 30.0mL of distilled/deionized water. Add NH_4OH. Mix thoroughly until dissolved. Bring volume to 100.0mL with distilled/deionized water. Store at 5° C.

Solution 6:
Composition per 10mL:
DL-Thioctic acid 0.01g
Ethanol (95% solution) 10.0mL

Preparation for Solution 6: Add DL-thioctic acid to 10.0mL of ethanol. Mix thoroughly. Store at –20°C.

Solution 7:
Composition per 10mL:
Coenzyme A 0.01g
$Na_2S \cdot 5H_2O$ (0.05% solution) 0.2mL

Preparation for Solution 7: Prepare $Na_2S \cdot 5H_2O$ solution in freshly boiled distilled/deionized water. Add coenzyme A to 9.8mL of distilled/deionized water. Mix thoroughly. Add freshly prepared $Na_2S \cdot 5H_2O$ solution. Mix thoroughly. Store the solution at –20°C.

Solution 8:
Composition per 100mL:
Oleic acid 0.10g

Preparation for Solution 8: Add oleic acid to 50.0mL of distilled/deionized water. Adjust pH to 9.0 with NaOH. Gently heat until dissolved. Bring volume to 100.0mL with distilled/deionized water. Store at 5°C.

Preparation of Medium: Add components—except potato starch, and solution 8—to distilled/deionized water and bring volume to 400.0mL. Mix thoroughly. Adjust pH to 6.5 with 20% KOH solution. Filter sterilize. In a separate flask add potato starch to 50.0mL of distilled/deionized water. Add the starch solution to 450.0mL of boiling distilled/deionized water. Add 1.0mL of solution 8. Mix thoroughly. Autoclave for 15 min at 15 psi pressure–121°C. Cool to 70°C. Aseptically combine the two sterile solutions. Pour into sterile Petri dishes or distribute into sterile tubes.

Use: For cultivation of *Histoplasma capsulatum* in the yeast phase. Also for the cultivation of *Histoplasma duboisii, Blastomyces dermatitidis* and *Sporothrix schenkii.*

Horse Blood Agar

Composition per liter:

Beef heart, infusion from 500.0g
Agar 15.0g
Tryptose 10.0g
NaCl 5.0g
Horse blood, defibrinated 50.0mL

pH 6.8 ± 0.2 at 25°C

Preparation of Medium: Add components, except horse blood, to distilled/deionized water and bring volume to 950.0mL. Mix thoroughly. Gently heat and bring to boiling. Autoclave for 15 min at 15 psi pressure–121°C. Cool to 45° to 50°C. Aseptically add sterile horse blood. Mix thoroughly. Pour into sterile Petri dishes or distribute into sterile tubes.

Use: For the cultivation and maintenance of *Yersinia pseudotuberculosis.*

Horse Serum Agar

Composition per liter:

Agar 15.0g
Pancreatic digest of gelatin 5.0g
Beef extract 3.0g
Horse serum 200.0mL

pH 6.8 ± 0.2 at 25°C

Preparation of Medium: Add components, except horse serum, to distilled/deionized water and bring volume to 800.0mL. Mix thoroughly. Gently heat and bring to boiling. Autoclave for 15 min at 15 psi pressure–121°C. Cool to 45° to 50°C. Aseptically add sterile horse serum. Mix thoroughly. Pour into sterile Petri dishes or distribute into sterile tubes.

Use: For the cultivation and maintenance of *Pseudomonas aeruginosa* and *Streptobacillus moniliformis.*

Horse Serum Broth

Composition per liter:

Pancreatic digest of gelatin 5.0g
Beef extract 3.0g
Horse serum 200.0mL

pH 6.8 ± 0.2 at 25°C

Source: This medium is available as a premixed powder from Difco Laboratories.

Preparation of Medium: Add components, except horse serum, to distilled/deionized water and bring volume to 800.0mL. Mix thoroughly. Gently heat and bring to boiling. Autoclave for 15 min at 15 psi pressure–121°C. Cool to 45° to 50°C. Aseptically add sterile horse serum. Mix thoroughly. Aseptically distribute into sterile tubes or flasks.

Use: For the cultivation and maintenance of *Pseudomonas aeruginosa* and *Streptobacillus moniliformis.*

Hoyle Medium Base

Composition per liter:

Agar 15.0g
Beef extract 10.0g
Peptone 10.0g
NaCl 5.0g
Blood, laked 50.0mL
Tellurite solution 10.0mL

pH 7.8 ± 0.2 at 25°C

Source: This medium is available as a premixed powder from Oxoid Unipath.

Tellurite Solution:

Composition per 100mL:

K_2TeO_3 .... 3.5g

Preparation of Tellurite Solution: Add K_2TeO_3 to distilled/deionized water and bring volume to 100.0mL. Mix thoroughly. Filter sterilize.

Caution: Potassium tellurite is toxic.

Preparation of Medium: Add components, except laked blood, to distilled/deionized water and bring volume to 940.0mL. Mix thoroughly. Gently heat and bring to boiling. Autoclave for 15 min at 15 psi pressure–121°C. Cool to 45° to 50°C. Aseptically add sterile components. Mix thoroughly. Pour into sterile Petri dishes or distribute into sterile tubes.

Use: For the isolation and differentiation of *Corynebacterium diphtheriae.*

HR Antifungal Assay Medium Buffered with MOPS

Composition per liter:

MOPS (3-N-Morpholino-
propanesulfonic acid) buffer 34.53g
Glucose 10.0g
$(NH_4)_2SO_4$ 2.5g
KH_2PO_4 .... 1.0g
$NaHCO_3$ 1.0g
Glutamine 0.58g
$MgSO_4 \cdot 7H_2O$ 0.5g
$CaCl_2 \cdot 2H_2O$ 0.1g
NaCl 0.1g
L-Lysine 0.07g
L-Isoleucine 0.05g
L-Leucine 0.05g
L-Threonine 0.05g
L-Valine 0.05g
L-Arginine 0.04g

L-Histidine 0.02g
L-Methionine 0.01g
L-Tryptophan 8.2mg
DL-Methionine 2.0mg
DL-Tryptophan 2.0mg
Inositol 2.0mg
L-Histidine·HCl 1.0mg
H_3BO_3 .... 0.5mg
Calciun pantothenate 0.4mg
$MnSO_4 \cdot H_2O$ 0.4mg
Niacin 0.4mg
Pyridoxine 0.4mg
Thiamine·HCl 0.4mg
$ZnSO_4 \cdot 7H_2O$ 0.4mg
p-Aminobenzoic acid 0.2mg
$FeCl_3$ 0.2mg
Riboflavin 0.2mg
Na_2MoO_3 .... 0.2mg
KI 0.1mg
$CuSO_4 \cdot 5H_2O$ 0.04mg
Biotin 2.0μg
Folic acid 2.0μg

pH 7.0 ± 0.2 at 25°C

Preparation of Medium: Add components, except $NaHCO_3$ and MOPS buffer, to distilled/deionized water and bring volume to 900.0mL. Mix thoroughly. Add $NaHCO_3$ and MOPS buffer. Mix thoroughly. Adjust pH to 7.0. Bring volume to 1.0L with distilled/deionized water. Filter sterilize.

Use: For the testing of the effectiveness of antifungal agents against clinical fungal isolates using the broth dilution susceptibility testing method.

Hugh–Leifson's Glucose Broth

Composition per liter:

NaCl 30.0g
Glucose 10.0g
Agar 3.0g
Peptone 2.0g
Yeast extract 0.5g
Bromcresol Purple 0.015g

pH 7.4 ± 0.2 at 25°C

Preparation of Medium: Add components to distilled/deionized water and bring volume to 1.0L. Mix thoroughly. Gently heat while stirring and bring to boiling. Adjust pH to 7.4. Distibute into tubes or flasks. Autoclave for 15 min at 15 psi pressure–121°C.

Use: For the cultivation and differentiation of bacteria based on their ability to ferment glucose. Bacteria that ferment glucose turn the medium yellow.

Indole Medium

Composition per 200mL:

K_2HPO_4 .... 3.13g
L-Tryptophan 1.0g
NaCl 1.0g
KH_2PO_4 .... 0.27g

pH 7.2 ± 0.2 at 25°C

Preparation of Medium: Add components to distilled/deionized water and bring volume to 200.0mL. Mix thoroughly. Distribute into tubes or flasks. Autoclave for 15 min at 15 psi pressure–121°C.

Use: For the differentiation of microorganisms by means of the indole production from tryptophan test.

Indole Medium

Composition per liter:

Pancreatic digest of casein 20.0g

pH 7.3 ± 0.2 at 25°C

Preparation of Medium: Add pancreatic digest of casein to distilled/deionized water and bring volume to 1.0L. Mix thoroughly. Distribute into tubes or flasks. Autoclave for 15 min at 15 psi pressure–121°C.

Use: For the differentiation of microorganisms by means of the indole test.

Indole Nitrite Medium (Trypticase™ Nitrate Broth)

Composition per liter:

Pancreatic digest of casein 20.0g
Na_2HPO_4 .... 2.0g
Agar 1.0g
Glucose 1.0g
KNO_3 .... 1.0g

pH 7.2 ± 0.2 at 25°C

Source: This medium is available as a premixed powder from BBL Microbiology Systems.

Preparation of Medium: Add components to distilled/deionized water and bring volume to 1.0L. Mix thoroughly. Gently heat and bring to boiling with frequent agitation. Distribute into tubes or flasks. Autoclave for 15 min at 15 psi pressure–121°C.

Use: For the identification of microorganisms by means of the nitrate reduction and indole tests.

Infusion Broth

Composition per liter:

Pancreatic digest of casein 13.0g
NaCl 5.0g

Yeast extract..5.0g
Heart muscle, solids from infusion....................2.0g

pH 7.4 ± 0.2 at 25°C

Source: This medium is available as a premixed powder from BBL Microbiology Systems.

Preparation of Medium: Add components to distilled/deionized water and bring volume to 1.0L. Mix thoroughly. Distribute into tubes or flasks. Autoclave for 15 min at 15 psi pressure–121°C.

Use: For the cultivation of a wide variety of microorganisms.

Inhibitory Mold Agar

Composition per liter:

Agar..15.0g
Glucose..5.0g
Yeast extract..5.0g
Pancreatic digest of casein....................................3.0g
Na_2HPO_4..2.0g
Peptic digest of animal tissue....................................2.0g
Starch..2.0g
Dextrin..1.0g
$MgSO_4 \cdot 7H_2O$..0.8g
Chloramphenicol..0.125g
$FeSO_4$..0.04g
NaCl..0.04g
$MnSO_4$..0.16g

pH 6.7 ± 0.2 at 25°C

Source: This medium is available as a premixed powder from BBL Microbiology Systems.

Preparation of Medium: Add components to distilled/deionized water and bring volume to 1.0L. Mix thoroughly. Gently heat and bring to boiling with frequent agitation. Distribute into tubes or flasks. Autoclave for 15 min at 15 psi pressure–121°C. Pour into sterile Petri dishes or leave in tubes.

Use: For the isolation of pathogenic fungi.

Inositol Brilliant Green Bile Salts Agar (IBB Agar) (*Pleisomonas* Differential Agar)

Composition per liter:

Agar..15.0g
meso-Inositol..10.0g
Proteose peptone..10.0g
Bile salts no. 3..8.5g
Meat extract..5.0g
NaCL..5.0g
Neutral Red (2% solution)....................................1.25mL
Brilliant Green (0.1% solution)....................................0.33mL

pH 7.2 ± 0.1 at 25°C

Preparation of Medium: Add components to distilled/deionized water and bring volume to 1.0L. Mix thoroughly. Gently heat and bring to boiling. Distribute into tubes or flasks. Autoclave for 15 min at 15 psi pressure–121°C. Pour into sterile Petri dishes or leave in tubes.

Use: For isolation of *Aeromonas* and *Plesiomonas* species.

Inositol Urea Caffeic Acid Medium

Composition per liter:

Agar solution..900.0mL
Base solution..100.0mL

Agar Solution:

Composition per 900mL:

Agar..15.0g

Preparation of Agar Solution: Add agar to distilled/deionized water and bring volume to 900.0mL. Mix thoroughly. Gently heat and bring to boiling. Autoclave for 15 min at 15 psi pressure–121°C. Cool to 45° to 50°C.

Base Solution:

Composition per 100mL:

Inositol..10.0g
Urea..5.0g
KH_2PO_4..1.0g
$MgSO_4 \cdot 7H_2O$..0.5g
Caffeic acid..0.2g
NaCl..0.1g
$CaCl_2 \cdot 2H_2O$..0.1g
Gentamicin sulfate..0.04g
H_3BO_3..0.5mg
$ZnSO_4 \cdot 7H_2O$..0.4mg
$MnSO_4 \cdot 4H_2O$..0.4mg
Thiamine·HCl..0.4mg
Pyroxidine·HCl..0.4mg
Niacin..0.4mg
Calcium pantothenate..0.4mg
p-Aminobenzoic acid..0.2mg
Riboflavin..0.2mg
$FeCl_3$..0.2mg
$Na_2MoO_4 \cdot 4H_2O$..0.2mg
KI..0.1mg
$CuSO_4 \cdot 5H_2O$..0.04mg
Folic acid..2.0μg
Biotin..2.0μg
Ferric citrate solution (1% solution)....................1.0mL

Preparation of Base Solution: Add components, except urea, to distilled/deionized water and bring volume to 100.0mL. Mix thoroughly. Gently heat just until components are dissolved. Cool to 75° to 80°C. Add urea. Mix thoroughly. Do not heat after addition of urea. Do not autoclave. Filter sterilize.

Preparation of Medium: Aseptically combine the cooled, sterile agar solution with the sterile base solution. Mix thoroughly. Pour into sterile Petri dishes.

Use: For the selective isolation and differentiation of *Cryptococcus* species based on inositol and urea utilization and pigment production from caffeic acid. On this medium, only *Cryptococcus* species utilize inositol as sole carbon source and urea as sole nitrogen source. *Cryptococcus neoformans* appears as dark brown colonies. Other *Cryptococcus* species are unpigmented.

Ion Agar for *Ureaplasma*

Composition per 101.45mL:

HEPES (N-[2-Hydroxyethyl] piperazine-N′-[2-ethane-sulfonic acid]) buffer	1.19g
Ionagar No. 2	0.75g
Pancreatic digest of casein	0.7g
NaCl	0.5g
Beef extract	0.3g
Yeast extract	0.3g
Beef heart, solids from infusion	0.2g
Yeast extract	0.1g
Horse serum, normal sterile	10.0mL
Ampicillin solution	1.0mL
Urea solution	0.25mL
Nystatin solution	0.1mL
Tripeptide solution	0.1mL

pH 7.2 ± 0.2 at 25°C

Ampicillin Solution:

Composition per 10mL:

Ampicillin	1.0g

Preparation of Ampicillin Solution: Add ampicillin to distilled/deionized water and bring volume to 10.0mL. Mix thoroughly. Filter sterilize.

Urea Solution:

Composition per 100mL:

Urea	10.0g

Preparation of Urea solution: Add urea to distilled/deionized water and bring volume to 100.0mL. Filter sterilize. Store at –20°C.

Nystatin Solution:

Composition per 1mL:

Nystatin	50,000U

Preparation of Nystatin Solution: Add nystatin to distilled/deionized water and bring volume to 1.0mL. Filter sterilize.

Tripeptide Solution:

Composition per 10mL:

Glycyl-L-histidyl-L-lysine acetate	0.2mg

Preparation of Tripeptide Solution: Add glycyl-L-histidyl-L-lysine acetate to distilled/deionized water and bring volume to 10.0mL. Mix thoroughly. Filter sterilize. Store at –20°C.

Preparation of Medium: Add components—except horse serum, ampicillin solution, urea solution, nystatin solution and tripeptide solution—to distilled/deionized water and bring volume to 90.0mL. Mix thoroughly. Gently heat and bring to boiling. Autoclave for 15 min at 15 psi pressure–121°C. Cool to 45° to 50°C. Aseptically add 10.0mL of sterile horse serum, 1.0mL of sterile ampicillin solution, 0.25mL of sterile urea solution, 0.1mL of sterile nystatin solution and 0.1mL of sterile tripeptide solution. Mix thoroughly. Pour into sterile Petri dishes.

Use: For the cultivation of *Ureaplasma* species from clinical specimens.

Irgasan® Ticarcillin Chlorate Broth (ITC Broth)

Composition per liter:

$MgCl_2 \cdot 6H_20$	60.0g
Pancreatic digest of casein	10.0g
NaCl	5.0g
$KClO_4$	1.0g
Yeast extract	1.0g
Malachite Green (0.2% solution)	5.0mL
Irgasan solution	1.0mL
Ticarcillin solution	1.0mL

pH 7.6 ± 0.2 at 25°C

Irgasan Solution:

Composition per 10mL:

Irgasan (triclosan)	1.0mg

Preparation of Irgasan Solution: Add Irgasan to distilled/deionized water and bring volume to 10.0mL. Mix thoroughly. Filter sterilize.

Ticarcillin Solution:

Composition per 10mL:

Ticarcillin	1.0 mg

Preparation of Ticarcillin Solution: Add ticarcillin to distilled/deionized water and bring volume to 10.0mL. Mix thoroughly. Filter sterilize.

Preparation of Medium: Add components, except Irgasan solution and ticarcillin solution, to distilled/deionized water and bring volume to 998.0mL. Mix thoroughly. Autoclave for 15 min at 15 psi pressure–121°C. Cool to 45° to 50°C. Adjust to pH 7.6. Aseptically add 1.0mL Irgasan solution and 1.0mL ticarcillin solution. Mix thoroughly. Aseptically distribute into sterile tubes or flasks.

Use: For the selective isolation and cultivation of *Yersinia* species.

Iso–Sensitest Agar

Composition per liter:

Casein, hydrolyzed	11.0g
Agar	8.0g
Peptones	3.0g
NaCl	3.0g
Na_2HPO_4	2.0g
Glucose	2.0g
Sodium acetate	1.0g
Soluble starch	1.0g
Magnesium glycerophosphate	0.2g
Calcium gluconate	0.1g
L-Cysteine·HCl	0.02g
L-Tryptophan	0.02g
Adenine	0.01g
Guanine	0.01g
Xanthine	0.01g
Uracil	0.01g
Nicotinamide	3.0mg
Pantothenate	3.0mg
Pyridoxine	3.0mg
$MnCl_2 \cdot 4H_2O$	2.0mg
$CoSO_4$	1.0mg
$CuSO_4 \cdot 5H_2O$	1.0mg
$FeSO_4 \cdot 7H_2O$	1.0mg
Menadione	1.0mg
Cyanocobalamin	1.0mg
$ZnSO_4 \cdot 7H_2O$	1.0mg
Biotin	0.3mg
Thiamine	.04mg

pH 7.4 ± 0.2 at 25°C

Source: This medium is available as a premixed powder from Oxoid Unipath.

Preparation of Medium: Add components to distilled/deionized water and bring volume to 1.0L. Mix thoroughly. Gently heat and bring to boiling. Distribute into tubes or flasks. Autoclave for 15 min at 15 psi pressure–121°C. Pour into sterile Petri dishes or leave in tubes.

Use: For antimicrobial susceptibility testing.

Iso–Sensitest Broth

Composition per liter:

Casein, hydrolysed	11.0g
Peptones	3.0g
NaCl	3.0g
Glucose	2.0g
Na_2HPO_4	2.0g
Sodium acetate	1.0g
Soluble starch	1.0g
Magnesium glycerophosphate	0.2g
Calcium gluconate	0.1g
L-Cysteine·HCl	0.02g
L-Tryptophan	0.02g
Adenine	0.01g
Guanine	0.01g
Xanthine	0.01g
Uracil	0.01g
Nicotinamide	3.0mg
Pantothenate	3.0mg
Pyridoxine	3.0mg
$MnCl_2 \cdot 4H_2O$	2.0mg
$CoSO_4$	1.0mg
$CuSO_4 \cdot 5H_2O$	1.0mg
$FeSO_4 \cdot 7H_2O$	1.0mg
Menadione	1.0mg
Cyanocobalamin	1.0mg
$ZnSO_4 \cdot 7H_2O$	1.0mg
Biotin	0.3mg
Thiamine	0.04mg

pH 7.4 ± 0.2 at 25°C

Source: This medium is available as a premixed powder from Oxoid Unipath.

Preparation of Medium: Add components to distilled/deionized water and bring volume to 1.0L. Mix thoroughly. Gently heat and bring to boiling. Distribute into tubes or flasks. Autoclave for 15 min at 15 psi pressure–121°C. Pour into sterile Petri dishes or leave in tubes.

Use: For antimicrobial susceptibility testing.

IsoVitaleX® Enrichment

Composition per liter:

Glucose	100.0g
L-Cysteine·HCl	25.9g
L-Glutamine	10.0g
Adenine	1.0g
Thiamine pyrophosphate	0.1g
Vitamin B_{12}	0.1g
Guanine·HCl	0.03g
$Fe(NO_3)_3 \cdot 9H_2O$	0.02g
p-Aminobenzoic acid	0.013g
Thiamine·HCl	0.003g

Preparation of IsoVitaleX® Enrichment: Add components to distilled/deionized water and bring volume to 1.0L. Mix thoroughly. Filter sterilize.

Use: As a nutrient supplement for the isolation and cultivation of fastidious microorganisms.

Jones–Kendrick Pertussis Transport Medium

Composition per liter:

Beef heart, solids from infusion	500.0g
Tryptose	10.0g
NaCl	5.0g
Agar	20.0g
Soluble starch	10.0g
Charcoal powder, activated	4.0g
Yeast extract	3.5g
Penicillin solution	10.0mL

pH 7.4 ± 0.2 at 25°C

Penicillin Solution:

Composition per 10mL:

Penicillin	300U

Preparation of Penicillin Solution: Add penicillin to distilled/deionized water and bring volume to 10.0mL. Mix thoroughly. Filter sterilize.

Preparation: Add components, except penicillin solution, starch, yeast extract, heart infusion, and agar to water. Boil to dissolve. Add charcoal, mix well and autoclave. Cool to 50°C, add penicillin and dispense into small bottles as slants. Cool and seal tightly. Store at 5°C. Stable for 2 to 3 months.

Use: For the cultivation and transport of *Bordetella pertussis* between clinical isolation and laboratory cultivation.

Kanamycin Vancomycin Blood Agar (KVBA)

Composition per liter:

Agar	17.5g
Pancreatic digest of casein	15.0g
Papaic digest of soybean meal	5.0g
NaCl	5.0g
Kanamycin	0.1g
Sheep blood, defibrinated	50.0mL
Vancomycin solution	10.0mL
Vitamin K_1 solution	1.0mL

Vancomycin Solution:

Composition per 10mL:

Vancomycin	7.5mg

Preparation of Vancomycin Solution: Add vancomycin to distilled/deionized water and bring volume to 10.0mL. Mix thoroughly. Filter sterilize.

Vitamin K_1 Solution:

Composition per 100mL:

Vitamin K_1	1.0g

Preparation of Vitamin K_1 Solution: Add vitamin K_1 to 99.0mL of absolute ethanol. Mix thoroughly. Filter sterilize.

Preparation of Medium: Add components, except sheep blood, vancomycin and vitamin K_1 solution, to distilled/deionized water and bring volume to 939.0mL. Mix thoroughly. Gently heat and bring to boiling. Autoclave for 15 min at 15 psi pressure–121°C. Cool to 45°–50°C. Aseptically add sheep blood, vancomycin solution and 1.0mL vitamin K_1 solution. Mix thoroughly. Pour into sterile Petri dishes or distribute into sterile tubes.

Use: For the selective isolation of anaerobes, particularly *Bacteroides*, from clinical specimens.

Kanamycin Vancomycin Laked Blood Agar

Composition per liter:

Agar	17.5g
Pancreatic digest of casein	15.0g
Papaic digest of soybean meal	5.0g
NaCl	5.0g
Kanamycin	0.075g
Sheep blood, laked	50.0mL
Vancomycin solution	10.0mL
Vitamin K_1 solution	1.0mL

Vancomycin Solution:

Composition per 10mL:

Vancomycin	7.5mg

Preparation of Vancomycin Solution: Add vancomycin to distilled/deionized water and bring volume to 10.0mL. Mix thoroughly. Filter sterilize.

Vitamin K_1 Solution:

Composition per 100mL:

Vitamin K_1	1.0g

Preparation of Vitamin K_1 Solution: Add vitamin K_1 to 99.0mL of absolute ethanol. Mix thoroughly. Filter sterilize.

Preparation of Medium: The blood is laked (hemolyzed) by freezing whole blood overnight and then thawing. Add components, except sheep blood, vancomycin and vitamin K_1 solution, to distilled/deionized water and bring volume to 939.0mL. Mix thoroughly. Gently heat and bring to boiling. Autoclave for 15 min at 15 psi pressure–121°C. Cool to 45° to 50°C. Aseptically add sheep blood, vancomycin solution and 1.0mL vitamin K_1 solution. Mix thoroughly. Pour into sterile Petri dishes or distribute into sterile tubes.

Use: For the isolation of *Prevotella melaninogenicus* group.

KCN Broth

Composition per liter:

Na_2HPO_4	5.64g
NaCl	5.0g
Peptone	3.0g
KH_2PO_4	0.225g
KCN (0.5% solution)	15.0mL

pH 7.6 ± 0.2 at 25°C

Caution: Cyanide is toxic.

Preparation of Medium: Add components, except KCN solution, to distilled/deionized water and bring volume to 985.0mL. Mix thoroughly. Autoclave for 15 min at 15 psi pressure–121°C. Cool to 25°C. Aseptically add KCN solution. Mix thoroughly. Aseptically distribute into sterile tubes. Stopper immediately.

Use: For the differentiation of Enterobacteriaceae based upon growth in the presence of potassium cyanide.

Kelly Medium, Nonselective Modified

Composition per 1430mL:

HEPES buffer (*N*-2-Hydroxyethylpiperazine-*N*-2-ethanesulfonic acid)	6.0g
Proteose peptone No. 2	5.0g
D-Glucose	3.0g
$NaHCO_3$	2.2g
Pancreatic digest of casein	1.0g
Yeast, autolyzed	1.0g
Sodium pyruvate	0.8g
Sodium citrate	0.7g
N-Acetylglucosamine	0.4g
$MgCl_2 \cdot 6H_2O$	0.3g
Gelatin solution	200.0mL
Bovine serum albumin	143.0mL
CMRL-1066 medium with glutamine, 10X	100.0mL
Rabbit serum, heat inactivated	86.0mL
Hemin solution	1.0mL

pH 7.2 ± 0.2 at 25°C

CMRL-1066 Medium with Glutamine, 10X:
Composition per liter:

NaCl	6.8g
$NaHCO_3$	2.2g
D-Glucose	1.0g
KCl	0.4g
L-Cysteine·HCl·H_2O	0.26g
$CaCl_2$, anhydrous	0.2g
$MgSO_4 \cdot 7H_2O$	0.2g
$NaH_2PO_4 \cdot H_2O$	0.14g
L-Glutamine	0.1g
Sodium acetate·$3H_2O$	0.083g
L-Glutamic acid	0.075g
L-Arginine·HCl	0.070g
L-Lysine·HCl	0.070g
L-Leucine	0.060g
Glycine	0.050g
Ascorbic acid	0.050g
L-Proline	0.040g
L-Tyrosine	0.040g
L-Aspartic acid	0.030g
L-Threonine	0.030g
L-Alanine	0.025g
L-Phenylalanine	0.025g
L-Serine	0.025g
L-Valine	0.025g
L-Cystine	0.020g
L-Histidine·HCl·H_2O	0.020g
L-Isoleucine	0.020g
Phenol Red	0.020g
L-Methionine	0.015g
Deoxyadenosine	0.010g
Deoxycytidine	0.010g
Deoxyguanosine	0.010g
Glutathione, reduced	0.010g
Thymidine	0.010g
Hydroxy-L-proline	0.010g
L-Tryptophan	0.010g
Nicotinamide adenine dinucleotide	7.0mg
Tween™ 80	5.0mg
Sodium glucuronate·H_2O	4.2mg
Coenzyme A	2.5mg
Cocarboxylase	1.0mg
Flavin adenine dinucleotide	1.0mg
Nicotinamide adenine dinucleotide phosphate	1.0mg
Uridine triphosphate	1.0mg
Choline chloride	0.50mg
Cholesterol	0.20mg
5-Methyldeoxycytidine	0.10mg
Inositol	0.05mg
p-Aminobenzoic acid	0.05mg
Niacin	0.025mg
Niacinamide	0.025mg
Pyridoxine	0.025mg

Pyridoxal·HCl 0.025mg
Biotin 0.01mg
D-Calcium pantothenate 0.01mg
Folic acid 0.01mg
Riboflavin 0.01mg
Thiamine·HCl 0.01mg

pH 7.2 ± 0.2 at 25°C

Source: This solution is available as a premixed powder from GIBCO BRL.

Preparation of CMRL-1066 Medium with Glutamine, 10X: Add components to distilled/deionized water and bring volume to 1.0L. Mix thoroughly. Adjust pH to 7.2. Filter sterilize.

Gelatin Solution:

Composition per 200mL:

Gelatin 14.0g

Preparation of Gelatin Solution: Add gelatin to distilled/deionized water and bring volume to 1.0L. Mix thoroughly. Gently heat and bring to boiling. Autoclave for 15 min at 15 psi pressure–121°C. Cool to 50°C.

Hemin Solution:

Composition per 100mL:

Hemin 1.0g
NaOH (1*N* solution) 20.0mL

Preparation of Hemin Solution: Add hemin to 20.0mL of 1*N* NaOH solution. Mix thoroughly. Bring volume to 100.0mL with distilled/deionized water.

Bovine Serum Albumin Solution:

Composition per 200mL:

Bovine serum albumin 70.0g

Preparation of Bovine Serum Albumin Solution: Add bovine serum albumin to distilled/deionized water and bring volume to 200.0mL. Filter sterilize.

Preparation of Medium: Add components, except gelatin solution, bovine serum albumin solution, and rabbit serum to distilled/deionized water and bring volume to 1001.0mL. Mix thoroughly. Bring pH to 7.6 with 5 *N* NaOH. Filter sterilize. Aseptically add 200.0mL sterile gelatin solution, 143.0mL sterile bovine serum albumin and 86.0mL sterile heat-inactivated rabbit serum. Mix thoroughly. Aseptically dispense into sterile tubes or flasks.

Use: For isolation of *Borrelia burgdorferi* and other spirochetes.

Kelly Medium, Selective Modified

Composition per liter:

Bovine serum albumin fraction V 50.0g
HEPES buffer (*N*-2-Hydroxyethylpiperazine-*N*-2-ethanesulfonic acid) 6.0g
Glucose 5.0g
Neopeptone 5.0g
$NaHCO_3$ 2.2g
Sodium pyruvate 0.8g
Sodium citrate 0.7g
N-Acetylglucosamine 0.4g
Kanamycin 8.0mg
5-Fluorouracil 2.3mg
Gelatin solution 200.0mL
CMRL-1066 medium with glutamine, 10X 100.0mL
Rabbit serum, partially hemolyzed 70.0mL

pH 7.7 ± 0.2 at 25°C

Gelatin Solution:

Composition per 200mL:

Gelatin 14.0g

Preparation of Gelatin Solution: Add gelatin to distilled/deionized water and bring volume to 1.0L. Mix thoroughly. Gently heat and bring to boiling. Autoclave for 15 min at 15 psi pressure–121°C. Cool to 50°C.

CMRL-1066 Medium with Glutamine, 10X:

Composition per liter:

NaCl 6.8g
$NaHCO_3$ 2.2g
D-Glucose 1.0g
KCl 0.4g
L-Cysteine·HCl·H_2O 0.26g
$CaCl_2$, anhydrous 0.2g
$MgSO_4 \cdot 7H_2O$ 0.2g
$NaH_2PO_4 \cdot H_2O$ 0.14g
L-Glutamine 0.1g
Sodium acetate·$3H_2O$ 0.083g
L-Glutamic acid 0.075g
L-Arginine·HCl 0.070g
L-Lysine·HCl 0.070g
L-Leucine 0.060g
Glycine 0.050g
Ascorbic acid 0.050g
L-Proline 0.040g
L-Tyrosine 0.040g
L-Aspartic acid 0.030g
L-Threonine 0.030g
L-Alanine 0.025g
L-Phenylalanine 0.025g
L-Serine 0.025g
L-Valine 0.025g

L-Cystine 0.020g
L-Histidine·HCl·H_2O 0.020g
L-Isoleucine 0.020g
Phenol Red 0.020g
L-Methionine 0.015g
Deoxyadenosine 0.010g
Deoxycytidine 0.010g
Deoxyguanosine 0.010g
Glutathione, reduced 0.010g
Thymidine 0.010g
Hydroxy-L-proline 0.010g
L-Tryptophan 0.010g
Nicotinamide adenine dinucleotide 7.0mg
Tween™ 80 .... 5.0mg
Sodium glucuronate·H_2O 4.2mg
Coenzyme A 2.5mg
Cocarboxylase 1.0mg
Flavin adenine dinucleotide 1.0mg
Nicotinamide adenine dinucleotide phosphate 1.0mg
Uridine triphosphate 1.0mg
Choline chloride 0.50mg
Cholesterol 0.20mg
5-Methyldeoxycytidine 0.10mg
Inositol 0.05mg
p-Aminobenzoic acid 0.05mg
Niacin 0.025mg
Niacinamide 0.025mg
Pyridoxine 0.025mg
Pyridoxal·HCl 0.025mg
Biotin 0.01mg
D-Calcium pantothenate 0.01mg
Folic acid 0.01mg
Riboflavin 0.01mg
Thiamine·HCl 0.01mg

pH 7.2 ± 0.2 at 25°C

Source: This solution is available as a premixed powder from GIBCO BRL.

Preparation of CMRL-1066 Medium with Glutamine, 10X: Add components to distilled/deionized water and bring volume to 1.0L. Mix thoroughly. Adjust pH to 7.2. Filter sterilize.

Preparation of Medium: Add components, except gelatin solution, partially hemolyzed rabbit serum solution, kanamycin, and 5-fluorouracil to distilled/deionized water and bring volume to 1.0mL. Mix thoroughly. Bring pH to 7.6 with 5 *N* NaOH. Filter sterilize. Aseptically add 200.0mL sterile gelatin solution, 70.0mL of partially hemolyzed rabbit serum, 8.0mg kanamycin, and 230.0mg 5-fluorouracil. Mix thoroughly. Aseptically distribute into sterile tubes or flasks.

Use: For the isolation of *Borrelia burgdorferi.*

Kimmig's Agar

Composition per liter:

Agar 15.0g
Glucose 10.0g
Pancreatic digest of gelatin 9.5g
Beef extract 5.5g
NaCl 5.0g
Peptone 5.0g
Glycerol 5.0mL

pH 6.9 ± 0.2 at 35°C

Preparation of Medium: Add glycerol and then other components to distilled/deionized water and bring volume to 1.0L. Mix thoroughly. Gently heat and bring to boiling. Distribute into tubes or flasks. Autoclave for 15 min at 15 psi pressure–121°C. Pour into sterile Petri dishes or leave in tubes.

Use: For the assay of fungistatic agents. For agar dilution test of antifungal agents. For cultivation and preservation of various fungi.

Kirchner's Enrichment Medium

Composition per liter:

$Na_2HPO_4 \cdot 12H_2O$ 19.0g
Asparagine 5.0g
KH_2PO_4 .... 2.5g
Sodium citrate 2.5g
$MgSO_4$ 0.6g
Serum 100.0mL
Glycerol 20.0mL
Penicillin solution 10.0mL
Phenol Red (0.4% solution) 3.0mL

pH 7.4–7.6 at 25°C

Penicillin Solution:

Composition per 10mL:

Penicillin 100,000U

Preparation of Penicillin Solution: Add penicillin to distilled/deionized water and bring volume to 10.0mL. Mix thoroughly. Filter sterilize.

Preparation of Medium: Add components, except serum and penicillin solution, to distilled/deionized water and bring volume to 890.0mL. Mix thoroughly. Gently heat and bring to boiling. Autoclave for 15 min at 15 psi pressure–121°C. Cool to 45° to 50°C. Aseptically add sterile serum and penicillin solution. Mix thoroughly. Aseptically distribute into sterile tubes or flasks.

Use: For the cultivation and enrichment of *Mycobacterium* species.

K–L Virulence Agar (Klebs–Loeffler Virulence Agar) (Elek Agar) (*Corynebacterium diphtheriae* Virulence Test Medium)

Composition per 1300mL:

K-L agar base 1.0L
Rabbit serum 200.0mL
K_2TeO_3 solution 100.0mL
K-L filter strips 100
pH 7.8 ± 0.2 at 25°C

Source: This medium is available as a premixed powder from Difco Laboratories.

K-L Agar Base:
Composition per liter:

Meat peptone 20.0g
Agar 15.0g
NaCl 2.5g

Preparation of K-L Agar Base: Add components to distilled/deionized water and bring volume to 1.0L. Mix thoroughly. Gently heat and bring to boiling. Autoclave for 15 min at 15 psi pressure–121°C. Cool to 50°C.

K_2TeO_3 Solution:
Composition per 100mL:

K_2TeO_3 .... 0.3g

Preparation of K_2TeO_3 Solution: Add K_2TeO_3 to distilled/deionized water and bring volume to 100.0mL. Mix thoroughly. Filter sterilize.

K-L Filter Strips:
Composition:

Whatman No. 3 filter paper .... as needed
Diphtheria toxin solution 10.0mL

Preparation of K-L Strips: Cut Whatman No. 3 filter paper into 1.5cm × 7cm strips. Autoclave for 15 min at 15 psi pressure–121°C. Aseptically dip each strip into a sterile solution containing 1000U of purified diphtheria toxin/mL. Drain off excess liquid. Store each strip in a sterile bottle.

Caution: Potassium tellurite is toxic.

Preparation of Medium: Filter sterilize rabbit serum. To 1.0L of cooled, sterile K-L agar base, aseptically add sterile rabbit serum and sterile K_2TeO_3 solution. Mix thoroughly. Pour into sterile Petri dishes in 13.0mL volumes. Before the agar solidifies, aseptically add one K-L filter strip across the diameter of the plate. Allow the filter strip to sink to the bottom of the plate or press it down with sterile forceps. Allow the agar to solidify. Dry the surface of the plates by incubating at 35°C with lid of plate ajar for 2 hr.

Use: For *in vitro* toxigenicity testing of *Corynebacterium diphtheriae* by the agar diffusion technique. *C. diphtheriae* that produce toxin form white precipitin lines at approximately 45° angles from the culture streak line.

K–L Virulence Agar (Klebs–Loeffler Virulence Agar)

Composition per 1250mL:

K-L agar base 1.0L
K-L enrichment 200.0mL
K_2TeO_3 solution 50.0mL
K-L filter strips 100
pH 7.8 ± 0.2 at 25°C

Source: This medium is available as a premixed powder from Difco Laboratories.

K-L Agar Base:
Composition per liter:

Meat peptone 20.0g
Agar 15.0g
NaCl 2.5g

Preparation of K-L Agar Base: Add components to distilled/deionized water and bring volume to 1.0L. Mix thoroughly. Gently heat and bring to boiling. Autoclave for 15 min at 15 psi pressure–121°C. Cool to 50°C.

K-L Enrichment:
Composition per 200mL:

Casamino acids 4.0g
Glycerol 100.0mL
Tween™ 80 .... 100.0mL

Preparation of K-L Enrichment: Combine components. Mix thoroughly. Filter sterilize.

K_2TeO_3 Solution:
Composition per 100mL:

K_2TeO_3 .... 1.0g

Preparation of K_2TeO_3 Solution: Add K_2TeO_3 to distilled/deionized water and bring volume to 100.0mL. Mix thoroughly. Filter sterilize.

K-L Filter Strips:
Composition:

Whatman No. 3 filter paper .... as needed
Diphtheria toxin solution .... as needed

Preparation of K-L Strips: Cut Whatman No. 3 filter paper into 1.5cm × 7cm strips. Autoclave for 15 min at 15 psi pressure–121°C. Aseptically dip each strip into a sterile solution containing 1000U of purified diphtheria toxin/mL. Drain off excess liquid. Store each strip in a sterile bottle.

Caution: Potassium tellurite is toxic.

Preparation of Medium: To 1.0L of cooled, sterile K-L agar base, aseptically add sterile K-L enrichment and sterile K_2TeO_3 solution. Mix thoroughly. Pour into sterile Petri dishes in 13.0mL volumes. Before the agar solidifies, aseptically add one K-L filter strip across the diameter of the plate. Allow the filter strip to sink to the bottom of the plate or press it down with sterile forceps. Allow the agar to solidify. Dry the surface of the plates by incubating at 35°C with lid of plate ajar for 2 hr.

Use: For *in vitro* toxigenicity testing of *Corynebacterium diphtheriae* by the agar diffusion technique. *C. diphtheriae* that produce toxin form white precipitin lines at approximately 45° angles from the culture streak line.

Klebsiella Selective Agar

Composition per liter:

Agar	26.0g
DL–Phenylalanine	10.0g
L-Ornithine·HCl	10.0g
Raffinose	7.0g
Pancreatic digest of casein	2.5g
Yeast extract	2.5g
K_2HPO_4	2.0g
Phenol Red solution	10.0mL
Carbenicillin solution	10.0mL

pH 5.6 ± 0.2 at 25°C

Phenol Red Solution:

Composition per 10mL:

Phenol Red	0.5g

Preparation of Phenol Red Solution: Add Phenol Red to 50% ethanol and bring volume to 10.0mL. Mix thoroughly.

Preparation of Medium: Add components, except carbenicillin solution, to distilled/deionized water and bring volume to 990.0mL. Mix thoroughly. Gently heat and bring to boiling. Autoclave for 15 min at 15 psi pressure–121°C. Cool to 45° to 50°C. Aseptically add 10.0mL carbenicillin solution. Mix thoroughly. Adjust pH to 5.6–5.7 with sterile 1 *N* HCl. Pour into sterile Petri dishes or distribute into sterile tubes.

Use: For the isolation and identification of *Klebsiella pneumoniae* from clinical specimens.

Kligler Iron Agar

Composition per liter:

Peptone	20.0g
Agar	12.0g
Lactose	10.0g
NaCl	5.0g
Beef extract	3.0g
Yeast extract	3.0g
Glucose	1.0g
Ferric citrate	0.3g
$Na_2S_2O_3$	0.3g
Phenol Red	0.05g

pH 7.4 ± 0.2 at 25°C

Source: This medium is available as a premixed powder from Difco Laboratories and Oxoid Unipath.

Preparation of Medium: Add components to distilled/deionized water and bring volume to 1.0L. Mix thoroughly. Gently heat and bring to boiling. Distribute into tubes. Autoclave for 15 min at 15 psi pressure–121°C. Pour into sterile Petri dishes or leave in tubes.

Use: For the differentiation and identification of Enterobacteriaceae based upon sugar fermentation and hydrogen sulfide production. Sugar fermentation is indicated by the medium turning yellow. H_2S production results in the medium turning black.

Kligler Iron Agar

Composition per liter:

Agar	15.0g
Lactose	10.0g
Pancreatic digest of casein	10.0g
Peptic digest of animal tissue	10.0g
NaCl	5.0g
Glucose	1.0g
Ferric ammonium citrate	0.5g
$Na_2S_2O_3$	0.5g
Phenol Red	0.025g

pH 7.4 ± 0.2 at 25°C

Source: This medium is available as a premixed powder from BBL Microbiology Systems.

Preparation of Medium: Add components to distilled/deionized water and bring volume to 1.0L. Mix thoroughly. Gently heat and bring to boiling. Distribute into tubes or flasks. Autoclave for 15 min at 15 psi pressure–121°C. Pour into sterile Petri dishes or leave in tubes.

Use: For the differentiation and identification of Enterobacteriaceae based upon sugar fermentation and hydrogen sulfide production. Sugar fermentation is indicated by the medium turning yellow. H_2S production results in the medium turning black.

Kligler Iron Agar (FDA M71)

Composition per liter:

Agar	15.0g
Lactose	20.0g

Pancreatic digest of casein ... 10.0g
Peptic digest of animal tissue ... 10.0g
NaCl ... 5.0g
Glucose ... 1.0g
Ferric ammonium citrate ... 0.5g
$Na_2S_2O_3$ ... 0.5g
Phenol Red ... 0.025g
pH 7.4 ± 0.2 at 25°C

Preparation of Medium: Add components to distilled/deionized water and bring volume to 1.0L. Mix thoroughly. Gently heat and bring to boiling. Distribute into tubes or flasks. Autoclave for 15 min at 15 psi pressure–121°C. Pour into sterile Petri dishes or leave in tubes.

Use: For the differentiation and identification of Enterobacteriaceae based upon sugar fermentation and hydrogen sulfide production. Sugar fermentation is indicated by the medium turning yellow. H_2S production results in the medium turning black.

Knisely Medium for *Bacillus anthracis*

Composition per liter:

Beef heart, solids from infusion ... 500.0g
Agar ... 15.0g
Pancreatic digest of casein ... 10.0g
NaCl ... 5.0g
EDTA ... 200.0mg
Lysozyme ... 40.0mg
Thallous acetate ... 40.0mg
Polymyxin ... 30,000U

Preparation of Medium: Add components, except EDTA, lysozyme, thallous acetate, and polymyxin, to distilled/deionized water and bring volume to 1.0mL. Mix thoroughly. Gently heat and bring to boiling. Adjust pH to 7.3. Autoclave for 15 min at 15 psi pressure–121°C. Cool to 45° to 50°C. Aseptically add sterile EDTA, lysozyme, thallous acetate, and polymyxin. Mix thoroughly. Pour into sterile Petri dishes or distribute into sterile tubes.

Use: For the cultivation and maintenance of *Bacillus anthracis*.

Korthof Medium

Composition per 1088mL:

NaCl ... 1.4g
$Na_2HPO_4 \cdot 2H_2O$ ... 0.88g
Peptone ... 0.8g
KH_2PO_4 ... 0.24g
$CaCl_2$ ... 0.04g
KCl ... 0.04g
$NaHCO_3$ ... 0.02g
Rabbit serum, inactivated ... 80.0mL
Rabbit hemoglobin solution ... 8.0mL
pH 7.2 ± 0.2 at 25°C

Rabbit Hemoglobin Solution:
Composition per 20mL:

Rabbit blood clot ... 10.0mL

Preparation of Rabbit Hemoglobin Solution: Add rabbit blood clot to 10.0mL of distilled/deionized water. Lyse the clot by freezing and thawing.

Preparation of Medium: Add components, except rabbit serum and rabbit hemoglobin solution, to distilled/deionized water and bring volume to 1.0L. Mix thoroughly. Gently heat and bring to boiling. Cool to 25°C. Filter through Whatman #1 filter paper. Distribute into flasks in 100.0mL volumes. Autoclave for 15 min at 15 psi pressure–121°C. Cool to 45° to 50°C. Aseptically add 8.0mL of rabbit serum and 0.8mL of rabbit hemoglobin solution to each flask. Mix thoroughly.

Use: For the cultivation of *Leptospira* species.

Korthof Medium, Modified

Composition per liter:

NaCl ... 1.40g
$Na_2HPO_4 \cdot 2H_2O$ ... 0.88g
Peptone ... 0.80g
KH_2PO_4 ... 0.24g
$CaCl_2$ ... 0.04g
KCl ... 0.04g
$NaHCO_3$ ... 0.02g
Rabbit serum, heat inactivated at 56°C ... 100.0mL
pH 7.2–7.6 at 25°C

Preparation of Medium: Add components, except rabbit serum, to distilled/deionized water and bring volume to 900.0L. Mix thoroughly. Gently heat and bring to boiling. Boil for 20 min. Cool overnight at 4°C. Filter through Whatman No. 2 filter paper. Distribute into tubes or flasks. Autoclave for 15 min at 15 psi pressure–121°C. Cool to50° to 56°C. Aseptically add 100.0mL rabbit serum. Mix thoroughly.

Use: For the cultivation of *Leptospira* species.

Kupferberg *Trichomonas* Base

Composition per liter:

Pancreatic digest of casein ... 20.0g
Cysteine·HCl·H_2O ... 1.5g
Agar ... 1.0g
Maltose ... 1.0g
Methylene Blue ... 3.0mg
Bovine serum ... 50.0mL
pH 6.0 ± 0.2 at 25°C

Source: This medium is available as a premixed powder from Difco Laboratories.

Preparation of Medium: Add components, except bovine serum, to distilled/deionized water and bring volume to 950.0mL. Mix thoroughly. Gently heat and bring to boiling. Autoclave for 15 min at 15 psi pressure–121°C. Cool to 45° to 50°C. Aseptically add 50.0mL bovine serum. If desired, additional selectivity can be obtained by aseptically adding 250,000U penicillin and 1.0g streptomycin or 1.0g chloramphenicol. Mix thoroughly. Pour into sterile Petri dishes or distribute into sterile tubes.

Use: For the cultivation of the *Trichomonas* species from clinical specimens.

Kupferberg *Trichomonas* Broth

Composition per liter:

Enzymatic digest of protein	20.0g
Cysteine·HCl·H_2O	1.5g
Agar	1.0g
Maltose	1.0g
Chloramphenicol	0.1g
Methylene Blue	3.0mg
Bovine serum	50.0mL

pH 6.0 ± 0.2 at 25°C

Source: This medium is available as a premixed powder from Difco Laboratories.

Preparation of Medium: Add components, except bovine serum, to distilled/deionized water and bring volume to 950.0mL. Mix thoroughly. Gently heat and bring to boiling. Autoclave for 15 min at 15 psi pressure–121°C. Cool to 45° to 50°C. Aseptically add bovine serum. If desired, additional selectivity can be obtained by aseptically adding 250,000U penicillin and 1.0g streptomycin or 1.0g chloramphenicol. Mix thoroughly. Pour into sterile Petri dishes or distribute into sterile tubes.

Use: For the cultivation of the *Trichomonas* species from clinical specimens.

Lactose Broth

Composition per liter:

Lactose	5.0g
Pancreatic digest of gelatin	5.0g
Beef extract	3.0g

pH 6.9 ± 0.2 at 25°C

Source: This medium is available as a premixed powder from BBL Microbiology Systems, Difco Laboratories, and Oxoid Unipath.

Preparation of Medium: Add components to distilled/deionized water and bring volume to 1.0L. Mix thoroughly. Distribute into tubes containing an inverted Durham tube in 10.0mL volumes. Autoclave for 12 min at 15 psi pressure–121°C. Cool broth quickly to 25°C. For testing water samples with 10.0mL volumes, prepare medium double strength.

Use: For detection of lactose-fermenting, Gram-negative coliforms, as a pre-enrichment broth for *Salmonella* species and in the study of lactose fermentation of bacteria in general.

Lash Serum Medium

Composition per liter:

Casamino acids	14.0g
NaCl	6.0g
Glucose	2.0g
Maltose	1.5g
Sodium lactate (60% solution)	0.5g
KCl	0.1g
$CaCl_2·2H_2O$	0.1g
Serum solution	500.0mL

pH 5.8 ± 0.2 at 25°C

Serum Solution:

Composition per500mL:

$NaHCO_3$	0.1g
Bovine serum	200.0mL

Preparation of Serum Solution: Add components to distilled/deionized water and bring volume to 500.0mL. Mix thoroughly. Filter sterilize.

Preparation of Medium: Add components, except serum solution, to distilled/deionized water and bring volume to 500.0mL. Mix thoroughly. Distribute into tubes in 5.0mL volumes. Autoclave for 15 min at 15 psi pressure–121°C. Cool to 25°C. Aseptically add 5.0mL of sterile serum solution to each tube. Mix thoroughly.

Use: For the cultivation of *Trichomonas vaginalis* from clinical specimens.

Lecithin Lactose Agar

Composition per liter:

Agar	15.0g
Pancreatic digest of casein	12.7g
Lactose	10.0g
NaCl	5.5g
Peptic digest of animal tissue	5.5g
Yeast extract	3.9g
Pancreatic digest of heart muscle	3.3g
Cornstarch	1.1g
Egg lecithin	0.66g
L-Cysteine·HCl·H_2O	0.5g
NaN_3	0.2g

Neomycin sulfate 0.15g
$CaCl_2$ 0.05g
Bromcresol Purple 0.02g
pH 6.8 ± 0.2 at 25°C

Source: Available as a prepared medium from BBL Microbiology Systems.

Caution: Sodium azide is toxic. Azides also react with metals and disposal must be highly diluted.

Preparation of Medium: Add components to distilled/deionized water and bring volume to 1.0L. Mix thoroughly. Gently heat and bring to boiling. Distribute into tubes or flasks. Autoclave for 15 min at 15 psi pressure–121°C. Pour into sterile Petri dishes.

Use: For the isolation, cultivation and differentiation of histolytic clostridia from clinical specimens based on lecithinase production and lactose fermentation. It is especially useful for the differentiation of *Clostridium perfringens, C. sordelli, C. novyi, C. septicum* and *C. histolyticum*. Bacteria that produce lecithinase appear as colonies surrounded by an opalescent zone. Bacteria that ferment lactose appear as colonies surrounded by a yellow zone.

Lecithin Lipase Anaerobic Agar

Composition per liter:
Pancreatic digest of casein 40.0g
Agar 25.0g
Yeast extract 5.0g
Glucose 2.0g
NaCl 2.0g
KH_2PO_4 ....1.0g
$Na_2HPO_4 \cdot 12H_2O$ 5.0g
$MgSO_4 \cdot 7H_2O$ 0.1g
Egg yolk emulsion 100.0mL
pH 7.6 ± 0.2 at 25°C

Egg Yolk Emulsion:
Composition:
Chicken egg yolks 11
Whole chicken egg 1

Preparation of Egg Yolk Emulsion: Soak eggs with 1:100 dilution of saturated mercuric chloride solution for 1 min. Crack eggs and separate yolks from whites. Mix egg yolks with 1 chicken egg. Filter sterilize.

Preparation of Medium: Add components, except egg yolk emulsion, to distilled/deionized water and bring volume to 900.0mL. Mix thoroughly. Gently heat and bring to boiling. Autoclave for 15 min at 15 psi pressure–121°C. Cool to 45° to 50°C. Aseptically add sterile egg yolk emulsion. Mix thoroughly. Pour into sterile Petri dishes or distribute into sterile tubes.

Use: For the isolation, cultivation and differentiation of *Clostridium* species based on lecithinase production and lipase production. Bacteria that produce lecithinase appear as colonies surrounded by a zone of insoluble precipitate. Bacteria that produce lipase appear as colonies with a pearly iridescent sheen.

Lecithin Tween™ Medium (LT Medium)

Composition per liter:
Tween™ 80 ....30.0g
Agar 15.0g
Pancreatic digest of casein 10.0g
Peptic digest of animal tissue 10.0g
NaCl 5.0g
Glucose 1.0g
Lecithin 5.0g
$Na_2S_2O_3 \cdot 5H_2O$ 5.0g
Glycerol 3.0g
Histidine, free base 1.0g
pH 7.5 ± 0.2 at 25°C

Antibiotic Solution:
Composition per 10mL:
5–Fluorocytosine 0.2g
Fosfomicin 0.1g
Ticarcillin 0.1g

Preparation of Antibiotic Solution: Add components to distilled/deionized water and bring volume to 10.0mL. Mix thoroughly. Filter sterilize.

Preparation of Medium: Add components, except antibiotic solution, to distilled/deionized water and bring volume to 990.0mL. Mix thoroughly. Gently heat and bring to boiling. Autoclave for 15 min at 15 psi pressure–121°C. Cool to 45° to 50°C. Aseptically add sterile antibiotic solution. Mix thoroughly. Pour into sterile Petri dishes in 20.0mL volumes.

Use: For the isolation and cultivation of multiresistant lipophilic *Corynebacterium* species, especially *Corynebacterium* Group JK found primarily in infections in immunocompromised hosts and patients with prosthetic valve endocarditis.

Legionella Agar Base (*Legionella* Medium) (BCYEα Agar, Modified)

Composition per liter:
Agar 17.0g
Yeast extract 10.0g
ACES buffer (N-2-acetamido-2-aminoethane sulfonic acid) 6.0g
Charcoal, activated 1.5g

KOH....1.5g
α-Ketoglutarate....1.0g

pH 6.85–7.0 at 25°C

Source: This medium is available as a prepared medium from Difco Laboratories.

Legionella Agar Enrichment:

Composition per 10mL:

L-Cysteine·HCl·H_2O....0.4g
$Fe_4(P_2O_7)_3$....0.25g

Preparation of *Legionella* Agar Enrichment: Add components to distilled/deionized water and bring volume to 10.0mL. Mix thoroughly. Filter sterilize.

Preparation of Medium: Add components, except *Legionella* agar enrichment, to distilled/deionized water and bring volume to 990.0mL. Mix thoroughly. Gently heat to boiling. Autoclave for 15 min at 15 psi pressure–121°C. Cool to 50° C. Add 10.0mL of sterile *Legionella* agar enrichment. Adjust pH to 6.9 at 50°C by adding 4.0–4.5mL of 1.0*N* KOH—this is a critical step. Mix thoroughly. Pour into sterile Petri dishes in 20.0mL volumes. Swirl medium while pouring to keep charcoal in suspension.

Use: For the preparation of *Legionella* agars. Also, for the isolation and cultivation of *Legionella* species from clinical specimens and in epidemiological investigations.

Legionella pneumophila Medium (Charcoal Yeast Extract Diphasic Blood Culture Medium) (Diphasic Blood Culture Buffered Charcoal Yeast Extract Medium) (CYE–DBCM)

Composition per liter:

Agar phase....500.0mL
Broth phase....500.0mL

pH 6.9–7.0 at 25°C

Agar Phase:

Composition per 500mL:

Agar....17.0g
Charcoal, activated....2.0g

Preparation of Agar Phase: Add components to distilled/deionized water and bring volume to 500.0mL. Mix thoroughly. Gently heat and bring to boiling. Distribute in 20.0mL volumes into 125.0mL serum bottles with aluminum crimp seals and rubber stoppers. Autoclave for 20 min at 15 psi pressure–121°C. Cool to 50°C. Swirl medium to put charcoal in suspension. Allow agar to solidify so that a slant with a 6.0cm height is formed.

Broth Phase:

Composition per 500mL:

Yeast extract....20.0g
L-Cysteine·HCl·H_2O solution....0.40g
$Fe(NO_3)_3 \cdot 9H_2O$ solution....0.10g

Preparation of Broth Phase: Add yeast extract to distilled/deionized water and bring volume to 480.0mL. Mix thoroughly. Autoclave for 15 min at 15 psi pressure–121°C. Cool to 25°C. Aseptically add sterile cysteine·HCl·H_2O solution and $Fe(NO_3)_3 \cdot 9H_2O$ solution. Mix thoroughly. Adjust pH to 6.9 with 6.0mL of sterile 1*N* KOH.

L-Cysteine·HCl·H_2O Solution:

Composition per 10mL:

L-Cysteine·HCl·H_2O....0.04g

Preparation of L-Cysteine·HCl·H_2O Solution: Add cysteine·HCl·H_2O to distilled/deionized water and bring volume to 10.0mL. Mix thoroughly. Filter sterilize.

$Fe(NO_3)_3 \cdot 9H_2O$ Solution:

Composition per 10mL:

$Fe(NO_3)_3 \cdot 9H_2O$....0.04g

Preparation of $Fe(NO_3)_3 \cdot 9H_2O$ Solution: Add $Fe(NO_3)_3 \cdot 9H_2O$ to distilled/deionized water and bring volume to 10.0mL. Mix thoroughly. Filter sterilize.

Preparation of Medium: Add 20.0mL of sterile broth phase to 125.0mL serum bottles containing 20.0mL of solidified agar phase. Seal bottles by crimping metal caps over rubber stoppers.

Use: For the isolation and cultivation of *Legionella pneumophila* from blood cultures.

Legionella Selective Agar

Composition per liter:

Agar....15.0g
ACES (2-[(2-amino-2-oxoethyl)-amino]ethane sulfonic acid) buffer....10.0g
Yeast extract....10.0g
Charcoal, activated....2.0g
α-Ketoglutarate....1.0g
L-Cysteine·HCl·H_2O solution....10.0mL
$Fe_4(P_2O_7)_3$ solution....10.0mL
Antibiotic solution....10.0mL

pH 6.85–7.0 at 25°C

Source: This medium is available as a prepared medium from BBL Microbiology Systems.

Cysteine·HCl·H_2O Solution:
Composition per 10mL:
L-Cysteine·HCl·H_2O 0.4g

Preparation of Cysteine·HCl·H_2O Solution: Add cysteine·HCl·H_2O to distilled/deionized water and bring volume to 10.0mL. Mix thoroughly. Filter sterilize.

$Fe_4(P_2O_7)_3$ Solution:
Composition per 10mL:
$Fe_4(P_2O_7)_3$ 0.25g

Preparation of $Fe_4(P_2O_7)_3$ Solution: Add $Fe_4(P_2O_7)_3$ to distilled/deionized water and bring volume to 10.0mL. Mix thoroughly. Filter sterilize.

Antibiotic Solution:
Composition per 10mL:
Anisomycin 10.0mg
Colistin 3.75mg
Vancomycin 2.0mg

Preparation of Antibiotic Solution: Add components to distilled/deionized water and bring volume to 10.0mL. Mix thoroughly. Filter sterilize.

Preparation of Medium: Add components—except cysteine·HCl·H_2O, $Fe_4(P_2O_7)_3$, and antibiotic solutions—to distilled/deionized water and bring volume to 970.0mL. Mix thoroughly. Gently heat and bring to boiling. Autoclave for 15 min at 15 psi pressure–121°C. Cool to 45° to 50°C. Aseptically add sterile cysteine·HCl·H_2O, $Fe_4(P_2O_7)_3$, and antibiotic solutions. Mix thoroughly. Pour into sterile Petri dishes. Swirl medium while pouring to keep charcoal in suspension.

Use: *Legionella* Selective Agar is used in qualitative procedures for the isolation of *Legionella* species from clinical specimens and in epidemiological investigations.

Leptospira Medium

Composition per liter:
$(NH_4)_2Fe(SO_4)_2·6H_2O$ 6.0g
NaH_2PO_4 .... 0.53g
L-Asparagine 0.5g
Glycerol 0.2g
Tween™ 60 .... 0.2g
$MgSO_4·7H_2O$ 0.15g
KH_2PO_4 .... 0.069g
Tween™ 80 .... 0.05g
EDTA 0.01g
$CaCO_3$ 4.0mg
Thiamine·HCl 1.0mg
Vitamin B_{12} 1.0μg
pH 7.4–7.6 at 25°C

Preparation of Medium: Add components, except thiamine·HCl to distilled/deionized water and bring volume to 990.0mL. Mix thoroughly. Gently heat and bring to boiling. Autoclave for 15 min at 15 psi pressure–121°C. Aseptically add 1.0mg of thiamine·HCl. Aseptically distribute into sterile tubes or flasks.

Use: For the cultivation of *Leptospira* species.

Leptospira Medium, EMJH (*Leptospira* Medium, Ellinghausen–McCullough/ Johnson–Harris)

Composition per liter:
Na_2HPO_4 .... 1.0g
NaCl 1.0g
KH_2PO_4 .... 0.3g
NH_4Cl 0.25g
Thiamine 5.0mg
Rabbit serum 100.0mL
pH 7.5 ± 0.2 at 25°C

Source: This medium is available as a premixed powder from Difco Laboratories.

Preparation of Medium: Add components, except rabbit serum, to distilled/deionized water and bring volume to 900.0mL. Mix thoroughly. Gently heat and bring to boiling. Autoclave for 15 min at 15 psi pressure–121°C. Cool to 25°C. Aseptically add sterile rabbit serum. Mix thoroughly. Aseptically distribute into sterile tubes or flasks.

Use: For the cultivation and maintenance of *Leptospira* species.

Leptospira Medium, Modified

Composition per liter:
Agar 1.5g
NaCl 0.5g
Peptone 0.3g
Beef extract 0.2g
Hemin solution 2.5mL
Sterile rabbit serum 100.0mL
pH 7.3 ± 0.1 at 25°C

Hemin Solution:
Composition per 100mL:
Hemin 0.05g
NaOH (1*N* solution) 1.0mL

Preparation of Hemin Solution: Add hemin to 1.0mL of 1*N* NaOH solution. Mix thoroughly. Bring volume to 100.0mL with distilled/deionized water. Autoclave for 15 min at 15 psi pressure–121°C. Cool to 45° to 50°C.

Preparation of Medium: Add components, except hemin solution and rabbit serum, to distilled/deionized water and bring volume to 897.5mL. Mix thoroughly. Gently heat and bring to boiling. Adjust pH to 7.4. Autoclave for 15 min at 15 psi pressure–121°C. Cool to 45° to 50°C. Aseptically add 2.5mL of sterile hemin solution and 100.0mL of sterile rabbit serum. Mix thoroughly. The pH of the medium should be 7.3. Store at 4°C for 24 hr. Inactivate medium at 56°C for 60 min. Aseptically distribute into sterile tubes or flasks.

Use: For the cultivation and maintenance of *Leptospira biflexa, Leptospira borgpetersenii, Leptospira interrogans, Leptospira meyeri, Leptospira noguchii, Leptospira santarosai,* and *Leptospira weili.*

Leptospira Protein–Free Medium (*Leptospira* PF Medium)

Composition per liter:

TES (*N*-tris[hydroxymethyl]methyl-2-aminoethane sulfonic acid) buffer....................1.2g
NaCl....................0.9g
Sodium pyruvate....................0.2g
CT-Tween™ 60....................12.0mL
CT-Tween™ 40....................3.0mL
$MgCl_2$-$CaCl_2$ solution....................1.0mL
Cyanocobalamin (0.02% solution)....................1.0mL
Glycerol (10% solution)....................1.0mL
KH_2PO_4 (1% solution)....................1.0mL
$MnSO_4 \cdot H_2O$ (0.1% solution)....................1.0mL
$ZnSO_4$ (0.4% solution)....................0.1mL

pH 7.6 ± 0.2 at 25°C

CT-Tween™ 60:

Composition per 200mL:

Charcoal, Norit A....................40.0g
Tween™ 60....................20.0g

Preparation of CT-Tween™ 60: Add Tween™ 60 to 200.0mL of distilled/deionized water. Mix thoroughly. While stirring, add charcoal. Stir mixture for 18 hr at 25°C. Allow charcoal to settle out of suspension for 18 hr at 4°C. Carefully decant the Tween™ solution off the sediment. Centrifuge the Tween™ solution at 10,000 × g for 1 hr. Decant supernatant solution. Pass Tween™ solution through a thin channel ultrafiltration XM 100 membrane. Store stock solution at –20°C.

CT-Tween™ 40:

Composition per 200mL:

Charcoal, Norit A....................40.0g
Tween™ 40....................20.0g

Preparation of CT-Tween™ 40: Add Tween™ 40 to 200.0mL of distilled/deionized water. Mix thoroughly. While stirring, add charcoal. Stir mixture for 18 hr at 25°C. Allow charcoal to settle out of suspension for 18 hr at 4°C. Carefully decant the Tween™ solution off the sediment. Centrifuge the Tween™ solution at 10,000 × g for 1 hr. Decant supernatant solution. Pass Tween™ solution through a thin channel ultrafiltration XM 100 membrane. Store stock solution at –20°C.

$MgCl_2$–$CaCl_2$ Solution:

Composition per 100mL:

$CaCl_2 \cdot 2H_2O$....................1.5g
$MgCl_2 \cdot 6H_2O$....................1.5g

Preparation of $MgCl_2$–$CaCl_2$ Solution: Add components to distilled/deionized water and bring volume to 100.0mL. Mix thoroughly.

Preparation of Medium: Add components to distilled/deionized water and bring volume to 1.0L. Mix thoroughly. Filter sterilize. Aseptically distribute into sterile tubes or flasks.

Use: For the cultivation of *Leptospira* species.

Letheen Agar

Composition per liter:

Agar....................15.0g
Tween™ 80....................7.0g
Pancreatic digest of casein....................5.0g
Beef extract....................3.0g
Glucose....................1.0g
Lecithin....................1.0g

pH 7.0 ± 0.2 at 25°C

Source: This medium is available as a premixed powder from BBL Microbiology Systems and Difco Laboratories.

Preparation of Medium: Add components to distilled/deionized water and bring volume to 1.0L. Mix thoroughly. Gently heat and bring to boiling. Distribute into tubes or flasks. Autoclave for 15 min at 15 psi pressure–121°C. Pour into sterile Petri dishes or leave in tubes.

Use: For the determination of the antimicrobial activity of quaternary ammonium compounds.

Letheen Agar, Modified

Composition per liter:

Agar....................20.0g
Thiotone....................10.0g
Pancreatic digest of casein....................10.0g
Tween™ 80....................7.0g
NaCl....................5.0g
Beef extract....................3.0g
Yeast extract....................2.0g

Glucose 1.0g
Lecithin 1.0g
$NaHSO_3$ 0.1g
pH 7.2 ± 0.2 at 25°C

Preparation of Medium: Add components to distilled/deionized water and bring volume to 1.0L. Mix thoroughly. Gently heat and bring to boiling. Distribute into tubes or flasks. Autoclave for 15 min at 15 psi pressure–121°C. Pour into sterile Petri dishes in 20.0mL.

Use: For the determination of the antimicrobial activity of quaternary ammonium compounds.

Letheen Broth

Composition per liter:

Peptic digest of animal tissue 10.0g
Beef extract 5.0g
NaCl 5.0g
Tween™ 80 .... 5.0g
Lecithin 0.7g
pH 7.0 ± 0.2 at 25°C

Source: This medium is available as a premixed powder from Difco Laboratories.

Preparation of Medium: Add components to distilled/deionized water and bring volume to 1.0L. Mix thoroughly. Distribute into tubes or flasks. Autoclave for 15 min at 15 psi pressure–121°C.

Use: For the determination of the antimicrobial activity of quaternary ammonium compounds.

Letheen Broth, Modified

Composition per liter:

Peptic digest of animal tissue 10.0g
Thiotone peptone 10.0g
Beef extract 5.0g
NaCl 5.0g
Tween™ 80 .... 5.0g
Pancreatic digest of casein 5.0g
Yeast extract 2.0g
Lecithin 0.7g
$NaHSO_3$ 0.1g
pH 7.2 ± 0.2 at 25°C

Preparation of Medium: Add components to distilled/deionized water and bring volume to 1.0L. Mix thoroughly. Distribute into screw-capped bottles in 90.0mL volumes. Autoclave for 15 min at 15 psi pressure–121°C.

Use: For the determination of the antimicrobial activity of quaternary ammonium compounds.

Levine EMB Agar (Levine Eosin Methylene Blue Agar) (Eosin Methylene Blue Agar, Levine) (LEMB Agar)

Composition per liter:

Agar 15.0g
Lactose 10.0g
Peptone 10.0g
K_2HPO_4 .... 2.0g
Eosin Y 0.4g
Methylene Blue 0.065mg
pH 7.1 ± 0.2 at 25°C

Source: This medium is available as a premixed powder from BBL Microbiology Systems and Difco Laboratories.

Preparation of Medium: Add components to distilled/deionized water and bring volume to 1.0L. Mix thoroughly. Gently heat and bring to boiling. Distribute into tubes or flasks. Autoclave for 15 min at 15 psi pressure–121°C. Pour into sterile Petri dishes or leave in tubes.

Use: For the isolation, cultivation and differentiation of Gram-negative enteric bacteria based on lactose fermentation. Bacteria that ferment lactose, especially the coliform bacterium *Escherichia coli*, appear as colonies with a green metallic sheen or blue-black to brown color. Bacteria that do not ferment lactose appear as colorless or transparent light purple colonies.

Litmus Milk

Composition per liter:

Skim milk 100.0g
Azolitmin 0.5g
Na_2SO_3 .... 0.5g
pH 6.5 ± 0.2 at 25°C

Source: This medium is available as a premixed powder from BBL Microbiology Systems, Difco Laboratories, and Oxoid Unipath.

Preparation of Medium: Add components to distilled/deionized water and bring volume to 1.0L. Mix thoroughly. Gently heat and bring to boiling. Distribute into tubes or flasks. Autoclave for 20 min at 10 psi pressure–115°C.

Use: For the maintenance of lactic acid bacteria and for the differentiation of several bacteria, especially *Clostridium* species, based on their action on milk. Bacteria that do not ferment carbohydrates, such as *Proteus vulgaris* or *Moraxella lacunata*, show no change in the litmus indicator. Bacteria that ferment lactose or glucose with the production of gas, such as *Clostridium perfringens*, turn the medium pink and frothy. Bacteria that proteolytically degrade lactalbumin turn the medium blue. Bacteria that coagulate casein form a curd or clot. Bacteria that peptonize casein, such as *Pseudomonas aeruginosa*, show a dissolution of the clot.

Littman Oxgall Agar

Composition per liter:

Agar	20.0g
Oxgall	15.0g
Glucose	10.0g
Peptone	10.0g
Crystal Violet	0.01g
Streptomycin solution	10.0mL

pH 6.5 ± 0.2 at 25°C

Source: This medium is available as a premixed powder from Difco Laboratories.

Streptomycin Solution:
Composition per 10mL:

Streptomycin	0.03g

Preparation of Streptomycin Solution: Add streptomycin to distilled/deionized water and bring volume to 10.0mL. Mix thoroughly. Filter sterilize.

Preparation of Medium: Add components, except streptomycin solution, to distilled/deionized water and bring volume to 990.0mL. Mix thoroughly. Gently heat and bring to boiling. Autoclave for 15 min at 15 psi pressure–121°C. Cool to 45° to 50°C. Aseptically add sterile streptomycin solution. Mix thoroughly. Pour into sterile Petri dishes or distribute into sterile tubes. Allow tubes to cool in a slanted position.

Use: For the selective isolation and cultivation of fungi, especially dermatophytes.

Liver Broth

Composition per liter:

Beef liver, fresh	453.0g
Pancreatic digest of casein	10.0g
K_2HPO_4	1.0g
Soluble starch	1.0g

pH 7.6 ± 0.2 at 25°C

Preparation of Medium: Remove the fat from fresh beef liver. Grind the liver. Add 1.0L of distilled/deionized water. Gently heat and bring to boiling. Continue boiling for 60 min. Adjust pH to 7.6. Filter through cheesecloth. Reserve meat. To filtrate, add pancreatic digest of casein, K_2HPO_4, and soluble starch. Bring volume to 1.0L with distilled/deionized water. Refilter solution. Add meat particles to test tubes to a depth of approximately 2 cm. Distribute broth into tubes with meat particles in 15.0mL volumes. Autoclave for 20 min at 15 psi pressure–121°C.

Use: For the isolation and cultivation of anaerobic microorganisms, especially *Clostridium botulinum*, from foods.

Liver Infusion Agar

Composition per liter:

Beef liver, infusion from	500.0g
Agar	20.0g
Proteose peptone	10.0g
NaCl	5.0g

pH 6.9 ± 0.2 at 25°C

Source: This medium is available as a premixed powder from Difco Laboratories.

Preparation of Medium: Add components to distilled/deionized water and bring volume to 1.0L. Mix thoroughly. Gently heat and bring to boiling. Distribute into tubes or flasks. Autoclave for 15 min at 15 psi pressure–121°C. Pour into sterile Petri dishes or leave in tubes.

Use: For the cultivation of *Brucella* species and other fastidious pathogenic bacteria.

Liver Infusion Broth

Composition per liter:

Beef liver, infusion from	500.0g
Proteose peptone	10.0g
NaCl	5.0g

pH 6.9 ± 0.2 at 25°C

Source: This medium is available as a premixed powder from Difco Laboratories.

Preparation of Medium: Add components to distilled/deionized water and bring volume to 1.0L. Mix thoroughly. Gently heat and bring to boiling. Distribute into tubes or flasks. Autoclave for 15 min at 15 psi pressure–121°C.

Use: For the cultivation of *Brucella* species and other fastidious pathogenic bacteria.

Liver Veal Egg Yolk Agar

Composition per 1080mL:

Liver veal agar	1.0L
Egg yolk emulsion, 50%	80.0mL

pH 7.3 ± 0.2 at 25°C

Liver Veal Agar:

Composition per liter:

Veal, infusion from	500.0g
Beef liver, infusion from	50.0g
Gelatin	20.0g
Proteose peptone	20.0g
Agar	15.0g
Soluble starch	10.0g
Glucose	5.0g
NaCl	5.0g
Casein	2.0g
$NaNO_3$	2.0g
Enzymatic digest of protein	1.3g
Pancreatic digest of casein	1.3g

Source: This medium is available as a premixed powder from Difco Laboratories.

Preparation of Liver Veal Agar: Add components to distilled/deionized water and bring volume to 1.0L. Mix thoroughly. Gently heat and bring to boiling. Distribute into tubes or flasks. Make sure that some liver and veal particles are transferred to each tube. Autoclave for 15 min at 15 psi pressure–121°C. Cool to 50°C.

Egg Yolk Emulsion, 50%:

Composition per 100mL:

Chicken egg yolks	11
Whole chicken egg	1
NaCl (0.9% solution)	50.0mL

Preparation of Egg Yolk Emulsion, 50%: Soak eggs with 1:100 dilution of saturated mercuric chloride solution for 1 min. Crack eggs and separate yolks from whites. Mix egg yolks with 1 chicken egg. Measure 50.0mL of egg yolk emulsion and add to 50.0mL of 0.9% NaCl solution. Mix thoroughly. Filter sterilize. Warm to 45° to 50°C.

Preparation of Medium: To 1.0L of cooled sterile liver veal agar, aseptically add 80.0mL of sterile egg yolk emulsion, 50%. Mix thoroughly. Pour into sterile Petri dishes. Dry plates at 35°C for 24 hr.

Use: For the cultivation of a variety of anaerobic organisms.

Loeffler Blood Serum Medium

Composition per liter:

Beef blood serum	750.0mL
Dextrose broth	250.0mL

pH 7.1 ± 0.2 at 25°C

Source: This medium is available as a premixed powder from Difco Laboratories.

Dextrose Broth:

Composition per liter:

Tryptose	10.0g
Glucose	5.0g
Sodium chloride	5.0g
Beef extract	3.0g

Preparation of Dextrose Broth: Add components to distilled/deionized water and bring volume to 1.0L. Mix thoroughly.

Preparation of Medium: Combine 750.0mL of beef blood serum with 250.0mL of dextrose broth. Mix thoroughly. Distribute into screw-capped tubes. Slant tubes in the autoclave. Close the autoclave door loosely. Autoclave for 10 min at 0 psi pressure–100°C. Close the autoclave door tightly. Autoclave for 15 min at 15 psi pressure–121°C.

Use: For the cultivation of *Corynebacterium diphtheriae*. Also used for the demonstration of pigment production and proteolysis by *Corynebacterium diphtheriae.*

Loeffler Blood Serum Medium

Composition per liter:

Beef blood serum	750.0mL
Dextrose broth	250.0mL

pH 7.1 ± 0.2 at 25°C

Dextrose Broth:

Composition per liter:

Enzymatic digest of protein	2.50g
Glucose	1.25g
NaCl	1.25g
Beef extract	0.75g

Preparation of Dextrose Broth: Add components to distilled/deionized water and bring volume to 1.0L. Mix thoroughly.

Preparation of Medium: Combine 750.0mL of beef blood serum with 250.0mL of dextrose broth. Mix thoroughly. Distribute into screw-capped tubes. Slant tubes in the autoclave. Close the autoclave door loosely. Autoclave for 10 min at 0 psi pressure–100°C. Close the autoclave door tightly. Autoclave for 15 min at 15 psi pressure–121°C.

Use: For the cultivation of *Corynebacterium diphtheriae*. Also used for the demonstration of pigment production and proteolysis by *Corynebacterium diphtheriae.*

Loeffler Medium

Composition per liter:

Beef serum	70.0g
Egg, dried	7.5g

Heart muscle, solids from infusion 0.72g
Glucose 0.71g
Peptic digest of animal tissue...........................0.71g
NaCl 0.36g

pH 7.6 ± 0.2 at 25°C

Source: This medium is available as a premixed powder from BBL Microbiology Systems.

Preparation of Medium: Add components to distilled/deionized water and bring volume to 1.0L. Mix thoroughly. Distribute into screw-capped tubes. Slant tubes in the autoclave. Close the autoclave door loosely. Autoclave for 10 min at 0 psi pressure–100°C. Close the autoclave door tightly. Autoclave for 15 min at 15 psi pressure–121°C.

Use: For the cultivation of *Corynebacterium diphtheriae.* For the demonstration of pigment production and proteolysis by *Corynebacterium diphtheriae.* Also used for the cultivation and maintenance of *Moraxella lacunata.*

Loeffler Slant

Composition per liter:

Tryptose 5.0g
Glucose 1.0g
Beef serum 750.0mL

Preparation of Medium: Add components to distilled/deionized water and bring volume to 1.0L. Mix thoroughly. Distribute into screw-capped tubes. Slant tubes in the autoclave. Close the autoclave door loosely. Autoclave for 10 min at 0 psi pressure–100°C. Close the autoclave door tightly. Autoclave for 15 min at 15 psi pressure–121°C.

Use: For the cultivation of *Corynebacterium diphtheriae.* For the demonstration of pigment production and proteolysis by *Corynebacterium diphtheriae.* Also used for the cultivation and maintenance of *Moraxella lacunata.*

Loeffler Slant, Modified

Composition per liter:

Peptone..0.5g
Glucose 1.0g
Beef serum 300.0mL

pH 7.6 ± 0.2 at 25°C

Preparation of Medium: Add peptone and glucose to distilled/deionized water and bring volume to 100.0mL. Mix thoroughly. Add beef serum. Mix thoroughly. Adjust pH to 7.6. Distribute into screw-capped tubes in 3.0mL volumes. Slant tubes in the autoclave. Autoclave for 30 min at 0 psi pressure–100°C.

Use: For the cultivation of *Corynebacterium diphtheriae.* For the demonstration of pigment production and proteolysis by *Corynebacterium diphtheriae.* Also used for the cultivation and maintenance of *Moraxella lacunata.*

Lombard–Dowell Agar (LD Agar)

Agar..20.0g
Pancreatic digest of casein 5.0g
Yeast extract...5.0g
NaCl 2.5g
L-Cystine 0.4g
L-Tryptophan 0.2g
Na_2SO_3 ....0.1g
Hemin.. 10.0mg
NaOH (1*N* NaOH) 5.0mL
Vitamin K_1 solution 1.0mL

pH 7.5 ± 0.2 at 25°C

Vitamin K_1 Solution:
Composition per 100mL:

Vitamin K_1 ....1.0g
Ethanol 99.0mL

Preparation of Vitamin K_1 Solution: Add vitamin K_1 to 99.0mL of absolute ethanol. Mix thoroughly.

Preparation of Medium: Add hemin and cystine to 5.0mL of NaOH. Mix thoroughly. Add remaining components. Bring volume to 1.0L with distilled/deionized water. Mix thoroughly. Gently heat and bring to boiling. Distribute into tubes or flasks. Autoclave for 15 min at 15 psi pressure–121°C. Pour into sterile Petri dishes.

Use: For the cultivation and identification of a variety of obligate anaerobic bacteria. For the cultivation of *Bacteroides species, Fusobacterium species, Clostridium* species, and non–spore–forming Gram–positive anaerobes.

Lombard–Dowell Bile Agar (LD Bile Agar)

Composition per liter:

Agar..20.0g
Oxgall...20.0g
Pancreatic digest of casein 5.0g
Yeast extract...5.0g
NaCl 2.5g
D-Glucose 1.0g
L-Cystine 0.4g
L-Tryptophan 0.2g
Na_2SO_3 ....0.1g

Hemin 10.0mg
NaOH (1*N* NaOH) 5.0mL
Vitamin K_1 solution 1.0mL
pH 7.5 ± 0.2 at 25°C

Vitamin K_1 Solution:
Composition per 100mL:
Vitamin K_1 .... 1.0g
Ethanol 99.0mL

Preparation of Vitamin K_1 Solution: Add vitamin K_1 to 99.0mL of absolute ethanol. Mix thoroughly.

Preparation of Medium: Add hemin and cystine to 5.0mL of NaOH. Mix thoroughly. Add remaining components. Bring volume to 1.0L with distilled/deionized water. Mix thoroughly. Gently heat and bring to boiling. Distribute into tubes or flasks. Autoclave for 15 min at 15 psi pressure–121°C. Pour into sterile Petri dishes.

Use: For the cultivation and identification of a variety of obligate anaerobic bacteria in the presence of 20% bile.

Lombard–Dowell Broth (LD Broth)

Composition per liter:
Pancreatic digest of casein 5.0g
Yeast extract 5.0g
Agar 0.7g
NaCl 2.5g
L-Tryptophan 0.2g
Na_2SO_3 .... 0.1g
NaOH (1*N* NaOH) 5.0mL
Hemin solution 1.0mL
Vitamin K_1 solution 1.0mL
pH 7.5 ± 0.2 at 25°C

Hemin Solution:
Composition per 100mL:
Hemin 1.0g
NaOH (1*N* solution) 20.0mL

Preparation of Hemin Solution: Add hemin to 20.0mL of 1*N* NaOH solution. Mix thoroughly. Bring volume to 100.0mL with distilled/deionized water.

Vitamin K_1 Solution:
Composition per 100mL:
Vitamin K_1 .... 1.0g
Ethanol 99.0mL

Preparation of Vitamin K_1 Solution: Add vitamin K_1 to 99.0mL of absolute ethanol. Mix thoroughly.

Preparation of Medium: Add tryptophan to 5.0mL of NaOH. Mix thoroughly. Add remaining components. Bring volume to 1.0L with distilled/deionized water. Mix thoroughly. Gently heat and bring to boiling. Adjust pH to 7.5. Distribute into screw-capped tubes in 7.0mL volumes. Autoclave for 15 min at 15 psi pressure–121°C. Cool tubes, with caps loose, under 85% N_2 + 10% H_2 + 5% CO_2. Tighten caps.

Use: For the cultivation of a wide variety of anaerobic bacteria.

Lombard–Dowell Egg Yolk Agar (LD Egg Yolk Agar) (Egg Yolk Agar, Lombard–Dowell)

Composition per 9100mL:
$Na_2HPO_4 \cdot 12H_2O$ 5.0g
Glucose 2.0g
LD Agar 9000.0mL
Egg yolk emulsion 100.0mL
$MgSO_4 \cdot 7H_20$ (5% solution) 0.2mL
pH 7.5 ± 0.2 at 25°C

LD Agar:
Composition per liter:
Agar 20.0g
Pancreatic digest of casein 5.0g
Yeast extract 5.0g
NaCl 2.5g
L-Cystine 0.4g
L-Tryptophan 0.2g
Na_2SO_3 .... 0.1g
Hemin 10.0mg
NaOH (1*N* NaOH) 5.0mL
Vitamin K_1 solution 1.0mL

Preparation of LD Agar: Add hemin and cystine to 5.0mL of NaOH. Mix thoroughly. Add remaining components. Mix thoroughly. Gently heat and bring to boiling.

Vitamin K_1 Solution:
Composition per 100mL:
Vitamin K_1 .... 1.0g
Ethanol 99.0mL

Preparation of Vitamin K_1 Solution: Add vitamin K_1 to 99.0mL of absolute ethanol. Mix thoroughly.

Egg Yolk Emulsion:
Composition:
Chicken egg yolks 11
Whole chicken egg 1

Preparation of Egg Yolk Emulsion: Soak eggs with 1:100 dilution of saturated mercuric chloride solution for 1 min. Crack eggs and separate yolks from whites. Mix egg yolks with 1 chicken egg.

Preparation of Medium: Combine components, except egg yolk emulsion. Mix thoroughly. Autoclave for 15 min at 15 psi pressure–121°C. Cool to 45° to 50°C. Aseptically add 100.0mL of egg yolk emulsion. Mix thoroughly. Pour into sterile Petri dishes.

Use: For the cultivation of a wide variety of anaerobic bacteria. For the differentiation of anaerobic bacteria based on lecithinase production, lipase production and proteolytic ability. Bacteria that produce lecithinase appear as colonies surrounded by a zone of insoluble precipitate. Bacteria that produce lipase appear as colonies with a pearly iridescent sheen. Bacteria that produce proteolytic activity appear as colonies surrounded by a clear zone.

Lombard–Dowell Esculin Agar (LD Esculin Agar) (Esculin Agar, Lombard–Dowell)

Composition per liter:

Agar	20.0g
Pancreatic digest of casein	5.0g
Yeast extract	5.0g
NaCl	2.5g
Esculin	1.0g
Ferric citrate	0.5g
L-Cystine	0.4g
L-Tryptophan	0.2g
Hemin	10.0mg
NaOH (1*N* NaOH)	5.0mL
Vitamin K_1 solution	1.0mL

pH 7.5 ± 0.2 at 25°C

Vitamin K_1 Solution:
Composition per 100mL:

Vitamin K_1	1.0g
Ethanol	99.0mL

Preparation of Vitamin K_1 Solution: Add vitamin K_1 to 99.0mL of absolute ethanol. Mix thoroughly.

Preparation of Medium: Add hemin and cystine to 5.0mL of NaOH. Mix thoroughly. Add remaining components. Bring volume to 1.0L with distilled/deionized water. Mix thoroughly. Gently heat and bring to boiling. Distribute into tubes or flasks. Autoclave for 15 min at 15 psi pressure–121°C. Pour into sterile Petri dishes.

Use: For the cultivation of a wide variety of anaerobic bacteria. For the differentiation of anaerobic bacteria based on esculin hydrolysis, H_2S production and catalase production. Bacteria that hydrolyze esculin appear as colonies surrounded by a red-brown to dark brown zone. Bacteria that produce H_2S appear as black colonies.

Lombard–Dowell Gelatin Agar (LD Gelatin Agar)

Composition per liter:

Agar	20.0g
Pancreatic digest of casein	5.0g
Yeast extract	5.0g
Gelatin	4.0g
NaCl	2.5g
Glucose	1.0g
L-Cystine	0.4g
L-Tryptophan	0.2g
Na_2SO_3	0.1g
Hemin	10.0mg
NaOH (1*N* NaOH)	5.0mL
Vitamin K_1 solution	1.0mL

pH 7.5 ± 0.2 at 25°C

Vitamin K_1 Solution:
Composition per 100mL:

Vitamin K_1	1.0g
Ethanol	99.0mL

Preparation of Vitamin K_1 Solution: Add vitamin K_1 to 99.0mL of absolute ethanol. Mix thoroughly.

Preparation of Medium: Add hemin and cystine to 5.0mL of NaOH. Mix thoroughly. Add remaining components, except agar and gelatin. Bring volume to 750.0mL with distilled/deionized water. Mix thoroughly. Gently heat and bring to boiling. In a separate flask, add gelatin to 100.0mL of cold distilled/deionized water. Gently heat and bring to 70°C. Add gelatin solution to the 750.0mL of basal medium. Mix thoroughly. Add agar. Bring volume to 1.0L with distilled/deionized water. Autoclave for 15 min at 15 psi pressure–121°C. Pour into sterile Petri dishes.

Use: For the cultivation of a wide variety of anaerobic bacteria. For the differentiation of anaerobic bacteria based on gelatinase production. After incubation of plates, gelatinase activity is determined by addition of Frazier's reagent. Bacteria that hydrolyze gelatin appear as colonies surrounded by a clear zone.

Lombard–Dowell Neomycin Agar (Egg Yolk Agar with Neomycin)

Composition per 9100mL:

$Na_2HPO_4 \cdot 12H_2O$	5.0g
Glucose	2.0g
Neomycin sulfate	0.1g
LD Agar	9000.0mL
Egg yolk emulsion	100.0mL
$MgSO_4 \cdot 7H_20$ (5% solution)	0.2mL

pH 7.5 ± 0.2 at 25°C

LD Agar:

Composition per liter:

Agar....................20.0g
Pancreatic digest of casein....................5.0g
Yeast extract....................5.0g
NaCl....................2.5g
L-Cystine....................0.4g
L-Tryptophan....................0.2g
Na_2SO_3....................0.1g
Hemin....................10.0mg
NaOH (1*N* NaOH)....................5.0mL
Vitamin K_1 solution....................1.0mL

Preparation of LD Agar: Add hemin and cystine to 5.0mL of NaOH. Mix thoroughly. Add remaining components. Mix thoroughly. Gently heat and bring to boiling.

Vitamin K_1 Solution:

Composition per 100mL:

Vitamin K_1....................1.0g
Ethanol....................99.0mL

Preparation of Vitamin K_1 Solution: Add vitamin K_1 to 99.0mL of absolute ethanol. Mix thoroughly.

Egg Yolk Emulsion:

Composition:

Chicken egg yolks....................11
Whole chicken egg....................1

Preparation of Egg Yolk Emulsion: Soak eggs with 1:100 dilution of saturated mercuric chloride solution for 1 min. Crack eggs and separate yolks from whites. Mix egg yolks with 1 chicken egg.

Preparation of Medium: Combine components, except egg yolk emulsion and neomycin sulfate. Mix thoroughly. Autoclave for 15 min at 15 psi pressure–121°C. Cool to 45° to 50°C. Aseptically add 100.0mL of egg yolk emulsion and neomycin sulfate. Mix thoroughly. Pour into sterile Petri dishes.

Use: For the selective cultivation of a wide variety of anaerobic bacteria. For the differentiation of anaerobic bacteria based on lecithinase production, lipase production and proteolytic ability. Bacteria that produce lecithinase appear as colonies surrounded by a zone of insoluble precipitate. Bacteria that produce lipase appear as colonies with a pearly iridescent sheen. Bacteria that produce proteolytic activity appear as colonies surrounded by a clear zone.

Low Iron YC Agar

Composition per 1033mL:

Solution 1....................1.0L
Solution 4....................30.0mL
Solution 2....................2.0mL
Solution 3....................1.0mL

pH 7.4 ± 0.2 at 25°C

Solution 1:

Composition per liter:

Yeast extract....................20.0g
Noble agar....................10.0g
Casamino acids....................10.0g
KH_2PO_4....................5.0g
$CaCl_2$....................1.0g
Tryptophan....................0.05g

Preparation of Solution 1: Add components to distilled/deionized water and bring volume to 1.0L. Mix thoroughly. Adjust pH to 7.4. Gently heat and bring to boiling. Filter through #40 ashless filter paper.

Solution 2:

Composition per 100mL:

$MgSO_4 \cdot 7H_2O$....................22.5g
$CuSO_4 \cdot 5H_2O$....................0.5g
$ZnSO_4 \cdot 5H_2O$....................0.2g
β-Alanine....................0.115g
Nicotinic Acid....................0.115g
$MnCl_2 \cdot 4H_2O$....................0.075g
Pimelic acid....................7.5mg
HCl, conc.....................3.0mL

Preparation of Solution 2: Add components to distilled/deionized water and bring volume to 100.0mL. Mix thoroughly.

Solution 3:

Composition per 100mL:

L-Cystine....................20.0g
HCl, concentrated....................20.0mL

Preparation of Solution 3: Add components to distilled/deionized water and bring volume to 100.0mL. Mix thoroughly.

Solution 4:

Composition per 100mL:

Maltose....................50.0g
$CaCl_2 \cdot 2H_2O$....................0.5g

Preparation of Solution 4: Add components to distilled/deionized water and bring volume to 100.0mL. Mix thoroughly. Autoclave for 10 min at 11 psi pressure–116°C. Cool to 45° to 50°C.

Preparation of Medium: To 1.0L of solution 1, add 2.0mL of solution 2 and 1.0mL of solution 3. Mix thoroughly. Autoclave for 15 min at 15 psi pressure–121°C. Cool to 45° to 50°C. Aseptically add 30.0mL of sterile solution 4. Mix thoroughly. Pour into sterile Petri dishes or distribute into sterile tubes.

Use: For the cultivation and maintenance of *Corynebacterium diphtheriae.*

Low Iron YC Broth

Composition per 1033mL:

Solution 1 .... 1.0L
Solution 4 .... 30.0mL
Solution 2 .... 2.0mL
Solution 3 .... 1.0mL
pH 7.4 ± 0.2 at 25°C

Solution 1:

Composition per liter:

Yeast extract 20.0g
Casamino acids 10.0g
KH_2PO_4 .... 5.0g
$CaCl_2 \cdot 2H_2O$ 1.0g
Tryptophan 0.05g

Preparation of Solution 1: Add components to distilled/deionized water and bring volume to 1.0L. Mix thoroughly. Adjust pH to 7.4. Gently heat and bring to boiling. Filter through #40 ashless filter paper.

Solution 2:

Composition per 100mL:

$MgSO_4 \cdot 7H_2O$ 22.5g
$CuSO_4 \cdot 5H_2O$ 0.5g
$ZnSO_4 \cdot 5H_2O$ 0.2g
β-Alanine 0.115g
Nicotinic Acid 0.115g
$MnCl_2 \cdot 4H_2O$ 0.075g
Pimelic acid 7.5mg
HCl, conc. 3.0mL

Preparation of Solution 2: Add components to distilled/deionized water and bring volume to 100.0mL. Mix thoroughly.

Solution 3:

Composition per 100mL:

L-Cystine 20.0g
HCl, concentrated 20.0mL

Preparation of Solution 3: Add components to distilled/deionized water and bring volume to 100.0mL. Mix thoroughly.

Solution 4:

Composition per 100mL:

Maltose 50.0g
$CaCl_2 \cdot 2H_2O$ 0.5g

Preparation of Solution 4: Add components to distilled/deionized water and bring volume to 100.0mL. Mix thoroughly. Autoclave for 10 min at 11 psi pressure–116°C. Cool to 25°C.

Preparation of Medium: To 1.0L of solution 1, add 2.0mL of solution 2 and 1.0mL of solution 3. Mix thoroughly. Autoclave for 15 min at 15 psi pressure–121°C. Cool to 25°C. Aseptically add 30.0mL of sterile solution 4. Mix thoroughly. Aseptically distribute into sterile tubes or flasks.

Use: For the cultivation and maintenance of *Corynebacterium diphtheriae*.

Lowenstein–Gruft Medium

Composition per 1600mL:

Potato starch 30.0g
Asparagine 3.6g
KH_2PO_4 .... 2.4g
Magnesium citrate 0.6g
Malachite Green 0.4g
$MgSO_4 \cdot 7H_2O$ 0.24g
Nalidixic acid 0.056g
Ribonucleic acid 0.08mg
Homogenized whole egg 1.0L
Glycerol 12.0mL
Penicillin 80,000U

Homogenized Whole Egg:

Composition per liter:

Whole eggs 18–24

Preparation of Homogenized Whole Egg: Use fresh eggs, less than 1 week old. Scrub the shells with soap. Let stand in a soap solution for 30 min. Rinse in running water. Soak eggs in 70% ethanol for 15 min. Break the eggs into a sterile container. Homogenize by shaking. Filter through four layers of sterile cheesecloth into a sterile graduated cylinder. Measure out 1.0L.

Preparation of Medium: Add glycerol to 600.0mL of distilled/deionized water. Mix thoroughly. Add remaining components, except fresh egg mixture. Mix thoroughly. Gently heat while stirring and bring to boiling. Autoclave for 15 min at 15 psi pressure–121°C. Cool to 50°C. Aseptically add 1.0L of homogenized whole egg. Mix thoroughly. Distribute into sterile screw-capped tubes. Place tubes in a slanted position. Inspissate at 85°C (moist heat) for 45 min.

Use: For the cultivation and differentiation of *Mycobacterium* species. *M. tuberculosis* appears as granular, rough, dry colonies. *M. kansasii* appears as smooth to rough photochromogenic colonies. *M. gordonae* appears as smooth yellow-orange colonies. *M. avium* appears as smooth, colorless colonies. *M. smegmatis* appears as wrinkled, creamy white colonies.

Lowenstein–Jensen Medium

Composition per 1600mL:

Potato starch 30.0g
Asparagine 3.6g

KH_2PO_4 .... 2.4g
Magnesium citrate 0.6g
Malachite Green 0.4g
$MgSO_4 \cdot 7H_2O$ 0.24g
Homogenized whole egg 1.0L
Glycerol 12.0mL

Source: Available as a prepared medium from BBL Microbiology Systems, Difco Laboratories and Oxoid Unipath.

Homogenized Whole Egg:
Composition per liter:
Whole eggs 18-24

Preparation of Homogenized Whole Egg: Use fresh eggs, less than 1 week old. Scrub the shells with soap. Let stand in a soap solution for 30 min. Rinse in running water. Soak eggs in 70% ethanol for 15 min. Break the eggs into a sterile container. Homogenize by shaking. Filter through four layers of sterile cheesecloth into a sterile graduated cylinder. Measure out 1.0L.

Preparation of Medium: Add glycerol to 600.0mL of distilled/deionized water. Mix thoroughly. Add remaining components, except fresh egg mixture. Mix thoroughly. Gently heat while stirring and bring to boiling. Autoclave for 15 min at 15 psi pressure–121°C. Cool to 50°C. Aseptically add 1.0L of homogenized whole egg. Mix thoroughly. Distribute into sterile screw-capped tubes. Place tubes in a slanted position. Inspissate at 85°C (moist heat) for 45 min.

Use: For the cultivation and differentiation of *Mycobacterium* species. *M. tuberculosis* appears as granular, rough, dry colonies. *M. kansasii* appears as smooth to rough photochromogenic colonies. *M. gordonae* appears as smooth yellow-orange colonies. *M. avium* appears as smooth, colorless colonies. *M. smegmatis* appears as wrinkled, creamy white colonies. Also used for the cultivation and maintenance of *Gordona* species, *Nocardia* species, *Rhodococcus* species, and *Tsukamurella paurometabolum*.

Lowenstein–Jensen Medium with NaCl

Composition per 1600mL:
NaCl 80.0g
Potato starch 30.0g
Asparagine 3.6g
KH_2PO_4 .... 2.4g
Magnesium citrate 0.6g
Malachite Green 0.4g
$MgSO_4 \cdot 7H_2O$ 0.24g
Homogenized whole egg 1.0L
Glycerol 12.0mL

Homogenized Whole Egg:
Composition per liter:
Whole eggs 18-24

Preparation of Homogenized Whole Egg: Use fresh eggs, less than 1 week old. Scrub the shells with soap. Let stand in a soap solution for 30 min. Rinse in running water. Soak eggs in 70% ethanol for 15 min. Break the eggs into a sterile container. Homogenize by shaking. Filter through four layers of sterile cheesecloth into a sterile graduated cylinder. Measure out 1.0L.

Preparation of Medium: Add glycerol to 600.0mL of distilled/deionized water. Mix thoroughly. Add remaining components, except fresh egg mixture. Mix thoroughly. Gently heat while stirring and bring to boiling. Autoclave for 15 min at 15 psi pressure–121°C. Cool to 50°C. Aseptically add 1.0L of homogenized whole egg. Mix thoroughly. Distribute into sterile screw-capped tubes. Place tubes in a slanted position. Inspissate at 85°C (moist heat) for 45 min.

Use: For the cultivation of *Mycobacterium smegmatis* and other salt-tolerant *Mycobacterium* species.

Lowenstein–Jensen Medium without Glycerol

Composition per liter:
Potato starch 30.0g
Asparagine 3.6g
KH_2PO_4 .... 2.4g
Magnesium citrate 0.6g
Malachite green 0.4g
$MgSO_4 \cdot 7H_2O$ 0.24g
Homogenized whole egg 1.0L

Homogenized Whole Egg:
Composition per liter:
Whole eggs 18–24

Preparation of Homogenized Whole Egg: Use fresh eggs, less than 1 week old. Scrub the shells with soap. Let stand in a soap solution for 30 min. Rinse in running water. Soak eggs in 70% ethanol for 15 min. Break the eggs into a sterile container. Homogenize by shaking. Filter through four layers of sterile cheesecloth into a sterile graduated cylinder. Measure out 1.0L.

Preparation of Medium: Add components, except fresh egg mixture, to 600.0mL of distilled/deionized water. Mix thoroughly. Gently heat while stirring and bring to boiling. Autoclave for 15 min at 15 psi pressure–121°C. Cool to 50°C. Aseptically add 1.0L of homogenized whole egg. Mix thoroughly. Distribute into sterile screw-capped tubes. Place tubes in a slanted position. Inspissate at 85°C (moist heat) for 45 min.

Use: For the cultivation and maintenance of *Mycobacterium* species, especially *M. bovis* and other species that are sensitive to glycerol.

Lowenstein–Jensen Medium with Streptomycin

Composition per 161..0mL:

Potato starch	30.0g
Asparagine	3.6g
KH_2PO_4	2.4g
Magnesium citrate	0.6g
Malachite Green	0.4g
$MgSO_4 \cdot 7H_2O$	0.24g
Homogenized whole egg	1.0L
Glycerol	12.0mL
Streptomycin solution	10.0mL

Homogenized Whole Egg:
Composition per liter:

Whole eggs	18-24

Preparation of Homogenized Whole Egg: Use fresh eggs, less than 1 week old. Scrub the shells with soap. Let stand in a soap solution for 30 min. Rinse in running water. Soak eggs in 70% ethanol for 15 min. Break the eggs into a sterile container. Homogenize by shaking. Filter through four layers of sterile cheesecloth into a sterile graduated cylinder. Measure out 1.0L.

Streptomycin Solution:
Composition per 10mL:

Streptomycin	0.1mg

Preparation of Streptomycin Solution: Add streptomycin to distilled/deionized water and bring volume to 10.0mL. Mix thoroughly. Filter sterilize.

Preparation of Medium: Add glycerol to 600.0mL of distilled/deionized water. Mix thoroughly. Add remaining components, except fresh egg mixture. Mix thoroughly. Gently heat while stirring and bring to boiling. Autoclave for 15 min at 15 psi pressure–121°C. Cool to 50°C. Aseptically add 1.0L of homogenized whole egg and 10.0mL of sterile streptomycin solution. Mix thoroughly. Distribute into sterile screw-capped tubes. Place tubes in a slanted position. Inspissate at 85°C (moist heat) for 45 min.

Use: For the cultivation and differentiation of *Mycobacterium* species. *M. tuberculosis* appears as granular, rough, dry colonies. *M. kansasii* appears as smooth to rough photochromogenic colonies. *M. gordonae* appears as smooth yellow-orange colonies. *M. avium* appears as smooth, colorless colonies. *M. smegmatis* appears as wrinkled, creamy white colonies. Also used for the cultivation and maintenance of *Gordona* species, *Nocardia* species, *Rhodococcus* species, and *Tsukamurella paurometabolum*.

LPM Agar (Lithium Chloride Phenylethanol Moxalactam Plating Agar)

Composition per liter:

Agar	15.0g
Glycine anhydride	10.0g
LiCl	5.0g
NaCl	5.0g
Pancreatic digest of casein	5.0g
Peptic digest of animal tissue	5.0g
Beef extract	3.0g
Phenylethyl alcohol	2.5g
Moxalactam solution	2.0mL

pH 7.3 ± 0.2 at 25°C

Source: This medium is available as a premixed powder from BBL Microbiology Systems and Difco Laboratories.

Moxalactam Solution:
Composition per 10mL:

Moxalactam	0.1g

Preparation of Moxalactam Solution: Add moxalactam to distilled/deionized water and bring volume to 10.0mL. Mix thoroughly. Filter sterilize.

Preparation of Medium: Add components, except moxalactam solution, to distilled/deionized water and bring volume to 998.0mL. Mix thoroughly. Gently heat while stirring and bring to boiling. Autoclave for 12 min at 15 psi pressure–121°C. Cool to 45° to 50°C. Aseptically add 2.0mL of sterile moxalactam solution. Mix thoroughly. Pour into sterile Petri dishes or distribute into sterile tubes.

Use: For the isolation and cultivation of *Listeria monocytogenes*.

LPM Agar with Esculin and Ferric Iron

Composition per liter:

Agar	15.0g
Glycine anhydride	10.0g
LiCl	5.0g
NaCl	5.0g
Pancreatic digest of casein	5.0g
Peptic digest of animal tissue	5.0g

Beef extract	3.0g
Phenylethyl alcohol	2.5g
Esculin	1.0g
Ferric ammonium citrate	0.5g
Moxalactam solution	2.0mL

pH 7.3 ± 0.2 at 25°C

Moxalactam Solution:
Composition per 10mL:

Moxalactam	0.1g

Preparation of Moxalactam Solution: Add moxalactam to distilled/deionized water and bring volume to 10.0mL. Mix thoroughly. Filter sterilize.

Preparation of Medium: Add components, except moxalactam solution, to distilled/deionized water and bring volume to 998.0mL. Mix thoroughly. Gently heat while stirring and bring to boiling. Autoclave for 12 min at 15 psi pressure–121°C. Cool to 45° to 50°C. Aseptically add 2.0mL of sterile moxalactam solution. Mix thoroughly. Pour into sterile Petri dishes or distribute into sterile tubes.

Use: For the isolation and cultivation of *Listeria monocytogenes*.

Lysine Arginine Iron Agar

Composition per liter:

Agar	15.0g
L-Arginine	10.0g
L-Lysine	10.0g
Peptone	5.0g
Yeast extract	3.0g
Glucose	1.0g
Ferric ammonium citrate	0.5g
Sodium thiosulfate	0.04g
Bromcresol Purple	0.02g

pH 6.8 ± 0.2 at 25°C

Preparation of Medium: Add components to distilled/deionized water and bring volume to 1.0L. Mix thoroughly. Gently heat and bring to boiling. Adjust pH to 6.8. Distribute into screw-capped tubes in 5.0mL volumes. Autoclave for 12 min at 15 psi pressure–121°C. Allow tubes to cool in a slanted position.

Use: For the cultivation and differentiation of bacteria based on their ability to decarboxylate lysine, decarboxylate arginine and produce H_2S. Bacteria that decarboxylate lysine or arginine turn the medium purple. Bacteria that produce H_2S appear as black colonies.

Lysine Decarboxylase Broth, Falkow

Composition per liter:

Peptone	5.0g
L-Lysine	5.0g
Yeast extract	3.0g
Glucose	1.0g
Bromcresol Purple	0.02g

pH 6.5–6.8 at 25°C

Preparation of Medium: Add components to distilled/deionized water and bring volume to 1.0L. Mix thoroughly. Gently heat and bring to boiling. Adjust pH to 6.5–6.8. Distribute into tubes in 5.0mL volumes. Autoclave for 15 min at 15 psi pressure–121°C.

Use: For the cultivation and differentiation of bacteria, especially *Salmonella*, based on their ability to decarboxylate lysine. Bacteria that decarboxylate lysine turn the medium turbid purple.

Lysine Decarboxylase Broth, Taylor Modification

Composition per liter:

L-Lysine	5.0g
Yeast extract	3.0g
Glucose	1.0g
Bromcresol Purple	0.02g

pH 6.1 ± 0.2 at 25°C

Source: This medium is available as a premixed powder from Oxoid Unipath.

Preparation of Medium: Add components to distilled/deionized water and bring volume to 1.0L. Mix thoroughly. Gently heat and bring to boiling. Adjust pH to 6.1. Distribute into tubes in 5.0mL volumes. Autoclave for 15 min at 15 psi pressure–121°C.

Use: For the cultivation and differentiation of bacteria, especially *Salmonella*, based on their ability to decarboxylate lysine. Bacteria that decarboxylate lysine turn the medium turbid purple.

Lysine Decarboxylase Broth, Taylor Modification (Lysine Decarboxylase Broth)

Composition per liter:

L-Lysine	5.0g
Peptone	5.0g
Yeast extract	3.0g
Glucose	1.0g
Bromcresol Purple	0.02g

pH 6.8 ± 0.2 at 25°C

Source: This medium is available as a premixed powder from Difco Laboratories.

Preparation of Medium: Add components to distilled/deionized water and bring volume to 1.0L. Mix thoroughly. Gently heat and bring to boiling. Adjust pH to 6.1. Distribute into tubes in 5.0mL volumes. Autoclave for 15 min at 15 psi pressure–121°C.

Use: For the cultivation and differentiation of bacteria, especially *Salmonella,* based on their ability to decarboxylate lysine. Bacteria that decarboxylate lysine turn the medium turbid purple.

Lysine Decarboxylase Medium

Composition per liter:

Glucose 0.5g
KH_2PO_4 .... 0.5g
L-Lysine·HCl 0.5g

pH 4.6 ± 0.2 at 25°C

Preparation of Medium: Add components to distilled/deionized water and bring volume to 1.0L. Mix thoroughly. Gently heat and bring to boiling. Adjust pH to 4.6. Autoclave for 15 min at 15 psi pressure–121°C. Aseptically distribute into sterile tubes in 1.0mL volumes.

Use: For the cultivation and differentiation of Gram-negative nonfermentative bacteria based on their ability to decarboxylate lysine. Bacteria that decarboxylate lysine turn the medium turbid purple.

Lysine Iron Agar

Composition per liter:

Agar 13.5g
L-Lysine 10.0g
Pancreatic digest of gelatin 5.0g
Yeast extract 3.0g
Glucose 1.0g
Ferric ammonium citrate 0.5g
$Na_2S_2O_3 \cdot 5H_2O$ 0.04g
Bromcresol Purple 0.02g

pH 6.7 ± 0.2 at 25°C

Source: This medium is available as a premixed powder from BBL Microbiology Systems, Difco Laboratories and Oxoid Unipath.

Preparation of Medium: Add components to distilled/deionized water and bring volume to 1.0L. Mix thoroughly. Gently heat while stirring and bring to boiling. Distribute into tubes in 10.0mL volumes. Autoclave for 12 min at 15 psi pressure–121°C. Allow tubes to cool in a slanted position.

Use: For the cultivation and differentiation of members of the Enterobacteriaceae based on their ability to decarboxylate lysine and to form H_2S. Bacteria that decarboxylate lysine turn the medium purple. Bacteria that produce H_2S appear as black colonies.

Lysine Ornithine Mannitol Agar (LOM Agar)

Composition per liter:

Agar 13.5g
L–Ornithine·HCl 6.5g
D–Mannitol 5.25g
L–Lysine·HCl 5.0g
NaCl 5.0g
Yeast extract 3.0g
Bromthymol Blue 0.3g
Vancomycin solution 10.0mL

pH 6.5 ± 0.2 at 25°C

Vancomycin Solution:
Composition per 10mL:

Vancomycin·HCl 0.03g

Preparation of Vancomycin Solution: Add vancomycin to distilled/deionized water and bring volume to 10.0mL. Mix thoroughly. Filter sterilize.

Preparation of Medium: Add components, except vancomycin solution, to distilled/deionized water and bring volume to 990.0mL. Mix thoroughly. Gently heat and bring to boiling. Autoclave for 15 min at 15 psi pressure–121°C. Cool to 45° to 50°C. Aseptically add sterile vancomycin solution. Mix thoroughly. Pour into sterile Petri dishes or distribute into sterile tubes.

Use: For the cultivation and differentiation of Gram-negative bacilli based on their ability to decarboxylate lysine or ornithine and mannitol fermentation. Especially useful for the identification of *Enterobacter agglomerans.* Bacteria that ferment mannitol appear as dark yellow colonies. Bacteria that decarboxylate lysine or ornithine appear as green-yellow colonies.

Lysozyme Broth

Composition per 1005mL:

Basal glycerol broth 1.0L
Lysozyme solution 5.0mL

Basal Glycerol Broth:
Composition per liter:

Peptone 5.0g
Beef extract 3.0g
Glycerol 70.0mL

Preparation of Basal Glycerol Broth: Add components to distilled/deionized water and bring volume to 1.0L. Mix thoroughly. Distribute 500.0mL of the broth into screw-capped tubes in 5.0mL volumes. Autoclave the tubes and the flask with the remaining broth for 15 min at 15 psi pressure–121°C. Cool to 25°C.

Lysozyme Solution:
Composition per 100mL:
Lysozyme 0.1g
HCl (0.01*N* solution) 100.0mL

Preparation of Lysozyme Solution: Add lysozyme to 100.0mL of HCl solution. Mix thoroughly. Filter sterilize. Store for up to 1 week at 4°C.

Preparation of Medium: Add 5.0mL of the sterile lysozyme solution to 95.0mL of the cooled, sterile basal glycerol broth. Mix thoroughly. Aseptically distribute into sterile screw-capped tubes in 5.0mL volumes.

Use: For the cultivation and differentiation of *Nocardia asteroides, Streptomyces griseus* and *Actinomadura madurae* based on sensitivity to lysozyme. *Nocardia asteroides* grows well in both the basal glycerol broth and the lysozyme broth. *Actinomadura madurae* and *Streptomyces griseus* grow well in the basal glycerol broth but not in the lysozyme broth.

M9 Medium with Casamino Acids

Composition per liter:
Na_2HPO_4 .... 6.0g
Casamino acids 5.0g
KH_2PO_4 .... 3.0g
NH_4Cl 1.0g
NaCl 0.5g
Glucose solution 10.0mL
$MgSO_4 \cdot 7H_2O$ solution 1.0mL
Thiamine·HCl solution 1.0mL
$CaCl_2$ solution 1.0mL
pH 7.0 ± 0.2 at 25°C

Glucose solution:
Composition per 100mL:
D-Glucose 20.0g

Preparation of Glucose Solution: Add glucose to distilled/deionized water and bring volume to 1.0L. Mix thoroughly. Autoclave for 15 min at 15 psi pressure–121°C.

$MgSO_4 \cdot 7H_2O$ Solution:
Composition per liter:
$MgSO_4 \cdot 7H_2O$ 246.5g

Preparation of $MgSO_4 \cdot 7H_2O$ Solution: Add $MgSO_4 \cdot 7H_2O$ to distilled/deionized water and bring volume to 1.0L. Mix thoroughly. Autoclave for 15 min at 15 psi pressure–121°C.

Thiamine·HCl Solution:
Composition per 10mL:
Thiamine·HCl 10.0mg

Preparation of Thiamine·HCl Solution: Add thiamine·HCl to distilled/deionized water and bring volume to 1.0L. Mix thoroughly. Filter sterilize.

$CaCl_2$ Solution:
Composition per liter:
$CaCl_2$ solution 14.7g

Preparation of $CaCl_2$ Solution: Add $CaCl_2$ solution to distilled/deionized water and bring volume to 1.0L. Mix thoroughly. Autoclave for 15 min at 15 psi pressure–121°C.

Preparation of Medium: Add components, except $MgSO_4 \cdot 7H_2O$ solution, glucose solution, thiamine·HCl solution, and $CaCl_2$ solution, to distilled/deionized water and bring volume to 987.0mL. Mix thoroughly. Adjust pH to 7.0. Autoclave for 15 min at 15 psi pressure–121°C. Cool to room temperature. Aseptically add sterile $MgSO_4 \cdot 7H_2O$ solution, sterile glucose solution, sterile thiamine·HCl solution, and sterile $CaCl_2$ solution. Mix thoroughly. Distribute into tubes or flasks.

Use: For the cultivation and maintenance of *Flavobacterium meningosepticum.*

MacConkey Agar

Composition per liter:
Pancreatic digest of gelatin 17.0g
Agar 13.5g
Lactose 10.0g
NaCl 5.0g
Bile salts 1.5g
Pancreatic digest of casein 1.5g
Peptic digest of animal tissue 1.5g
Neutral Red 0.03g
Crystal Violet 1.0mg
pH 7.1 ± 0.2 at 25°C

Source: This medium is available as a premixed powder from BBL Microbiology Systems and Difco Laboratories.

Preparation of Medium: Add components to distilled/deionized water and bring volume to 1.0L. Mix thoroughly. Gently heat while stirring until boiling. Autoclave for 15 min at 15 psi pressure–121°C. Pour into sterile Petri dishes or distribute into sterile tubes.

Use: For the selective isolation, cultivation and differentiation of coliforms and enteric pathogens based on the ability to ferment lactose. Lactose-fermenting organisms appear as red to pink colonies. Lactose-nonfermenting organisms appear as colorless or transparent colonies.

MacConkey Agar

Composition per liter:

Peptone	20.0g
Agar	12.0g
Lactose	10.0g
Bile salts	5.0g
NaCl	5.0g
Neutral Red	0.075g

pH 7.4 ± 0.2 at 25°C

Source: This medium is available as a premixed powder from Oxoid Unipath.

Preparation of Medium: Add components to distilled/deionized water and bring volume to 1.0L. Mix thoroughly. Gently heat while stirring until boiling. Autoclave for 15 min at 15 psi pressure–121°C. Pour into sterile Petri dishes or distribute into sterile tubes.

Use: For the selective isolation, cultivation and differentiation of coliforms and enteric pathogens based on the ability to ferment lactose. Lactose-fermenting organisms appear as red to pink colonies. Lactose-nonfermenting organisms appear as colorless or transparent colonies.

MacConkey Agar No. 3

Composition per liter:

Peptone	20.0g
Agar	15.0g
Lactose	10.0g
NaCl	5.0g
Bile salts No.3	1.5g
Neutral Red	0.03g
Crystal Violet	0.001g

pH 7.1 ± 0.2 at 25°C

Source: This medium is available as a premixed powder from Oxoid Unipath.

Preparation of Medium: Add components to distilled/deionized water and bring volume to 1.0L. Mix thoroughly. Gently heat while stirring until boiling. Autoclave for 15 min at 15 psi pressure–121°C. Pour into sterile Petri dishes or distribute into sterile tubes.

Use: For the selective isolation, cultivation and differentiation of enteric pathogens, especially *Salmonella* and *Shigella*, in clinical specimens and in foods.

MacConkey Agar without Crystal Violet

Composition per liter:

Agar	12.0g
Lactose	10.0g
Pancreatic digest of casein	10.0g
Peptic digest of animal tissue	10.0g
Bile salts	5.0g
NaCl	5.0g
Neutral Red	0.05g

pH 7.4 ± 0.2 at 25°C

Source: This medium is available as a premixed powder from BBL Microbiology Systems and Difco Laboratories.

Preparation of Medium: Add components to distilled/deionized water and bring volume to 1.0L. Mix thoroughly. Gently heat while stirring until boiling. Autoclave for 15 min at 15 psi pressure–121°C. Pour into sterile Petri dishes or distribute into sterile tubes.

Use: For the detection of members of the *Enterobacteriaceae* and enterococci as well as some staphylococci. Used for the isolation and detection of coliforms and enteric pathogens from water and wastewater. For susceptibility testing of *Mycobacterium fortuitum* and other rapidly-growing *Mycobacterium* species.

MacConkey Agar without Salt

Composition per liter:

Peptone	20.0g
Agar	12.0g
Lactose	10.0g
Bile salts	5.0g
Neutral Red	0.075g

pH 7.4 ± 0.2 at 25°C

Source: This medium is available as a premixed powder from Difco Laboratories and Oxoid Unipath.

Preparation of Medium: Add components to distilled/deionized water and bring volume to 1.0L. Mix thoroughly. Gently heat while stirring until boiling. Autoclave for 15 min at 15 psi pressure–121°C. Pour into sterile Petri dishes or distribute into sterile tubes. Dry the surface of plates before inoculation.

Use: Used for the isolation and detection of coliforms and enteric pathogens from urine. Provides a low electrolyte medium on which most *Proteus* species will not swarm and therefore avoids overgrow of the plate.

Malonate Broth

Composition per liter:

Sodium malonate	3.0g
NaCl	2.0g
$(NH_4)_2SO_4$	2.0g
K_2HPO_4	0.6g

KH_2PO_4 .... 0.4g
Bromthymol Blue 0.025g

pH 6.7 ± 0.2 at 25°C

Source: This medium is available as a premixed powder from Difco Laboratories.

Preparation of Medium: Add components to distilled/deionized water and bring volume to 1.0L. Mix thoroughly. Distribute into tubes or flasks. Autoclave for 15 min at 15 psi pressure–121°C. Avoid introduction of carbon and nitrogen from other sources.

Use: For the cultivation and differentiation of coliforms and other enteric organisms, particularly *Enterobacter* and *Escherichia* based on their ability to utilize malonate as the sole carbon source and ammonium sulfate as the sole nitrogen source. Malonate-utilizing organisms turn the medium blue.

Malonate Broth, Ewing Modified

Composition per liter:

Sodium malonate 3.0g
NaCl 2.0g
$(NH_4)_2SO_4$ 2.0g
Yeast extract 1.0g
Glucose 0.25g
K_2HPO_4 .... 0.6g
KH_2PO_4 .... 0.4g
Bromthymol Blue 0.025g

pH 6.7 ± 0.2 at 25°C

Source: This medium is available as a premixed powder from BBL Microbiology Systems and Difco Laboratories.

Preparation of Medium: Add components to distilled/deionized water and bring volume to 1.0L. Mix thoroughly. Distribute into tubes or flasks. Autoclave for 15 min at 15 psi pressure–121°C.

Use: For the cultivation and differentiation of coliforms and other enteric organisms, particularly *Enterobacter* and *Escherichia* based on their ability to utilize malonate as a carbon source and ammonium sulfate as a nitrogen source. The small amount of yeast extract and glucose encourages the growth of some organisms that may be distressed or fail to respond. Malonate-utilizing organisms turn the medium blue.

Malt Extract Agar

Composition per liter:

Glucose 20.0g
Malt extract 20.0g
Agar 20.0g
Peptone 1.0g

pH 5.4 ± 0.2 at 25°C

Preparation of Medium: Add components to distilled/deionized water and bring volume to 1.0L. Mix thoroughly. Gently heat while stirring until boiling. Autoclave for 15 min at 15 psi pressure–121°C. Do not overheat or agar will not harden. Pour into sterile Petri dishes or distribute into sterile tubes.

Use: For the maintenance of stock fungal cultures including species of *Aspergillus, Penicillium, Sporothrix schenkii*, and most yeasts.

Mannitol Salt Agar

Composition per liter:

NaCl 75.0g
Agar 15.0g
D-Mannitol 10.0g
Pancreatic digest of casein 5.0g
Peptic digest of animal tissue 5.0g
Beef extract 1.0g
Phenol Red 0.025g

pH 7.4 ± 0.2 at 25°C

Source: This medium is available as a premixed powder from BBL Microbiology Systems and Difco Laboratories and Oxoid Unipath.

Preparation of Medium: Add components to distilled/deionized water and bring volume to 1.0L. Mix thoroughly. Gently heat while stirring and bring to boiling. Distribute into tubes or flasks. Autoclave for 15 min at 15 psi pressure–121°C. Pour into sterile Petri dishes or leave in tubes.

Use: For the selective isolation, cultivation and enumeration of staphylococci from clinical specimens and in epidemiological investigations. Mannitol-utilizing organisms turn the medium yellow.

Martin–Lewis Agar

Composition per liter:

Agar 12.0g
Hemoglobin 10.0g
Pancreatic digest of casein 7.5g
Selected meat peptone 7.5g
NaCl 5.0g
K_2HPO_4 .... 4.0g
Cornstarch 1.0g
KH_2PO_4 .... 1.0g
Supplement solution 10.0mL
VCAT inhibitor 10.0mL

pH 7.2 ± 0.22 at 25°C

Source: Available as a prepared medium from BBL Microbiology Systems.

Supplement Solution:

Composition per liter:

Glucose 100.0g
L-Cysteine·HCl 25.9g
L-Glutamine 10.0g
L-Cystine 1.1g
Adenine 1.0g
Nicotinamide adenine dinucleotide 0.25g
Vitamin B_{12} 0.1g
Thiamine pyrophosphate 0.1g
Guanine·HCl 0.03g
$Fe(NO_3)_3 \cdot 6H_2O$ 0.02g
p-Aminobenzoic acid 0.013g
Thiamine·HCl 3.0mg

Source: The supplement solution IsoVitaleX® enrichment is available from BBL Microbiology Laboratories. This enrichment may be replaced by supplement VX from Difco Laboratories.

Preparation of Supplement Solution: Add components to distilled/deionized water and bring volume to 1.0L. Mix thoroughly. Filter sterilize.

VCAT Inhibitor:

Composition per 10mL:

Vancomycin 4.0mg
Colistin 7.5mg
Anisomycin 0.02g
Trimethoprim lactate 5.0mg

Preparation of VCAT Inhibitor: Add components to distilled/deionized water and bring volume to 10.0mL. Mix thoroughly. Filter sterilize.

Preparation of Medium: Add components, except supplement solution enrichment and VCAT inhibitor, to distilled/deionized water and bring volume to 980.0mL. Gently heat while stirring and bring to boiling. Autoclave for 15 min at 15 psi pressure–121°C. Cool to 45° to 50°C. Aseptically add sterile supplement solution enrichment and sterile VCAT inhibitor. Mix thoroughly. Pour into sterile Petri dishes.

Use: For the isolation and cultivation of pathogenic *Neisseria* from specimens containing mixed flora of bacteria and fungi. Vancomycin-sensitive strains of *N. gonorrhoeae* have been reported.

Martin–Lewis Agar, Enriched

Composition per liter:

Agar 12.0g
Pancreatic digest of casein 7.5g
Selected meat peptone 7.5g
NaCl 5.0g
K_2HPO_4 .... 4.0g
Cornstarch 1.0g
KH_2PO_4 .... 1.0g
Sarcina lutea suspension 20.0mL
Horse serum, inactivated 20.0mL
Supplement solution 10.0mL
PCAT inhibitor 10.0mL

pH 7.2 ± 0.22 at 25°C

***Sarcina lutea* Suspension:**

Composition per 20mL:

Sarcina lutea FDA 1001 .... 10^6–10^7 cells

Preparation of *Sarcina lutea* Suspension: Aseptically wash the growth of 24-hr cultures of *Sarcina lutea* FDA 1001 cells from Thayer-Martin plates with sterile soybean casein digest broth. Standardize the suspension by adding additional sterile tryptic soy broth to yield 40% light transmission at 530nm wavelength.

Soybean Casein Digest Broth:

Composition per liter:

Pancreatic digest of casein 17.0g
NaCl 5.0g
Papaic digest of soybean meal 3.0g
K_2HPO_4 .... 2.5g
Glucose 2.5g

pH 7.3 ± 0.2 at 25°C

Preparation of Soybean Casein Digest Broth: Add components to distilled/deionized water and bring volume to 1.0L. Mix thoroughly. Distribute into tubes or flasks. Autoclave for 15 min at 15 psi pressure–121°C.

Supplement Solution:

Composition per liter:

Glucose 100.0g
L-Cysteine·HCl 25.9g
L-Glutamine 10.0g
L-Cystine 1.1g
Adenine 1.0g
Nicotinamide adenine dinucleotide 0.25g
Vitamin B_{12} 0.1g
Thiamine pyrophosphate 0.1g
Guanine·HCl 0.03g
$Fe(NO_3)_3 \cdot 6H_2O$ 0.02g
p-Aminobenzoic acid 0.013g
Thiamine·HCl 3.0mg

Source: The supplement solution IsoVitaleX® enrichment is available from BBL Microbiology Laboratories. This enrichment may be replaced by supplement VX from Difco Laboratories.

Preparation of Supplement Solution: Add components to distilled/deionized water and bring volume to 1.0L. Mix thoroughly. Filter sterilize.

PCAT Inhibitor:

Composition per 10mL:

Anisomycin 0.02g
Colistin 7.5mg

Trimethoprim lactate 5.0mg
Penicillin G 25,000U

Preparation of PCAT Inhibitor: Add components to distilled/deionized water and bring volume to 10.0mL. Mix thoroughly. Filter sterilize.

Preparation of Medium: Add components—except *Sarcina lutea* suspension, horse serum, supplement solution, and PCAT inhibitor—to distilled/deionized water and bring volume to 940.0mL. Gently heat while stirring and bring to boiling. Autoclave for 15 min at 15 psi pressure–121°C. Cool to 45° to 50°C. Aseptically add 20.0mL of sterile *Sarcina lutea* suspension, 20.0mL of sterile horse serum, 10.0mL of supplement solution, and 10.0mL of sterile PCAT inhibitor. Mix thoroughly. Pour into sterile Petri dishes.

Use: For the isolation and cultivation of pathogenic *Neisseria*, especially penicillinase-producing strains, from specimens containing mixed flora of bacteria and fungi.

McCarthy Agar

Composition per liter:

Cornstarch 10.0g
Naladixic acid 0.015g
Colistin 0.010g
GC Agar base 1.0L

pH 7.2 ± 0.2 at 25°C

GC Agar Base:

Composition per liter:

Agar 10.0g
Pancreatic digest of casein 7.5g
Peptic digest of animal tissue 7.5g
NaCl 5.0g
K_2HPO_4 .... 4.0g
Cornstarch 1.0g
KH_2PO_4 .... 1.0g

Preparation of GC Agar Base: Add components to distilled/deionized water and bring volume to 1.0L. Mix thoroughly.

Preparation of Medium: To 1.0L of GC Agar base add the cornstarch. Gently heat while stirring to dissolve. Add the naladixic acid and colistin. Mix thoroughly. Distribute into tubes or flasks. Autoclave for 15 min at 15 psi pressure–121°C. Pour into sterile Petri dishes or leave in tubes.

Use: For the isolation and differentiation of *Gardnerella vaginalis (Haemophilus vaginalis, Corynebacterium vaginale*) from genitourinary specimens. Bacteria which can utilize starch appear as colonies surrounded by a clear zone.

Middlebrook 7H9 Broth, Supplemented

Composition per liter:

Na_2HPO_4 .... 2.5g
KH_2PO_4 .... 1.0g
Monosodium glutamate 0.5g
$(NH_4)_2SO_4$ 0.5g
Tween™ 80 .... 0.5g
Sodium citrate 0.1g
$MgSO_4 \cdot 7H_2O$ 0.05g
Ferric ammonium citrate 0.04g
Mycobactin J, Allied Laboratories, Inc. 2.0mg
$CuSO_4 \cdot 5H_2O$ 1.0mg
Pyridoxine 1.0mg
$ZnSO_4 \cdot 7H_2O$ 1.0mg
Biotin 0.5mg
$CaCl_2 \cdot 2H_2O$ 0.5mg
Dubos oleic albumin complex 100.0mL
Glycerol 2.0mL

pH 6.6 ± 0.2 at 25°C

Dubos Oleic Albumin Complex:

Composition per 100mL:

Bovine serum albumin, fraction V 5.0g
Oleic acid, sodium salt 0.05g
NaCl (0.85% solution) 100.0mL

Preparation of Dubos Oleic Albumin Complex: Add bovine serum albumin and oleic acid to 100.0mL of NaCl solution. Mix thoroughly. Filter sterilize.

Preparation of Medium: Add components, except Dubos oleic albumin complex, to distilled/deionized water and bring volume to 900.0mL. Mix thoroughly. Gently heat and bring to boiling. Autoclave for 15 min at 15 psi pressure–121°C. Cool to 45° to 50°C. Aseptically add sterile Dubos oleic albumin complex. Mix thoroughly. Pour into sterile Petri dishes or distribute into sterile tubes.

Use: For the cultivation and maintenance of *Mycobacterium avium.*

Middlebrook 7H9 Broth with Middlebrook ADC Enrichment

Composition per liter:

Na_2HPO_4 .... 2.5g
KH_2PO_4 .... 1.0g
Monosodium glutamate 0.5g
$(NH_4)_2SO_4$ 0.5g
Sodium citrate 0.1g
$MgSO_4 \cdot 7H_2O$ 0.05g
Ferric ammonium citrate 0.04g
$CuSO_4 \cdot 5H_2O$ 1.0mg
Pyridoxine 1.0mg

$ZnSO_4 \cdot 7H_2O$ 1.0mg
Biotin 0.5mg
$CaCl_2 \cdot 2H_2O$ 0.5mg
Middlebrook ADC enrichment 100.0mL
Glycerol 2.0mL

pH 6.6 ± 0.2 at 25°C

Source: This medium is available as a premixed powder from BBL Microbiology Systems and Difco Laboratories.

Middlebrook ADC Enrichment:

Composition per 100mL:

Bovine albumin fraction V 5.0g
Glucose 2.0g
Catalase 3.0mg

Source: This enrichment is available as a prepared enrichment from Difco Laboratories.

Preparation of Middlebrook ADC Enrichment: Add components to distilled/deionized water and bring volume to 100.0mL. Mix thoroughly. Filter sterilize.

Preparation of Medium: Add glycerol to 900.0mL of distilled/deionized water and add remaining components, except Middlebrook ADC enrichment. Mix thoroughly. Gently heat and bring to boiling. Autoclave for 15 min at 15 psi pressure–121°C. Cool to 50° to 55°C. Aseptically add 100.0mL of sterile Middlebrook ADC enrichment. Mix thoroughly. Distribute into sterile tubes or flasks.

Use: For the isolation, cultivation and maintenance of *Mycobacterium* species, including *M. tuberculosis.* Also used for determining the antimicrobial susceptibility of mycobacteria.

Middlebrook 7H9 Broth with Middlebrook OADC Enrichment

Composition per liter:

Na_2HPO_4 .... 2.5g
KH_2PO_4 .... 1.0g
Monosodium glutamate 0.5g
$(NH_4)_2SO_4$ 0.5g
Sodium citrate 0.1g
$MgSO_4 \cdot 7H_2O$ 0.05g
Ferric ammonium citrate 0.04g
$CuSO_4 \cdot 5H_2O$ 1.0mg
Pyridoxine 1.0mg
$ZnSO_4 \cdot 7H_2O$ 1.0mg
Biotin 0.5mg
$CaCl_2 \cdot 2H_2O$ 0.5mg
Middlebrook OADC enrichment 100.0mL
Glycerol 2.0mL

pH 6.6 ± 0.2 at 25°C

Source: This medium is available as a premixed powder from BBL Microbiology Systems and Difco Laboratories.

Middlebrook OADC Enrichment:

Composition per 100mL:

Bovine albumin fraction V 5.0g
Glucose 2.0g
NaCl 0.85g
Oleic acid 0.05g
Catalase 4.0mg

Source: Available as a prepared enrichment from Difco Laboratories.

Preparation of Middlebrook OADC Enrichment: Add components to distilled/deionized water and bring volume to 100.0mL. Mix thoroughly. Filter sterilize.

Preparation of Medium: Add glycerol to 900.0mL of distilled/deionized water and add remaining components, except Middlebrook OADC enrichment. Mix thoroughly. Gently heat and bring to boiling. Autoclave for 15 min at 15 psi pressure–121°C. Cool to 50° to 55°C. Aseptically add 100.0mL of sterile Middlebrook OADC enrichment. Mix thoroughly. Distribute into sterile tubes or flasks.

Use: For the isolation, cultivation and maintenance of *Mycobacterium* species, including *M. tuberculosis.* Also used for determining the antimicrobial susceptibility of mycobacteria.

Middlebrook 7H9 Broth with Middlebrook OADC Enrichment and Triton WR 1339

Composition per liter:

Na_2HPO_4 .... 2.5g
KH_2PO_4 .... 1.0g
Monosodium glutamate 0.5g
$(NH_4)_2SO_4$ 0.5g
Sodium citrate 0.1g
$MgSO_4 \cdot 7H_2O$ 0.05g
Ferric ammonium citrate 0.04g
$CuSO_4 \cdot 5H_2O$ 1.0mg
Pyridoxine 1.0mg
$ZnSO_4 \cdot 7H_2O$ 1.0mg
Biotin 0.5mg
$CaCl_2 \cdot 2H_2O$ 0.5mg
Middlebrook OADC enrichment
with Triton WR 1339 .... 100.0mL
Glycerol 2.0mL

pH 6.6 ± 0.2 at 25°C

Source: This medium is available as a premixed powder from BBL Microbiology Systems.

Middlebrook OADC Enrichment with Triton WR 1339:

Composition per 100mL:

Bovine albumin fraction V	5.0g
Glucose	2.0g
NaCl	0.85g
Triton WR-1339	0.25g
Oleic acid	0.05g
Catalase	4.0mg

Source: Available as a prepared enrichment from Difco Laboratories.

Preparation of Middlebrook OADC Enrichment with Triton WR 1339: Add components to distilled/deionized water and bring volume to 100.0mL. Mix thoroughly. Filter sterilize.

Preparation of Medium: Add glycerol to 900.0mL of distilled/deionized water and add remaining components, except Middlebrook OADC enrichment with Triton WR-1339. Mix thoroughly. Gently heat and bring to boiling. Autoclave for 15 min at 15 psi pressure–121°C. Cool to 50° to 55°C. Aseptically add 100.0mL of sterile Middlebrook OADC enrichment with Triton WR-1339. Mix thoroughly. Distribute into sterile tubes or flasks.

Use: For the isolation, cultivation and maintenance of *Mycobacterium* species, including *M. tuberculosis*. Also used for determining the antimicrobial susceptibility of mycobacteria.

Middlebrook 7H10 Agar with Middlebrook ADC Enrichment

Composition per liter:

Agar	15.0g
Na_2HPO_4	1.5g
KH_2PO_4	1.5g
$(NH_4)_2SO_4$	0.5g
L-Glutamic Acid	0.5g
Sodium citrate	0.4g
Ferric ammonium citrate	0.04g
$MgSO_4 \cdot 7H_2O$	0.025g
$ZnSO_4 \cdot 7H_2O$	1.0mg
$CuSO_4 \cdot 5H_2O$	1.0mg
Pyridoxine	1.0mg
Biotin	0.5mg
$CaCl_2 \cdot 2H_2O$	0.5mg
Malachite Green	.0.25mg
Middlebrook ADC enrichment	100.0mL
Glycerol	5.0mL

pH 6.6 ± 0.2 at 25°C

Source: Available as a premixed powder from BBL Microbiology Systems and Difco Laboratories.

Middlebrook ADC Enrichment:

Composition per 100mL:

Bovine albumin fraction V	5.0g
Glucose	2.0g
Catalase	0.003g
Distilled water	100.0mL

Source: Available as a prepared enrichment from Difco Laboratories.

Preparation of Middlebrook ADC Enrichment: Add components to distilled/deionized water and bring volume to 100.0mL. Mix thoroughly. Filter sterilize.

Preparation of Medium: Add glycerol to 900.0mL of distilled/deionized water and add remaining components, except Middlebrook ADC enrichment. Mix thoroughly. Gently heat and bring to boiling. Autoclave for 15 min at 15 psi pressure–121°C. Cool to 50° to 55°C. Aseptically add 100.0mL of sterile Middlebrook ADC enrichment. Mix thoroughly. Pour into sterile Petri dishes or distribute into sterile tubes.

Use: For the isolation, cultivation and maintenance of *Mycobacterium* species, including *M. tuberculosis*. Also used for determining the antimicrobial susceptibility of mycobacteria.

Middlebrook 7H10 Agar with Middlebrook OADC Enrichment (Middlebrook and Cohn 7H10 Agar)

Composition per liter:

Agar	15.0g
Na_2HPO_4	1.5g
KH_2PO_4	1.5g
$(NH_4)_2SO_4$	0.5g
L-Glutamic Acid	0.5g
Sodium citrate	0.4g
Ferric ammonium citrate	0.04g
$MgSO_4 \cdot 7H_2O$	0.025g
$ZnSO_4 \cdot 7H_2O$	1.0mg
$CuSO_4 \cdot 5H_2O$	1.0mg
Pyridoxine	1.0mg
Biotin	0.5mg
$CaCl_2 \cdot 2H_2O$	0.5mg
Malachite Green	.0.25mg
Middlebrook OADC enrichment	100.0mL
Glycerol	5.0mL

pH 6.6 ± 0.2 at 25°C

Source: This medium is available as a premixed powder from BBL Microbiology Systems and Difco Laboratories.

Middlebrook OADC Enrichment:
Composition per 100mL:

Bovine albumin fraction V	5.0g
Glucose	2.0g
NaCl	0.85g
Oleic acid	0.05g
Catalase	4.0mg

Source: Available as a prepared enrichment from Difco Laboratories.

Preparation of Middlebrook OADC Enrichment: Add components to distilled/deionized water and bring volume to 100.0mL. Mix thoroughly. Filter sterilize.

Preparation of Medium: Add glycerol to 900.0mL of distilled/deionized water and add remaining components, except Middlebrook OADC enrichment. Mix thoroughly. Gently heat and bring to boiling. Autoclave for 15 min at 15 psi pressure–121°C. Cool to 50° to 55°C. Aseptically add 100.0mL of sterile Middlebrook OADC enrichment. Mix thoroughly. Pour into sterile Petri dishes or distribute into sterile tubes.

Use: For the isolation, cultivation and maintenance of *Mycobacterium* species, including *M. tuberculosis*. Also used for determining the antimicrobial susceptibility of mycobacteria.

Middlebrook 7H10 Agar with Middlebrook OADC Enrichment and Hemin (Hemin Medium for *Mycobacterium*)

Composition per liter:

Agar	15.0g
Na_2HPO_4	1.5g
KH_2PO_4	1.5g
$(NH_4)_2SO_4$	0.5g
L-Glutamic Acid	0.5g
Sodium citrate	0.4g
Ferric ammonium citrate	0.04g
$MgSO_4 \cdot 7H_2O$	0.025g
$ZnSO_4 \cdot 7H_2O$	1.0mg
$CuSO_4 \cdot 5H_2O$	1.0mg
Pyridoxine	1.0mg
Biotin	0.5mg
$CaCl_2 \cdot 2H_2O$	0.5mg
Malachite Green	0.25mg
Middlebrook OADC enrichment	100.0mL
Glycerol	5.0mL
Hemin solution	3.9mL

pH 6.6 ± 0.2 at 25°C

Source: This medium is available as a premixed powder from BBL Microbiology Systems and Difco Laboratories.

Middlebrook OADC Enrichment:
Composition per 100mL:

Bovine albumin fraction V	5.0g
Glucose	2.0g
NaCl	0.85g
Oleic acid	0.05g
Catalase	4.0mg

Preparation of Middlebrook OADC Enrichment: Add components to distilled/deionized water and bring volume to 100.0mL. Mix thoroughly. Filter sterilize.

Hemin Solution:
Composition per 100mL:

Hemin	1.0g
NaOH (1*N* solution)	20.0mL

Preparation of Hemin Solution: Add hemin to 20.0mL of 1*N* NaOH solution. Mix thoroughly. Bring volume to 100.0mL with distilled/deionized water.

Preparation of Medium: Add glycerol to 891.1mL of distilled/deionized water and add remaining components, except Middlebrook OADC enrichment. Mix thoroughly. Gently heat and bring to boiling. Autoclave for 15 min at 15 psi pressure–121°C. Cool to 50° to 55°C. Aseptically add 100.0mL of sterile Middlebrook OADC enrichment. Mix thoroughly. Pour into sterile Petri dishes or distribute into sterile tubes.

Use: For the isolation, cultivation and maintenance of *Mycobacterium* species, including *Mycobacterium tuberculosis*. For the cultivation and maintenance of *Mycobacterium haemophilum*. Also used for determining the antimicrobial susceptibility of mycobacteria.

Middlebrook 7H10 Agar with Middlebrook OADC Enrichment and Triton WR 1339

Composition per liter:

Agar	15.0g
Na_2HPO_4	1.5g
KH_2PO_4	1.5g
$(NH_4)_2SO_4$	0.5g
L-Glutamic Acid	0.5g
Sodium citrate	0.4g
Ferric ammonium citrate	0.04g
$MgSO_4 \cdot 7H_2O$	0.025g
$ZnSO_4 \cdot 7H_2O$	1.0mg
$CuSO_4 \cdot 5H_2O$	1.0mg

Pyridoxine 1.0mg
Biotin 0.5mg
$CaCl_2 \cdot 2H_2O$ 0.5mg
Malachite Green 0.25mg
Middlebrook OADC enrichment with Triton WR 1339 .... 100.0mL
Glycerol 5.0mL

pH 6.6 ± 0.2 at 25°C

Source: Available as a premixed powder from BBL Microbiology Systems and Difco Laboratories.

Middlebrook OADC Enrichment with Triton WR 1339:

Composition per 100mL:

Bovine albumin fraction V 5.0g
Glucose 2.0g
NaCl 0.85g
Triton WR-1339 .... 0.25g
Oleic acid 0.05g
Catalase 4.0mg

Source: Available as a prepared enrichment from Difco Laboratories.

Preparation of Middlebrook OADC Enrichment with Triton WR 1339: Add components to distilled/deionized water and bring volume to 100.0mL. Mix thoroughly. Filter sterilize.

Preparation of Medium: Add glycerol to 900.0mL of distilled/deionized water and add remaining components, except Middlebrook OADC enrichment with Triton WR-1339. Mix thoroughly. Gently heat and bring to boiling. Autoclave for 15 min at 15 psi pressure–121°C. Cool to 50° to 55°C. Aseptically add 100.0mL of sterile Middlebrook OADC enrichment with Triton WR-1339. Mix thoroughly. Pour into sterile Petri dishes or distribute into sterile tubes.

Use: For the isolation, cultivation and maintenance of *Mycobacterium* species, including *M. tuberculosis*. Also used for determining the antimicrobial susceptibility of mycobacteria.

Middlebrook 7H10 Agar with Streptomycin

Composition per liter:

Agar 15.0g
Na_2HPO_4 .... 1.5g
KH_2PO_4 .... 1.5g
$(NH_4)_2SO_4$ 0.5g
L-Glutamic Acid 0.5g
Sodium citrate 0.4g
Ferric ammonium citrate 0.04g
$MgSO_4 \cdot 7H_2O$ 0.025g
$ZnSO_4 \cdot 7H_2O$ 1.0mg
$CuSO_4 \cdot 5H_2O$ 1.0mg
Pyridoxine 1.0mg
Biotin 0.5mg
$CaCl_2 \cdot 2H_2O$ 0.5mg
Malachite Green 0.25mg
Glycerol 5.0mL
Streptomycin 100.0mg

pH 6.6 ± 0.2 at 25°C

Source: This medium is available as a premixed powder from BBL Microbiology Systems and Difco Laboratories.

Preparation of Medium: Add glycerol to 1.0L of distilled/deionized water and add remaining components. Mix thoroughly. Gently heat and bring to boiling. Autoclave for 15 min at 15 psi pressure–121°C. Cool to 50° to 55°C. Aseptically add streptomycin. Mix thoroughly. Pour into sterile Petri dishes or distribute into sterile tubes.

Use: For the isolation, cultivation and maintenance of *Mycobacterium kansasii*.

Middlebrook 7H11 Agar, Selective

Composition per liter:

Agar 15.0g
Na_2HPO_4 .... 1.5g
KH_2PO_4 .... 1.5g
Pancreatic digest of casein 1.0g
$(NH_4)_2SO_4$ 0.5g
L-Glutamic acid 0.5g
Sodium citrate 0.4g
$MgSO_4 \cdot 7H_2O$ 0.05g
Ferric ammonium citrate 0.04g
Pyridoxine 1.0mg
$ZnSO_4 \cdot 7H_2O$ 1.0mg
$CuSO_4 \cdot 5H_2O$ 1.0mg
$CaCl_2 \cdot 2H_2O$ 0.5mg
Malachite Green 0.25mg
D-Biotin 0.5μg
Middlebrook OADC enrichment 100.0mL
Antibiotic solution 10.0mL
Glycerol 5.0mL

pH 6.6 ± 0.2 at 25°C

Middlebrook OADC Enrichment:

Composition per 100mL:

Bovine albumin fraction V 5.0g
Glucose 2.0g
NaCl 0.85g
Oleic acid 0.05g
Catalase 4.0mg

Source: Available as a prepared enrichment from Difco Laboratories.

Preparation of Middlebrook OADC Enrichment: Add components to distilled/deionized water and bring volume to 100.0mL. Mix thoroughly. Filter sterilize.

Antibiotic Solution:
Composition per 10mL:

Carbenicillin	0.050mg
Trimethoprim lactate	0.020mg
Amphotericin B	0.010mg
Polymyxin B	200,000U

Preparation of Antibiotic Solution: Add components to distilled/deionized water and bring volume to 10.0mL. Mix thoroughly. Filter sterilize.

Preparation of Medium: Add glycerol to 890.0mL of distilled/deionized water and add remaining components, except Middlebrook OADC enrichment and antibiotic solution. Mix thoroughly. Gently heat and bring to boiling. Autoclave for 15 min at 15 psi pressure–121°C. Cool to 50° to 55°C. Aseptically add 100.0mL of sterile Middlebrook OADC enrichment and 10.0mL of sterile antibiotic solution. Mix thoroughly. Pour into sterile Petri dishes or distribute into sterile tubes.

Use: For the selective isolation and cultivation of pathogenic mycobacteria from specimens potentially contaminated with bacteria and fungi.

Middlebrook 7H11 Agar with Middlebrook ADC Enrichment (Mycobacteria 7H11 Agar with Middlebrook ADC Enrichment)

Composition per liter:

Agar	15.0g
Na_2HPO_4	1.5g
KH_2PO_4	1.5g
Pancreatic digest of casein	1.0g
$(NH_4)_2SO_4$	0.5g
L–Glutamic acid	0.5g
Sodium citrate	0.4g
$MgSO_4 \cdot 7H_2O$	0.05g
Ferric ammonium citrate	0.04g
Pyridoxine	1.0mg
Malachite Green	0.25mg
D–Biotin	0.5μg
Middlebrook ADC enrichment	100.0mL
Glycerol	5.0mL

pH 6.6 ± 0.2 at 25°C

Source: This medium is available as a premixed powder from BBL Microbiology Systems and Difco Laboratories.

Middlebrook ADC Enrichment:
Composition per 100mL:

Bovine albumin fraction V	5.0g
Glucose	2.0g
Catalase	0.003g
Distilled water	100.0mL

Source: Available as a prepared enrichment from Difco Laboratories.

Preparation of Middlebrook ADC Enrichment: Add components to distilled/deionized water and bring volume to 100.0mL. Mix thoroughly. Filter sterilize.

Preparation of Medium: Add glycerol to 900.0mL of distilled/deionized water and add remaining components, except Middlebrook ADC enrichment. Mix thoroughly. Gently heat and bring to boiling. Autoclave for 15 min at 15 psi pressure–121°C. Cool to 50° to 55°C. Aseptically add 100.0mL of sterile Middlebrook ADC enrichment. Mix thoroughly. Pour into sterile Petri dishes or distribute into sterile tubes.

Use: For the cultivation of drug resistant (isoniazid [INH]) strains of *M. tuberculosis*. For the cultivation of particularly fastidious strains of tubercle bacilli which occur following treatment of tuberculosis patients with secondary anti-tubercular drugs. Generally these strains fail to grow on 7H10 medium.

Middlebrook 7H11 Agar with Middlebrook OADC Enrichment (Mycobacteria 7H11 Agar with Middlebrook OADC Enrichment)

Composition per liter:

Agar	15.0g
Na_2HPO_4	1.5g
KH_2PO_4	1.5g
Pancreatic digest of casein	1.0g
$(NH_4)_2SO_4$	0.5g
L–Glutamic acid	0.5g
Sodium citrate	0.4g
$MgSO_4 \cdot 7H_2O$	0.05g
Ferric ammonium citrate	0.04g
Pyridoxine	1.0mg
Malachite Green	0.25mg
D–Biotin	0.5μg
Middlebrook OADC enrichment	100.0mL
Glycerol	5.0mL

pH 6.6 ± 0.2 at 25°C

Source: This medium is available as a premixed powder from BBL Microbiology Systems and Difco Laboratories.

Middlebrook OADC Enrichment:

Composition per 100mL:

Bovine albumin fraction V....5.0g
Glucose....2.0g
NaCl....0.85g
Oleic acid....0.05g
Catalase....4.0mg

Source: Available as a prepared enrichment from Difco Laboratories.

Preparation of Middlebrook OADC Enrichment: Add components to distilled/deionized water and bring volume to 100.0mL. Mix thoroughly. Filter sterilize.

Preparation of Medium: Add glycerol to 900.0mL of distilled/deionized water and add remaining components, except Middlebrook OADC enrichment. Mix thoroughly. Gently heat and bring to boiling. Autoclave for 15 min at 15 psi pressure–121°C. Cool to 50° to 55°C. Aseptically add 100.0mL of sterile Middlebrook OADC enrichment. Mix thoroughly. Pour into sterile Petri dishes or distribute into sterile tubes.

Use: For the cultivation of drug resistant (isoniazid [INH]) strains of *M. tuberculosis*. For the cultivation of particularly fastidious strains of tubercle bacilli which occur following treatment of tuberculosis patients with secondary anti-tubercular drugs. Generally these strains fail to grow on 7H10 medium.

Middlebrook 7H11 Agar with Middlebrook OADC Enrichment and Triton WR 1339 (Mycobacteria 7H11 Agar with Middlebrook OADC Enrichment and Triton WR 1339)

Composition per liter:

Agar....15.0g
Na_2HPO_4....1.5g
KH_2PO_4....1.5g
Pancreatic digest of casein....1.0g
$(NH_4)_2SO_4$....0.5g
L–Glutamic acid....0.5g
Sodium citrate....0.4g
$MgSO_4 \cdot 7H_2O$....0.05g
Ferric ammonium citrate....0.04g
Pyridoxine....1.0mg
Malachite Green....0.25mg
D–Biotin....0.5μg
Middlebrook OADC enrichment with Triton WR 1339....100.0mL
Glycerol....5.0mL

pH 6.6 ± 0.2 at 25°C

Source: This medium is available as a premixed powder from BBL Microbiology Systems and Difco Laboratories.

Middlebrook OADC Enrichment with Triton WR 1339:

Composition per 100mL:

Bovine albumin fraction V....5.0g
Glucose....2.0g
NaCl....0.85g
Triton WR-1339....0.25g
Oleic acid....0.05g
Catalase....4.0mg

Source: Available as a prepared enrichment from Difco Laboratories.

Preparation of Middlebrook OADC Enrichment with Triton WR 1339: Add components to distilled/deionized water and bring volume to 100.0mL. Mix thoroughly. Filter sterilize.

Preparation of Medium: Add glycerol to 900.0mL of distilled/deionized water and add remaining components, except Middlebrook OADC enrichment with Triton WR-1339. Mix thoroughly. Gently heat and bring to boiling. Autoclave for 15 min at 15 psi pressure–121°C. Cool to 50° to 55°C. Aseptically add 100.0mL of sterile Middlebrook OADC enrichment with Triton WR-1339. Mix thoroughly. Pour into sterile Petri dishes or distribute into sterile tubes.

Use: For the cultivation of drug resistant (isoniazid [INH]) strains of *M. tuberculosis*. For the cultivation of particularly fastidious strains of tubercle bacilli which occur following treatment of tuberculosis patients with secondary anti-tubercular drugs. Generally these strains fail to grow on 7H10 medium.

Middlebrook 7H12 Medium

Composition per 102.5mL:

Bovine serum albumin....0.5g
Casein hydrolyslate....0.1g
Catalase....4,800U
^{14}C-Palmitic acid....100μCi
Middlebrook 7H9 broth....100.0mL
Antibiotic solution....2.5mL

pH 6.8 ± 0.1 at 25°C

Middlebrook 7H9 Broth:

Composition per liter:

Na_2HPO_4....2.5g
KH_2PO_4....1.0g
Monosodium glutamate....0.5g
$(NH_4)_2SO_4$....0.5g
Sodium citrate....0.1g
$MgSO_4 \cdot 7H_2O$....0.05g

Ferric ammonium citrate....0.04g
$CuSO_4 \cdot 5H_2O$....1.0mg
Pyridoxine....1.0mg
$ZnSO_4 \cdot 7H_2O$....1.0mg
Biotin....0.5mg
$CaCl_2 \cdot 2H_2O$....0.5mg
Glycerol....2.0mL

Preparation of Middlebrook 7H9 Broth: Add components to distilled/deionized water and bring volume to 1.0L. Mix thoroughly.

Antibiotic Solution:
Composition per 5mL:

Nalidixic acid....0.2g
Azlocillin....0.1g
Amphotericin B....0.050g
Trimethoprim....0.050g
Polymyxin B....500,000U

Preparation of Antibiotic Solution: Add components to distilled/deionized water and bring volume to 5.0mL. Mix thoroughly. Filter sterilize.

Preparation of Medium: To 100.0mL of Middlebrook 7H9 broth, add remaining components, except antibiotic solution. Mix thoroughly. Filter sterilize. Aseptically distribute into bottles in 4.0mL volumes. Prior to inoculation, aseptically add 0.1mL of antibiotic solution to each bottle. Mix thoroughly.

Use: For the cultivation of *Mycobacterium* species from the blood of patients suspected of having mycobacteremia.

Middlebrook 13A Medium

Composition per 112.5mL:

Casein hydrolysate....0.1g
Tween™ 80....0.02g
Sodium polyanetholesulfonate....0.025g
Catalase....36,000U
^{14}C-substrate....125μCi(185kBq)
Middlebrook 7H9 broth....100.0mL
Middlebrook 13A enrichment....12.5mL

pH 6.6 ± 0.2 at 25°C

Middlebrook 7H9 Broth:
Composition per liter:

Na_2HPO_4....2.5g
KH_2PO_4....1.0g
Monosodium glutamate....0.5g
$(NH_4)_2SO_4$....0.5g
Sodium citrate....0.1g
$MgSO_4 \cdot 7H_2O$....0.05g
Ferric ammonium citrate....0.04g
$CuSO_4 \cdot 5H_2O$....1.0mg
Pyridoxine....1.0mg
$ZnSO_4 \cdot 7H_2O$....1.0mg
Biotin....0.5mg
$CaCl_2 \cdot 2H_2O$....0.5mg
Glycerol....2.0mL

Preparation of Middlebrook 7H9 Broth: Add components to distilled/deionized water and bring volume to 1.0L. Mix thoroughly.

Middlebrook 13A Enrichment:
Composition per 20mL:

Bovine serum albumin....3.0g

Preparation of Middlebrook 13A Enrichment: Add bovine serum albumin to distilled/deionized water and bring volume to 20.0mL. Mix thoroughly. Filter sterilize.

Preparation of Medium: To 100.0mL of Middlebrook 7H9 broth, add remaining components, except Middlebrook 13A enrichment. Mix thoroughly. Filter sterilize. Aseptically distribute into bottles in 4.0mL volumes. Prior to inoculation, aseptically add 0.5mL of Middlebrook 13A enrichment to each bottle. Mix thoroughly.

Use: For the cultivation of *Mycobacterium* species from the blood of patients suspected of having mycobacteremia.

Middlebrook ADC Enrichment (Middlebrook Albumin Dextrose Catalase Enrichment)

Composition per 100mL:

Bovine albumin fraction V....5.0g
Glucose....2.0g
Catalase....0.003g

Source: Available as a prepared enrichment from Difco Laboratories.

Preparation of Enrichment: Add components to distilled/deionized water and bring volume to 100.0mL. Mix thoroughly. Filter sterilize.

Use: For use as a supplement to other Middlebrook media for the isolation, cultivation and maintenance of *Mycobacterium* species. Also used as a supplement to other Middlebrook media for determining the antimicrobial susceptibility of mycobacteria.

Middlebrook OADC Enrichment (Middlebrook Oleic Albumin Dextrose Catalase Enrichment)

Composition per 100mL:

Bovine albumin fraction V....5.0g
Glucose....2.0g
NaCl....0.85g

Oleic acid 0.05g
Catalase 4.0mg

Source: Available as a prepared enrichment from Difco Laboratories.

Preparation of Enrichment: Add components to distilled/deionized water and bring volume to 100.0mL. Mix thoroughly. Filter sterilize.

Use: For use as a supplement to other Middlebrook media for the isolation, cultivation and maintenance of *Mycobacterium* species. Also used as a supplement to other Middlebrook media for determining the antimicrobial susceptibility of mycobacteria.

Middlebrook OADC Enrichment with Triton WR 1339 (Middlebrook Oleic Albumin Dextrose Catalase Enrichment with Triton WR 1339)

Composition per 100mL:

Bovine albumin fraction V 5.0g
Glucose 2.0g
NaCl 0.85g
Triton WR 1339 .... 0.25g
Oleic acid 0.05g
Catalase 4.0mg

Source: Available as a prepared enrichment from Difco Laboratories.

Preparation of Enrichment: Add components to distilled/deionized water and bring volume to 100.0mL. Mix thoroughly. Filter sterilize.

Use: For use as a supplement to other Middlebrook media for the isolation, cultivation and maintenance of *Mycobacterium* species. Also used as a supplement to other Middlebrook media for determining the antimicrobial susceptibility of mycobacteria.

MIL Medium (Motility Indole Lysine Medium)

Composition per liter:

Peptone 10.0g
Pancreatic digest of casein 10.0g
L-Lysine·HCl 10.0g
Yeast extract 3.0g
Agar 2.0g
Dextrose 1.0g
Ferric ammonium citrate 0.5g
Bromcresol Purple 0.02g

pH 6.6 ± 0.2 at 25°C

Source: Available as a premixed powder and prepared medium from Difco Laboratories.

Preparation of Medium: Add components to distilled/deionized water and bring volume to 1.0L. Mix thoroughly. Gently heat and bring to boiling. Distribute into tubes in 5.0mL volumes. Autoclave for 15 min at 15 psi pressure–121°C.

Use: For the cultivation and differentiation of members of the Enterobacteriaceae on the basis of motility, lysine decarboxylase activity, lysine deaminase activity and indole production.

Mitis–Salivarius Agar

Composition per liter:

Sucrose 50.0g
Agar 15.0g
Enzymatic digest of protein 10.0g
Proteose peptone 10.0g
K_2HPO_4 .... 4.0g
Dextrose 1.0g
Trypan Blue 0.08g
Crystal Violet 0.8mg
Na_2TeO_3 solution 1.0mL

pH 7.0 ± 0.2 at 25°C

Source: This medium is available as a premixed powder from Difco Laboratories.

Na_2TeO_3 Solution:

Composition per 10mL:

Na_2TeO_3 .... 0.1g

Preparation of Na_2TeO_3 Solution: Add Na_2TeO_3 to 10.0mL of distilled/deionized water. Mix thoroughly. Filter sterilize.

Caution: Potassium tellurite is toxic.

Preparation of Medium: Add components to distilled/deionized water and bring volume to 999.0mL. Mix thoroughly. Gently heat and bring to boiling. Autoclave for 15 min at 15 psi pressure–121°C. Cool medium to 50–55°C. Aseptically add 1.0mL of the sterile Na_2TeO_3 solution to the cooled basal medium. Mix thoroughly. Pour into sterile Petri dishes or distribute into sterile tubes.

Use: For the selective isolation of *Streptococcus mitis, Streptococcus salivarius,* other viridans streptococci and enterococci.

Møller Decarboxylase Broth

Composition per liter:

Amino acid 10.0g
Peptic digest of animal tissue 5.0g
Beef extract 5.0g
Glucose 0.5g

Bromcresol Purple 0.01g
Cresol Red 5.0mg
Pyridoxal 5.0mg
pH 6.0 ± 0.2 at 25°C

Source: Available as a premixed powder from BBL Microbiology Systems and Difco Laboratories.

Preparation of Medium: Add components to distilled/deionized water and bring volume to 1.0L. Use L-lysine, L-arginine or L-ornithine. Mix thoroughly. Gently heat until dissolved. Distribute into screw-capped tubes in 5.0mL volumes. Autoclave for 15 min at 15 psi pressure–121°C. A slight precipitate may form in the ornithine broth.

Use: For the differentiation of Gram-negative enteric bacteria based on the production of arginine dihydrolase, lysine decarboxylase or ornithine decarboxylase.

Møller KCN Broth Base

Composition per liter:
Na_2HPO_4 .... 5.64g
NaCl 5.0g
Pancreatic digest of casein 1.5g
Peptic digest of animal tissue 1.5g
KH_2PO_4 .... 0.225g
KCN solution 0.15mL
pH 7.6 ± 0.2 at 25°C

Source: This medium is available as a premixed powder from BBL Microbiology Systems.

KCN Solution:
Composition per100mL:
KCN 0.5g

Preparation of KCN Solution: Add KCN to 100.0mL of cold distilled/deionized water. Mix thoroughly and cap. Do not mouth pipette.

Caution: Cyanide is toxic.

Preparation of Medium: Add components, except KCN solution, to distilled/deionized water and bring volume to 1.0L. Mix thoroughly. Autoclave for 15 min at 15 psi pressure–121°C. Cool to room temperature. Prior to use, add 0.15mL of KCN solution. Mix thoroughly. Aseptically distribute into sterile tubes.

Use: For the differentiation of Gram-negative enteric bacteria on the basis of their ability to grow in the presence of cyanide.

Molybdate Agar

Composition per101.5mL:
Base 100.0mL
Phosphomolybdic acid solution 1.5mL
pH 5.3 ± 0.2 at 25°C

Base:
Composition per liter:
Sucrose 40.0g
Agar 15.0g
Meat peptone 10.0g

Preparation of Base: Add components to distilled/deionized water and bring volume to 1.0L. Mix thoroughly. Adjust pH to 7.6. Gently heat and bring to boiling. Autoclave for 15 min at 15 psi pressure–121°C. Cool to 45° to 50°C.

Phosphomolybdic Acid Solution:
Composition per 100mL:
$P_2O_5{\cdot}2OMoO_3$ 12.5g

Preparation of Base: Add $P_2O_5{\cdot}2OMoO_3$ (phospho-12-molybdic acid, 12–molybdophosphoric acid, or PMA) to sterile distilled/deionized water. Mix thoroughly. Do not adjust pH.

Preparation of Medium: To 100.0mL of cooled sterile base add 1.5mL of phosphomolybdic acid solution. Mix thoroughly. Pour into sterile Petri dishes or distribute into sterile tubes.

Use: For the isolation and presumptive identification of yeast, especially *Candida* species. *Candida albicans* appears as smooth, medium olive colonies with medium olive bottoms. *Candida stellatoidea* appears as shiny, light gray colonies with light gray bottoms. *Candida tropicalis* appears as smooth, shiny, dark blue/gray colonies with dark blue/gray bottoms. *Candida krusei* appears as smooth, dull white colonies with white bottoms. *Saccharomyces cerevisiae* appears as smooth, shiny light blue/dark blue colonies with dark blue/green bottoms.

Monsur Agar (Taurocholate Tellurite Gelatin Agar)

Composition per liter:
Gelatin 30.0g
Agar 15.0g
Casein peptone 10.0g
NaCl 10.0g
Sodium taurocholate 5.0g
$Na_2CO_3{\cdot}H_2O$ 1.0g
K_2TeO_3 solution 10.0mL
pH 8.5 ± 0.2 at 25°C

K_2TeO_3 Solution:
Composition per 10mL:
K_2TeO_3 .... 0.02g

Preparation of K_2TeO_3 Solution: Add K_2TeO_3 to 10.0mL of distilled/deionized water. Mix thoroughly. Filter sterilize.

Caution: Potassium tellurite is toxic.

Preparation of Medium: Add components, except K_2TeO_3 solution, to distilled/deionized water and bring volume to 990.0mL. Mix thoroughly. Gently heat and bring to boiling. Autoclave for 15 min at 15 psi pressure–121°C. Cool to 45° to 50°C. Add 10.0mL of sterile K_2TeO_3 solution. Mix thoroughly. Pour into sterile Petri dishes or distribute into sterile tubes.

Use: For the isolation of *Vibrio cholerae* from fecal specimens.

Motility Indole Ornithine Medium (MIO Medium)

Composition per liter:

Pancreatic digest of gelatin	10.0g
Pancreatic digest of casein	9.5g
L-Ornithine·HCl	5.0g
Yeast extract	3.0g
Agar	2.0g
Glucose	1.5g
Bromcresol Purple	0.02g

pH 6.6 ± 0.2 at 25°C

Source: This medium is available as a premixed powder from BBL Microbiology Systems and Difco Laboratories.

Preparation of Medium: Add components to distilled/deionized water and bring to 1.0L. Mix thoroughly. Gently heat and bring to boiling. Distribute into tubes or flasks. Autoclave for 15 min at 15 psi pressure–121°C.

Use: For the differentiation of Gram-negative enteric bacteria based on their motility, indole production and ornithine decarboxylase activity.

Motility Nitrate Agar

Composition per liter:

Beef heart, solids from infusion	100.0g
Tryptose	12.0g
Agar	3.0g
NaCl	1.0g
KNO_3	1.0g
Glucose	0.5g

pH 7.4 ± 0.2 at 25°C

Preparation of Medium: Add components to distilled/deionized water and bring volume to 1.0L. Mix thoroughly. Gently heat and bring to boiling. Distribute into tubes in 4.0mL volumes. Autoclave for 15 min at 15 psi pressure–121°C.

Use: For the cultivation and observation of motility and nitrate reduction in some Gram-negative bacteria.

Motility Sulfide Medium

Composition per liter:

Gelatin	80.0g
Proteose peptone	10.0g
NaCl	5.0g
Agar	4.0g
Beef extract	3.0g
Sodium citrate	2.0g
L-Cystine	0.2g
Ferrous ammonium citrate	0.2g

pH 7.3 ± 0.2 at 25°C

Source: This medium is available as a premixed powder from Difco Laboratories.

Preparation of Medium: Add components to distilled/deionized water and bring volume to 1.0L. Mix thoroughly. Gently heat while stirring and bring to boiling. Distribute into tubes in 4–5.0mL volumes. Autoclave for 15 min at 10 psi pressure–116°C.

Use: For the determination of bacterial motility and the ability of bacteria to produce H_2S from L-cystine. Used for the differentiation of Gram-negative bacteria of the Enterobacteriaceae.

Motility Test and Maintenance Medium

Composition per liter:

Peptone	10.0g
NaCl	5.0g
Agar	4.0g
Beef extract	3.0g
2,3,5-triphenyltetrazolium chloride	0.05g

Preparation of Medium: Add components to distilled/deionized water and bring volume to 1.0L. Mix thoroughly. Distribute into screw-capped tubes in 8.0mL volumes. Autoclave for 15 min at 15 psi pressure–121°C.

Use: For the cultivation, maintenance and observation of motility of *Listeria monocytogenes*.

Motility Test and Maintenance Medium

Composition per liter:

Agar	9.0g
Tryptose	8.0g
NaCl	5.0g
Pancreatic digest of gelatin	2.5g
Beef extract	1.5g

pH 7.2 ± 0.1 at 25°C

Preparation of Medium: Add components to distilled/deionized water and bring volume to 1.0L. Mix thoroughly. Gently heat and bring to boiling. Distribute into tubes in 7.0mL volumes. Autoclave for 15 min at 15 psi pressure–121°C. Cool to 45° to 50°C. Pass the cooled tubes into an anaerobic chamber containing 85% N_2 + 10% H_2 + 5% CO_2.

Use: For the cultivation, maintenance and observation of motility in a variety of anaerobic bacteria.

Motility Test and Maintenance Medium

Composition per liter:

Peptone	10.0g
NaCl	5.0g
Agar	4.0g
Beef extract	3.0g

pH 7.4 ± 0.1 at 25°C

Preparation of Medium: Add components to distilled/deionized water and bring volume to 1.0L. Mix thoroughly. Distribute into screw-capped tubes in 8.0mL volumes. Autoclave for 15 min at 15 psi pressure–121°C.

Use: For the cultivation, maintenance and observation of motility in members of the *Enterobacteriaceae*.

Motility Test and Maintenance Medium, Gilardi

Composition per liter:

Pancreatic digest of casein	10.0g
NaCl	5.0g
Agar	3.0g
Yeast extract	3.0g

pH 7.2 ± 0.1 at 25°C

Preparation of Medium: Add components to distilled/deionized water and bring volume to 1.0L. Mix thoroughly. Distribute into screw-capped tubes in 3.5mL volumes. Autoclave for 15 min at 15 psi pressure–121°C.

Use: For the cultivation, maintenance and observation of motility in nonfermenting Gram–negative bacteria.

Motility Test and Maintenance Medium, Tatum

Composition per liter:

Tryptose	8.0g
NaCl	5.0g
Agar	4.0g
Pancreatic digest of gelatin	2.5g
Beef extract	1.5g

pH 6.9 ± 0.2 at 25°C

Preparation of Medium: Add components to distilled/deionized water and bring volume to 1.0L. Mix thoroughly. Distribute into screw-capped tubes in 8.0mL volumes. Autoclave for 15 min at 15 psi pressure–121°C.

Use: For the cultivation, maintenance and observation of motility in nonfermenting Gram–negative bacteria.

Motility Test Medium

Composition per liter:

Pancreatic digest of gelatin	10.0g
NaCl	5.0g
Agar	4.0g
Beef extract	3.0g

pH 7.3 ± 0.2 at 25°C

Source: This medium is available as a premixed powder from BBL Microbiology Systems.

2,3,5-Triphenyltetrazolium Chloride Solution:
Composition per 10mL:

2,3,5-triphenyltetrazolium chloride	0.1g

Preparation of 2,3,5-Triphenyltetrazolium Chloride Solution: Add 2,3,5-triphenyltetrazolium chloride to distilled/deionized water and bring volume to 10.0mL. Mix thoroughly. Filter sterilize.

Preparation of Medium: Add components to distilled/deionized water and bring volume to 995.0mL. Mix thoroughly. Gently heat while stirring and bring to boiling. Autoclave for 15 min at 15 psi pressure–121°C. Cool to 45° to 50°C. Aseptically add 5.0mL of sterile 2,3,5-triphenyltetrazolium chloride solution. Mix thoroughly. Aseptically distribute into sterile tubes.

Use: For the detection of motility of Gram-negative enteric bacteria.

Motility Test Medium

Composition per liter:

Tryptose	10.0g
NaCl	5.0g
Agar	5.0g

pH 7.2 ± 0.2 at 25°C

Source: This medium is available as a premixed powder from Difco Laboratories.

Preparation of Medium: Add components to distilled/deionized water and bring volume to 1.0L. Mix thoroughly. Gently heat while stirring and bring to boiling. Distribute into tubes in 4–5.0mL volumes. Autoclave for 15 min at 15 psi pressure–121°C. Cool tubes quickly in an upright position.

Use: For the determination of bacterial motility.

Motility Test Medium, Semisolid

Composition per liter:

Peptone..10.0g
NaCl...5.0g
Agar..4.0g
Beef extract...3.0g

pH 7.4 ± 0.2 at 25°C

Preparation of Medium: Add components to distilled/deionized water and bring volume to 1.0L. Mix thoroughly. Gently heat while stirring and bring to boiling. Distribute into screw-capped tubes in 8.0mL or 20.0mL volumes. Autoclave for 15 min at 15 psi pressure–121°C. Pour into sterile Petri dishes in 20.0mL volumes or leave in tubes.

Use: For the cultivation and observation of motility in a variety of bacteria, especially *Salmonella* species.

MRVP Broth (Methyl Red–Voges–Proskauer Broth)

Composition per liter:

Glucose 5.0g
KH_2PO_4..5.0g
Pancreatic digest of casein 3.5g
Peptic digest of animal tissue..............................3.5g

pH 6.9 ± 0.2 at 25°C

Source: Available as a premixed powder from BBL Microbiology Systems and as a prepared medium from Difco Laboratories.

Preparation of Medium: Add components to distilled/deionized water and bring volume to 1.0L. Mix thoroughly. Distribute into tubes or flasks. Autoclave for 15 min at 15 psi pressure–121°C.

Use: For the differentiation of bacteria based on acid production (methyl red test) and acetoin production (Voges-Proskauer reaction).

MRVP Medium (Methyl Red Voges–Proskauer Medium)

Composition per liter:

Glucose 5.0g
Peptone..5.0g
Phosphate buffer 5.0g

pH 7.5 ± 0.2 at 25°C

Source: This medium is available as a premixed powder from Oxoid Unipath.

Preparation of Medium: Add components to distilled/deionized water and bring volume to 1.0L. Mix thoroughly. Distribute into tubes or flasks. Autoclave for 15 min at 15 psi pressure–121°C.

Use: For the differentiation of bacteria based on acid production (methyl red test) and acetoin production (Voges-Proskauer reaction).

Mucate Broth

Composition per liter:

Mucic acid...10.0g
Peptone...10.0g
Bromthymol Blue...0.024g

pH 7.4 ± 0.1 at 25°C.

Preparation of Medium: Add components to distilled/deionized water and bring volume to 1.0L. Mix thoroughly. Add 5*N* NaOH while stirring until mucic acid dissolves. Distribute into screw-capped tubes in 5.0mL volumes. Autoclave for 10 min at 15 psi pressure–121°C.

Use: For the isolation and cultivation of enterovirulent *Escherichia coli* and *Shigella* species.

Mucate Control Broth

Composition per liter:

Peptone...10.0g
Bromthymol Blue...0.024g

pH 7.4 ± 0.1 at 25°C

Preparation of Medium: Add components to distilled/deionized water and bring volume to 1.0L. Mix thoroughly. Distribute into screw-capped tubes in 5.0mL volumes. Autoclave for 10 min at 15 psi pressure–121°C.

Use: For the isolation and cultivation of enterovirulent *Escherichia coli* and *Shigella* species.

Mueller–Hinton Agar

Composition per liter:

Beef infusion..300.0g
Acid hydrolysate of casein.................................17.5g
Agar...17.0g
Starch..1.5g

pH 7.4 ± 0.2 at 25°C

Source: This medium is available as a premixed powder from Difco Laboratories and Oxoid Unipath.

Preparation of Medium: Add components to distilled/deionized water and bring to 1.0L. Mix thoroughly. Gently heat and bring to boiling. Distribute into tubes or flasks. Autoclave for 15 min at 15 psi pressure–121°C. Pour into sterile Petri dishes or leave in tubes.

Use: For antimicrobial susceptibility testing of a variety of nonfastidious, rapidly-growing microorganisms.

Mueller–Hinton Agar with IsoVitaleX® and Hemoglobin

Composition per liter:

Component A 490.0mL
Component B 490.0mL
IsoVitaleX® enrichment 20.0mL

pH 6.9 ± 0.2 at 25°C

Component A:

Composition per 490mL:

Beef infusion 300.0g
Acid hydrolysate of casein 17.5g
Agar 17.0g
Starch 1.5g

Preparation of Component A: Add components to distilled/deionized water and bring to 490.0mL. Mix thoroughly. Gently heat and bring to boiling. Autoclave for 15 min at 15 psi pressure–121°C. Cool to 45° to 50°C.

Component B:

Composition per 490mL:

Hemoglobin 10.0g

Preparation of Component B: Add hemoglobin to distilled/deionized water and bring to 490.0mL. Mix thoroughly. Gently heat and bring to boiling. Autoclave for 15 min at 15 psi pressure–121°C. Cool to 45° to 50°C.

IsoVitaleX® Enrichment:

Composition per liter:

Glucose 100.0g
L-Cysteine·HCl 25.9g
L-Glutamine 10.0g
L-Cystine 1.1g
Adenine 1.0g
Nicotinamide adenine dinucleotide 0.25g
Vitamin B_{12} 0.1g
Thiamine pyrophosphate 0.1g
Guanine·HCl 0.03g
$Fe(NO_3)_3 \cdot 6H_2O$ 0.02g
p-Aminobenzoic acid 0.013g
Thiamine·HCl 3.0mg

Source: The supplement solution IsoVitaleX® enrichment is available from BBL Microbiology Laboratories. This enrichment may be replaced by supplement VX from Difco Laboratories.

Preparation of IsoVitaleX® Enrichment: Add components to distilled/deionized water and bring volume to 1.0L. Mix thoroughly. Filter sterilize. Warm to 45° to 50°C.

Preparation of Medium: Aseptically combine 490.0mL of component A, 490.0mL of component B, and 20.0mL of IsoVitaleX® enrichment. Mix thoroughly. Adjust pH to 6.9. Pour into sterile Petri dishes in 20.0mL volumes.

Use: For the isolation and cultivation of fastidious microorganisms including *Legionella pneumophila.*

Mueller–Hinton Broth

Composition per liter:

Acid hydrolysate of casein 17.5g
Beef extract 3.0g
Starch 1.5g

pH 7.3 ± 0.1 at 25°C

Source: This medium is available as a premixed powder from BBL Microbiology Systems, Difco Laboratories and Oxoid Unipath.

Preparation of Medium: Add components to distilled/deionized water and bring to 1.0L. Mix thoroughly. Gently heat and bring to boiling. Distribute into tubes or flasks. Autoclave for 10 min at 10 psi pressure–115°C. Do not overheat.

Use: For the cultivation of a wide variety of nonfastidious microorganisms. For antimicrobial susceptibility testing.

Mueller–Hinton Chocolate Agar

Beef infusion 300.0g
Acid hydrolysate of casein 17.5g
Agar 17.0g
Starch 1.5g
Sheep blood 50.0mL

pH 7.4 ± 0.2 at 25°C

Preparation of Medium: Add components, except sheep blood, to distilled/deionized water and bring volume to 950.0mL. Mix thoroughly. Gently heat and bring to boiling. Autoclave for 15 min at 15 psi pressure–121°C. Cool to 45° to 50°C. Aseptically add sterile sheep blood. Mix thoroughly. Gently heat to 70°C for 10 min. Pour into sterile Petri dishes or distribute into sterile tubes.

Use: For the cultivation and maintenance of *Neisseria gonorrhoeae* and *Neisseria meningitidis.* For antimicrobial susceptibility testing of fastidious microorganisms.

Mueller–Hinton II Agar

Composition per liter:

Acid hydrolysate of casein 17.5g
Agar 17.0g

Beef extract 2.0g
Starch 1.5g

pH 7.3 ± 0.1 at 25°C

Source: This medium is available as a premixed powder from BBL Microbiology Systems.

Preparation of Medium: Add components to distilled/deionized water and bring to 1.0L. Mix thoroughly. Gently heat and bring to boiling. Distribute into tubes or flasks. Autoclave for 15 min at 15 psi pressure–121°C. Pour into sterile Petri dishes or leave in tubes.

Use: For antimicrobial disc diffusion susceptibility testing by the Bauer-Kirby method of a variety of bacteria. This medium supplemented with 5% sheep blood is recommended for use in antimicrobial susceptibility testing of *Streptococcus pneumoniae* and related organisms.

Mueller–Hinton Medium with Rabbit Serum

Composition per liter:

Beef infusion...300.0g
Acid hydrolysate of casein..............................17.5g
Agar...17.0g
Starch 1.5g
Rabbit serum 100.0mL

pH 7.4 ± 0.2 at 25°C

Preparation of Medium: Add components, except rabbit serum, to distilled/deionized water and bring volume to 900.0mL. Mix thoroughly. Gently heat and bring to boiling. Autoclave for 15 min at 15 psi pressure–121°C. Cool to 45° to 50°C. Aseptically add sterile rabbit serum. Pour into sterile Petri dishes or distribute into sterile tubes.

Use: For the cultivation and maintenance of *Corynebacterium* species.

Mueller Tellurite Medium

Composition per liter:

Casamino acids 20.0g
Agar...20.0g
Casein...5.0g
KH_2PO_4...0.3g
$MgSO_4 \cdot 7H_2O$ 0.1g
L-Tryptophan 0.05g
Mueller tellurite serum................................25.0mL

pH 7.4 ± 0.1 at 25°C

Mueller Tellurite Serum:
Composition per 100mL:

K_2TeO_3 solution 0.4g
Calcium pantothenate.................................... 0.2mg
Horse or beef serum, sterile 50.0mL
Sodium lactate solution................................40.0mL
Ethyl alcohol 10.0mL

Preparation of Mueller Tellurite Serum: Add calcium pantothenate to 1.0mL of distilled/deionized water. Autoclave for 15 min at 15 psi pressure–121°C. Add K_2TeO_3 to 1.0mL of sterile distilled/deionized water. To 40.0mL of cooled, sterile sodium lactate solution add filter-sterilized ethanol, sterile calcium pantothenate solution, sterile serum and K_2TeO_3 solution. Mix thoroughly. Store at 4° to 8°C.

Sodium Lactate Solution:
Composition per 100mL:

Lactic acid (85% solution)..............................50mL
Phenol Red solution (0.2g in 50% ethanol)....0.1mL

Preparation of Sodium Lactate Solution: Add lactic acid to distilled/deionized water and bring volume to 100.0mL. Add 0.1mL of Phenol Red solution. Add enough 40% NaOH solution to adjust pH to 7.0. Gently heat and bring to boiling for 5 min. Add more NaOH solution to retain red color, if necessary. Autoclave for 15 min at 15 psi pressure–121°C. Cool to 50°C.

Caution: Potassium tellurite is toxic.

Preparation of Medium: Add components to distilled/deionized water and bring volume to 975.0mL. Gently heat and bring to boiling. Autoclave for 15 min at 15 psi pressure–121°C. Cool quickly to 50°C. Aseptically add 25.0mL Mueller tellurite serum. Mix thoroughly. Distribute into sterile Petri dishes. Allow the surface of the plates to dry by partially removing the covers during solidification.

Use: For isolation, cultivation and differentiation of *Corynebacterium diphtheriae.*

Muller–Kauffmann Tetrathionate Broth

Composition per 1028mL:

$Na_2S_2O_3$ 40.7g
$CaCO_3$ 25.0g
Pancreatic digest of casein 7.0g
Ox bile..4.75g
Soya peptone..2.3g
NaCl 2.3g
Iodine solution 19.0mL
Brilliant Green solution..................................9.5mL

Iodine Solution:
Composition per 100mL:

Iodine 20.0g
KI 25.0g

Preparation of Iodine Solution: Add the KI to approximately 5.0mL of distilled/deionized water. Mix thoroughly. Add the iodine. Gently heat to dissolve. Bring volume to 100.0mL with distilled/deionized water. Filter sterilize.

Brilliant Green Solution:
Composition per 100mL:

Brilliant Green 0.1g

Preparation of Brilliant Green Solution: Add the Brilliant Green to distilled/deionized water and bring volume to 100.0mL. Mix thoroughly. Gently heat while stirring and bring to boiling. Continue boiling for 30 min while stirring until dye has dissolved. Filter sterilize. Store protected from light.

Preparation of Medium: Add components, except iodine solution and Brilliant Green solution, to distilled/deionized water and bring volume to 1.0L. Mix thoroughly. Gently heat and bring to boiling. Do not autoclave. Cool to 45°C. Prior to use, add 19.0mL of iodine solution and 9.5mL of Brilliant Green solution. Mix thoroughly. Aseptically distribute into sterile tubes or flasks.

Use: For the isolation and cultivation of *Salmonella* species from specimens with a mixed flora.

MWY Medium (Wadowsky and Yee Medium, Modified)

Composition per liter:

Agar...13.0g
Yeast extract...10.0g
Glycine..3.0g
ACES buffer (2-[(2-Amino-2-oxoethyl)-amino]-ethane sulfonic acid).......................2.0g
Charcoal, activated..2.0g
α-Ketoglutarate...0.2g
$Fe_4(P_2O_7)_3 \cdot 9H_2O$..0.05g
Bromcresol Purple 0.01g
Bromcresol Blue 0.01g
Antibiotic inhibitor....................................... 10.0mL
L-Cysteine·HCl·H_2O solution....................... 10.0mL

pH 6.9 ± 0.2 at 25°C

Antibiotic Inhibitor:
Composition per 10mL:

Anisomycin 0.16g
Cefamandole 4.0mg
Vancomycin.. 1.0mg
Polymyxin B 130,000U

Preparation of Antibiotic Inhibitor: Add components to distilled/deionized water and bring volume to 10.0mL. Mix thoroughly. Filter sterilize.

L-Cysteine·HCl·H_2O Solution:
Composition per 10mL:

L-Cysteine·HCl·H_2O...0.08g

Preparation of L-Cysteine·HCl·H_2O Solution: Add L-Cysteine·HCl·H_2O to distilled/deionized water and bring volume to 10.0mL. Mix thoroughly. Filter sterilize.

Preparation of Medium: Add components, except cysteine and antibiotic inhibitor, to distilled/deionized water and bring volume to 980.0mL. Mix thoroughly. Adjust medium to pH 6.9 with 1*N* KOH. Heat gently and bring to boiling for 1 min. Autoclave for 15 min at 15 psi pressure–121°C. Cool to 50° to 55°C. Add 10.0mL of the sterile L-cysteine·HCl·H_2O solution and 10.0mL of the sterile antibiotic solution. Mix thoroughly. Pour into sterile Petri dishes with constant agitation to keep charcoal in suspension.

Use: For the selective isolation and cultivation of *Legionella pneumophila* and other *Legionella* species.

Mycobactin Medium

Serum Agar Medium:
Composition per liter:

Noble agar..15.0g
Casamino acids 2.5g
Na_2HPO_4 (anhydrous).......................................2.5g
Sodium citrate 1.5g
KH_2PO_4...1.0g
$MgSO_4 \cdot 7H_2O$ 0.6g
Asparagine 0.3g
Crude mycobactin 0.16g
Chloramphenicol...0.05g
Primaricine (myprozine) 0.05g
Penicillin 100,000U
Bovine serum , 56°C-inactivated...............200.0mL
Tween™ 80 (1% solution)............................50.0mL
Glycerol..25.0mL

pH 7.2 ± 0.2 at 25°C

Preparation of Crude Mycobactin: Grow *Mycobacterium phlei* in 600.0mL of Mycobactin Production Broth for 2 weeks at 37°C. Autoclave the culture for 15 min at 15 psi pressure–121°C. Filter the cells and wash with distilled/deionized water. Dry cells under $CaCl_2$. Treat 100.0g of dried culture with 3 successive acetone extractions—500.0mL of acetone for 30 min in a liter flask fitted with a reflux condenser. Evaporate the acetone to dryness. Extract the residue in a Soxhlet apparatus with petroleum ether for 18–20 hrs at 40° to 60°C. A hard red residue will remain. Dissolve the residue in warm absolute ethanol. Centrifuge for 30 min at 2,250 rpm to remove debris. Evaporate the supernatant to dryness. Grind the residue to a powder of crude mycobactin.

Source: Purified mycobactin is available from Allied Labs, Inc., 2520 Hunt St., Ames, IA 50010.

Mycobactin Production Broth:
Composition per 600mL:

Solution B 500.0mL
Solution A 100.0mL

Preparation of Mycobactin Production Broth: Aseptically mix the cooled sterile solution A and solution B.

Solution A:
Composition per 100mL:

L-Asparagine 5.0g
Na_2HPO_4 .... 2.0g
KH_2PO_4 .... 1.0g
Glycerol 30.0mL

Preparation of Solution A: Add components to distilled/deionized water and bring volume to 100.0mL. Mix thoroughly. Autoclave for 15 min at 15 psi pressure–121°C. Cool to 45° to 50°C.

Solution B:
Composition per 500mL:

Glucose 10.0g
$MgSO_4 \cdot 7H_2O$ 0.2g

Preparation of Solution B: Add components to distilled/deionized water and bring volume to 500.0mL. Mix thoroughly. Autoclave for 15 min at 15 psi pressure–121°C. Cool to 45° to 50°C.

Preparation of Medium: Add components, except penicillin, chloramphenicol, primaricine and bovine serum, to distilled/deionized water and bring volume to 800.0mL. Mix thoroughly. Gently heat with a minimum of heat. Autoclave for 15 min at 10 psi pressure–116°C. Cool to 50°C. Aseptically add penicillin, chloramphenicol, primaricine and sterile bovine serum. Mix thoroughly. Adjust pH to 7.2. Distribute into sterile tubes or flasks.

Use: For the cultivation and maintenance of *Mycobacterium avium* and *Mycobacterium paratuberculosis.*

Mycobactosel™ Agar

Composition per liter:

Agar 13.5g
Bovine albumin fraction V 5.0g
Glucose 2.0g
Na_2HPO_4 .... 1.5g
KH_2PO_4 .... 1.5g
Pancreatic digest of casein 1.0g
NaCl 0.85g
$(NH_4)_2SO_4$ 0.5g
Monosodium glutamate 0.5g
Sodium citrate 0.4g
$MgSO_4 \cdot 7H_2O$ 0.05g
Ferric ammonium citrate 0.04g
Pyridoxine 1.0mg
$ZnSO_4 \cdot 7H_2O$ 1.0mg
$CuSO_4 \cdot 5H_2O$ 1.0mg
Biotin 0.5mg
$CaCl_2 \cdot H_2O$ 0.5mg
Malachite Green 0.25mg
Antibiotic solution 10.0mL
Catalase solution 10.0mL
Glycerol 5.0mL
Oleic Acid 0.06mL

pH 6.6 ± 0.2 at 25°C

Source: Available as a prepared medium from BBL Microbiology Systems.

Antibiotic Solution:
Composition per 10mL:

Cycloheximide 0.36g
Nalidixic acid 0.02g
Lincomycin 2.0mg

Preparation of Antibiotic Solution: Add components to distilled/deionized water and bring volume to 10.0mL. Mix thoroughly. Filter sterilize.

Catalase Solution:
Composition per 10mL:

Catalase 3.0mg

Preparation of Catalase Solution: Add catalase to distilled/deionized water and bring volume to 10.0mL. Mix thoroughly. Filter sterilize.

Preparation of Medium: Add components, except antibiotic solution and catalase solution, to distilled/deionized water and bring volume to 980.0mL. Mix thoroughly. Gently heat and bring to boiling. Autoclave for 15 min at 15 psi pressure–121°C. Cool to 45° to 50°C. Aseptically add sterile antibiotic solution and sterile catalase solution. Mix thoroughly. Pour into sterile Petri dishes or distribute into sterile tubes.

Use: For the selective isolation of mycobacteria from specimens containing mixed flora.

Mycobactosel™ L–J Medium

Composition per liter:

Potato flour 30.0g
L-Asparagine 3.6g
KH_2PO_4, anhydrous 2.5g
Sodium citrate 0.6g
$MgSO_4 \cdot 7H_2O$ 0.24g
Homogenized whole egg 1.0L
Malachite Green solution 20.0mL
Glycerol 12.0mL
Antibiotic solution 10.0mL

pH 7.0 ± 0.2 at 25°C

Source: Available as a prepared medium from BBL Microbiology Systems.

Homogenized Whole Egg:
Composition per liter:
Whole eggs...18–24

Preparation of Whole Egg: Use fresh eggs, less than 1 week old. Scrub the shells with soap. Let stand in a soap solution for 30 min. Rinse in running water. Soak eggs in 70% ethanol for 15 min. Break the eggs into a sterile container. Homogenize by shaking. Filter through four layers of sterile cheesecloth into a sterile graduated cylinder. Measure out 1.0L.

Malachite Green Solution:
Composition per 20mL:
Malachite Green...0.4g

Preparation of Malachite Green Solution: Add Malachite Green to sterile distilled/deionized water and bring volume to 20.0mL in a sterile container. Mix thoroughly.

Antibiotic Solution:
Composition per 10mL:
Cycloheximide...0.64g
Nalidixic acid..0.056g
Lincomycin 3.2mg

Preparation of Antibiotic Solution: Add components to distilled/deionized water and bring volume to 10.0mL. Mix thoroughly. Filter sterilize.

Preparation of Medium: Add components—except whole egg, Malachite Green solution, and antibiotic solution—to distilled/deionized water and bring volume to 600.0mL. Mix thoroughly. Autoclave for 30 min at 15 psi pressure–121°C. Cool to room temperature. Add the homogenized whole egg, Malachite Green solution, and antibiotic solution. Distribute into sterile tubes in 8.0mL volumes. Coagulate medium in a slanted position at 85°C (moist heat) for 50 min.

Use: For the isolation and cultivation of *Mycobacterium* species from clinical specimens.

Mycobiotic Agar (Cycloheximide Chloramphenicol Agar)

Composition per liter:
Agar...15.0g
Enzymatic hydrolysate of soybean meal...........10.0g
Glucose 10.0g
Cycloheximide..0.5g
Chloramphenicol..0.05g
pH 6.5 ± 0.2 at 25°C

Source: This medium is available as a premixed powder from Difco Laboratories.

Preparation of Medium: Add components to distilled/deionized water and bring volume to 1.0L. Mix thoroughly. Gently heat and bring to boiling. Distribute into tubes or flasks. Autoclave for 15 min at 15 psi pressure–121°C. Cool tubes quickly in a slanted position.

Use: For the selective isolation and cultivation of pathogenic fungi.

Mycological Agar

Composition per liter:
Agar...15.0g
Enzymatic hydrolysate of soybean meal...........10.0g
Glucose 10.0g
pH 7.0 ± 0.2 at 25°C

Source: This medium is available as a premixed powder from Difco Laboratories.

Preparation of Medium: Add components to distilled/deionized water and bring volume to 1.0L. Mix thoroughly. Gently heat and bring to boiling. Distribute into tubes or flasks. Autoclave for 15 min at 15 psi pressure–121°C.

Use: For the selective isolation, cultivation and maintenance of pathogenic fungi.

Mycological Agar with Low pH

Composition per liter:
Agar...15.0g
Enzymatic hydrolysate of soybean meal...........10.0g
Glucose 10.0g
pH 4.8 ± 0.2 at 25°C

Source: This medium is available as a premixed powder from Difco Laboratories.

Preparation of Medium: Add components to distilled/deionized water and bring volume to 1.0L. Mix thoroughly. Gently heat and bring to boiling. Distribute into tubes or flasks. Autoclave for 15 min at 15 psi pressure–121°C.

Use: For the selective isolation, cultivation and maintenance of pathogenic fungi.

Mycophil™ Agar

Composition per liter:
Agar...16.0g
Papaic digest of soybean meal 10.0g
Glucose 10.0g
pH 7.0 ± 0.2 at 25°C

Source: This medium is available as a premixed powder from BBL Microbiology Systems.

Preparation of Medium: Add components to distilled/deionized water and bring volume to 1.0L. Mix thoroughly. Gently heat and bring to boiling. Distribute into tubes or flasks. Autoclave for 15 min at 15 psi pressure–121°C. Pour into sterile Petri dishes or distribute into sterile tubes.

Use: For the cultivation, maintenance, and enumeration of fungi. For the demonstration of pigment production in fungal species. Also used for the cultivation and maintenance of *Bacillus* species.

Mycophil™ Agar with Low pH

Composition per liter:

Agar	18.0g
Papaic digest of soybean meal	10.0g
Glucose	10.0g

pH 4.7 ± 0.2 at 25°C

Source: This medium is available as a premixed powder from BBL Microbiology Systems.

Preparation of Medium: Add components to distilled/deionized water and bring volume to 1.0L. Mix thoroughly. Gently heat and bring to boiling. Distribute into tubes or flasks. Autoclave for 15 min at 15 psi pressure–121°C. Adjust pH to 4.7 by adding approximately 10.0mL of sterile 10% lactic acid. Mix thoroughly. Pour into sterile Petri dishes or distribute into sterile tubes.

Use: For the isolation and enumeration of yeasts and molds.

Mycophil™ Broth

Composition per liter:

Glucose	40.0g
Pancreatic digest of casein	5.0g
Peptic digest of animal tissue	5.0g

pH 7.0 ± 0.2 at 25°C

Source: This medium is available as a premixed powder from BBL Microbiology Systems.

Preparation of Medium: Add components to distilled/deionized water and bring volume to 1.0L. Mix thoroughly. Gently heat and bring to boiling. Distribute into tubes or flasks. Autoclave for 15 min at 15 psi pressure–118°C. Do not overheat.

Use: For the isolation and cultivation of a wide variety of fungi.

Mycoplasma Agar Base (PPLO Agar Base)

Composition per 1300mL:

Agar	14.0g
Pancreatic digest of casein	7.0g
NaCl	5.0g
Beef extract	3.0g
Yeast extract	3.0g
Beef heart, infusion from (solids)	2.0g
Horse serum	260.0mL
Fresh yeast extract solution	65.0mL

pH 7.8 ± 0.2 at 25°C

Source: This medium is available as a premixed powder from BBL Microbiology Systems.

Fresh Yeast Extract Solution:
Composition per 100mL:

Baker's yeast, live, pressed, starch-free	25.0g

Preparation of Fresh Yeast Extract Solution: Add the live Baker's yeast to 100mL of distilled/deionized water. Autoclave for 90 min at 15 psi pressure–121°C. Allow to stand. Remove supernatant solution. Adjust pH to 6.6–6.8.

Preparation of Medium: Add components, except horse serum and special yeast extract, to distilled/deionized water and bring volume to 1.0L. Mix thoroughly. Gently heat and bring to boiling. Distribute into tubes or flasks. Autoclave for 15 min at 15 psi pressure–121°C. Cool to 50°C. To each 75.0mL of cooled, sterile basal medium add 20.0mL of sterile horse serum and 5.0mL of special yeast extract. Mix thoroughly. Pour into sterile Petri dishes or distribute into sterile tubes.

Use: For the preparation of media for cultivation of *Mycoplasma.*

Mycoplasma pneumoniae Isolation Medium

Composition per 1200mL:

Beef heart for infusion, Difco	50.0g
Peptone	10.0g
NaCl	5.0g
Water	900.0mL
Yeast extract solution	100.0mL
Agamma horse serum, unheated	200.0mL

pH 7.6–7.8 at 25°C

Yeast Extract Solution:
Composition per 10mL:

Yeast, active dry Baker's	250.0g

Preparation of Yeast Extract Solution: Add yeast to 1.0L of distilled/deionized water. Mix thoroughly. Gently heat and bring to boiling. Filter through Whatman #2 filter paper. Adjust the pH of the filtrate to 8.0 with NaOH. Distribute into tubes in 10.0mL volumes. Autoclave for 15 min at 15 psi pressure–121°C. Store at –20°C.

Preparation of Medium: Add components, except yeast extract solution and agamma horse serum, to distilled/deionized water and bring volume to 990.0mL. Mix thoroughly. Gently heat and bring to boiling. Autoclave for 15 min at 15 psi pressure–121°C. Cool to 45° to 50°C. Aseptically add sterile yeast extract solution and agamma horse serum. Mix thoroughly. Aseptically distribute into sterile tubes.

Use: For the isolation and cultivation of *Mycoplasma pneumoniae*.

Mycoplasmal Agar

Composition per liter:

Papaic digest of soy meal	20.0g
Agarose	10.0g
NaCl	5.0g
Phenol Red (2% solution)	1.0mL

pH 7.3 ± 0.2 at 25°C

Preparation of Medium: Add components, except agarose, to distilled/deionized water and bring volume to 1.0L. Mix thoroughly. Adjust pH to 7.3 with 1*N* NaOH. Add agarose. Mix thoroughly. Gently heat and bring to boiling. Distribute into tubes or flasks. Autoclave for 15 min at 15 psi pressure–121°C. Pour into sterile Petri dishes or leave in tubes.

Use: For isolation and cultivation of human mycoplasmas and ureaplasmas.

Mycosel™ Agar (Cycloheximide Chloramphenicol Agar)

Composition per liter:

Agar	15.5g
Papaic digest of soybean meal	10.0g
Glucose	10.0g
Cycloheximide	0.4g
Chloramphenicol	0.05g

pH 6.9 ± 0.2 at 25°C

Source: This medium is available as a premixed powder from BBL Microbiology Systems.

Preparation of Medium: Add components to distilled/deionized water and bring volume to 1.0L. Mix thoroughly. Gently heat while stirring and bring to boiling. Autoclave for 15 min at 14 psi pressure–118°C. Avoid overheating. Pour into sterile Petri dishes or distribute into sterile tubes.

Use: For the selective isolation of pathogenic fungi from specimens contaminated with bacteria and saprophytic fungi.

Neisseria meningitidis Medium

Composition per liter:

Beef infusion	300.0g
Acid hydrolysate of casein	17.5g
Agar	17.0g
Starch	1.5g
Antibiotic solution	10.0mL

pH 7.4 ± 0.2 at 25°C

Antibiotic Solution:

Composition per 10mL:

Vancomycin	3.0mg
Colistin	7.5mg
Nystatin	12,500U

Preparation of Antibiotic Solution: Add components to distilled/deionized water and bring volume to 10.0mL. Mix thoroughly. Filter sterilize.

Preparation of Medium: Add components, except antibiotic solution, to distilled/deionized water and bring volume to 990.0mL. Mix thoroughly. Gently heat and bring to boiling. Autoclave for 15 min at 15 psi pressure–121°C. Cool to 45°–50°C. Aseptically add sterile antibiotic solution. Mix thoroughly. Pour into sterile Petri dishes or distribute into sterile tubes.

Use: For the selective isolation and cultivation of *Neisseria meningitidis*.

Nelson Medium for *Naegleria fowleri*

Composition per liter:

Glucose	1.0g
Ox liver digest	1.0g
Page's amoeba saline	1.0L
Fetal calf serum, inactivated	20.0mL

Page's Amoeba Saline:

Composition per liter:

Na_2HPO_4	0.142g
KH_2PO_4	0.136g
NaCl	0.12g
$MgSO_4 \cdot 7H_2O$	4.0mg
$CaCl_2 \cdot 2H_2O$	4.0mg

Preparation of Page's Amoeba Saline: Add components to distilled/deionized water and bring volume to 10.0mL. Mix thoroughly.

Preparation of Medium: Add the glucose and ox liver digest to 1.0L of Page's amoeba saline. Mix thoroughly. Distribute into screw-capped tubes in 10.0mL volumes. Autoclave for 15 min at 15 psi pressure–121°C. Cool to 25°C. Aseptically add 0.2mL of sterile fetal calf serum to each tube. Mix thoroughly.

Use: For the cultivation of *Naegleria fowleri*.

Neomycin Agar, Modified

Composition per liter:

Agar	15.0g
Peptone	6.0g
Pancreatic digest of casein	4.0g
Yeast extract	3.0g
Beef extract	1.5g
Glucose	1.0g
Methanol	20.0mL
Neomycin solution	10.0mL

pH 7.0 ± 0.2 at 25°C

Neomycin Solution:
Composition per 10mL:

Neomycin sulfate	1.0g

Preparation of Neomycin Solution: Add neomycin sulfate to distilled/deionized water and bring volume to 10.0mL. Mix thoroughly. Filter sterilize.

Preparation of Medium: Add components, except methanol and neomycin solution, to distilled/deionized water and bring volume to 970.0mL. Mix thoroughly. Gently heat and bring to boiling. Autoclave for 15 min at 15 psi pressure–121°C. Cool to 45° to 50°C. Filter sterilize methanol. To cooled, sterile basal medium aseptically add sterile methanol and sterile neomycin solution. Mix thoroughly. Pour into sterile Petri dishes or distribute into sterile tubes.

Use: For the cultivation and maintenance of *Bordetella bronchiseptica*.

Neomycin Blood Agar

Composition per liter:

Pancreatic digest of casein	14.5g
Agar	14.0g
Papaic digest of soybean meal	5.0g
NaCl	5.0g
Growth factors	1.5g
Neomycin solution	10.0mL
Sheep blood, defibrinated	50.0mL

pH 7.3 ± 0.2 at 25°C

Source: This medium is available as a premixed powder from BBL Microbiology Systems.

Neomycin Solution:
Composition per 10mL:

Neomycin sulfate	0.030g

Preparation of Neomycin Solution: Add neomycin sulfate to distilled/deionized water and bring volume to 10.0mL. Mix thoroughly. Filter sterilize.

Preparation of Medium: Add components, except sheep blood and neomycin solution, to distilled/deionized water and bring volume to 940.0mL. Mix thoroughly. Gently heat and bring to boiling. Autoclave for 15 min at 15 psi pressure–121°C. Cool to 45° to 50°C. Aseptically add sterile sheep blood and sterile neomycin solution. Mix thoroughly. Pour into sterile Petri dishes or distribute into sterile tubes.

Use: For the isolation and cultivation of Group A streptococci (*Streptococcus pyogenes)* and Group B streptococci (*S. agalactiae*) from throat cultures and other clinical specimens.

Neopeptone Glucose Agar

Composition per liter:

Agar	20.0g
Glucose	10.0g
Neopeptone	5.0g

pH 6.5 ± 0.2 at 25°C

Preparation of Medium: Add components to distilled/deionized water and bring volume to 1.0L. Mix thoroughly. Gently heat and bring to boiling. Distribute into tubes or flasks. Autoclave for 15 min at 15 psi pressure–121°C. Adjust pH to 6.5. Pour into sterile Petri dishes or leave in tubes.

Use: For the maintenance of stock cultures of a variety of microorganisms.

Neopeptone Glucose Rose Bengal Aureomycin® Agar

Composition per liter:

Agar	20.0g
Neopeptone	5.0g
Glucose	1.0g
Tetracycline solution	5.0mL
Rose Bengal solution	3.5mL

pH 6.5 ± 0.2 at 25°C

Tetracycline Solution:
Composition per 150mL:

Tetracycline	1.0g

Preparation of Tetracycline Solution: Add tetracycline to distilled/deionized water and bring volume to 150.0mL. Mix thoroughly. Filter sterilize.

Rose Bengal Solution:

Composition per 100mL:

Rose Bengal 1.0g

Preparation of Rose Bengal Solution: Add Rose Bengal to distilled/deionized water and bring volume to 100.0mL. Mix thoroughly. Filter sterilize.

Preparation of Medium: Add components, except tetracycline solution, to distilled/deionized water and bring volume to 995.0mL. Mix thoroughly. Gently heat and bring to boiling. Autoclave for 15 min at 15 psi pressure–121°C. Cool to 45° to 50°C. Aseptically add 5.0mL of sterile tetracycline solution. Mix thoroughly. Pour into sterile Petri dishes or distribute into sterile tubes.

Use: For the isolation and cultivation of a wide variety of fungal species.

Neopeptone Infusion Agar

Composition per liter:

Beef heart, infusion from 500.0g
Neopeptone 20.0g
Agar 20.0g
NaCl 5.0g
Sheep blood, defibrinated 50.0mL

pH 7.4 ± 0.2 at 25°C

Preparation of Medium: Add components, except sheep blood, to distilled/deionized water and bring volume to 950.0mL. Mix thoroughly. Gently heat and bring to boiling. Autoclave for 15 min at 15 psi pressure–121°C. Cool to 45° to 50°C. Aseptically add sterile sheep blood. Mix thoroughly. Pour into sterile Petri dishes or distribute into sterile tubes.

Use: For the cultivation of a wide variety of fastidious microorganisms.

New York City Medium

Composition per liter:

NYC basal medium 640.0mL
Horse blood cells 200.0mL
Horse plasma, citrated 120.0mL
Glucose solution 10.0mL
Yeast dialysate 25.0mL
Antibiotic VCNT solution 5.0mL

pH 7.4 ± 0.2 at 25°C

NYC Basal Medium:

Composition per 640mL:

Solution 1 .... 400.0mL
Solution 3 .... 200.0mL
Solution 2 .... 40.0mL

Preparation of NYC Basal Medium: Combine solution 1, solution 2 and solution 3. Mix thoroughly. Autoclave for 15 min at 15 psi pressure–121°C. Cool to 45° to 50°C.

Solution 1:

Composition per 400mL:

Agar 20.0g

Preparation of Solution 1: Add agar to distilled/deionized water and bring volume to 400.0mL. Mix thoroughly. Melt agar in autoclave for 10 min at 0 psi pressure–100°C. Cool to 45° to 50°C.

Solution 2:

Composition per 40mL:

Cornstarch 1.0g

Preparation of Solution 2: Add cornstarch to distilled/deionized water and bring volume to 40.0mL. Mix thoroughly. Warm to 45° to 50°C.

Solution 3:

Composition per 200mL:

Proteose peptone No. 3 .... 15.0g
NaCl 5.0g
K_2HPO_4 .... 4.0g
KH_2PO_4 .... 1.0g

Preparation of Solution 3: Add components to distilled/deionized water and bring volume to 1.0L. Mix thoroughly. Gently heat and bring to boiling. Cool to 45° to 50°C.

Horse Blood Cells:

Composition per 200mL:

Horse blood cells, sedimented 6.0mL

Preparation of Horse Blood Cells: Cow blood may be used instead of horse blood but do not use sheep blood. Use cells freshly packed by sedimentation. Do not pack by centrifugation. Aseptically add 6.0mL of sedimented blood cells to 200.0mL of sterile distilled/deionized water. Mix thoroughly.

Horse Plasma, Citrated:

Composition per 6 liters:

Horse blood 5400.0mL
Citrate solution 600.0mL

Preparation of Horse Plasma, Citrated: Place 600.0mL of sterile citrate solution into a receiving bottle. Draw horse blood to the 6.0L mark. Allow cells to sediment out. Aseptically remove plasma.

Citrate Solution:

Composition per liter:

Sodium citrate 150.0g
NaCl 81.13g

Preparation of Citrate Solution: Add components to distilled/deionized water and bring volume to 1.0L. Mix thoroughly. Filter sterilize.

Glucose Solution:
Composition per 10mL:

D-Glucose	5.0g

Preparation of Glucose Solution: Add D-glucose to distilled/deionized water and bring volume to 10.0mL. Mix thoroughly. Autoclave for 10 min at 10 psi pressure–115°C. Cool to 45° to 50°C.

Yeast Dialysate:
Composition per 2500mL:

Baker's yeast, fresh	908.0g

Preparation of Yeast Dialysate: Add fresh baker's yeast to 2500.0mL of distilled/deionized water. Mix thoroughly. Autoclave for 10 min at 15 psi pressure–121°C. Cool to 25°C. Put into dialysis tubing. Dialyze against 2.0L of distilled/deionized water for 48 hr at 4°C.

Antibiotic VCNT Solution:
Composition per 5mL:

Colistin	7.5mg
Trimethorprim lactate	3.0mg
Vancomycin·HCl	2.0mg
Nystatin	12.5U

Preparation of Antibiotic Solution: Add components to distilled/deionized water and bring volume to 5.0mL. Mix thoroughly. Filter sterilize.

Preparation of Medium: Have all solutions prepared and at 45° to 50°C. Aseptically combine components. Mix thoroughly. Pour into sterile Petri dishes.

Use: For the isolation and cultivation of pathogenic *Neisseria* species. Used as a transport medium for urogenital and other clinical specimens. For the isolation and presumptive identification of Mycoplasmatales, including large-colony species (*Mycoplasma pneumoniae*) and T–mycoplasmas from urogenital specimens.

New York City Medium, Modified

Composition per liter:

NYC basal medium	840.0mL
Agamma horse serum (Flow Labs)	120.0mL
Glucose solution	10.0mL
Yeast dialysate	25.0mL
Antibiotic LCNT solution	5.0mL

pH 7.4 ± 0.2 at 25°C

NYC Basal Medium:
Composition per 840mL:

Horse blood	5400.0mL
Solution 1	600.0mL
Solution 3	200.0mL
Solution 2	40.0mL

Preparation of NYC Basal Medium: Combine solution 1, solution 2 and solution 3. Mix thoroughly. Autoclave for 15 min at 15 psi pressure–121°C. Cool to 45° to 50°C.

Solution 1:
Composition per 600mL:

Agar	20.0g

Preparation of Solution 1: Add agar to distilled/deionized water and bring volume to 600.0mL. Mix thoroughly. Melt agar in autoclave for 10 min at 0 psi pressure–100°C. Cool to 45° to 50°C.

Solution 2:
Composition per 40mL:

Cornstarch	1.0g

Preparation of Solution 2: Add cornstarch to distilled/deionized water and bring volume to 40.0mL. Mix thoroughly. Warm to 45° to 50°C.

Solution 3:
Composition per 200mL:

Proteose peptone No. 3	15.0g
NaCl	5.0g
K_2HPO_4	4.0g
KH_2PO_4	1.0g

Preparation of Solution 3: Add components to distilled/deionized water and bring volume to 200.0mL. Mix thoroughly. Gently heat and bring to boiling. Cool to 45° to 50°C.

Glucose Solution:
Composition per 10mL:

D-Glucose	5.0g

Preparation of Glucose Solution: Add glucose to distilled/deionized water and bring volume to 10.0mL. Mix thoroughly. Autoclave for 10 min at 10 psi pressure–115°C. Cool to 45° to 50°C.

Yeast Dialysate:
Composition per 2500mL:

Baker's yeast, fresh	908.0g

Preparation of Yeast Dialysate: Add fresh baker's yeast to 2500.0mL of distilled/deionized water. Mix thoroughly. Autoclave for 10 min at 15 psi pressure–121°C. Cool to 25°C. Put into dialysis tubing. Dialyze against 2.0L of distilled/deionized water for 48 hr at 4°C.

Antibiotic LCNT Solution:
Composition per 5mL:

Colistin	7.5mg
Lincomycin·HCl	4.0mg
Trimethorprim lactate	3.0mg
Nystatin	12.5U

Preparation of Antibiotic LCNT Solution: Add the components to distilled/deionized water and bring volume to 5.0mL. Mix thoroughly. Filter sterilize the solution.

Preparation of Medium: Have all solutions prepared and at 45° to 50°C. Aseptically combine components. Mix thoroughly. Pour into sterile Petri dishes.

Use: For the isolation and cultivation of pathogenic *Neisseria* species. Used as a transport medium for urogenital and other clinical specimens. For the isolation and presumptive identification of Mycoplasmatales, including large-colony species (*Mycoplasma pneumoniae*) and T–mycoplasmas from urogenital specimens.

Nitrate Broth

Composition per liter:

Pancreatic digest of gelatin 20.0g
KNO_3 .... 2.0g

pH 7.2 ± 0.2 at 25°C

Source: This medium is available as a premixed powder from BBL Microbiology Systems.

Preparation of Medium: Add components to distilled/deionized water and bring volume to 1.0L. Mix thoroughly. Distribute into tubes or flasks. Autoclave for 15 min at 15 psi pressure–121°C.

Use: For the differentiation of aerobic and facultative Gram-negative microorganisms based on their ability to reduce nitrate. Test for nitrates with sulfanilic acid and α-naphthylamine reagents. Bacteria that reduce nitrate to nitrite turn the reagents red or pink.

Nitrate Broth, *Campylobacter*

Composition per liter:

Beef heart, solids from infusion 500.0g
Tryptose 10.0g
NaCl 5.0g
KNO_3 .... 2.0g

pH 7.0 ± 0.2 at 25°C

Preparation of Medium: Add components to distilled/deionized water and bring volume to 1.0L. Mix thoroughly. Adjust pH to 7.0. Distribute 4.0mL volumes into test tubes that contain inverted Durham tubes. Autoclave for 15 min at 15 psi pressure–121°C.

Use: For the cultivation and differentiation of *Campylobacter* species based on their ability to reduce nitrate.

Nitrate Broth, Enriched

Composition per liter:

Pancreatic digest of casein 13.0g
NaCl 5.0g
Yeast extract 5.0g
Heart muscle, solids from infusion 2.0g
KNO_3 .... 2.0g

pH 7.3 ± 0.2 at 25°C

Preparation of Medium: Add components to distilled/deionized water and bring volume to 1.0L. Mix thoroughly. Distribute into test tubes that contain an inverted Durham tube. Autoclave for 15 min at 15 psi pressure–121°C.

Use: For the differentiation of aerobic and facultative Gram-negative microorganisms based on their ability to reduce nitrate to nitrite or form N_2 gas. Test for nitrates with sulfanilic acid and α-naphthylamine reagents. Bacteria that reduce nitrate to nitrite turn the reagents red or pink.

Nitrate Reduction Broth

Composition per liter:

Pancreatic digest of casein 13.0g
NaCl 5.0g
Yeast extract 5.0g
Heart muscle, solids from infusion 2.0g
KNO_3 or $NaNO_3$ 2.0g

pH 7.4 ± 0.2 at 25°C

Preparation of Medium: Add components to distilled/deionized water and bring volume to 1.0L. Mix thoroughly. Distribute into test tubes that contain an inverted Durham tube. Autoclave for 15 min at 15 psi pressure–121°C.

Use: For the differentiation of a variety of Gram-negative bacteria based on their ability to reduce nitrate to nitrite or form N_2 gas. Test for nitrates with sulfanilic acid and α-naphthylamine reagents. Bacteria that reduce nitrate to nitrite turn the reagents red or pink.

Nitrate Reduction Broth for *Pseudomonas* and Related Genera

Composition per liter:

Peptone 5.0g
NaCl 5.0g
Yeast extract 2.0g
Beef extract 1.0g
$NaNO_3$ 0.1g

pH 7.4 ± 0.2 at 25°C

Preparation of Medium: Add components to distilled/deionized water and bring volume to 1.0L. Mix thoroughly. Distribute into test tubes that contain an inverted Durham tube. Autoclave for 15 min at 15 psi pressure–121°C.

Use: For the differentiation of members of the Pseudomonadaceae based on their ability to reduce nitrate to nitrite or form N_2 gas. Test for nitrates with sulfanilic acid and α-naphthylamine reagents. Bacteria that reduce nitrate to nitrite turn the reagents red or pink.

NNN Medium (Novy, MacNeal and Nicole Medium)

Composition per liter:

Agar....7.0g
NaCl....3.0g
Rabbit blood, defibrinated....150.0mL

Preparation of Medium: Add components, except rabbit blood, to distilled/deionized water and bring volume to 850.0mL. Mix thoroughly. Gently heat and bring to boiling. Autoclave for 15 min at 15 psi pressure–121°C. Cool to 50°C. Aseptically add sterile rabbit blood. Mix thoroughly. Aseptically distribute into sterile tubes in 5.0mL volumes. Allow tubes to cool in a slanted position at 4°C.

Use: For the cultivation and maintenance of *Leishmania* species and *Trypanosoma cruzi*.

Nutrient Agar

Composition per liter:

Agar....15.0g
Peptone....5.0g
NaCl....5.0g
Yeast extract....2.0g
Beef extract....1.0g

pH 7.4 ± 0.2 at 25°C

Source: This medium is available as a premixed powder from Oxoid Unipath.

Preparation of Medium: Add components to distilled/deionized water and bring volume to 1.0L. Mix thoroughly. Gently heat and bring to boiling. Distribute into tubes or flasks. Autoclave for 15 min at 15 psi pressure–121°C. Pour into sterile Petri dishes or leave in tubes.

Use: For the cultivation and maintenance of a wide variety of microorganisms.

Nutrient Agar, 1.5% (ATCC Medium 105)

Composition per liter:

Agar....15.0g
NaCl....8.0g
Pancreatic digest of gelatin....5.0g
Beef extract....3.0g

pH 7.3 ± 0.2 at 25°C

Source: This medium is available as a premixed powder from BBL Microbiology Systems and Difco Laboratories.

Preparation of Medium: Add components to distilled/deionized water and bring volume to 1.0L. Mix thoroughly. Gently heat while stirring and bring to boiling. Distribute into tubes or flasks. Autoclave for 15 min at 15 psi pressure–121°C. Pour into sterile Petri dishes or leave in tubes.

Use: For the cultivation and maintenance of a variety of nonfastidious bacteria.

Nutrient Agar with 0.5% NaCl

Composition per liter:

Agar....15.0g
NaCl....5.0g
Pancreatic digest of gelatin....5.0g
Beef extract....3.0g

pH 6.8 ± 0.2 at 25°C

Source: Nutrient Agar is available as a premixed powder from Difco Laboratories.

Preparation of Medium: Add components to distilled/deionized water and bring volume to 1.0L. Mix thoroughly. Gently heat and bring to boiling. Distribute into tubes or flasks. Autoclave for 15 min at 15 psi pressure–121°C. Pour into sterile Petri dishes or leave in tubes.

Use: For the cultivation and maintenance of *Agrobacterium tumefaciens*, *Escherichia coli*, *Pseudomonas aeruginosa*, *Salmonella choleraesuis*, *Shigella dysenteriae*, *Shigella flexneri*, *Vibrio* species, and *Yersinia* species.

Nutrient Broth

Composition per liter:

Pancreatic digest of gelatin....5.0g
Beef extract....3.0g

pH 6.9 ± 0.2 at 25°C

Source: This medium is available as a premixed powder from BBL Microbiology Systems and Difco Laboratories.

Preparation of Medium: Add components to distilled/deionized water and bring volume to 1.0L. Mix thoroughly. Distribute into tubes or flasks. Autoclave for 15 min at 15 psi pressure–121°C.

Use: For the cultivation of a wide variety of nonfastidious microorganisms.

Nutrient Broth No. 2

Composition per liter:

Beef extract	10.0g
Peptone	10.0g
NaCl	5.0g

pH 7.5 ± 0.2 at 25°C

Source: This medium is available as a premixed powder from Oxoid Unipath.

Preparation of Medium: Add components to distilled/deionized water and bring volume to 1.0L. Mix thoroughly. Distribute into tubes or flasks. Autoclave for 15 min at 15 psi pressure–121°C.

Use: For the cultivation of a variety of fastidious and nonfastidious microorganisms.

Nutrient Gelatin

Composition per liter:

Gelatin	120.0g
Pancreatic digest of gelatin	5.0g
Beef extract	3.0g

pH 6.8 ± 0.2 at 25°C

Source: This medium is available as a premixed powder from BBL Microbiology Systems and Difco Laboratories and Oxoid Unipath.

Preparation of Medium: Add components to distilled/deionized water and bring volume to 1.0L. Mix thoroughly. Gently heat while stirring to 50°C. Distribute into tubes. Autoclave for 15 min at 15 psi pressure–121°C.

Use: For the cultivation and differentiation of bacteria based on their ability to liquefy gelatin.

Ogawa TB Medium

Composition per 300mL:

KH_2PO_4	1.0g
Homogenized whole egg	200.0mL
Glycerol	6.0mL
Malachite Green (2% solution)	6.0mL

pH 6.5 ± 0.2 at 25°C

Homogenized Whole Egg:

Composition per liter:

Whole eggs	18–24

Preparation of Homogenized Whole Egg: Use fresh eggs, less than 1 week old. Scrub the shells with soap. Let stand in a soap solution for 30 min. Rinse in running water. Soak eggs in 70% ethanol for 15 min. Break the eggs into a sterile container. Homogenize by shaking. Filter through four layers of sterile cheesecloth into a sterile graduated cylinder. Measure out 1.0L.

Preparation of Medium: Add components, except homogenized whole egg, to distilled/deionized water and bring volume to 100.0mL. Mix thoroughly. Autoclave for 15 min at 15 psi pressure–121°C. Cool to 45° to 50°C. Aseptically add 200.0mL of sterile homogenized whole egg. Mix thoroughly. Aseptically distribute into sterile screw-capped tubes in 7.0mL volumes. Inspissate at 85° to 90°C (moist heat) for 60 min.

Use: For the isolation and cultivation of *Mycobacterium* species except for *M. leprae*.

Oleic Albumin Complex

Composition per liter:

Bovine albumin fraction V	50.0g
NaCl	8.5g
Oleic acid	0.6mL

Preparation of Medium: Add components to distilled/deionized water and bring volume to 1.0L. Mix thoroughly. Filter sterilize.

Use: For use in media employed for the cultivation of mycobacteria.

ONPG Broth

Composition per liter:

Peptone water	750.0mL
ONPG solution	250.0mL

pH 7.2–7.4 at 25°C

ONPG Solution:

Composition per 250mL:

ONPG (*o*-nitrophenyl-β-D-galactopyranoside)	1.5g
Sodium phosphate buffer (0.01M, pH 7.5)	250.0mL

Preparation of ONPG Solution: Add ONPG to 250.0mL of sodium phosphate buffer. Mix thoroughly. Filter sterilize.

Peptone Water:

Composition per 750mL:

Peptone	7.5g
NaCl	3.75g

Preparation of Peptone Water: Add components to distilled/deionized water and bring volume to 750.0mL. Mix thoroughly. Gently heat and bring to boiling. Adjust pH to 8.0–8.4. Continue boiling for 10 min. Filter through Whatman #1 filter paper. Readjust pH of filtrate to 7.2–7.4. Autoclave for 20 min at 10 psi pressure–115°C. Cool to 25°C.

Preparation of Medium: Aseptically combine the sterile ONPG solution with the cooled, sterile peptone water. Mix thoroughly. Aseptically distribute into tubes in 2.5–3.0.0mLvolumes. Store at 4°C for up to one month.

Use: For the differentiation of a variety of Gram-negative bacteria based on production of β-galactosidase. For the differentiation of lactose-delayed bacteria from lactose-negative bacteria. For the differentiation of *Pseudomonas cepacia* (positive) and *Pseudomonas maltophila* (positive) from other *Pseudomonas* species (negative). Bacteria that produce β-galactosidase turn the medium yellow.

OR Indicator Agar (Oxidation–Reduction Indicator Agar)

Composition per liter:

Agar	15.0g
Sodium glycerol phosphate	10.0g
Sodium thioglycollate	1.7g
$CaCl_2 \cdot 2H_2O$	0.1g
Methylene Blue	6.0mg

Preparation of Medium: Add components to distilled/deionized water and bring volume to 1.0L. Mix thoroughly. Gently heat and bring to boiling. Distribute into tubes or flasks. Autoclave for 15 min at 15 psi pressure–121°C. Pour into sterile Petri dishes or leave in tubes.

Use: For use as an indicator of oxygen free conditions in anaerobic culture chambers.

Oxford Agar (*Listeria* Selective Agar, Oxford)

Composition per liter:

Special peptone	23.0g
LiCl	15.0g
Agar	10.0g
NaCl	5.0g
Cornstarch	1.0g
Esculin	1.0g
Ferric ammonium citrate	0.5g
Antibiotic inhibitor	10.0mL

pH 7.0 ± 0.2 at 25°C

Source: This medium is available as a premixed powder from Oxoid Unipath.

Antibiotic Inhibitor:

Composition per 10mL:

Cycloheximide	0.4g
Colistin sulphate	0.02g
Fosfomycin	0.01g
Acriflavine	5.0mg
Cefotetan	2.0mg
Ethanol (50% solution)	10.0mL

Preparation of Antibiotic Inhibitor: Add antibiotics to 10.0mL of ethanol. Mix thoroughly. Filter sterilize.

Preparation of Medium: Add components, except antibiotic inhibitor, to distilled/deionized water and bring volume to 990.0mL. Mix thoroughly. Gently heat and bring to boiling. Autoclave for 15 min at 15 psi pressure–121°C. Cool to 45° to 50°C. Aseptically add 10.0mL of sterile antibiotic inhibitor. Mix thoroughly. Pour into sterile Petri dishes or distribute into sterile tubes.

Use: For the isolation and cultivation of *Listeria monocytogenes* from specimens containing a mixed bacterial flora.

Oxford Agar, Modified (*Listeria* Selective Agar, Modified Oxford) (MOX Agar)

Composition per liter:

Special peptone	23.0g
LiCl	15.0g
Agar	12.0g
NaCl	5.0g
Cornstarch	1.0g
Esculin	1.0g
Ferric ammonium citrate	0.5g
Antibiotic inhibitor	10.0mL

pH 7.0 ± 0.2 at 25°C

Antibiotic Inhibitor:

Composition per 10mL:

Moxalactam	0.015g
Colistin sulfate	0.010g

Preparation of Antibiotic Inhibitor: Add components to distilled /deionized water and bring volume to 10.0mL. Mix thoroughly. Filter sterilize.

Preparation of Medium: Add components, except antibiotic inhibitor, to distilled/deionized water and bring volume to 990.0mL. Mix thoroughly. Gently heat and bring to boiling. Autoclave for 10 min at 15 psi pressure–121°C. Cool to 45° to 50°C. Aseptically add 10.0mL of sterile antibiotic inhibitor. Mix thoroughly. Pour into sterile Petri dishes or distribute into sterile tubes.

Use: For the isolation and cultivation of *Listeria monocytogenes* from specimens containing a mixed bacterial flora.

Oxidative Fermentative Glucose Medium, Semisolid (OF Glucose Medium, Semisolid)

Composition per liter:

Glucose	10.0g
NaCl	5.0g
Agar	2.0g
Pancreatic digest of casein	2.0g
K_2HPO_4	0.3g
Bromthymol Blue dye	0.08g

pH 6.8 ± 0.2 at 25°C

Preparation of Medium: Add components to distilled/deionized water and bring volume to 1.0L. Mix thoroughly. Gently heat and bring to boiling. Distribute into tubes or flasks. Autoclave for 15 min at 15 psi pressure–121°C. Pour into sterile Petri dishes or leave in tubes.

Use: For differentiating Gram-negative bacteria based upon determining the oxidative and fermentative metabolism of glucose. Bacteria that ferment glucose turn the medium yellow.

Oxidation Fermentation Medium (OF Medium)

Composition per liter:

NaCl	5.0g
Agar	2.5g
Pancreatic digest of casein	2.0g
K_2HPO_4	0.3g
Bromthymol Blue	0.03g
Carbohydrate solution	100.0mL

pH 6.8 ± 0.1 at 25°C

Source: This medium is available as a premixed powder from BBL Microbiology Systems.

Carbohydrate Solution:
Composition per 100mL:

Carbohydrate	10.0g

Preparation of Carbohydrate Solution: Add carbohydrate to distilled/deionized water and bring volume to 100.0mL. Mix thoroughly. Filter sterilize.

Preparation of Medium: Add components, except carbohydrate solution, to distilled/deionized water and bring volume to 900.0mL. Mix thoroughly. Gently heat and bring to boiling. Autoclave for 15 min at 15 psi pressure–121°C. Cool to 45° to 50°C. Aseptically add 100.0mL sterile carbohydrate solution. Mix thoroughly. Pour into sterile Petri dishes or distribute into sterile tubes.

Use: For differentiating Gram-negative bacteria based upon determining the oxidative and fermentative metabolism of carbohydrates.

Oxidation Fermentation Medium, Hugh–Leifson's (Hugh–Leifson's Oxidation Fermentation Medium)

Composition per liter:

NaCl	5.0g
Agar	3.0g
Peptone	2.0g
K_2HPO_4	0.3g
Carbohydrate solution	100.0mL
Bromthymol Blue solution (0.2%)	15.0mL

pH 7.1 ± 0.2 at 25°C

Carbohydrate Solution:
Composition per 100mL:

Carbohydrate	10.0g

Preparation of Carbohydrate Solution: Add carbohydrate to distilled/deionized water and bring volume to 100.0mL. Mix thoroughly. Filter sterilize.

Preparation of Medium: Add components, except carbohydrate solution, to distilled/deionized water and bring volume to 900.0mL. Mix thoroughly. Gently heat and bring to boiling. Autoclave for 15 min at 15 psi pressure–121°C. Cool to 45° to 50°C. Aseptically add 100.0mL sterile carbohydrate solution. Mix thoroughly. Pour into sterile Petri dishes or distribute into sterile tubes.

Use: For differentiating Gram-negative bacteria, such as *Vibrio* species, based upon determining the oxidative and fermentative metabolism of carbohydrates. Bacteria that ferment the carbohydrate turn the medium yellow.

Oxidation Fermentation Medium, King's (King's OF Medium)

Composition per liter:

Base:

Agar	3.0g
Pancreatic digest of casein	2.0g
Carbohydrate solution	100.0mL
Phenol Red (1.5% solution)	2.0mL

pH to 7.3 ± 0.2

Carbohydrate Solution:
Composition per 100mL:
Carbohydrate 10.0g

Preparation of Carbohydrate Solution: Add carbohydrate to distilled/deionized water and bring volume to 100.0mL. Mix thoroughly. Filter sterilize.

Preparation of Medium: Add components, except carbohydrate solution, to distilled/deionized water and bring volume to 900.0mL. Mix thoroughly. Gently heat and bring to boiling. Autoclave for 15 min at 15 psi pressure–121°C. Cool to 45° to 50°C. Aseptically add 100.0mL sterile carbohydrate solution. Mix thoroughly. Pour into sterile Petri dishes or distribute into sterile tubes.

Use: For differentiating bacteria based upon determining the oxidative and fermentative metabolism of carbohydrates. Bacteria that ferment the carbohydrate turn the medium yellow.

Oxidative–Fermentative Glucose Medium, Semisolid, with NaCl (OF Glucose Medium, Semisolid with NaCl)

Composition per liter:
NaCl 20.0g
Glucose 10.0g
Agar 2.0g
Pancreatic digest of casein 2.0g
K_2HPO_4 .... 0.3g
Bromthymol Blue dye 0.08g
pH 6.8 ± 0.2 at 25°C

Preparation of Medium: Add components to distilled/deionized water and bring volume to 1.0L. Mix thoroughly. Gently heat and bring to boiling. Distribute into tubes or flasks. Autoclave for 15 min at 15 psi pressure–121°C. Pour into sterile Petri dishes or leave in tubes.

Use: For differentiating halophilic *Vibrio* species based upon determining the oxidative and fermentative metabolism of glucose. Bacteria that ferment glucose turn the medium yellow.

Oxidative–Fermentative Medium (OF Medium)

Composition per liter:
NaCl 5.0g
Agar 2.0g
Pancreatic digest of casein 2.0g
K_2HPO_4 .... 0.3g
Bromothymol Blue 0.08g
Carbohydrate solution 100.0mL
pH 6.8 ± 0.1 at 25°C

Source: This medium is available as a premixed powder from Difco Laboratories and Oxoid Unipath.

Carbohydrate Solution:
Composition per 100mL:
Carbohydrate 10.0g

Preparation of Carbohydrate Solution: Add carbohydrate to distilled/deionized water and bring volume to 100.0mL. Mix thoroughly. Filter sterilize.

Preparation of Medium: Add components, except carbohydrate solution, to distilled/deionized water and bring volume to 900.0mL. Mix thoroughly. Gently heat and bring to boiling. Autoclave for 15 min at 15 psi pressure–121°C. Cool to 45° to 50°C. Aseptically add 100.0mL sterile carbohydrate solution. Mix thoroughly. Pour into sterile Petri dishes or distribute into sterile tubes.

Use: For differentiating bacteria based upon determining the oxidative and fermentative metabolism of carbohydrates. Bacteria that ferment the carbohydrate turn the medium yellow.

Oxidative–Fermentative Test Medium (OF Test Medium)

Composition per liter:
NaCl 5.0g
Agar 3.0g
Peptone 2.0g
K_2HPO_4 .... 0.3g
Bromthymol Blue 0.03g
Carbohydrate solution 100.0mL

Carbohydrate Solution:
Composition per 100mL:
Carbohydrate 10.0g

Preparation of Carbohydrate Solution: Add carbohydrate to distilled/deionized water and bring volume to 100.0mL. Mix thoroughly. Filter sterilize.

Preparation of Medium: Add components, except carbohydrate solution, to distilled/deionized water and bring volume to 1.0L. Mix thoroughly. Gently heat and bring to boiling. Distribute into tubes in 3.0mL volumes. Autoclave for 15 min at 15 psi pressure–121°C. Cool to 45° to 50°C. Aseptically add 0.3mL of sterile carbohydrate solution to each tube. Mix thoroughly.

Use: For the cultivation and differentiation of a variety of microorganisms based on their ability to ferment a specific carbohydrate. Bacteria that ferment the specific carbohydrate turn the medium yellow.

Pagano Levin Agar

Composition per liter:

Glucose	40.0g
Agar	15.0g
Peptone	10.0g
Yeast extract	1.0g
Neomycin	0.5g
2,3,5-Triphenyltetrazolium chloride	0.1g

pH 6.0 ± 0.1 at 25°C

Source: This medium is available as a premixed powder from Difco Laboratories.

Preparation of Medium: Add components, except neomycin and 2,3,5-triphenyltetrazolium chloride, to distilled/deionized water and bring volume to 1.0L. Mix thoroughly. Gently heat and bring to boiling. Autoclave for 15 min at 15 psi pressure–121°C. Cool to 45° to 50°C. Aseptically add neomycin and 2,3,5-triphenyltetrazolium chloride. Mix thoroughly. Pour into sterile Petri dishes or distribute into sterile tubes. Allow tubes to cool in a slanted position.

Use: For the isolation, cultivation and differentiation of *Candida* species. *Candida albicans* appears as smooth, shiny cream-light pink colonies.

Pasteurella haemolytica Selective Medium

Composition per 1010mL:

Tryptose agar with peptic digest of blood	1.0L
Antibiotic solution	10.0mL

pH 7.2 ± 0.2 at 25°C

Tryptose Agar With Peptic Digest of Blood:
Composition per liter:

Agar	15.0g
Pancreatic digest of casein	10.0g
Peptic digest of animal tissue	10.0g
NaCl	5.0g
Glucose	1.0g
Peptic digest of blood	50.0mL

Preparation of Tryptose Agar With Peptic Digest of Blood: Add components to distilled/deionized water and bring volume to 950.0mL. Mix thoroughly. Gently heat and bring to boiling. Autoclave for 15 min at 15 psi pressure–121°C. Cool to 45° to 50°C. Aseptically add peptic digest of blood. Mix thoroughly.

Antibiotic Solution:
Composition per 10mL:

Actidione (cycloheximide)	0.1g
Novobiocin	2.0mg
Neomycin	1.5mg

Preparation of Antibiotic Solution: Add components to distilled/deionized water and bring volume to 10.0mL. Mix thoroughly. Filter sterilize.

Preparation of Medium: To 1.0L of cooled, sterile tryptose agar with peptic digest of blood, aseptically add 10.0mL of sterile antibiotic solution. Mix thoroughly. Pour into sterile Petri dishes or distribute into sterile tubes.

Use: For the selective cultivation of *Pasteurella haemolytica*.

Pasteurella multocida Selective Medium

Composition per 1020mL:

Tryptose agar with peptic digest of blood	1.0L
Antibiotic solution	10.0mL
K_2TeO_3 solution	10.0mL

pH 7.2 ± 0.2 at 25°C

Tryptose Agar With Peptic Digest of Blood:
Composition per liter:

Agar	15.0g
Pancreatic digest of casein	10.0g
Peptic digest of animal tissue	10.0g
NaCl	5.0g
Glucose	1.0g
Peptic digest of blood	50.0mL

Preparation of Tryptose Agar With Peptic Digest of Blood: Add components to distilled/deionized water and bring volume to 950.0mL. Mix thoroughly. Gently heat and bring to boiling. Autoclave for 15 min at 15 psi pressure–121°C. Cool to 45° to 50°C. Aseptically add peptic digest of blood. Mix thoroughly.

Antibiotic Solution:
Composition per 10mL:

Actidione (cycloheximide)	0.1g
Novobiocin	0.01g
Erythrocin	5.0mg

Preparation of Antibiotic Solution: Add components to distilled/deionized water and bring volume to 10.0mL. Mix thoroughly. Filter sterilize.

K_2TeO_3 Solution:
Composition per 10mL:

K_2TeO_3	5.0mg

Preparation of K_2TeO_3 Solution: Add K_2TeO_3 to distilled/deionized water and bring volume to 10.0mL. Mix thoroughly. Filter sterilize.

Caution: Potassium tellurite is toxic.

Preparation of Medium: To 1.0L of cooled, sterile tryptose agar with peptic digest of blood, aseptically add 10.0mL of sterile antibiotic solution and 10.0mL of sterile K_2TeO_3 solution. Mix thoroughly. Pour into sterile Petri dishes or distribute into sterile tubes.

Use: For the selective cultivation of *Pasteurella multocida*.

Pectin Medium

Composition per liter:

Pectin....30.0g
Yeast extract....5.0g
Bromthymol Blue (0.1% solution)....1.0mL
$CaCl_2 \cdot 2H_2O$ (10.0% solution)....0.6mL

pH 7.3 ± 0.2 at 25°C

Preparation of Medium: Add 100.0mL of distilled/deionized water to a 2-liter flask and place on a magnetic stirrer with no heat. While stirring, slowly add $CaCl_2 \cdot 2H_2O$, Bromthymol Blue solution, yeast extract, and pectin. Add slowly to ensure each particle is wetted. Gently heat and bring to almost boiling. Adjust pH to 7.3 with 1N NaOH. Do not overshoot pH 7.3.

Use: For the isolation and presumptive identification of *Yersinia enterocolitica* and *Yersinia pseudotuberculosis* from other fermenting Gram–negative bacilli such as *Enterobacter agglomerans* which is often confused with *Yersinia*. *E. agglomerans* does not produce pectinase. *Y. enterocolitica* is strongly positive for pectinase activity. *Y. pseudotuberculosis* is weakly positive for pectinase activity. *Yersinia pestis* is negative for pectinase activity. Also, use for the differentiation of *Klebsiella oxytoca* which is pectinase positive.

Peptone Starch Dextrose Agar (PSD Agar) (Dunkelberg Agar)

Composition per liter:

Proteose peptone No. 3....20.0g
Agar....15.0g
Soluble starch....10.0g
Glucose....2.0g
Na_2HPO_4....1.0g
NaH_2PO_4....1.0g

pH 6.8 ± 0.2 at 25°C

Preparation of Medium: Add starch to approximately 100.0mL of cold distilled/deionized water. Mix thoroughly. Add starch solution to 400.0mL of boiling distilled/deionized water. Add remaining components. Mix thoroughly. Bring volume to 1.0L with distilled/deionized water. Autoclave for 12 min at 8 psi pressure–112°C. Pour into sterile Petri dishes or distribute into screw-capped tubes.

Use: For the selective isolation and cultivation of *Gardnerella vaginalis*.

Petragnani Medium

Composition per 2398mL:

Skim milk....100.0g
Potato flour....36.4g
L-Asparagine....5.1g
Pancreatic digest of casein....5.1g
Malachite Green....1.2g
Whole egg....1277.0mL
Egg yolk....121.0mL
Glycerol....60.0mL

pH 7.0 ± 0.2 at 25°C

Source: Available as a prepared medium from BBL Microbiology Systems.

Preparation of Medium: Add components—except whole egg, egg yolk, and glycerol—to distilled/deionized water and bring volume to 940.0mL. Mix thoroughly. Add glycerol. Gently heat while stirring and bring to boiling. Autoclave for 15 min at 15 psi pressure–121°C. Cool to 45° to 50°C. Scrub the eggshells with soap. Let stand in a soap solution for 30 min. Rinse in running water. Soak eggs in 70% ethanol for 15 min. Break the eggs into a sterile container. Homogenize by shaking. Filter through four layers of sterile cheesecloth into a sterile graduated cylinder. Measure out 1277.0mL. Add separated egg yolks to another sterile container. Measure out 121.0mL. Aseptically add homogenized whole egg and egg yolk to cooled sterile basal medium. Mix thoroughly. Aseptically distribute into sterile tubes. Inspissate at 85° to 90°C (moist heat) for 45 min.

Use: For the isolation and cultivation of *Mycobacterium* species obtained from clinical specimens. For the cultivation and maintenance of *Mycobacterium smegmatis*.

Petragnani Medium

Composition per 2285mL:

Potato....500.0g
Potato flour....36.0g
Malachite Green....1.2g
Whole egg....1200.0mL
Whole milk....900.0mL
Egg yolk....115.0mL
Glycerol....70.0mL

pH 7.2 ± 0.2 at 25°C

Source: Available as a prepared medium from Difco Laboratories.

Preparation of Medium: Peel and dice potato. Add potato to 500.0mL of distilled/deionized water. Gently heat and bring to boiling. Continue boiling for 30 min. Filter solids through two layers of cheesecloth. Combine potato solids with remaining components—except whole egg, egg yolk, and glycerol. Mix thoroughly. Add glycerol. Gently heat while stirring and bring to boiling. Autoclave for 15 min at 15 psi pressure–121°C. Cool to 45° to 50°C. Scrub the eggshells with soap. Let stand in a soap solution for 30 min. Rinse in running water. Soak eggs in 70% ethanol for 15 min. Break the eggs into a sterile container. Homogenize by shaking. Filter through four layers of sterile cheesecloth into a sterile graduated cylinder. Measure out 1200.0mL. Add separated egg yolks to another sterile container. Measure out 115.0mL. Aseptically add homogenized whole egg and egg yolk to cooled sterile basal medium. Mix thoroughly. Aseptically distribute into sterile tubes. Inspissate at 85° to 90°C (moist heat) for 45 min.

Use: For the isolation and cultivation of *Mycobacterium* species from clinical specimens. For the cultivation and maintenance of *Mycobacterium smegmatis.*

Pfizer Selective *Enterococcus* Agar (PSE Agar)

Peptone C	17.0g
Agar	15.0g
Bile	10.0g
NaCl	5.0g
Yeast extract	5.0g
Peptone B	3.0g
Esculin	1.0g
Sodium citrate	1.0g
Ferric ammonium citrate	0.5g
NaN_3	0.25g

pH 7.1 ± 0.2 at 25°C

Caution: Sodium azide is toxic. Azides also react with metals and disposal must be highly diluted.

Preparation of Medium: Add components to distilled/deionized water and bring volume to 1.0L. Mix thoroughly. Gently heat and bring to boiling. Distribute into tubes or flasks. Autoclave for 15 min at 15 psi pressure–121°C. Pour into sterile Petri dishes or leave in tubes.

Use: For the selective isolation, cultivation and enumeration of *Enterococcus* species by the multiple tube technique.

PGT Medium

Composition per liter:

Casamino acids	30.0g
L-Glutamic acid	0.5g
$MgSO_4 \cdot 7H_2O$	0.45g
Maltose	0.2g
L-Cystine	0.2g
DL-Tryptophan	0.1g
Solution 3	100.0mL
Solution 2	2.0mL
Calcium pantothenate (0.1% solution)	0.5mL

pH 6.8 ± 0.2 at 25°C

Solution 3:

Composition per 500mL:

Maltose	200.0g
$CaCl_2$	1.5g
Calcium pantothenate (0.1% solution)	3.0mL
$FeSO_4$ (1% in 1*N* HCl)	0.2mL

Preparation of Solution 3: Add components to distilled/deionized water and bring volume to 500.0mL. Mix thoroughly. Autoclave for 15 min at 7 psi pressure–111°C. Cool to 45° to 50°C.

Solution 2:

Composition per 100mL:

β-Alanine	0.115g
Nicotinic acid	0.115g
$CuSO_4 \cdot 5H_2O$	0.05g
$ZnSO_4 \cdot 7H_2O$	0.045g
$MnCl_2 \cdot 4H_2O$	0.015g
Pimelic acid	7.5mg
HCl, concentrated	3.0mL

Preparation of Solution 2: Add components to distilled/deionized water and bring volume to 100.0mL. Mix thoroughly.

Preparation of Medium: Add components, except solution 3, to distilled/deionized water and bring volume to 900.0mL. Mix thoroughly. Adjust pH to 6.8 with 50% KOH. Gently heat and bring to boiling. Autoclave for 15 min at 15 psi pressure–121°C. Cool to 45° to 50°C. Aseptically add sterile solution 3. Mix thoroughly. Aseptically distribute into sterile tubes or flasks.

Use: For the cultivation and maintenance of *Corynebacterium diphtheriae.*

Phenethyl Alcohol Agar (Phenylethanol Agar) (Phenylethyl Alcohol Agar)

Composition per liter:

Agar	15.0g
Pancreatic digest of casein	15.0g

NaCl 5.0g
Papaic digest of soybean meal 5.0g
β-Phenethyl alcohol 2.5g
Blood 50.0mL

pH 7.3 ± 0.2 at 25°C

Source: This medium is available as a premixed powder from BBL Microbiology Systems.

Preparation of Medium: Add components, except blood, to distilled/deionized water and bring volume to 950.0mL. Mix thoroughly. Gently heat and bring to boiling. Autoclave for 15 min at 13 psi pressure–118°C. Cool to 45° to 50°C. Aseptically add sterile defibrinated blood. Mix thoroughly. Pour into sterile Petri dishes or distribute into sterile tubes.

Use: For the selective isolation of Gram-positive bacteria, particularly Gram-positive cocci, from specimens with a mixed flora. Do not use for the observation of hemolytic reactions.

Phenol Red Agar

Composition per liter:

Agar 15.0g
Pancreatic digest of casein 10.0g
NaCl 5.0g
Phenol Red 0.018g
Carbohydrate solution 20.0mL

pH 7.4 ± 0.2 at 25°C

Source: This medium is available as a premixed powder from BBL Microbiology Systems.

Carbohydrate Solution:
Composition per 20mL:

Carbohydrate 5.0–10.0g

Preparation of Carbohydrate Solution: Add carbohydrate to distilled/deionized water and bring volume to 20.0mL. Mix thoroughly. Filter sterilize.

Preparation of Medium: Add components, except carbohydrate solution, to distilled/deionized water and bring volume to 980.0mL. Mix thoroughly. Adjust pH to 7.4 if necessary. Autoclave for 15 min at 15 psi pressure–121°C. Cool to 45° to 50°C. Aseptically add 20.0mL of sterile carbohydrate solution. Pour into sterile Petri dishes or distribute into sterile tubes. Allow tubes to cool in a slanted position.

Use: For the determination of fermentation reactions. Bacteria that can ferment the added carbohydrate turn the medium yellow.

Phenol Red Agar

Composition per liter:

Agar 15.0g
Proteose peptone No. 3 ....10.0g
NaCl 5.0g
Beef extract 1.0g
Phenol Red 0.025g
Carbohydrate solution 20.0mL

pH 7.4 ± 0.2 at 25°C

Source: This medium is available as a premixed powder from Difco Laboratories.

Carbohydrate Solution:
Composition per 20mL:

Carbohydrate 5.0–10.0g

Preparation of Carbohydrate Solution: Add carbohydrate to distilled/deionized water and bring volume to 20.0mL. Mix thoroughly. Filter sterilize.

Preparation of Medium: Add components, except carbohydrate solution, to distilled/deionized water and bring volume to 980.0mL. Mix thoroughly. Adjust pH to 7.4 if necessary. Autoclave for 15 min at 15 psi pressure–121°C. Cool to 45° to 50°C. Aseptically add 20.0mL of sterile carbohydrate solution. Pour into sterile Petri dishes or distribute into sterile tubes. Allow tubes to cool in a slanted position.

Use: For the determination of fermentation reactions. Bacteria that can ferment the added carbohydrate turn the medium yellow.

Phenol Red Broth

Composition per liter:

Pancreatic digest of casein 10.0g
NaCl 5.0g
Phenol Red 0.018g
Carbohydrate solution 20.0mL

pH 7.4 ± 0.2 at 25°C

Source: This medium is available as a premixed powder from BBL Microbiology Systems.

Carbohydrate Solution:
Composition per 20mL:

Carbohydrate 5.0–10.0g

Preparation of Carbohydrate Solution: Add carbohydrate to distilled/deionized water and bring volume to 20.0mL. Mix thoroughly. Filter sterilize.

Preparation of Medium: Add components, except carbohydrate solution, to distilled/deionized water and bring volume to 980.0mL. Mix thoroughly. Adjust pH to 7.4 if necessary. Distribute into tubes containing an inverted Durham tube. Fill each tube with 9.8mL of medium. Autoclave for 15 min at 13 psi pressure–118°C. Cool to 45° to 50°C. Aseptically add 0.2mL of sterile carbohydrate solution to each tube.

Use: For the determination of fermentation reactions in the differentiation of microorganisms. Fermentation is determined by the production of acid—broth turns yellow—and formation of gas—bubble trapped in Durham tube.

Phenol Red Glucose Broth

Composition per liter:

Pancreatic digest of casein 10.0g
Glucose 5.0g
NaCl 5.0g
Phenol Red 0.018g

pH 7.3 ± 0.2 at 25°C

Source: This medium is available as a premixed powder from BBL Microbiology Systems.

Preparation of Medium: Add components to distilled/deionized water and bring volume to 1.0L. Mix thoroughly. Adjust pH to 7.3 if necessary. Distribute into tubes containing an inverted Durham tube. Fill each tube with 10.0mL of medium. Autoclave for 15 min at 13 psi pressure–118°C.

Use: For the determination of the ability of a microorganism to ferment glucose. Fermentation is determined by the production of acid—broth turns yellow—and formation of gas—bubble trapped in Durham tube.

Phenol Red Lactose Agar

Composition per liter:

Agar 15.0g
Lactose 10.0g
Proteose peptone No. 3 .... 10.0g
NaCl 5.0g
Beef extract 1.0g
Phenol Red 25.0mg

pH 7.4 ± 0.2 at 25°C

Source: This medium is available as a premixed powder from Difco Laboratories.

Preparation of Medium: Add components to distilled/deionized water and bring volume to 1.0L. Mix thoroughly. Gently heat and bring to boiling. Distribute into tubes or flasks. Autoclave for 15 min at 13 psi pressure–118°C. Pour into sterile Petri dishes or leave in tubes. Allow tubes to cool in a slanted position.

Use: For the determination of the ability of a microorganism to ferment lactose. Fermentation is determined by the production of acid—medium turns yellow.

Phenol Red Lactose Broth

Composition per liter:

Pancreatic digest of casein 10.0g
Lactose 5.0g
NaCl 5.0g
Phenol Red 0.018g

pH 7.3 ± 0.2 at 25°C

Source: This medium is available as a premixed powder from BBL Microbiology Systems.

Preparation of Medium: Add components to distilled/deionized water and bring volume to 1.0L. Mix thoroughly. Adjust pH to 7.3 if necessary. Distribute into tubes containing an inverted Durham tube. Fill each tube with 10.0mL of medium. Autoclave for 15 min at 13 psi pressure–118°C.

Use: For the determination of the ability of a microorganism to ferment lactose. Fermentation is determined by the production of acid—broth turns yellow—and formation of gas—bubble trapped in Durham tube.

Phenol Red Mannitol Agar

Composition per liter:

Agar 15.0g
Mannitol 10.0g
Proteose peptone No. 3 .... 10.0g
NaCl 5.0g
Beef extract 1.0g
Phenol Red 25.0mg

pH 7.4 ± 0.2 at 25°C

Source: This medium is available as a premixed powder from Difco Laboratories.

Preparation of Medium: Add components to distilled/deionized water and bring volume to 1.0L. Mix thoroughly. Gently heat and bring to boiling. Distribute into tubes or flasks. Autoclave for 15 min at 13 psi pressure–118°C. Pour into sterile Petri dishes or leave in tubes. Allow tubes to cool in a slanted position.

Use: For the determination of the ability of a microorganism to ferment mannitol. Fermentation is determined by the production of acid—medium turns yellow.

Phenol Red Mannitol Broth

Composition per liter:

Pancreatic digest of casein 10.0g
D-Mannitol 5.0g
NaCl 5.0g
Phenol Red 0.018g

pH 7.3 ± 0.2 at 25°C

Source: This medium is available as a premixed powder from BBL Microbiology Systems.

Preparation of Medium: Add components to distilled/deionized water and bring volume to 1.0L. Mix

thoroughly. Adjust pH to 7.3 if necessary. Distribute into tubes containing an inverted Durham tube. Fill each tube with 10.0mL of medium. Autoclave for 15 min at 13 psi pressure–118°C.

Use: For the determination of the ability of a microorganism to ferment mannitol. Fermentation is determined by the production of acid—broth turns yellow—and formation of gas—bubble trapped in Durham tube.

Phenol Red Sucrose Broth

Composition per liter:

Pancreatic digest of casein 10.0g
NaCl 5.0g
Sucrose 5.0g
Phenol Red 0.018g

pH 7.3 ± 0.2 at 25°C

Source: This medium is available as a premixed powder from BBL Microbiology Systems.

Preparation of Medium: Add components to distilled/deionized water and bring volume to 1.0L. Mix thoroughly. Adjust pH to 7.3 if necessary. Distribute into tubes containing an inverted Durham tube. Fill each tube with 10.0mL of medium. Autoclave for 15 min at 13 psi pressure–118°C.

Use: For the determination of the ability of a microorganism to ferment sucrose. Fermentation is determined by the production of acid—broth turns yellow—and formation of gas—bubble trapped in Durham tube.

Phenol Red Tartrate Agar

Composition per liter:

Agar 15.0g
Peptone 10.0g
Potassium tartrate 10.0g
NaCl 5.0g
Phenol Red 0.024g

pH 7.6 ± 0.2 at 25°C

Source: This medium is available as a premixed powder from Difco Laboratories.

Preparation of Medium: Add components to cold distilled/deionized water and bring volume to 1.0L. Mix thoroughly. Gently heat and bring to boiling. Distribute into tubes or flasks. Autoclave for 15 min at 13 psi pressure–118°C. Pour into sterile Petri dishes or leave in tubes. Allow tubes to cool in an upright position.

Use: For the differentiation of Gram-negative bacteria of the intestinal groups, particularly members of the *Salmonella* (paratyphoid) group based on their ability to ferment tartrate.

Phenol Red Tartrate Broth

Composition per liter:

Pancreatic digest of casein 10.0g
Potassium tartrate 10.0g
Agar 5.0g
NaCl 5.0g
Phenol Red 0.024g

pH 7.6 ± 0.2 at 25°C

Preparation of Medium: Add components to distilled/deionized water and bring volume to 1.0L. Mix thoroughly. Distribute into tubes or flasks. Autoclave for 15 min at 15 psi pressure–121°C.

Use: For the differentiation of Gram-negative bacteria of the intestinal groups, particularly members of the *Salmonella* (paratyphoid) group based on their ability to ferment tartrate.

Phenylalanine Agar (Phenylalanine Deaminase Medium)

Composition per liter:

Agar 12.0g
NaCl 5.0g
Yeast extract 3.0g
DL-Phenylalanine 2.0g
Na_2HPO_4 .... 1.0g

pH 7.3 ± 0.2 at 25°C

Source: This medium is available as a premixed powder from BBL Microbiology Systems, and Difco Laboratories.

Preparation of Medium: Add components to distilled/deionized water and bring volume to 1.0L. Mix thoroughly. Gently heat while stirring and bring to boiling. Distribute into tubes or flasks. Autoclave for 10 min at 15 psi pressure–121°C. Pour into sterile Petri dishes or leave in tubes.

Use: For the differentiation of enteric Gram-negative bacilli on the basis of their ability to produce phenylpyruvic acid from phenylalanine. After appropriate incubation of bacteria on phenylalanine agar, ferric chloride reagent is added on the agar. Formation of a green color in 1–5 min indicates the production of phenylpyruvic acid.

Phenylalanine Malonate Broth

Composition per liter:

Sodium malonate 3.0g
DL-Phenylalanine 2.0g
NaCl 2.0g
$(NH_4)_2SO_4$ 2.0g

Yeast extract..1.0g
K_2HPO_4..0.6g
KH_2PO_4..0.4g
Bromthymol Blue....................................0.025g
pH 7.3 ± 0.2 at 25°C

Source: This medium is available as a premixed powder from Difco Laboratories.

Preparation of Medium: Add components to distilled/deionized water and bring volume to 1.0L. Mix thoroughly. Distribute into tubes or flasks. Autoclave for 10 min at 10 psi pressure–115°C.

Use: For the differentiation of Gram-negative enteric bacilli on the basis of malonate utilization and formation of pyruvic acid from phenylalanine.

Phenylethanol Agar

Composition per liter:

Agar..15.0g
Tryptose..10.0g
NaCl..5.0g
Beef extract..3.0g
Phenylethanol..2.5g
pH 7.3 ± 0.2 at 25°C

Source: This medium is available as a premixed powder from Difco Laboratories.

Preparation of Medium: Add components to distilled/deionized water and bring volume to 1.0L. Mix thoroughly. Gently heat and bring to boiling. Distribute into tubes or flasks. Autoclave for 15 min at 15 psi pressure–121°C. Pour into sterile Petri dishes or leave in tubes.

Use: For the isolation of staphylococci and streptococci from specimens containing a mixed flora.

Phenylethanol Blood Agar

Composition per liter:

Agar..15.0g
Tryptose..10.0g
NaCl..5.0g
Beef extract..3.0g
Phenylethanol..2.5g
Blood, defibrinated....................................50.0mL
pH 7.3 ± 0.2 at 25°C

Preparation of Medium: Add components, except blood, to distilled/deionized water and bring volume to 950.0mL. Mix thoroughly. Gently heat and bring to boiling. Autoclave for 15 min at 13 psi pressure–118°C. Cool to 45° to 50°C. Aseptically add sterile defibrinated blood. Mix thoroughly. Pour into sterile Petri dishes or distribute into sterile tubes.

Use: For the isolation of staphylococci and streptococci from specimens containing a mixed flora.

Phytone™ Yeast Extract Agar

Composition per liter:

Glucose..40.0g
Agar..17.0g
Papaic digest of soybean meal....................................10.0g
Yeast extract..5.0g
Chloramphenicol..0.05g
Streptomycin..0.03g
pH 6.6 ± 0.2 at 25°C

Source: This medium is available as a premixed powder from BBL Microbiology Systems.

Preparation of Medium: Add components to distilled/deionized water and bring volume to 1.0L. Mix thoroughly. Gently heat and bring to boiling. Distribute into tubes or flasks. Autoclave for 15 min at 15 psi pressure–121°C. Pour into sterile Petri dishes or leave in tubes.

Use: For the selective isolation of dermatophytes, particularly *Trichophyton verrucosum* and other pathogenic fungi, from clinical specimens.

Pike Streptococcal Broth

Composition per liter:

Pancreatic digest of casein....................................10.0g
Tryptose..10.0g
Yeast extract..10.0g
Glucose..0.2g
NaN_3..0.065g
Crystal Violet..2.0mg
Rabbit blood, defibrinated....................................50.0mL
pH 7.4 ± 0.2 at 25°C

Caution: Sodium azide is toxic. Azides also react with metals and disposal must be highly diluted.

Preparation of Medium: Add components, except rabbit blood, to distilled/deionized water and bring volume to 950.0mL. Mix thoroughly. Gently heat and bring to boiling. Distribute into flasks in 100.0mL volumes. Autoclave for 15 min at 15 psi pressure–121°C. Cool to 45° to 50°C. Aseptically add 5.0mL of sterile rabbit blood to each flask. Mix thoroughly.

Use: For the isolation and enrichment of hemolytic streptococci from throat swabs and other clinical specimens. After incubation of bacteria for 18–24 hr in this medium, they may be isolated by streaking the culture onto blood agar plates.

PKU Test Agar (Phenylketonuria Test Agar)

Composition per liter:

Agar	15.0g
K_2HPO_4	15.0g
Glucose	10.0g
KH_2PO_4	5.0g
$(NH_4)Cl$	2.5g
$(NH_4)NO_3$	0.5g
Asparagine	0.5g
DL-Alanine	0.5g
L-Glutamic acid	0.5g
Na_2SO_4	0.5g
$MgSO_4 \cdot 7H_2O$	0.05g
$FeCl_3$	5.0mg
$MnCl_2 \cdot 4H_2O$	5.0mg
$CaCl_2 \cdot 2H_2O$	2.5mg
β-2-Thienylalanine	3.3mg
Bacillus subtilis spore suspension	10.0mL

pH 7.0 ± 0.2 at 25°C

Source: This medium is available as a premixed powder from Difco Laboratories.

Preparation of Medium: Add components to distilled/deionized water and bring volume to 1.0L. Mix thoroughly. Gently heat and bring to boiling. Continue boiling for 5 min. Do not autoclave. Cool to 50°C. Add 10.0mL of a suspension of *Bacillus subtilis* ATCC 6633 spores. Mix thoroughly. Pour into sterile Petri dishes or other containers.

Use: For the determination of phenylalanine concentrations in serum or urine. Use in the Guthrie-modified bacterial-inhibition assay procedure for screening newborn infants for phenylketonuria (PKU).

PKU Test Agar (Phenylketonuria Test Agar)

Composition per liter:

K_2HPO_4	15.0g
Agar	13.5g
Glucose	5.0g
KH_2PO_4	5.0g
$(NH_4)Cl$	2.5g
L-Asparagine	0.5g
L-Glutamic acid	0.5g
Na_2SO_4	0.5g
$(NH_4)NO_3$	0.5g
L-Alanine	0.25g
$MgSO_4 \cdot 7H_2O$	0.05g
$FeCl_3$	0.005g
$MnSO_4$	0.005g
$CaCl_2 \cdot 2H_2O$	2.5mg
β-2-Thienylalanine solution	1.0mL

pH 6.9 ± 0.2 at 25°C

Source: This medium is available as a premixed powder from BBL Microbiology Systems.

β-2-Thienylalanine Solution:

Composition per 100mL:

β-2-Thienylalanine	0.33g

Preparation of β-2-Thienylalanine Solution: Add β-2-thienylalanine to distilled/deionized water and bring volume to 100.0mL. Mix thoroughly. Filter sterilize.

Preparation of Medium: Add components, except β-2-thienylalanine solution, to distilled/deionized water and bring volume to 1.0L. Mix thoroughly. Gently heat and bring to boiling. Continue boiling for 5 min. Do not autoclave. Cool to 50°C. Add 10.0mL of a suspension of *Bacillus subtilis* ATCC 6633 spores and 1.0mL of sterile β-2-thienylalanine solution. Mix thoroughly. Pour into sterile Petri dishes or other containers.

Use: For the determination of phenylalanine concentrations in serum or urine. Used in the Guthrie-modified bacterial-inhibition assay procedure for screening newborn infants for phenylketonuria (PKU).

PLET Agar

Composition per liter:

Beef heart, solids from infusion	500.0g
Agar	15.0g
Tryptose	10.0g
NaCl	5.0g
Ethylenediamine tetracetic acid,EDTA	0.3g
Thallous acetate	0.04g
Antibiotic inhibitor	10.0mL

Antibiotic Inhibitor:

Composition per 10mL:

Lysozyme	300,000U
Polymyxin	30,000U

Preparation of Antibiotic Inhibitor: Add components to distilled/deionized water and bring volume to 10.0mL. Mix thoroughly. Filter sterilize.

Preparation of Medium: Add components, except antibiotic inhibitor, to distilled/deionized water and bring volume to 990.0mL. Mix thoroughly. Gently heat and bring to boiling. Autoclave for 15 min at 15 psi pressure–121°C. Cool to 50°C. Aseptically add sterile antibiotic inhibitor. Mix thoroughly. Pour into sterile Petri dishes or distribute into sterile tubes.

Use: For the selective isolation and cultivation of *Bacillus anthracis*.

Polymyxin *Staphylococcus* Medium

Composition per liter:

Agar	15.0g
Pancreatic digest of gelatin	5.0g
Beef extract	3.0g
Lecithin	0.7g
Polymyxin	0.075g
Tween™ 80	10.2mL

pH 6.8 ± 0.2 at 25°C

Preparation of Medium: Add components to distilled/deionized water and bring volume to 1.0L. Mix thoroughly. Gently heat and bring to boiling. Distribute into tubes or flasks. Autoclave for 15 min at 15 psi pressure–121°C. Pour into sterile Petri dishes.

Use: For the selective isolation and cultivation of pathogenic, coagulase-positive *Staphylococcus aureus. Proteus* species will grow on this medium but appear as translucent colonies.

Potassium Cyanide Broth

Composition per liter:

Na_2HPO_4	5.64g
NaCl	5.0g
Proteose peptone No. 3	3.0g
KH_2PO_4	0.225g
KCN solution	15.0mL

pH 7.6 ± 0.2 at 25°C

KCN Solution:

Composition per 100mL:

KCN	0.5g

Preparation of KCN Solution: Add KCN to distilled/deionized water and bring volume to 100.0mL. Mix thoroughly.

Caution: Cyanide is toxic.

Preparation of Medium: Add components, except KCN solution, to distilled/deionized water and bring volume to 985.0mL. Mix thoroughly. Gently heat and bring to boiling. Autoclave for 15 min at 15 psi pressure–121°C. Cool to 25°C. Aseptically add 15.0mL of KCN solution. Mix thoroughly. Distribute into sterile screw-capped tubes or flasks in 1.0–1.5mL volumes. Close caps tightly.

Use: For the cultivation and differentiation of urease negative Gram-negative enteric bacteria. *Salmonella* species and *Shigella* species are nonmotile in this medium. *Proteus* species are motile in this medium.

Potassium Tellurite Agar

Composition per liter:

Beef heart, solids from infusion	500.0g
Agar	15.0g
Tryptose	10.0g
NaCl	5.0g
K_2TeO_3 solution	20.0mL
Blood, defibrinated	50.0mL

pH 6.0 ± 0.2 at 25°C

K_2TeO_3 Solution:

Composition per 20mL:

K_2TeO_3	0.5g

Preparation of K_2TeO_3 Solution: Add K_2TeO_3 to distilled/deionized water and bring volume to 10.0mL. Mix thoroughly. Filter sterilize.

Caution: Potassium tellurite is toxic.

Preparation of Medium: Add components, except K_2TeO_3 solution, to distilled/deionized water and bring volume to 930.0mL. Mix thoroughly. Gently heat and bring to boiling. Autoclave for 15 min at 15 psi pressure–121°C. Cool to 45° to 50°C. Aseptically add sterile K_2TeO_3 solution and 50.0mL of blood. Rabbit or sheep blood may be used. Mix thoroughly. Pour into sterile Petri dishes or distribute into sterile tubes. Allow tubes to cool in a slanted position.

Use: For the cultivation and differentiation of *Enterococcus faecalis. E. faecalis* appears as black colonies.

Potato Dextrose Agar

Composition per liter:

Glucose	20.0g
Agar	15.0g
Potato, infusion from	4.0g
Tartaric acid solution	14.0mL

pH 5.6 ± 0.2 at 25°C

Source: This medium is available as a premixed powder from BBL Microbiology Systems and Oxoid Unipath.

Tartaric Acid Solution:

Composition per 50mL:

Tartaric acid	5.0g

Preparation of Tartaric Acid Solution: Add tartaric acid to distilled/deionized water and bring volume to 50.0mL. Mix thoroughly. Filter sterilize.

Preparation of Medium: Add components to distilled/deionized water and bring volume to 986.0mL. Mix thoroughly. Gently heat and bring to boiling. Distribute into tubes or flasks. Autoclave for 15 min at 15 psi pressure–121°C. Cool to 45° to 50°C. Aseptically add 14.0mL of sterile tartaric acid solution. Mix thoroughly. If medium is to be used for the enumeration of yeasts and molds in butter, adjust pH to 3.5. Pour into sterile Petri dishes or distribute into sterile tubes.

Use: For the cultivation and enumeration of yeasts and molds. For the enumeration of yeasts and molds in butter by the plate count method.

Potato Dextrose Agar (PDA)

Composition per liter:

Potato, peeled and diced	200.0g
Agar	20.0g
Glucose	10.0g

pH 5.6 ± 0.2 at 25°C

Source: This medium is available as a premixed powder from BBL Microbiology Systems and Difco Laboratories.

Preparation of Medium: Add potato to 1.0L of distilled/deionized water. Bring to boiling and cook potato until soft. Add glucose and agar. Mix thoroughly. Gently heat and bring to boiling. Filter through gauze and restore the volume to 1000mL with distilled/deionized water. Autoclave for 15 min at 15 psi pressure–121°C. Pour into sterile Petri dishes or distribute into sterile tubes.

Use: For the preservation of sporulation or pigmentation of a variety of pathogenic or saprophytic fungi.

Potato Flakes Agar

Composition per liter:

Potato flakes	20.0g
Agar	15.0g
Glucose	10.0g

pH 5.6 ± 0.2 at 25°C

Source: This medium is available as a premixed powder from BBL Microbiology Systems.

Preparation of Medium: Add components to distilled/deionized water and bring volume to 1.0L. Mix thoroughly. Gently heat and bring to boiling. Autoclave for 15 min at 15 psi pressure–121°C. Pour into sterile Petri dishes or distribute into sterile tubes.

Use: For the isolation and cultivation of pathogenic and saprophytic fungi. For the induction of sporulation and enhancement of morphological structures used for the identification of many pathogenic and opportunistic fungi.

Potato Flakes CC Agar

Composition per liter:

Potato flakes	20.0g
Agar	15.0g
Glucose	10.0g
Chloramphenical	0.5g
Cycloheximide	0.5g

pH 5.6 ± 0.2 at 25°C

Source: This medium is available as a premixed powder from BBL Microbiology Systems.

Preparation of Medium: Add components to distilled/deionized water and bring volume to 1.0L. Mix thoroughly. Gently heat and bring to boiling. Autoclave for 15 min at 15 psi pressure–121°C. Pour into sterile Petri dishes or distribute into sterile tubes.

Use: For the isolation and cultivation of pathogenic and saprophytic fungi. For the induction of sporulation and enhancement of morphological structures used for the identification of many pathogenic and opportunistic fungi. The addition of chloramphenical and cycloheximide enhances the selectivity of the medium for more effective isolation and identification of pathogenic and medically-significant fungi.

PPLO Agar

Composition per liter:

Beef heart, infusion from	50.0g
Agar	14.0g
Peptone	10.0g
NaCl	5.0g
Bovine serum	100.0mL

pH 7.8 ± 0.2 at 25°C

Source: This medium is available as a premixed powder from Difco Laboratories.

Preparation of Medium: Add components, except bovine serum, to distilled/deionized water and bring volume to 900.0mL. Mix thoroughly. Gently heat and bring to boiling. Autoclave for 15 min at 15 psi pressure–121°C. Cool to 45° to 50°C. Aseptically add sterile bovine serum. Mix thoroughly. Pour into sterile Petri dishes or distribute into sterile tubes.

Use: For the isolation and cultivation of *Mycoplasma* species (pleuro-pneumonia-like organisms).

PPLO Agar

Composition per liter:

Beef heart, infusion from	50.0g
Agar	14.0g
Peptone	10.0g
NaCl	5.0g
Ascitic fluid	250.0mL

pH 7.8 ± 0.2 at 25°C

Source: This medium is available as a premixed powder from Difco Laboratories.

Preparation of Medium: Add components, except ascitic fluid, to distilled/deionized water and bring volume to 750.0mL. Mix thoroughly. Gently heat and bring to boiling. Autoclave for 15 min at 15 psi pressure–121°C. Cool to 45° to 50°C. Aseptically add sterile ascitic fluid. Mix thoroughly. Pour into sterile Petri dishes or distribute into sterile tubes.

Use: For the isolation and cultivation of *Mycoplasma* species (pleuro-pneumonia-like organisms).

PPLO Broth with Crystal Violet

Composition per liter:

Beef heart, infusion from	50.0g
Peptone	10.0g
NaCl	5.0g
Crystal Violet	0.01g
Ascitic fluid	250.0mL
Chapman tellurite solution	2.85mL

pH 7.8 ± 0.2 at 25°C

Source: This medium is available as a premixed powder from Difco Laboratories.

Chapman Tellurite Solution:
Composition per 100mL:

K_2TeO_3	1.0g

Preparation of Chapman Tellurite Solution: Add K_2TeO_3 to distilled/deionized water and bring volume to 100.0mL. Mix thoroughly. Filter sterilize.

Caution: Potassium tellurite is toxic.

Preparation of Medium: Add components, except ascitic fluid and Chapman tellurite solution, to distilled/deionized water and bring volume to 747.15mL. Mix thoroughly. Autoclave for 15 min at 15 psi pressure–121°C. Cool to less than 37°C. Aseptically add sterile ascitic fluid and 2.85mL of Chapman tellurite solution. Mix thoroughly. Aseptically distribute into sterile tubes or flasks.

Use: For the isolation of *Mycoplasma* species from clinical specimens.

PPLO Broth without Crystal Violet

Composition per liter:

Beef heart, infusion from	50.0g
Peptone	10.0g
NaCl	5.0g
Ascitic fluid	250.0mL

pH 7.8 ± 0.2 at 25°C

Source: This medium is available as a premixed powder from Difco Laboratories.

Preparation of Medium: Add components, except ascitic fluid, to distilled/deionized water and bring volume to 750.0mL. Mix thoroughly. Autoclave for 15 min at 15 psi pressure–121°C. Cool to less than 37°C. Aseptically add sterile ascitic fluid. If desired, 0.5g of thallium acetate or 100,000U of penicillin may be added for a more selective medium. Mix thoroughly. Aseptically distribute into sterile tubes or flasks.

Use: For the enrichment of pleuro-pneumonia-like organisms (PPLOs) and *Mycoplasma* species from clinical specimens.

PPNG Selective Medium (Penicillinase-Producing *Neisseria gonorrhoeae* Medium)

Composition per plate:

Quadrant I	10.0mL
Quadrant II	10.0mL

Quadrant I:
Composition per 10mL:

Martin Lewis agar	10.0mL

Quadrant II:
Composition per 10mL:

Martin Lewis agar, enriched	10.0mL

Use: For the differentiation and presumptive identification of penicillinase-producing strains of *Neisseria gonorrhoeae*. The PPNG selective medium is a two-sectored plate each containing a different medium.

Presumpto Media

Composition per plate:

Quadrant I	5.0mL
Quadrant II	5.0mL
Quadrant III	5.0mL
Quadrant IV	5.0mL

Quadrant I:
Composition per 5mL:

Lombard-Dowell agar	5.0mL

Quadrant II:
Composition per 5mL:

Lombard-Dowell bile agar	5.0mL

Quadrant III:
Composition per 5mL:

Lombard-Dowell egg yolk agar	5.0mL

Quadrant IV:
Composition per 5mL:

Lombard-Dowell esculin agar	5.0mL

Use: For the differentiation and presumptive identification of anaerobic bacteria. The Presumpto media is a four-sectored plate each containing a different medium. See Lombard-Dowell agars for additional information.

Proskauer–Beck Medium for *Mycobacterium*

Composition per liter:

Asparagine	5.0g
KH_2PO_4	5.0g
Magnesium citrate	2.5g
$MgSO_4 \cdot 7H_2O$	0.6g
Glycerol	20.0mL

pH 7.4 ± 0.2 at 25°C

Preparation of Medium: Add components one at a time to distilled/deionized water and bring volume to 1.0L. Mix thoroughly. Make sure one salt is totally dissolved before the next one is added. Adjust pH to 7.8 with 40% NaOH. Autoclave for 15 min at 15 psi pressure–121°C. The pH of the medium after autoclaving should be 7.4. Filter through Whatman #1 filter paper to remove any precipitate. Distribute into tubes or flasks. Autoclave again for 15 min at 15 psi pressure–121°C.

Use: For the cultivation and maintenance of *Mycobacterium tuberculosis.*

Proteose No. 3 Agar

Composition per 1010mL:

Proteose peptone No. 3	20.0g
Agar	15.0g
Na_2HPO_4	5.0g
NaCl	5.0g
Glucose	0.5g
Hemoglobin solution	500.0mL
Supplement B (Difco)	10.0mL

pH 7.3 ± 0.2 at 25°C

Source: This medium is available as a premixed powder from Difco Laboratories.

Hemoglobin Solution:

Composition per 500mL:

Hemoglobin	10.0g

Preparation of Hemoglobin Solution: Add hemoglobin to distilled/deionized water and bring volume to 500.0mL. Mix thoroughly. Autoclave for 15 min at 15 psi pressure–121°C. Cool to 45° to 50°C.

Supplement B:

Composition per 10mL:

Supplement B contains yeast concentrate, glutamine, coenzyme, cocarboxylase, hematin and growth factors.

Preparation of Supplement B: Add components to distilled/deionized water and bring volume to 10.0mL. Mix thoroughly. Filter sterilize.

Preparation of Medium: Add components, except hemoglobin solution and supplement B, to distilled/deionized water and bring volume to 500.0mL. Mix thoroughly. Gently heat and bring to boiling. Autoclave for 15 min at 15 psi pressure–121°C. Cool to 50° to 60°C. Aseptically add 500.0mL of sterile hemoglobin solution and 10.0mL of sterile supplement B. Mix thoroughly. Pour into sterile Petri dishes or distribute into sterile tubes.

Use: For the isolation and cultivation of *Neisseria* species, *Haemophilus* species, and other fastidious bacteria. For the cultivation and maintenance of *Escherichia coli.*

Proteose No. 3 Agar

Composition per 1010mL:

Proteose peptone No. 3	20.0g
Agar	15.0g
Na_2HPO_4	5.0g
NaCl	5.0g
Glucose	0.5g
Hemoglobin solution	500.0mL
Supplement VX (Difco)	10.0mL

pH 7.3 ± 0.2 at 25°C

Source: This medium is available as a premixed powder from Difco Laboratories.

Hemoglobin Solution:

Composition per 500mL:

Hemoglobin	10.0g

Preparation of Hemoglobin Solution: Add hemoglobin to distilled/deionized water and bring volume to 500.0mL. Mix thoroughly. Autoclave for 15 min at 15 psi pressure–121°C. Cool to 45° to 50°C.

Supplement VX:

Composition per 10mL:

Supplement B contains essential growth factors.

Preparation of Supplement VX: Add components to distilled/deionized water and bring volume to 10.0mL. Mix thoroughly. Filter sterilize.

Preparation of Medium: Add components, except hemoglobin solution and supplement VX, to distilled/deionized water and bring volume to 500.0mL. Mix thoroughly. Gently heat and bring to boiling. Autoclave for 15 min at 15 psi pressure–121°C. Cool to 50° to 60°C. Aseptically add 500.0mL of sterile hemoglobin solution and 10.0mL of sterile supplement VX. Mix thoroughly. Pour into sterile Petri dishes or distribute into sterile tubes.

Use: For the isolation and cultivation of *Neisseria* species, *Haemophilus* species, and other fastidious bacteria. For the cultivation and maintenance of *Escherichia coli.*

Pseudomonas Agar F

Composition per liter:

Agar....15.0g
Glycerol....10.0g
Proteose peptone No. 3....10.0g
Pancreatic digest of casein....10.0g
K_2HPO_4....1.5g
$MgSO_4 \cdot 7H_2O$....1.5g

pH 7.0 ± 0.2 at 25°C

Source: This medium is available as a premixed powder from Difco Laboratories.

Preparation of Medium: Add components to distilled/deionized water and bring volume to 1.0L. Mix thoroughly. Gently heat and bring to boiling. Distribute into tubes or flasks. Autoclave for 15 min at 15 psi pressure–121°C. Pour into sterile Petri dishes or leave in tubes.

Use: For the isolation, cultivation and differentiation of *Pseudomonas aeruginosa* on the basis of pigment (fluorscein) production.

Pseudomonas Agar P

Composition per liter:

Proteose peptone No. 3....20.0g
Agar....15.0g
Glycerol....10.0g
K_2HPO_4....10.0g
$MgCl_2 \cdot 6H_2O$....1.4g

pH 7.0 ± 0.2 at 25°C

Source: This medium is available as a premixed powder from Difco Laboratories.

Preparation of Medium: Add components to distilled/deionized water and bring volume to 1.0L. Mix thoroughly. Gently heat and bring to boiling. Distribute into tubes or flasks. Autoclave for 15 min at 15 psi pressure–121°C. Pour into sterile Petri dishes or leave in tubes.

Use: For the isolation, cultivation and differentiation of *Pseudomonas aeruginosa* on the basis of pigment (pyocyanin) production.

Pseudomonas CN Agar

Composition per liter:

Pancreatic digest of gelatin....16.0g
Agar....11.0g
Pancreatic digest of casein....10.0g
K_2SO_4....10.0g
$MgCl_2 \cdot 6H_2O$....1.4g
CN selective supplement....10.0mL
Glycerol....10.0mL

CN Selective Supplement:

Cetrimide....0.1g
Sodium nalidixate....7.5mg

pH 7.1 ± 0.2 at 25°C

Source: This medium is available as a premixed powder from Oxoid Unipath.

Preparation of Medium: Add components, except CN selective supplement, to distilled/deionized water and bring volume to 990.0mL. Mix thoroughly. Gently heat and bring to boiling. Autoclave for 15 min at 15 psi pressure–121°C. Cool to 45° to 50°C. Aseptically add sterile CN selective supplement. Mix thoroughly. Pour into sterile Petri dishes or distribute into sterile tubes.

Use: For the selective isolation and cultivation of *Pseudomonas* species.

Pseudomonas Isolation Agar

Composition per liter:

Peptone....20.0g
Agar....13.6g
K_2SO_4....10.0g
$MgCl_2 \cdot 6H_2O$....1.4g
Irgasan® (triclosan)....0.025g
Glycerol....20.0mL

pH 7.0 ± 0.2 at 25°C

Source: This medium is available as a premixed powder from BBL Microbiology Systems and Difco Laboratories.

Preparation of Medium: Add components to distilled/deionized water and bring volume to 1.0L. Mix thoroughly. Gently heat and bring to boiling. Distribute into tubes or flasks. Autoclave for 15 min at 15 psi pressure–121°C. Pour into sterile Petri dishes or leave in tubes.

Use: For the isolation and cultivation of *Pseudomonas* species.

Purple Agar

Composition per liter:

Agar....15.0g
Proteose peptone No. 3....10.0g
NaCl....5.0g
Beef extract....1.0g
Bromcresol Purple....0.02g
Carbohydrate solution....20.0mL

pH 6.8 ± 0.2 at 25°C

Source: This medium is available as a premixed powder from Difco Laboratories.

Carbohydrate Solution:
Composition per 20mL:
Carbohydrate..10.0g

Preparation of Carbohydrate Solution: Add carbohydrate to distilled/deionized water and bring volume to 20.0mL. For expensive carbohydrates, 5.0g may be used instead of 10.0g. Mix thoroughly. Filter sterilize.

Preparation of Medium: Add components, except carbohydrate solution, to distilled/deionized water and bring volume to 980.0mL. Mix thoroughly. Gently heat and bring to boiling. Distribute into tubes in 9.8mL volumes. Autoclave for 15 min at 15 psi pressure–121°C. Cool to 45° to 50°C. Aseptically add 0.2mL of sterile carbohydrate solution to each tube. Mix thoroughly. Allow tubes to cool in a slanted position.

Use: For the preparation of carbohydrate media used in fermentation studies for the identification of bacteria, especially members of the Enterobacteriaceae. Bacteria that can ferment the carbohydrate turn the medium yellow.

Purple Broth (Purple Carbohydrate Broth)

Composition per liter:
Proteose peptone No. 3....................................10.0g
NaCl...5.0g
Beef extract..1.0g
Bromcresol Purple...0.015g
Carbohydrate solution..................................20.0mL
pH 6.8 ± 0.2 at 25°C

Source: This medium is available as a premixed powder from Difco Laboratories.

Carbohydrate Solution:
Composition per 20mL:
Carbohydrate..10.0g

Preparation of Carbohydrate Solution: Add carbohydrate to distilled/deionized water and bring volume to 20.0mL. For expensive carbohydrates, 5.0g may be used instead of 10.0g. Mix thoroughly. Filter sterilize.

Preparation of Medium: Add components, except carbohydrate solution, to distilled/deionized water and bring volume to 980.0mL. Mix thoroughly. Gently heat and bring to boiling. Distribute into tubes in 9.8mL volumes. Autoclave for 15 min at 15 psi pressure–121°C. Cool to 25°C. Aseptically add 0.2mL of sterile carbohydrate solution to each tube. Mix thoroughly.

Use: For the preparation of carbohydrate media used in fermentation studies for the identification of bacteria, especially members of the Enterobacteriaceae. Bacteria that can ferment the carbohydrate turn the medium yellow.

Purple Broth

Composition per liter:
Pancreatic digest of gelatin...............................10.0g
NaCl...5.0g
Bromcresol Purple...0.02g
Carbohydrate solution..................................20.0mL
pH 6.8 ± 0.2 at 25°C

Source: This medium is available as a premixed powder from BBL Microbiology Systems.

Carbohydrate Solution:
Composition per 20mL:
Carbohydrate..10.0g

Preparation of Carbohydrate Solution: Add carbohydrate to distilled/deionized water and bring volume to 20.0mL. For expensive carbohydrates, 5.0g may be used instead of 10.0g. Mix thoroughly. Filter sterilize.

Preparation of Medium: Add components, except carbohydrate solution, to distilled/deionized water and bring volume to 980.0mL. Mix thoroughly. Adjust pH to 7.4 if necessary. Distribute into tubes containing an inverted Durham tube. Fill each tube with 9.8mL of medium. Autoclave for 15 min at 15 psi pressure–121°C. Aseptically add 0.2mL of sterile carbohydrate solution to each tube.

Use: For the preparation of liquid fermentation media. Bacteria that can ferment the carbohydrate turn the medium yellow.

Purple Lactose Agar

Composition per liter:
Agar..10.0g
Lactose...10.0g
Peptone...5.0g
Beef extract..3.0g
Bromcresol Purple...0.025g
pH 6.8 ± 0.1 at 25°C

Source: This medium is available as a premixed powder from Difco Laboratories.

Preparation of Medium: Add components to distilled/deionized water and bring volume to 1.0L. Mix thoroughly. Gently heat and bring to boiling. Distribute into tubes or flasks. Autoclave for 15 min at 15 psi pressure–121°C. Pour into sterile Petri dishes or leave in tubes. Allow tubes to cool in a slanted position.

Use: For the detection and differentiation of members of the Enterobacteriaceae. Bacteria that can ferment lactose turn the medium yellow.

Purple Serum Agar Base

Composition per liter:

Agar	20.0g
Lactose	20.0g
Peptone	20.0g
NaCl	5.0g
Bromcresol Purple	0.030g
Phenol Eed	0.024g

pH 7.6 ± 0.2 at 25°C

Preparation of Medium: Add components to distilled/deionized water and bring volume to 1.0L. Mix thoroughly. Gently heat and bring to boiling. Distribute into tubes or flasks. Autoclave for 15 min at 15 psi pressure–121°C. Pour into sterile Petri dishes or leave in tubes.

Use: For the cultivation and differentiation of Gram-negative bacteria isolated from the urinary tract. Bacteria that can ferment lactose turn the medium yellow.

PV Blood Agar (Paromomycin Vancomycin Blood Agar)

Composition per liter:

Agar	20.0g
Pancreatic digest of casein	15.0g
NaCl	5.0g
Papaic digest of soybean meal	5.0g
Yeast extract	5.0g
L-Cystine	0.4g
Paromomycin	0.1g
Vancomycin	7.5mg
Hemin	5.0mg
Sheep blood, defibrinated	50.0mL
Vitamin K_1 solution	10.0mL

pH 7.5 ± 0.2 at 25°C

Vitamin K_1 Solution:
Composition per 10mL:

Vitamin K_1	0.01g
Ethanol	10.0mL

Preparation of Vitamin K_1 Solution: Add vitamin K_1 to 10.0mL of absolute ethanol. Mix thoroughly. Filter sterilize.

Preparation of Medium: Add components—except vitamin K_1 solution, sheep blood, paromomycin and vancomycin—to distilled/deionized water and bring volume to 940.0mL. Mix thoroughly. Gently heat and bring to boiling for 1 min. Autoclave for 15 min at 15 psi pressure–121°C. Cool to 50–55°C. Aseptically add the sterile Vitamin K_1 solution, sheep blood, vancomycin and paromomycin. Mix thoroughly. Pour into sterile Petri dishes.

Use: For the selective cultivation of fastidious anaerobic bacteria.

PYG Broth (Peptone Yeast Extract Glucose Broth)

Composition per liter:

Peptone	20.0g
D-Glucose	10.0g
Yeast extract	10.0g
Cysteine·HCl·H_2O	0.5g
VPI salt solution	40.0mL
Resazurin solution	4.0mL

pH 7.2 ± 0.2 at 25°C

Resazurin Solution:
Composition per 44mL:

Resazurin	0.044g

Preparation of Resazurin Solution: Add resazurin to distilled/deionized water and bring volume to 44.0mL. Mix thoroughly.

VPI Salt Solution:
Composition per 40mL:

$CaCl_2$	0.2g
$MgSO_4$	0.2g
K_2HPO_4	1.0g
KH_2PO_4	1.0g

Preparation of VPI Salt Solution: Add $CaCl_2$ and $MgSO_4$ to 300.0mL of distilled/deionized water. Mix thoroughly until dissolved. Bring volume to 800.0mL with distilled/deionized water. Add remaining components while stirring. Bring volume to 1.0L. Mix thoroughly. Store at 4°C.

Preparation of Medium: Add components to distilled/deionized water and bring volume to 1.0L. Mix thoroughly. Distribute into screw-capped tubes in 7.0mL volumes. Autoclave for 15 min at 15 psi pressure–121°C. Cool to 45° to 50°C under 100% N_2.

Use: For the cultivation of a wide variety of anaerobic bacteria.

Pyrazinamidase Agar (Pyrazinamide Medium)

Composition per liter:

Pancreatic digest of casein	15.0g
Agar	15.0g

Papaic digest of soybean meal 5.0g
NaCl 5.0g
Yeast extract 3.0g
Pyrazinecarboxamide 1.0g
Tris(hydroxymethyl)amino-
methane maleate buffer (0.2M, pH 6.0) 1.0L
pH 6.0 ± 0.2 at 25°C

Preparation of Medium: Combine components. Mix thoroughly. Gently heat and bring to boiling. Distribute into tubes in 5.0mL volumes. Autoclave for 15 min at 15 psi pressure–121°C. Allow tubes to cool in a slanted position.

Use: For the cultivation, differentiation and maintenance of pathogenic *Yersinia* species. Bacteria that produce pyrazinamidase turn the medium pink.

Pyrazinamide Medium

Composition per liter:

Agar 15.0g
Na_2HPO_4 .... 2.5g
Sodium pyruvate 2.0g
L-Asparagine 2.0g
KH_2PO_4 .... 1.0g
Pancreatic digest of casein 0.5g
Tween™ 80 .... 0.2g
Pyrazinamide 0.1g
$CaCl_2 \cdot 2H_2O$ 0.5mg
$CuSO_4 \cdot 5H_2O$ 0.1mg
$ZnSO_4 \cdot 7H_2O$ 0.1mg
Ferric ammonium citrate 0.05g
$MgSO_4 \cdot 7H_2O$ 0.01g
pH 6.6 ± 0.2 at 25°C

Preparation of Medium: Combine components. Mix thoroughly. Gently heat and bring to boiling. Distribute into tubes in 5.0mL volumes. Autoclave for 15 min at 15 psi pressure–121°C. Allow tubes to cool in a slanted position.

Use: For cultivation and differentiation of *Corynebacterium* species and related organisms. Bacteria that produce pyrazinamidase turn the medium pink.

Pyruvic Acid Egg Medium

Composition per 1640mL:

KH_2PO_4 .... 11.4g
Na_2HPO_4 .... 6.0g
Pyruvic acid 3.0g
$MgSO_4 \cdot 7H_2O$ 0.3g
Malachite Green 0.125g
D-Glucose 10.0g
Egg, homogenized whole 1.0L
Penicillin solution 10.0mL

Source: This medium is available as a prepared medium from Oxoid Unipath.

Penicillin Solution:
Composition per 10mL:

Penicillin G 100,000U

Preparation of Penicillin Solution: Add penicillin G to distilled/deionized water and bring volume to 10.0mL. Mix thoroughly. Filter sterilize.

Homogenized Whole Egg:
Composition per liter:

Whole eggs 18-24

Preparation of Homogenized Whole Egg: Use fresh eggs, less than 1 week old. Scrub the shells with soap. Let stand in a soap solution for 30 min. Rinse in running water. Soak eggs in 70% ethanol for 15 min. Break the eggs into a sterile container. Homogenize by shaking. Filter through four layers of sterile cheesecloth into a sterile graduated cylinder. Measure out 1.0L.

Preparation of Medium: Add components, except homogenized whole egg and penicillin solution, to distilled/deionized water and bring volume to 630.0mL. Mix thoroughly. Autoclave for 15 min at 15 psi pressure–121°C. Cool to 45° to 50°C. Aseptically add homogenized whole egg and penicillin solution to cooled sterile basal medium. Mix thoroughly. Aseptically distribute into sterile tubes. Inspissate at 85° to 90°C (moist heat) for 45 min.

Use: For the isolation and cultivation of *Mycobacterium* species, especially ones that are drug-resistant and difficult to grow.

R Medium

Composition per liter:

Agar 30.0g
$NaHCO_3$ 8.0g
K_2HPO_4 .... 3.0g
Glucose 2.5g
Glutamine 0.61g
Serine 0.24g
Leucine 0.23g
Lysine 0.23g
Asparagine 0.18g
Valine 0.17g
Isoleucine 0.17g
Tyrosine 0.14g
Arginine·HCl 0.125g
Phenylalanine 0.125g
Threonine 0.12g
Methionine 0.073g
Glycine 0.065g
Histidine·HCl 0.055g

Proline 0.043g
Tryptophan 0.035g
Cystine 0.025g
$MgSO_4 \cdot H_2O$ 9.9mg
$CaCl_2 \cdot 2H_2O$ 7.4mg
Adenine sulfate 2.1mg
Uracil 1.4mg
Thiamine·HCl 1.0mg
$MnSO_4 \cdot H_2O$ 0.9mg

pH 8.0 ± 0.2 at 25°C

Preparation of Medium: Add components, except agar, to distilled/deionized water and bring volume to 500.0mL. Mix thoroughly. Filter sterilize. Warm to 45° to 50°C. Add agar to distilled/deionized water and bring volume to 500.0mL. Mix thoroughly. Autoclave for 15 min at 15 psi pressure–121°C. Cool to 45° to 50°C. Aseptically combine both solutions. Mix thoroughly. Pour into sterile Petri dishes or distribute into sterile tubes.

Use: For the cultivation of *Bacillus anthracis* especially for the production of toxins.

Rabbit Laked Blood Agar

Composition per liter:

Agar 15.0g
Pancreatic digest of casein 10.0g
Peptic digest of animal tissue 10.0g
NaCl 5.0g
Yeast extract 2.0g
Glucose 1.0g
$NaHSO_3$ 0.1g
Hemin solution 1.0mL
Vitamin K_1 solution 1.0mL
Rabbit blood, laked 50.0mL

pH 7.0 ± 0.2 at 25°C

Hemin Solution:

Composition per 100mL:

Hemin 0.5g
NaOH (1*N* solution) 10.0mL

Preparation of Hemin Solution: Add hemin to 10.0mL of NaOH solution. Mix thoroughly.

Vitamin K_1 Solution:

Composition per 20mL:

Vitamin K_1 (phytomenadione) 0.2g
Ethanol (95% solution) 20.0mL

Preparation of Vitamin K_1 Solution: Add vitamin K_1 to 20.0mL of ethanol. Mix thoroughly.

Preparation of Medium: Add components, except vitamin K_1 solution and laked rabbit blood, to distilled/deionized water and bring volume to 849.0mL. Mix thoroughly. Gently heat and bring to boiling. Autoclave for 15 min at 15 psi pressure–121°C. Cool to 45° to 50°C. Aseptically add 1.0mL of sterile vitamin K_1 solution and 50.0mL of sterile laked rabbit blood. Laked blood is prepared by freezing whole blood overnight and thawing to room temperature. Mix thoroughly. Pour into sterile Petri dishes or distribute into sterile tubes.

Use: For the cultivation and enhancement of pigment production of a variety of anaerobic bacteria.

Rabbit Serum Medium (Rabbit Serum Bovine Serum Albumin Tween™ 80 Medium) (Rabbit Serum BSA Tween™ 80 Medium)

Composition per liter:

Basal medium 900.0mL
Rabbit serum with supplements 100.0mL

pH 7.4 ± 0.2 at 25°C

Basal Medium:

Composition per 900mL:

Na_2HPO_4 ....1.0g
NaCl 1.0g
KH_2PO_4 ....0.3g
Glycerol (10% solution) 1.0mL
NH_4Cl (25% solution) 1.0mL
Sodium pyruvate (10% solution) 1.0mL
Thiamine (0.5% solution 1.0mL

Preparation of Basal Medium: Add components to distilled/deionized water and bring volume to 900.0mL. Mix thoroughly. Gently heat and bring to boiling. Autoclave for 20 min at 15 psi pressure–121°C. Cool to 25°C.

Rabbit Serum with Supplements :

Composition per 106mL:

Rabbit serum 100.0mL
L-Asparagine (3% solution) 5.0mL
$MgCl_2$-$CaCl_2$ solution 1.0mL

Preparation of Rabbit Serum with Supplements: Combine the three solutions. Mix thoroughly. Filter sterilize.

$MgCl_2$–$CaCl_2$ Solution:

Composition per 100mL:

$CaCl_2 \cdot 2H_2O$ 1.5g
$MgCl_2 \cdot 6H_2O$ 1.5g

Preparation of $MgCl_2$–$CaCl_2$ Solution: Add components to distilled/deionized water and bring volume to 100.0mL. Mix thoroughly.

Preparation of Medium: Aseptically combine 900.0mL of cooled sterile basal medium and

100.0mL of sterile rabbit serum with supplements. Mix thoroughly. Aseptically distribute into sterile tubes or flasks.

Use: For the cultivation of *Leptospira* species.

Rapid Fermentation Medium

Composition per liter:

Pancreatic digest of casein 20.0g
NaCl 5.0g
Agar 3.5g
Cystine 0.5g
Na_2SO_3 .... 0.5g
Phenol Red 0.017g

pH 7.3 ± 0.2 at 25°C

Source: This medium is available as a premixed powder from BBL Microbiology Systems.

Preparation of Medium: Add components to distilled/deionized water and bring volume to 1.0L. Mix thoroughly. Distribute into tubes or flasks. Autoclave for 15 min at 15 psi pressure–121°C.

Use: For the differentiation of *Neisseria* species isolated from clinical specimens.

Rappaport Broth, Modified (Rap Broth, Modified)

Composition per 250.2mL:

Solution A 155.0mL
Solution C 53.0mL
Solution B 40.0mL
Solution D 1.6mL
Solution E 0.6mL

Solution A:

Composition per liter:

Pancreatic digest of casein 10.0g

Preparation of Solution A: Add pancreatic digest of casein to distilled/deionized water and bring volume to 1.0L. Mix thoroughly.

Solution B:

Composition per liter:

Na_2HPO_4 .... 9.5g

Preparation of Solution B: Add Na_2HPO_4 to distilled/deionized water and bring volume to 1.0L. Mix thoroughly.

Solution C:

Composition per 100mL:

$MgCl_2 \cdot 6H_2O$ 40.0g

Preparation of Solution C: Add $MgCl_2 \cdot 6H_2O$ to distilled/deionized water and bring volume to 100.0mL. Mix thoroughly. Autoclave for 15 min at 15 psi pressure–121°C. Cool to 25°C.

Solution D:

Composition per 100mL:

Malachite Green 0.2g

Preparation of Solution D: Add Malachite Green to sterile distilled/deionized water and bring volume to 100.0mL. Mix thoroughly. Do not sterilize.

Solution E:

Composition per 10mL:

Carbenicillin 0.010g

Preparation of Solution E: Add carbenicillin to distilled/deionized water and bring volume to 10.0mL. Mix thoroughly. Filter sterilize.

Preparation of Medium: Combine 155.0mL of solution A and 40.0mL of solution B. Mix thoroughly. Autoclave for 15 min at 15 psi pressure–121°C. Cool to 45° to 50°C. Aseptically add 53.0mL of sterile solution C, 1.6mL of solution D and 0.6mL of sterile solution E. Mix thoroughly. Aseptically distribute into sterile tubes or flasks.

Use: For the isolation and cultivation of *Yersinia enterocolitica* from foods.

Reduced Salt Solution Medium (RSS Medium)

Composition per liter:

$CaCl_2 \cdot H_2O$ 20.0g
$NaHCO_3$ 10.0g
Dithiothreitol 2.0g
$MgSO_4 \cdot 7H_2O$ 2.0g
K_2HPO_4 .... 1.0g
KH_2PO_4 .... 1.0g
NaCl 0.2g

pH 9.2 ± 0.2 at 25°C

Preparation of Medium: Add components to distilled/deionized water and bring volume to 1.0L. Mix thoroughly. Distribute into screw-capped tubes or flasks. Autoclave for 15 min at 15 psi pressure–121°C.

Use: For the transport and isolation of bacteria from dental plaque, especially, *Streptococcus mutans*, *Streptococcus sanguis*, and *Lactobacillus* species.

Reduced Transport Fluid

Composition per liter:

$(NH_4)_2SO_4$ 9.0g
NaCl 9.0g
K_2HPO_4 .... 4.5g
KH_2PO_4 .... 4.5g
Na_2CO_3 .... 4.0g
EDTA (ethylenediamine tetraacetic acid) 3.8g

Dithiothreitol..2.0g
$MgSO_4 \cdot 7H_2O$..1.8g

pH 8.0 ± 0.2 at 25°C

Preparation of Medium: Add components to distilled/deionized water and bring volume to 1.0L. Mix thoroughly. Filter sterilize. Aseptically distribute into sterile tubes with rubber stoppers.

Use: For the transport and isolation of bacteria from dental plaque, especially *Streptococcus mutans* and *Streptococcus sanguis*. Also, used for the cultivation of a variety of Gram-positive bacteria from the oral cavity, especially streptococci, actinomycetes, lactobacilli, clostridia, *Bacteroides* species, *Fusobacterium* species, and *Veillonella* species.

Reduced Transport Fluid

Composition per liter:

Stock mineral salt solution No. 1..................75.0mL
Stock mineral salt solution No. 2..................75.0mL
Ethylenediamine tetraacetic acid
(1M solution)..10.0mL
Na_2CO_3 (8% solution)....................................5.0mL
Dithiothreitol (1% solution)..........................20.0mL
Resazurin (0.1% solution)..............................1.0mL

pH 8.0 ± 0.2 at 25°C

Stock Mineral Salt Solution No. 1:
Composition per 100mL:

K_2HPO_4..0.6g

Preparation of Stock Mineral Salt Solution No. 1: Add K_2HPO_4 to distilled/deionized water and bring volume to 100.0mL. Mix thoroughly.

Stock Mineral Salt Solution No. 2:
Composition per 100mL:

NaCl...1.2g
$(NH_4)_2SO_4$..1.2g
K_2HPO_4..0.6g
$MgSO_4 \cdot 7H_2O$...0.25g

Preparation of Stock Mineral Salt Solution No. 2: Add components to distilled/deionized water and bring volume to 100.0mL. Mix thoroughly.

Preparation of Medium: Add components to distilled/deionized water and bring volume to 1.0L. Mix thoroughly. Filter sterilize. Aseptically distribute into sterile tubes with rubber stoppers.

Use: For the transport and isolation of bacteria from dental plaque, especially *Streptococcus mutans* and *Streptococcus sanguis*. Also, used for the cultivation of a variety of Gram-positive bacteria from the oral cavity, especially streptococci, actinomycetes, lactobacilli, clostrida, *Bacteroides*, Fusobacteria, and *Veillonella*.

Regan–Lowe Charcoal Agar (Regan–Lowe Medium)

Composition per liter:

Agar..12.0g
Beef extract..10.0g
Pancreatic digest of gelatin..............................10.0g
Soluble starch...10.0g
NaCl..5.0g
Charcoal..4.0g
Niacin...0.01g
Horse blood, defibrinated...........................100.0mL
Cephalexin solution.....................................10.0mL

pH 7.4 ± 0.2 at 25°C

Source: This medium is available as a premixed powder from BBL Microbiology Systems.

Cephalexin Solution:
Composition per 10mL:

Cephalexin..0.04g

Preparation of Cephalexin Solution: Add cephalexin to distilled/deionized water and bring volume to 10.0mL. Mix thoroughly. Filter sterilize.

Preparation of Medium: Add components, except horse blood and cephalexin solution, to distilled/deionized water and bring volume to 890.0mL. Mix thoroughly. Gently heat and bring to boiling. Autoclave for 15 min at 15 psi pressure–121°C. Cool to 45° to 50°C. Aseptically add sterile horse blood and sterile cephalexin solution. Mix thoroughly. Pour into sterile Petri dishes or distribute into sterile tubes. Swirl medium while dispensing to keep charcoal in suspension.

Use: For the selective isolation and cultivation of *Bordetella pertussis* and *Bordetella parapertussis* from clinical specimens.

Regan–Lowe Semisolid Transport Medium

Composition per liter:

Agar..6.0g
Beef extract..5.0g
Pancreatic digest of gelatin................................5.0g
Soluble starch...5.0g
NaCl..2.5g
Charcoal..2.0g
Niacin...0.01g
Horse blood, defibrinated...........................100.0mL
Cephalexin solution.....................................10.0mL

pH 7.4 ± 0.2 at 25°C

Cephalexin Solution:
Composition per 10mL:

Cephalexin..0.04g

Preparation of Cephalexin Solution: Add cephalexin to distilled/deionized water and bring volume to 10.0mL. Mix thoroughly. Filter sterilize.

Preparation of Medium: Add components, except horse blood and cephalexin solution, to distilled/deionized water and bring volume to 890.0mL. Mix thoroughly. Gently heat and bring to boiling. Autoclave for 15 min at 15 psi pressure–121°C. Cool to 45° to 50°C. Aseptically add sterile horse blood and sterile cephalexin solution. Mix thoroughly. Aseptically distribute into small, sterile, screw-capped tubes. Fill tubes half-full. Swirl medium while dispensing to keep charcoal in suspension.

Use: For the transport of *Bordetella pertussis* and *Bordetella parapertussis* isolated from clinical specimens.

Rimler–Shotts Medium (RS Medium)

Composition per liter:

Agar	13.5g
$Na_2S_2O_3 \cdot 5H_2O$	6.8g
L-Ornithine·HCl	6.5g
NaCl	5.0g
L-Lysine·HCl	5.0g
Maltose	3.5g
Yeast extract	3.0g
Sodium deoxycholate	1.0g
Ferric ammonium citrate	0.8g
L-Cysteine·HCl	0.3g
Bromthymol Blue	0.03g
Novobiocin solution	10.0mL

pH 7.0 ± 0.2 at 25°C

Novobiocin Solution:

Composition per 10mL:

Novobiocin	5.0mg

Preparation of Novobiocin Solution: Add novobiocin to distilled/deionized water and bring volume to 10.0mL. Mix thoroughly. Filter sterilize.

Preparation of Medium: Add components, except novobiocin solution, to distilled/deionized water and bring volume to 990.0mL. Mix thoroughly. Gently heat and bring to boiling. Autoclave for 15 min at 15 psi pressure–121°C. Cool to 45° to 50°C. Aseptically add sterile components. Mix thoroughly. Pour into sterile Petri dishes or distribute into sterile tubes.

Use: For the selective isolation, cultivation and presumptive identification of *Aeromonas hydrophila* and other Gram-negative bacteria based on their ability to decarboxylate lysine and ornithine, ferment maltose and produce H_2S. Maltose-fermenting bacteria appear as yellow colonies. Bacteria that produce lysine or ornithine decarboxylase turn the medium greenish-yellow to yellow. Bacteria that produce H_2S appear as colonies with black centers.

RIOT Agar (Rice Infusion Oxgall Tween™ 80 Agar)

Composition per 1010mL:

Agar	10.0g
Oxgall	10.0g
Rice extract	1.0L
Tween™ 80	10.0mL

pH 7.3 ± 0.2 at 25°C

Rice Extract:

Composition per liter:

Cream of rice cereal	10.0g

Preparation of Rice Extract: Add cream of rice cereal to 1.0L of boiling tap water. Mix thoroughly. Filter quickly through cheesecloth. Bring volume of filtrate to 1.0L with tap water.

Preparation of Medium: Combine components. Mix thoroughly. Gently heat and bring to boiling. Distribute into tubes or flasks. Autoclave for 15 min at 15 psi pressure–121°C. Pour into sterile Petri dishes or leave in tubes.

Use: For the cultivation and differentiation of *Candida albicans* and *Candida stellatoidea* from other *Candida* species based on chlamydospore formation.

RPMI 1640 Medium with L-Glutamine

Composition per liter:

NaCl	6.0g
$NaHCO_3$	2.0g
D-Glucose	2.0g
$Na_2HPO_4 \cdot 7H_2O$	1.5g
KCl	0.4g
L-Glutamine	0.3g
L-Arginine	0.2g
$Ca(NO_3)_2 \cdot 4H_2O$	0.1g
$MgSO_4 \cdot 7H_2O$	0.1g
L-Asparagine	0.05g
L-Cystine	0.05g
L-Isoleucine, allo free	0.05g
L-Leucine, methionine free	0.05g
L-Lysine·HCl	0.04g
i-Inositol	0.035g
L-Serine	0.03g
L-Aspartic acid	0.02g
L-Glutamic acid	0.02g
L-Hydroxyproline	0.02g

L-Proline, hydroxy-L-proline free....................0.02g
L-Threonine, allo free....................0.02g
L-Tyrosine....................0.02g
L-Valine....................0.02g
L-Histidine, free base....................0.015g
L-Methionine....................0.015g
L-Phenylalanine....................0.015g
Glycine....................0.01g
L-Tryptophan....................5.0mg
Phenol Red....................5.0mg
Choline chloride....................3.0mg
Glutathione, reduced....................1.0mg
p-Aminobenzoic acid....................1.0mg
Folic acid....................1.0mg
Nicotinamide....................1.0mg
Pyridoxine·HCl....................1.0mg
Thiamine·HCl....................1.0mg
D-Calcium pantothenate....................0.25mg
Biotin....................0.20mg
Riboflavin....................0.20mg
Vitamin B_{12}....................5.0µg

pH 7.3 ± 0.2 at 25°C

Preparation of Medium: Add components to distilled/deionized water and bring volume to 1.0L. Adjust pH to 7.3 with 1N HCl or 1N NaOH. Filter sterilize. Aseptically distribute into sterile tubes or flasks.

Use: For the cultivation of mammalian cells in tissue culture. Culture medium for human immunodeficiency viruses.

SABHI Agar (Sabouraud Glucose and Brain Heart Infusion Agar)

Composition per liter:

Glucose....................21.0g
Agar....................15.0g
Pancreatic digest of casein....................10.5g
Peptic digest of animal tissue....................5.0g
Brain heart, solids from infusion....................4.0g
NaCl....................2.5g
Na_2HPO_4....................1.25g

pH 6.8 ± 0.2 at 25°C

Source: This medium is available as a premixed powder from BBL Microbiology Systems.

Preparation of Medium: Add components to distilled/deionized water and bring volume to 1.0L. Mix thoroughly. Gently heat and bring to boiling. Distribute into tubes or flasks. Autoclave for 15 min at 15 psi pressure–121°C. Pour into sterile Petri dishes in 20.0mL volumes or leave in tubes.

Use: For the cultivation of dermatophytes and other pathogenic and nonpathogenic fungi from clinical specimens and in epidemiological investigations.

SABHI Agar

Composition per liter:

Beef heart, infusion from....................125.0g
Calf brains, infusion from....................100.0g
Glucose....................21.0g
Agar....................15.0g
Neopeptone....................5.0g
Proteose peptone....................5.0g
NaCl....................2.5g
Na_2HPO_4....................1.25g
Chloromycetin solution....................1.0mL

pH 7.0 ± 0.2 at 25°C

Source: This medium is available as a premixed powder from Difco Laboratories.

Chloromycetin Solution:

Composition per 10mL:

Chloromycetin....................1.0g

Preparation of Chloromycetin Solution: Add chloromycetin to distilled/deionized water and bring volume to 10.0mL. Mix thoroughly. Filter sterilize.

Preparation of Medium: Add components, except chloromycetin solution, to distilled/deionized water and bring volume to 999.0mL. Mix thoroughly. Gently heat and bring to boiling. Autoclave for 15 min at 15 psi pressure–121°C. Cool to 45° to 50°C. Aseptically add 1.0mL of sterile chloromycetin solution. Mix thoroughly. Aseptically distribute into sterile tubes in 5.0mL volumes.

Use: For the cultivation of dermatophytes and other pathogenic and nonpathogenic fungi from clinical specimens and in epidemiological investigations.

SABHI Agar, Modified

Composition per liter:

Beef heart, infusion from....................62.5g
Calf brain, infusion from....................50.0g
Glucose....................20.5g
Brain heart infusion broth....................18.6g
Agar....................7.5g
Neopeptone....................5.0g
Pancreatic digest of gelatin....................2.50g
NaCl....................1.25g
Na_2HPO_4....................0.625g

pH 6.8 ± 0.2 at 25°C

Preparation of Medium: Dissolve, autoclave at 121°C for 15 min. Cool to 50°C and add 1.0mL of sterile chloramphenicol solution (100.0mg/mL). Mix well and dispense into sterile tubes. Slant and allow to harden. Refrigerate until needed.

Use: For the cultivation of dermatophytes and other pathogenic and nonpathogenic fungi from clinical specimens and in epidemiological investigations.

SABHI Blood Agar

Composition per liter:

Beef heart, infusion from 125.0g
Calf brains, infusion from 100.0g
Glucose 21.0g
Agar 15.0g
Neopeptone 5.0g
Proteose peptone 5.0g
NaCl 2.5g
Na_2HPO_4 .... 1.25g
Blood 100.0mL
Chloromycetin solution 1.0mL

pH 7.0 ± 0.2 at 25°C

Source: This medium is available as a premixed powder from Difco Laboratories.

Chloromycetin Solution:
Composition per 10mL:

Chloromycetin 1.0g

Preparation of Chloromycetin Solution: Add chloromycetin to distilled/deionized water and bring volume to 10.0mL. Mix thoroughly. Filter sterilize.

Preparation of Medium: Add components, except blood and chloromycetin solution, to distilled/deionized water and bring volume to 899.0mL. Mix thoroughly. Gently heat and bring to boiling. Autoclave for 15 min at 15 psi pressure–121°C. Cool to 45° to 50°C. Aseptically add 100.0mL of sterile blood and 1.0mL of sterile chloromycetin solution. Sheep blood or human blood may be used. Mix thoroughly. Aseptically distribute into sterile tubes in 5.0mL volumes.

Use: For the cultivation of dermatophytes and other pathogenic and nonpathogenic fungi from clinical specimens and in epidemiological investigations. Blood enhances the recovery of *Blastomyces dermatitidis* and *Histoplasma capsulatum* and their conversion to the yeast phase.

Sabouraud Agar with CCG and 3% NaCl

Composition per 3031.5mL:

Glucose 120.0g
NaCl 90.0g
Agar 45.0g
Peptone 30.0g
Chloramphenicol solution 15.0mL
Cycloheximide solution 15.0mL
Gentamicin solution 1.5mL

Chloramphenicol Solution:
Composition per 15mL:

Chloramphenicol 0.15g

Preparation of Chloramphenicol Solution: Add chloramphenicol to distilled/deionized water and bring volume to 15.0mL. Mix thoroughly. Filter sterilize.

Cycloheximide Solution:
Composition per 15mL:

Cycloheximide 0.3g

Preparation of Cycloheximide Solution: Add cycloheximide to distilled/deionized water and bring volume to 15.0mL. Mix thoroughly. Filter sterilize.

Gentamicin Solution:
Composition per 10mL:

Gentamicin 0.4g

Preparation of Gentamicin Solution: Add gentamicin to distilled/deionized water and bring volume to 10.0mL. Mix thoroughly. Filter sterilize.

Preparation of Medium: Add components—except chloramphenicol solution, cycloheximide solution, and gentamicin solution—to distilled/deionized water and bring volume to 3.0L. Mix thoroughly. Gently heat and bring to boiling. Autoclave for 15 min at 15 psi pressure–121°C. Cool to 45° to 50°C. Aseptically add 15.0mL of sterile chloramphenicol solution, 15.0mL of sterile cycloheximide solution, and 1.5mL of sterile gentamicin solution. Mix thoroughly. Aseptically distribute into sterile tubes. Allow tubes to cool in a slanted position.

Use: For the selective isolation and cultivation of fungi from specimens with a mixed flora.

Sabouraud Agar with CCG and 5% NaCl

Composition per 3031.5mL:

NaCl 150.0g
Glucose 120.0g
Agar 45.0g
Peptone 30.0g
Chloramphenicol solution 15.0mL
Cycloheximide solution 15.0mL
Gentamicin solution 1.5mL

Chloramphenicol Solution:
Composition per 15mL:

Chloramphenicol 0.15g

Preparation of Chloramphenicol Solution: Add chloramphenicol to distilled/deionized water and bring volume to 15.0mL. Mix thoroughly. Filter sterilize.

Cycloheximide Solution:
Composition per 15mL:

Cycloheximide..0.3g

Preparation of Cycloheximide Solution: Add cycloheximide to distilled/deionized water and bring volume to 15.0mL. Mix thoroughly. Filter sterilize.

Gentamicin Solution:
Composition per 10mL:

Gentamicin...0.4g

Preparation of Gentamicin Solution: Add gentamicin to distilled/deionized water and bring volume to 10.0mL. Mix thoroughly. Filter sterilize.

Preparation of Medium: Add components—except chloramphenicol solution, cycloheximide solution, and gentamicin solution—to distilled/deionized water and bring volume to 3.0L. Mix thoroughly. Gently heat and bring to boiling. Autoclave for 15 min at 15 psi pressure–121°C. Cool to 45° to 50°C. Aseptically add 15.0mL of sterile chloramphenicol solution, 15.0mL of sterile cycloheximide solution, and 1.5mL of sterile gentamicin solution. Mix thoroughly. Aseptically distribute into sterile tubes. Allow tubes to cool in a slanted position.

Use: For the selective isolation and cultivation of fungi from specimens with a mixed flora.

Sabouraud Glucose Agar

Composition per liter:

Glucose 40.0g
Agar...15.0g
Pancreatic digest of casein...................................5.0g
Peptic digest of animal tissue.................................5.0g

pH 5.6 ± 0.2 at 25°C

Source: This medium is available as a premixed powder from BBL Microbiology Systems, Difco Laboratories and Oxoid Unipath.

Preparation of Medium: Add components to distilled/deionized water and bring volume to 1.0L. Mix thoroughly. Gently heat and bring to boiling. Distribute into tubes or flasks. Autoclave for 15 min at 15 psi pressure–121°C. Pour into sterile Petri dishes or leave in tubes.

Use: For the cultivation of pathogenic and nonpathogenic fungi, especially dermatophytes. The medium may be made more selective for fungi by the addition of specific antibiotics such as chloramphenicol. For the cultivation of yeast and filamentous fungi.

Sabouraud Glucose Agar with Olive Oil

Composition per liter:

Glucose 40.0g
Agar...15.0g
Pancreatic digest of casein...................................5.0g
Peptic digest of animal tissue.................................5.0g
Olive oil 20.0mL
Tween™ 80 ....2.0mL

pH 5.6 ± 0.2 at 25°C

Preparation of Medium: Add components to distilled/deionized water and bring volume to 1.0L. Mix thoroughly. Gently heat and bring to boiling. Distribute into tubes or flasks. Autoclave for 15 min at 15 psi pressure–121°C. Allow tubes to cool in a slanted position.

Use: For the cultivation and maintenance of *Malassezia* species.

Sabouraud Glucose Agar, Emmons

Composition per liter:

Glucose 20.0g
Agar...17.0g
Pancreatic digest of casein...................................5.0g
Peptic digest of animal tissue.................................5.0g

pH 6.9 ± 0.2 at 25°C

Source: This medium is available as a premixed powder from BBL Microbiology Systems.

Preparation of Medium: Add components to distilled/deionized water and bring volume to 1.0L. Mix thoroughly. Gently heat and bring to boiling. Distribute into tubes or flasks. Autoclave for 15 min at 13 psi pressure–118°C. Pour into sterile Petri dishes or leave in tubes.

Use: For the cultivation of dermatophytes and other pathogenic and nonpathogenic fungi from clinical specimens and in epidemiological investigations.

Sabouraud Glucose Broth

Composition per liter:

Glucose 20.0g
Neopeptone..10.0g

pH 5.6 ± 0.2 at 25°C

Source: This medium is available as a premixed powder from Difco Laboratories.

Preparation of Medium: Add components to distilled/deionized water and bring volume to 1.0L. Mix thoroughly. Distribute into tubes or flasks. Autoclave for 15 min at 15 psi pressure–121°C. Avoid overheating.

Use: For the cultivation of pathogenic and nonpathogenic fungi, especially dermatophytes. The medium may be made more selective for fungi by the addition of specific antibiotics such as chloramphenicol.

Sabouraud Maltose Agar

Composition per liter:

Maltose	40.0g
Agar	15.0g
Pancreatic digest of casein	5.0g
Peptic digest of animal tissue	5.0g

pH 5.6 ± 0.2 at 25°C

Source: This medium is available as a premixed powder from BBL Microbiology Systems, Difco Laboratories and Oxoid Unipath.

Preparation of Medium: Add components to distilled/deionized water and bring volume to 1.0L. Mix thoroughly. Gently heat and bring to boiling. Distribute into tubes or flasks. Autoclave for 15 min at 15 psi pressure–121°C. Avoid overheating. Pour into sterile Petri dishes or leave in tubes.

Use: For the cultivation and maintenance of a variety of fungi.

Sabouraud Maltose Broth

Composition per liter:

Maltose	40.0g
Neopeptone	10.0g

pH 5.6 ± 0.2 at 25°C

Source: This medium is available as a premixed powder from Difco Laboratories.

Preparation of Medium: Add components to distilled/deionized water and bring volume to 1.0L. Mix thoroughly. Distribute into tubes or flasks. Autoclave for 15 min at 15 psi pressure–121°C. Avoid overheating.

Use: For the cultivation of a variety of fungi.

Salmonella Shigella Agar (SS Agar)

Composition per liter:

Agar	13.5g
Lactose	10.0g
Bile salts	8.5g
$Na_2S_2O_3$	8.5g
Sodium citrate	8.5g
Beef extract	5.0g
Pancreatic digest of casein	2.5g
Peptic digest of animal tissue	2.5g
Ferric citrate	1.0g
Neutral Red	0.025g
Brilliant Green	0.330mg

pH 7.0 ± 0.2 at 25°C

Source: This medium is available as a premixed powder from BBL Microbiology Systems, Difco Laboratories and Oxoid Unipath.

Preparation of Medium: Add components to distilled/deionized water and bring volume to 1.0L. Mix thoroughly. Gently heat while stirring and bring to boiling. Do not autoclave. Cool to 45° to 50°C. Pour into sterile Petri dishes in 20.0mL volumes. Allow the surface of the plates to dry before inoculation.

Use: For the selective isolation and differentiation of pathogenic enteric bacilli, especially those belonging to the genus *Salmonella.* This medium is not recommended for the primary isolation of *Shigella* species. Lactose-fermenting bacteria such as *Escherichia coli* or *Klebsiella pneumoniae* appear as small pink or red colonies. Lactose-nonfermenting bacteria—such as *Salmonella* species, *Proteus* species and *Shigella* species—appear as colorless colonies. Production of H_2S by *Salmonella* species turns the center of the colonies black.

Salmonella Shigella Agar, Modified (SS Agar, Modified)

Composition per liter:

Agar	12.0g
Lactose	10.0g
Sodium citrate	10.0g
$Na_2S_2O_3$	8.5g
Bile salts	5.5g
Beef extract	5.0g
Peptone	5.0g
Ferric citrate	1.0g
Neutral Red	0.025g
Brilliant Green	0.330mg

pH 7.3 ± 0.2 at 25°C

Source: This medium is available as a premixed powder from Oxoid Unipath.

Preparation of Medium: Add components to distilled/deionized water and bring volume to 1.0L. Mix thoroughly. Gently heat while stirring and bring to boiling. Do not autoclave. Cool to 45° to 50°C. Pour into sterile Petri dishes in 20.0mL volumes. Allow the surface of the plates to dry before inoculation.

Use: For the selective isolation and differentiation of pathogenic enteric bacilli, especially those belonging to the genus *Salmonella.* This medium provides better growth of *Shigella* species. Lactose-fermenting bacteria such as *Escherichia coli* or *Klebsiella pneu-*

moniae appear as small pink or red colonies. Lactose-nonfermenting bacteria—such as *Salmonella* species, *Proteus* species and *Shigella* species—appear as colorless colonies. Production of H_2S by *Salmonella* species turns the center of the colonies black.

Salt Broth, Modified

Composition per liter:

NaCl 65.0g
Enzymatic digest of animal tissue 10.0g
Heart digest 10.0g
Glucose 1.0g
Bromcresol Purple 0.016g
pH 7.2 ± 0.2 at 25°C

Source: Available as a prepared medium from BBL Microbiology Systems.

Preparation of Medium: Add components to distilled/deionized water and bring volume to 1.0L. Mix thoroughly. Distribute into tubes or flasks. Autoclave for 15 min at 15 psi pressure–121°C.

Use: For the cultivation and differentiation of the enterococcal group D streptococci from nonenterococcal group D streptococci based on salt tolerance.

Salt Meat Broth

Composition per liter:

NaCl 100.0g
Neutral ox-heart tissue 30.0g
Beef extract 10.0g
Peptone 10.0g
pH 7.6 ± 0.2 at 25°C

Source: This medium is available as tablets from Oxoid Unipath.

Preparation of Medium: Add components to distilled/deionized water and bring volume to 1.0L. Mix thoroughly. Distribute into tubes or flasks. Autoclave for 15 min at 15 psi pressure–121°C.

Use: For the isolation and cultivation of staphylococci from specimens with a mixed flora such as fecal specimens, especially during the investigation of staphylococcal food poisoning.

Salt Tolerance Medium

Composition per liter:

Beef heart, infusion from 500.0g
NaCl 65.0g
Tryptose 10.0g
Glucose 1.0g
Indicator solution 1.0mL
pH 7.4 ± 0.2 at 25°C

Indicator Solution:
Composition per 100mL:

Bromcresol Purple 1.6g
Ethanol (95% solution) 100.0mL

Preparation of Indicator Solution: Add Bromcresol Purple to ethanol. Mix thoroughly.

Preparation of Medium: Add components to distilled/deionized water and bring volume to 1.0L. Mix thoroughly. Distribute into tubes or flasks. Autoclave for 15 min at 15 psi pressure–121°C.

Use: For the cultivation of salt-tolerant *Streptococcus* species and other salt-tolerant Gram-positive cocci. For the differentiation of Gram-positive cocci based on salt tolerance.

Salt Tolerance Medium

Composition per liter:

NaCl 60.0g
Peptone 5.0g
Yeast extract 2.0g
Beef extract 1.0g
pH 7.4 ± 0.2 at 25°C

Preparation of Medium: Add components to distilled/deionized water and bring volume to 1.0L. Mix thoroughly. Distribute into tubes or flasks. Autoclave for 15 min at 15 psi pressure–121°C.

Use: For the cultivation and differentiation of *Aeromonas* and *Plesiomonas* species based on salt tolerance.

Salt Tolerance Medium

Composition per liter:

Beef heart, solids from infusion 500.0g
NaCl 65.0g
Tryptose 10.0g
pH 7.4 ± 0.2 at 25°C

Preparation of Medium: Add components to distilled/deionized water and bring volume to 1.0L. Mix thoroughly. Distribute into tubes or flasks. Autoclave for 15 min at 15 psi pressure–121°C.

Use: For testing the salt tolerance of a variety of microorganisms.

Salt Tolerance Medium, Gilardi

Composition per liter:

NaCl 65.0g
Pancreatic digest of casein 15.0g
Agar 15.0g
Papaic digest of soybean meal 5.0g
pH 7.3 ± 0.2 at 25°C

Preparation of Medium: Add components to distilled/deionized water and bring volume to 1.0L. Mix thoroughly. Gently heat and bring to boiling. Distribute into tubes or flasks. Autoclave for 15 min at 15 psi pressure–121°C. Do not overheat. Pour into sterile Petri dishes or leave in tubes.

Use: For the cultivation and maintenance of salt-tolerant, nonfermenting Gram-negative bacteria. For the differentiation of nonfermenting Gram-negative bacteria based on salt tolerance.

Salt Tolerance Medium, Tatum

Composition per liter:

NaCl 65.0g
Peptone 5.0g
Yeast extract 2.0g
Beef extract 1.0g

pH 7.4 ± 0.2 at 25°C

Preparation of Medium: Add components to distilled/deionized water and bring volume to 1.0L. Mix thoroughly. Distribute into tubes or flasks. Autoclave for 15 min at 15 psi pressure–121°C.

Use: For the cultivation of salt-tolerant, nonfermenting Gram-negative bacteria. For the differentiation of nonfermenting Gram-negative bacteria based on salt tolerance.

Sauton's Medium

Composition per liter:

L-Asparagine 4.0g
Citric acid 2.0g
K_2HPO_4 .... 0.5g
$MgSO_4$ 0.5g
Triton® WR 1339 .... 0.25g
Ferric ammonium citrate 0.05g
Glycerol 40.0mL

Preparation of Medium: Add components to distilled/deionized water and bring volume to 1.0L. Mix thoroughly. Distribute into tubes or flasks. Autoclave for 15 min at 15 psi pressure–121°C.

Use: For the cultivation of *Mycobacterium tuberculosis* Bacille Calmette-Guèrin (BCG) for vaccine production.

SBG Enrichment Broth (Selenite Brilliant Green Enrichment Broth)

Composition per liter:

D-Mannitol 5.0g
Peptone 5.0g
Yeast extract 5.0g
$Na_2SeO_3 \cdot 5H_2O$ 4.0g
K_2HPO_4 .... 2.65g
KH_2PO_4 .... 1.02g
Sodium taurocholate 1.0g
Brilliant Green 5.0mg

pH 7.2 ± 0.2 at 25°C

Source: This medium is available as a premixed powder from Difco Laboratories.

Preparation of Medium: Add components to distilled/deionized water and bring volume to 1.0L. Mix thoroughly. Gently heat and bring to boiling. Continue boiling for 5–10 min. Do not autoclave. Distribute into sterile tubes or flasks.

Use: For the selective isolation of *Salmonella* species, especially from eggs and egg products.

SBG Sulfa Enrichment

Composition per liter:

D-Mannitol 5.0g
Peptone 5.0g
Yeast extract 5.0g
$Na_2SeO_3 \cdot 5H_2O$ 4.0g
K_2HPO_4 .... 2.65g
KH_2PO_4 .... 1.02g
Sodium taurocholate 1.0g
Sodium sulfapyridine 0.5g
Brilliant Green 5.0mg

pH 7.2 ± 0.2 at 25°C

Source: This medium is available as a premixed powder from Difco Laboratories.

Preparation of Medium: Add components to distilled/deionized water and bring volume to 1.0L. Mix thoroughly. Gently heat and bring to boiling. Continue boiling for 5–10 min. Do not autoclave. Distribute into sterile tubes or flasks.

Use: For the selective isolation of *Salmonella* species, especially from eggs and egg products.

Schaedler Agar (Schaedler Anaerobic Agar)

Composition per liter:

Agar 13.5g
Glucose 5.83g
Pancreatic digest of casein 5.7g
Proteose peptone No. 3 .... 5.0g
Yeast extract 5.0g
Tris(hydroxymethyl)aminomethane buffer 3.0g
NaCl 1.65g
Papaic digest of soybean meal 1.0g
K_2HPO_4 .... 0.83g

L-Cystine 0.4g
Hemin 0.01g

pH 7.6 ± 0.2 at 25°C

Source: This medium is available as a premixed powder from Difco Laboratories and Oxoid Unipath.

Preparation of Medium: Add components to distilled/deionized water and bring volume to 1.0L. Mix thoroughly. Gently heat and bring to boiling. Distribute into tubes or flasks. Autoclave for 15 min at 15 psi pressure–121°C. Pour into sterile Petri dishes or leave in tubes.

Use: For the isolation, cultivation and enumeration of anaerobic and aerobic microorganisms.

Schaedler Agar

Composition per liter:

Agar 13.5g
Pancreatic digest of casein 8.2g
Glucose 5.8g
Yeast extract 5.0g
Tris(hydroxymethyl)aminomethane buffer 3.0g
Peptic digest of animal tissue 2.5g
NaCl 1.7g
Papaic digest of soybean meal 1.0g
K_2HPO_4 .... 0.8g
L-Cystine 0.4g
Hemin 0.01g

pH 7.6 ± 0.2 at 25°C

Source: This medium is available as a premixed powder from BBL Microbiology Systems.

Preparation of Medium: Add components to distilled/deionized water and bring volume to 1.0L. Mix thoroughly. Gently heat and bring to boiling. Distribute into tubes or flasks. Autoclave for 15 min at 15 psi pressure–121°C. Pour into sterile Petri dishes or leave in tubes.

Use: For the isolation, cultivation and enumeration of anaerobic and aerobic microorganisms.

Schaedler Agar with Vitamin K_1 and Sheep Blood

Composition per liter:

Agar 13.5g
Pancreatic digest of casein 8.2g
Glucose 5.8g
Yeast extract 5.0g
Tris(hydroxymethyl)aminomethane buffer 3.0g
Peptic digest of animal tissue 2.5g
Papaic digest of soybean meal 1.0g
NaCl 1.7g
K_2HPO_4 .... 0.8g
L-Cystine 0.4g
Hemin 0.01g
Sheep blood, defibrinated 50.0mL
Vitamin K_1 solution 1.0mL

pH 7.6 ± 0.2 at 25°C

Vitamin K_1 Solution:

Composition per 10mL:

Vitamin K_1 .... 5.0g
Ethanol, absolute 10.0mL

Preparation of Vitamin K_1 Solution: Add vitamin K_1 to ethanol. Mix thoroughly.

Preparation of Medium: Add components, except sheep blood, to distilled/deionized water and bring volume to 950.0mL. Mix thoroughly. Gently heat and bring to boiling. Autoclave for 15 min at 15 psi pressure–121°C. Cool to 45° to 50°C. Aseptically add sterile sheep blood. Mix thoroughly. Pour into sterile Petri dishes or distribute into sterile tubes.

Use: For the recovery of fastidious anaerobic bacteria such as *Bacteroides* species.

Schaedler Broth (Schaedler Anaerobic Broth)

Composition per liter:

Pancreatic digest of casein 8.2g
Glucose 5.8g
Yeast extract 5.0g
Tris(hydroxymethyl)aminomethane buffer 3.0g
Peptic digest of animal tissue 2.5g
NaCl 1.7g
Papaic digest of soybean meal 1.0g
K_2HPO_4 .... 0.8g
L-Cystine 0.4g
Hemin 0.01g

pH 7.6 ± 0.2 at 25°C

Source: This medium is available as a premixed powder from BBL Microbiology Systems, Difco Laboratories and Oxoid Unipath.

Preparation of Medium: Add components to distilled/deionized water and bring volume to 1.0L. Mix thoroughly. Distribute into tubes or flasks. Autoclave for 15 min at 15 psi pressure–121°C.

Use: For the cultivation and maintenance of *Eubacterium combesii, Eubacterium contortum* and a variety of other anaerobic bacteria.

Schaedler CNA Agar with Vitamin K_1 and Sheep Blood

Composition per liter:

Agar 13.5g
Pancreatic digest of casein 8.2g

Glucose 5.8g
Yeast extract 5.0g
Tris(hydroxymethyl)aminomethane buffer 3.0g
Peptic digest of animal tissue 2.5g
Papaic digest of soybean meal 1.0g
NaCl 1.7g
K_2HPO_4 .... 0.8g
L-Cystine 0.4g
Hemin 0.01g
Colistin 0.01g
Nalidixic acid 0.01g
Sheep blood, defibrinated 50.0mL
Vitamin K_1 solution 1.0mL

pH 7.6 ± 0.2 at 25°C

Vitamin K_1 Solution:
Composition per 10mL:

Vitamin K_1 .... 5.0g
Ethanol, absolute 10.0mL

Preparation of Vitamin K_1 Solution: Add vitamin K_1 to ethanol. Mix thoroughly.

Preparation of Medium: Add components, except sheep blood, to distilled/deionized water and bring volume to 950.0mL. Mix thoroughly. Gently heat and bring to boiling. Autoclave for 15 min at 15 psi pressure–121°C. Cool to 45° to 50°C. Aseptically add sterile sheep blood. Mix thoroughly. Pour into sterile Petri dishes or distribute into sterile tubes.

Use: For the selective isolation of anaerobic, Gram-positive cocci, especially *Peptococcus* species and *Peptostreptococcus* species.

Selenite Broth (Selenite Broth, Lactose) (Selenite F Enrichment Medium) (Sodium Biselenite Medium) (Sodium Hydrogen Selenite Medium)

Composition per liter:

Na_2HPO_4 .... 10.0g
Pancreatic digest of casein 5.0g
Lactose 4.0g
$NaHSeO_3 \cdot 5H_2O$ 4.0g

pH 7.0 ± 0.2 at 25°C

Source: This medium is available as a premixed powder from Difco Laboratories and a prepared media from Oxoid Unipath.

Caution: Sodium biselenite is toxic and a potential teratogen and care must be taken to avoid inhalation of the powdered dye, contact with the skin, or ingestion, especially in pregnant laboratory workers.

Preparation of Medium: Add components to distilled/deionized water and bring volume to 1.0L. Mix thoroughly. Gently heat and bring to boiling. Do not autoclave. Distribute into sterile tubes in 10.0mL volumes.

Use: For the isolation and enrichment of *Salmonella* species from clinical specimens and food products.

Selenite Broth Base, Mannitol

Composition per liter:

Na_2HPO_4 .... 10.0g
Peptone 5.0g
Mannitol 4.0g
$NaHSeO_3 \cdot 5H_2O$ 4.0g

pH 7.1 ± 0.2 at 25°C

Source: This medium is available as a premixed powder from Oxoid Unipath.

Caution: Sodium selenite is toxic and a potential teratogen and care must be taken to avoid inhalation of the powdered dye, contact with the skin, or ingestion, especially in pregnant laboratory workers.

Preparation of Medium: Add components to distilled/deionized water and bring volume to 1.0L. Mix thoroughly. Gently heat. Do not autoclave. Distribute into sterile tubes in 10.0mL volumes. Sterilize for 10 min at 0 psi pressure–100°C.

Use: For the isolation and cultivation of *Salmonella* species.

Selenite Cystine Broth

Composition per liter:

Na_2HPO_4 .... 10.0g
Pancreatic digest of casein 5.0g
Lactose 4.0g
$Na_2SeO_3 \cdot 5H_2O$ 4.0g
L-Cystine 0.02g

pH 7.0 ± 0.2 at 25°C

Source: This medium is available as a premixed powder from BBL Microbiology Systems, Difco Laboratories and Oxoid Unipath.

Caution: Sodium selenite is toxic and a potential teratogen and care must be taken to avoid inhalation of the powdered dye, contact with the skin, or ingestion, especially in pregnant laboratory workers.

Preparation of Medium: Add components to distilled/deionized water and bring volume to 1.0L. Mix thoroughly. Gently heat. Do not autoclave. Distribute into sterile tubes in 10.0mL volumes. Sterilize for 15 min at 0 psi pressure–100°C.

Use: For the isolation and cultivation of *Salmonella* species from feces and other specimens.

Selenite F Broth

Composition per liter:

KH_2PO_4	7.0g
Pancreatic digest of casein	5.0g
Lactose	4.0g
$Na_2SeO_3 \cdot 5H_2O$	4.0g
Na_2HPO_4	3.0g

pH 7.0 ± 0.2 at 25°C

Source: This medium is available as a premixed powder from BBL Microbiology Systems.

Caution: Sodium selenite is toxic and a potential teratogen and care must be taken to avoid inhalation of the powdered dye, contact with the skin, or ingestion, especially in pregnant laboratory workers.

Preparation of Medium: Add components to distilled/deionized water and bring volume to 1.0L. Mix thoroughly. Gently heat. Do not autoclave. Distribute into sterile tubes in 10.0mL volumes. Sterilize for 30 min at 0 psi pressure–100°C.

Use: For the isolation and cultivation of *Salmonella* species from feces and other specimens.

Sellers Agar (Sellers Differential Agar)

Composition per 1015mL:

Pancreatic digest of gelatin	20.0g
Agar	13.5g
D-Mannitol	2.0g
NaCl	2.0g
$MgSO_4 \cdot 7H_2O$	1.5g
K_2HPO_4	1.0g
L-Arginine	1.0g
$NaNO_3$	1.0g
Yeast extract	1.0g
$NaNO_3$	0.35g
Bromthymol Blue	0.04g
Phenol Red	8.0mg
Glucose solution	15.0mL

pH 6.7 ± 0.2 at 25°C

Source: This medium is available as a premixed powder from BBL Microbiology Systems.

Glucose Solution:

Composition per 10mL:

D-Glucose	5.0g

Preparation of Glucose Solution: Add glucose to distilled/deionized water and bring volume to 10.0mL. Mix thoroughly. Filter sterilize.

Preparation of Medium: Add components, except glucose solution, to distilled/deionized water and bring volume to 1.0L. Mix thoroughly. Gently heat and bring to boiling. Distribute into tubes in 10.0mL volumes. Autoclave for 15 min at 15 psi pressure–121°C. Allow tubes to cool in a slanted position to form a 3 inch slant with a 1.5 inch butt. Immediately prior to inoculation aseptically add 0.15mL of sterile glucose solution to each tube. Let the glucose solution run down the side of the tube opposite the slant.

Use: For the cultivation and differentiation of nonfermentative Gram-negative bacilli, especially *Pseudomonas aeruginosa, Herellea vaginicola (Acinetobacter calcoaceticus) Mima polymorpha (Acinetobacter lwoffii), Alcaligenes faecalis* and *Bacterium anitratum (Acinetobacter calcoaceticus).*

Semisolid Medium for Motility

Composition per liter:

Biosate	5.0g
Polypeptone™	5.0g
NaCl	5.0g
Agar	4.0g
Myosate	1.5g
Triphenyltetrazolium chloride solution	2.5mL

pH 6.9–7.1at 25°C

Triphenyltetrazolium Chloride Solution:

Composition per 10mL:

Triphenyltetrazolium chloride	0.1g
Ethanol (95% solution)	10.0mL

Preparation of Triphenyltetrazolium Chloride Solution: Add triphenyltetrazolium chloride to 10.0mL of ethanol. Mix thoroughly.

Preparation of Medium: Add components, except triphenyltetrazolium chloride solution, to distilled/deionized water and bring volume to 997.5mL. Mix thoroughly. Gently heat and bring to boiling. Add 2.5mL of triphenyltetrazolium chloride solution. Mix thoroughly. Distribute into tubes in 10.0mL volumes. Autoclave for 15 min at 15 psi pressure–121°C.

Use: For the differentiation of bacteria based on motility.

Sensitest Agar

Composition per liter:

Pancreatic digest of casein	11.0g
Agar	8.0g
Buffer salts	3.3g
Peptone	3.0g

NaCl 3.0g
Glucose 2.0g
Starch 1.0g
Nucleoside bases 0.02g
Thiamine 0.02mg
pH 7.4 ± 0.2 at 25°C

Source: This medium is available as a premixed powder from Oxoid Unipath.

Preparation of Medium: Add components to distilled/deionized water and bring volume to 1.0L. Mix thoroughly. Gently heat and bring to boiling. Distribute into tubes or flasks. Autoclave for 15 min at 15 psi pressure–121°C. Pour into sterile Petri dishes.

Use: For performance of antibiotic sensitivity assays.

Serratia Differential Medium (SD Medium)

Composition per 102mL:

Solution A 92.0mL
Solution B 10.0mL
pH 6.7 ± 0.2 at 25°C

Solution A:

Composition per 92mL:

Yeast extract 1.0g
L-Ornithine 1.0g
NaCl 0.5g
Agar 0.4g
Irgasan inhibitor 1.0mL
Indicator solution 1.0mL

Preparation of Solution A: Add components to distilled/deionized water and bring volume to 92.0mL. Mix thoroughly. Adjust pH to 6.7 with 1*N* NaOH.

Irgasan Inhibitor:

Composition per 100mL:

Irgasan-DP-300 (4,2′, 4′-trichloro-2-hydroxydiphenylether) 0.1g
NaOH (1*N* solution) 10.0mL

Preparation of Irgasan Inhibitor: Add irgasan to 10.0mL of NaOH solution. Mix thoroughly. Gently heat to dissolve. Bring volume to 100.0mL with distilled/deionized water.

Indicator Solution:

Composition per 100mL:

Bromthymol Blue 0.2g
Phenol Red 0.1g

Preparation of Indicator Solution: Add components to 50.0mL of distilled/deionized water. Mix thoroughly for 1 hr. Bring volume to 100.0mL with distilled/deionized water.

Solution B:

Composition per 10mL:

L-Arabinose 1.0g

Preparation of Solution B: Add arabinose to distilled/deionized water and bring volume to 10.0mL. Mix thoroughly.

Preparation of Medium: Combine 92.0mL of solution A with 10.0mL of solution B. Mix thoroughly. Distribute into tubes. Autoclave for 15 min at 15 psi pressure–121°C. Allow tubes to cool in an upright position.

Use: For the cultivation and differentiation of *Serratia* species based on fermentation of arabinose and production of ornithine decarboxylase. *S. marcescens* changes the medium to purple throughout the tube. *S. liquefaciens* changes the medium to a band of purple at the top of the tube with a green/yellow butt. *S. rubidaea* changes the medium to yellow throughout the tube.

Serum Glucose Agar

Composition per 1060mL:

Agar 15.0g
Peptone 10.0g
Beef extract 5.0g
NaCl 5.0g
Serum-glucose solution 60.0mL
pH 7.3 ± 0.2 at 25°C

Serum-Glucose Solution:

Composition per 60mL:

D-Glucose 10.0g
Serum (inactivated at 56°C, 30 min) 50.0mL

Preparation of Serum-Glucose Solution: Add glucose to 50.0mL of heat-inactivated serum. Horse serum or ox serum may be used. Mix thoroughly. Filter sterilize.

Preparation of Medium: Add components, except serum-glucose solution, to distilled/deionized water and bring volume to 1.0L. Mix thoroughly. Gently heat and bring to boiling. Autoclave for 15 min at 10 psi pressure–115°C. Cool to 50°C. Aseptically add 60.0mL of sterile serum-glucose solution. Mix thoroughly. Pour into sterile Petri dishes or distribute into sterile tubes. Allow tubes to cool in a slanted position.

Use: For the cultivation and maintenance of *Brucella* species.

Serum Glucose Agar, Farrell Modified

Composition per 1086.9mL:

Agar....15.0g
Peptone....10.0g
Beef extract....5.0g
NaCl....5.0g
Serum-glucose solution....60.0mL
Bacitracin solution....12.5mL
Cycloheximide solution....10.0mL
Nystatin solution....2.0mL
Polymyxin B solution....1.0mL
Nalidixic acid solution....1.0mL
Vancomycin solution....0.4mL

pH 7.3 ± 0.2 at 25°C

Serum-Glucose Solution:

Composition per 60mL:

D-Glucose....10.0g
Serum (inactivated at 56°C, 30 min)....50.0mL

Preparation of Serum-Glucose Solution: Add glucose to 50.0mL of heat-inactivated serum. Horse serum or ox serum may be used. Mix thoroughly. Filter sterilize.

Bacitracin Solution:

Composition per 12.5mL:

Bacitracin....25,000U

Preparation of Bacitracin Solution: Add Bacitracin to distilled/deionized water and bring volume to 12.5mL. Mix thoroughly. Filter sterilize.

Cycloheximide Solution:

Composition per 100mL:

Cycloheximide....1.0g
Acetone....5.0mL

Preparation of Cycloheximide Solution: Add cycloheximide to 5.0mL of acetone. Mix thoroughly. Bring volume to 100.0mL with distilled/deionized water. Mix thoroughly. Filter sterilize.

Nystatin Solution:

Composition per 5mL:

Nystatin....250,000U

Preparation of Nystatin Solution: Add nystatin to distilled/deionized water and bring volume to 5.0mL. Mix thoroughly. Filter sterilize.

Polymyxin B Solution:

Composition per 2mL:

Polymyxin B....10,000U

Preparation of Polymyxin B Solution: Add polymyxin B to distilled/deionized water and bring volume to 2.0mL. Mix thoroughly. Filter sterilize.

Nalidixic Acid Solution:

Composition per 2mL:

Nalidixic acid....0.1g
NaOH (0.5*N* solution)....2.0mL

Preparation of Nalidixic Acid Solution: Add nalidixic acid to 2.0mL of NaOH solution. Mix thoroughly. Immediately before use add 1.0mL of this stock solution to 9.0mL of distilled/deionized water. Mix thoroughly. Filter sterilize.

Vancomycin Solution:

Composition per 1mL:

Vancomycin....0.05g

Preparation of Vancomycin Solution: Add vancomycin to distilled/deionized water and bring volume to 1.0mL. Mix thoroughly. Filter sterilize.

Preparation of Medium: Add components—except serum-glucose solution, bacitracin solution, cycloheximide solution, nystatin solution, polymyxin B solution, nalidixic acid solution, and vancomycin solution—to distilled/deionized water and bring volume to 1.0L. Mix thoroughly. Gently heat and bring to boiling. Autoclave for 15 min at 10 psi pressure–115°C. Cool to 50°C. Aseptically add 60.0mL of sterile serum-glucose solution, 12.5mL of sterile bacitracin solution, 10.0mL of sterile cycloheximide solution, 2.0mL of sterile nystatin solution, 1.0mL of sterile polymyxin B solution, 1.0mL of sterile nalidixic acid solution, and 0.4mL of sterile vancomycin solution. Mix thoroughly. Pour into sterile Petri dishes or distribute into sterile tubes. Allow tubes to cool in a slanted position.

Use: For the selective isolation and cultivation of *Brucella* species.

Serum Potato Infusion Agar

Composition per 1120mL:

Agar....15.0g
Peptone....10.0g
Meat extract....5.0g
NaCl....5.0g
Potato infusion....1.0L
Horse serum, heat inactivated....100.0mL
Glycerol....20.0mL

pH 6.8 ± 0.2 at 25°C

Potato Infusion:

Composition per 10mL:

Potatoes....250.0g

Preparation of Potato Infusion: Add peeled, thinly sliced potatoes to 1.0L of distilled/deionized water. Infuse overnight at 60°C. Filter through Whatman #1 filter paper. Bring volume to 1.0L with distilled/deionized water.

Preparation of Medium: Combine components, except horse serum. Mix thoroughly. Gently heat and bring to boiling. Autoclave for 15 min at 15 psi pressure–121°C. Cool to 45° to 50°C. Aseptically add 100.0mL of sterile horse serum. Mix thoroughly. Pour into sterile Petri dishes or distribute into sterile tubes.

Use: For the cultivation of *Brucella* species.

Serum Tellurite Agar

Composition per liter:

Agar	20.0g
Pancreatic digest of casein	10.0g
Peptic digest of animal tissue	10.0g
NaCl	5.0g
Glucose	2.0g
Lamb serum	50.0mL
Chapman tellurite solution	10.0mL

pH 7.5 ± 0.2 at 25°C

Source: This medium is available as a premixed powder from BBL Microbiology Systems.

Chapman Tellurite Solution:

Composition per 100mL:

K_2TeO_3	1.0g

Preparation of Chapman Tellurite Solution: Add K_2TeO_3 to distilled/deionized water and bring volume to 100.0mL. Mix thoroughly. Filter sterilize.

Preparation of Medium: Add components, except lamb serum and Chapman tellurite solution, to distilled/deionized water and bring volume to 940.0mL. Mix thoroughly. Gently heat and bring to boiling. Autoclave for 15 min at 15 psi pressure–121°C. Cool to 45° to 50°C. Aseptically add sterile lamb serum and 10.0mL of sterile Chapman tellurite solution. Mix thoroughly. Pour into sterile Petri dishes or distribute into sterile tubes.

Use: For the isolation and cultivation of *Corynebacterium* species, especially in the laboratory diagnosis of diphtheria.

Seven H11 Agar (Selective 7H11 Agar)

Composition per 1010mL:

Agar	13.5g
KH_2PO_4	1.5g
Na_2HPO_4	1.5g
Pancreatic digest of casein	1.0g
NaCl	0.85g
Monosodium glutamate	0.5g
$(NH_4)_2SO_4$	0.5g
Sodium citrate	0.4g
$MgSO_4 \cdot 7H_2O$	0.05g
Ferric ammonium citrate	0.04g
$CuSO_4 \cdot 5H_2O$	1.0mg
Pyridoxine	1.0mg
$ZnSO_4 \cdot 7H_2O$	1.0mg
Biotin	0.5mg
$CaCl_2 \cdot 2H_2O$	0.5mg
Malachite Green	0.25mg
Glycerol	5.0mL
Middlebrook OADC enrichment	100.0mL
Antibiotic inhibitor	10.0mL

pH 6.6 ± 0.2 at 25°C

Source: Available as a prepared medium from BBL Microbiology Systems.

Middlebrook OADC Enrichment:

Bovine albumin, fraction V	5.0g
Glucose	2.0g
NaCl	0.85g
Catalase	3.0mg
Oleic acid	0.06mL

Preparation of Middlebrook OADC Enrichment: Add components to distilled/deionized water and bring volume to 100.0mL. Mix thoroughly. Filter sterilize.

Antibiotic Inhibitor:

Composition per 10mL:

Carbenicillin	0.05g
Trimethoprim lactate	0.02g
Amphotericin B	0.01g
Polymyxin B	200,000U

Preparation of Antibiotic Inhibitor: Add components to distilled/deionized water and bring volume to 10.0mL. Mix thoroughly. Filter sterilize.

Preparation of Medium: Add glycerol to 900.0mL of distilled/deionized water. Mix thoroughly. Add remaining components, except Middlebrook OADC enrichment and antibiotic inhibitor. Mix thoroughly. Gently heat. Do not boil. Autoclave for 10 min at 15 psi pressure–121°C. Cool to 50° to 55°C. Aseptically add 100.0mL of sterile Middlebrook OADC enrichment and 10.0mL of sterile antibiotic solution. Mix thoroughly. Pour into sterile Petri dishes or distribute into sterile tubes.

Use: For the isolation and cultivation of *Mycobacterium* species from specimens with a mixed flora.

SF Broth (*Streptococcus faecalis* Broth) (*Enterococcus faecalis* Broth)

Composition per liter:

Pancreatic digest of casein	20.0g
Glucose	5.0g

NaCl....................5.0g
K_2HPO_4....................4.0g
KH_2PO_4....................1.5g
NaN_3....................0.5g
Bromcresol Purple....................0.032g
pH 6.9 ± 0.2 at 25°C

Source: This medium is available as a premixed powder from BBL Microbiology Systems and Difco Laboratories.

Preparation of Medium: Add components to distilled/deionized water and bring volume to 1.0L. Mix thoroughly. Distribute into tubes or flasks. Autoclave for 15 min at 15 psi pressure–121°C.

Use: For the cultivation and differentiation of Group D enterococci (*Enterococcus faecalis* and *Enterococcus faecium*) from Group D nonenterococci and from other *Streptococcus* species. Group D enterococci turn the medium turbid and yellow-brown.

SIM Medium

Composition per liter:

Peptone....................30.0g
Agar....................3.0g
Beef extract....................3.0g
Peptonized iron (Difco)....................0.2g
$Na_2S_2O_3{\cdot}5H_2O$....................0.025g
pH 7.3 ± 0.2 at 25°C

Source: This medium is available as a premixed powder from Difco Laboratories.

Preparation of Medium: Add components to distilled/deionized water and bring volume to 1.0L. Mix thoroughly. Gently heat and bring to boiling. Distribute into tubes in 15.0mL volumes. Autoclave for 15 min at 15 psi pressure–121°C. Allow tubes to cool in an upright position.

Use: For the differentiation of members of the Enterobacteriaceae, based on H_2S production, indole production and motility.

SIM Medium

Composition per liter:

Pancreatic digest of casein....................20.0g
Peptic digest of animal tissue....................6.1g
Agar....................3.5g
$Fe(NH_4)_2(SO_4)_2{\cdot}6H_2O$....................0.2g
$Na_2S_2O_3{\cdot}5H_2O$....................0.2g
pH 7.3 ± 0.2 at 25°C

Source: This medium is available as a premixed powder from BBL Microbiology Systems and Oxoid Unipath.

Preparation of Medium: Add components to distilled/deionized water and bring volume to 1.0L. Mix thoroughly. Gently heat and bring to boiling. Distribute into tubes in 15.0mL volumes. Autoclave for 15 min at 15 psi pressure–121°C. Allow tubes to cool in an upright position.

Use: For the differentiation of members of the Enterobacteriaceae, based on H_2S production, indole production and motility.

Simmons' Citrate Agar (Citrate Agar)

Composition per liter:

Agar....................15.0g
NaCl....................5.0g
Sodium citrate....................2.0g
K_2HPO_4....................1.0g
$(NH_4)H_2PO_4$....................1.0g
$MgSO_4{\cdot}7H_2O$....................0.2g
Bromthymol Blue....................0.08g
pH 6.9 ± 0.2 at 25°C

Source: This medium is available as a premixed powder from BBL Microbiology Systems, Difco Laboratories and Oxoid Unipath.

Preparation of Medium: Add components to distilled/deionized water and bring volume to 1.0L. Mix thoroughly. Gently heat while stirring and bring to boiling. Distribute into tubes or flasks. Autoclave for 15 min at 15 psi pressure–121°C. Pour into sterile Petri dishes or leave in tubes.

Use: For the differentiation of Gram-negative bacteria on the basis of citrate utilization. Bacteria which can utilize citrate as sole carbon source turn the medium blue.

Skirrow *Brucella* Medium

Composition per liter:

Blood agar base No. 2....................940.0mL
Horse blood, lysed defibrinated....................50.0mL
Antibiotic solution....................10.0mL
pH 7.4 ± 0.2 at 25°C

Blood Agar Base No. 2

Composition per 940mL:

Proteose peptone....................15.0g
Agar....................12.0g
NaCl....................5.0g
Yeast extract....................5.0g
Liver digest....................2.5g
pH 7.4 ± 0.2 at 25°C

Preparation of Blood Agar Base No. 2: Add components to distilled/deionized water and bring volume to 940.0mL. Mix thoroughly. Gently heat while stirring and bring to boiling. Autoclave for 15 min at 15 psi–121°C. Cool to 45° to 50°C.

Antibiotic Solution:
Composition per 10mL:

Vancomycin..0.01g
Trimethoprim 5.0mg
Polymyxin B 2,500U

Preparation of Antibiotic Solution: Add components to distilled/deionized water and bring volume to 10.0mL. Mix thoroughly. Filter sterilize.

Preparation of Medium: To 940.0mL of sterile cooled blood agar base No. 2, aseptically add 50.0mL of sterile, lysed defibrinated horse blood and 10.0mL of sterile antibiotic solution. Pour into sterile Petri dishes or distribute into sterile tubes.

Use: For the selective isolation and cultivation of *Campylobacter* species.

Sodium Chloride Broth, 6.5

Composition per liter:

Beef heart, solids from infusion.....................500.0g
NaCl 65.0g
Tryptose 10.0g

pH 7.4 ± 0.2 at 25°C

Preparation of Medium: Add components to distilled/deionized water and bring volume to 1.0L. Mix thoroughly. Distribute into tubes or flasks. Autoclave for 15 min at 15 psi pressure–121°C.

Use: For the cultivation of enterococci and other salt-tolerant organisms. Use for the differentiation of microorganisms based on salt tolerance.

Sodium Hippurate Broth (Hippurate Broth)

Composition per liter:

Beef heart, solids from infusion.....................500.0g
Tryptose 10.0g
Sodium hippurate..10.0g
NaCl 5.0g

pH 7.4 ± 0.2 at 25°C

Source: Heart infusion broth is available as a premixed powder from Difco Laboratories and BBL Microbiology Systems.

Preparation of Medium: Add components to distilled/deionized water and bring volume to 1.0L. Mix thoroughly. Gently heat and bring to boiling. Distribute into screw-capped tubes or flasks. Autoclave for 15 min at 15 psi pressure–121°C. Tighten caps to prevent drying.

Use: For the identification and differentiation of β-hemolytic streptococci based on hippurate hydrolysis. After inoculation and incubation, tubes are treated with $FeCl_3$ reagent. A heavy precipitate remaining after 10–15 min indicates that hippurate has been hydrolyzed.

Sorbitol MacConkey Agar (MacConkey Agar with Sorbitol)

Composition per liter:

Peptone...20.0g
Agar..15.0g
Sorbitol...10.0g
NaCl 5.0g
Bile salts No.3 ....1.5g
Neutral Red 0.03g
Crystal Violet 1.0mg

pH 7.1 ± 0.2 at 25°C

Source: This medium is available as a premixed powder from Difco Laboratories and Oxoid Unipath.

Preparation of Medium: Add components to distilled/deionized water and bring volume to 1.0L. Mix thoroughly. Gently heat and bring to boiling. Distribute into tubes or flasks. Autoclave for 15 min at 15 psi pressure–121°C. Pour into sterile Petri dishes or leave in tubes.

Use: For the isolation and cultivation of pathogenic *Escherichia coli* including strain O157:H7 which does not utilize sorbitol.

Specimen Preservative Medium

Composition per liter:

NaCl 5.0g
Sodium citrate ·2 H_2O...5.0g
$(NH_4)_2HPO_4$..4.0g
KH_2PO_4..2.0g
Yeast extract..1.0g
Sodium deoxycholate..0.5g
$MgSO_4 \cdot 7H_2O$ 0.4g
Glycerol...300.0mL

pH 7.0 ± 0.2 at 25°C

Preparation of Medium: Add components, except glycerol, to distilled/deionized water and bring volume to 700.0mL. Mix thoroughly. Gently heat and bring to boiling. Add 300.0mL glycerol. Mix thoroughly. Distribute into tubes or flasks. Autoclave for 10 min at 11 psi pressure–116°C.

Use: For the preservation of viable microorganisms in stool specimens. For transport of fecal material.

Spore Strip Broth

Composition per liter:

Spore strip broth	9.0g

Preparation of Medium: Add 9.0g of spore strip broth powder (a mixture of glucose, buffer salts, growth factors and Bromthymol Blue) to distilled/deionized water and bring volume to 1.0L. Mix thoroughly. Distribute into tubes or flasks. Autoclave for 15 min at 15 psi pressure–121°C.

Use: For the recovery of spores of *Bacillus stearothermophilus* on spore strips used to determine the sterilization efficiency of autoclaves.

Sporulation Broth

Composition per liter:

Polypeptone™	15.0g
Na_2HPO_4	11.0g
Starch, soluble	3.0g
Yeast extract	3.0g
Sodium thioglycollate	1.0g
$MgSO_4$, anhydrous	0.1g

pH 7.8 ± 0.1 at 25°C

Preparation of Medium: Add components to distilled/deionized water and bring volume to 1.0L. Mix thoroughly. Distribute into tubes in 15mL volumes. Autoclave for 15 min at 15 psi pressure–121°C.

Use: For the cultivation and observation of sporulation of *Clostridium perfringens*.

Standard Fluid Medium 10B (Shepard's M10 Medium)

Composition per 102.5mL:

Base solution	70.0mL
Horse serum, unheated	20.0mL
Fresh yeast extract solution	10.0mL
Penicillin solution	1.0mL
CVA enrichment	0.5mL
L-Cysteine·HCl·H_2O solution	0.5mL
Urea solution	0.4mL
Phenol Red	0.1mL

pH 6.0 ± 0.2 at 25°C

Base Solution:

Composition per 70mL:

Beef heart (solids from infusion)	5.0g
Peptone	1.0g
NaCl	0.5g

Preparation of Base Solution: Add components to distilled/deionized water and bring volume to 70.0mL. Mix thoroughly. Adjust pH to 5.5 with 2*N* HCl. Autoclave for 15 min at 15 psi pressure–121°C. Cool to 45° to 50°C.

Fresh Yeast Extract Solution:

Composition per 100mL:

Baker's yeast, live, pressed, starch-free,	25.0g

Preparation of Fresh Yeast Extract Solution: Add the live Baker's yeast to 100.0mL of distilled/deionized water. Autoclave for 90 min at 15 psi pressure–121°C. Allow to stand. Remove supernatant solution. Adjust pH to 6.6–6.8.

Penicillin Solution:

Composition per 10mL:

Penicillin G	1,000,000U

Preparation of Penicillin Solution: Add penicillin to distilled/deionized water and bring volume to 10.0mL. Mix thoroughly. Filter sterilize.

CVA Enrichment:

Composition per liter:

Glucose	100.0g
L-Cysteine·HCl·H_2O	25.9g
L-Glutamine	10.0g
L-Cystine·2HCl	1.0g
Adenine	1.0g
Nicotinamide adenine dinucleotide	0.25g
Cocarboxylase	0.1g
Guanine·HCl	0.03g
$Fe(NO_3)_3$	0.02g
p-Aminobenzoic acid	0.013g
Vitamin B_{12}	0.01g
Thiamine·HCl	3.0mg

Preparation of CVA Enrichment: Add components to distilled/deionized water and bring volume to 1.0L. Mix thoroughly. Filter sterilize.

Cysteine·HCl·H_2O Solution:

Composition per 10mL:

Cysteine·HCl·H_2O	0.2g

Preparation of Cysteine·HCl·H_2O Solution: Add cysteine·HCl·H_2O to distilled/deionized water and bring volume to 10.0mL. Mix thoroughly. Filter sterilize.

Urea Solution:

Composition per 10mL:

Urea	1.0g

Preparation of Urea Solution: Add urea to distilled/deionized water and bring volume to 10.0mL. Mix thoroughly. Filter sterilize.

Phenol Red Solution:

Composition per 10mL:

Phenol Red	0.1g

Preparation of Phenol Red Solution: Add Phenol Red to distilled/deionized water and bring volume to 10.0mL. Mix thoroughly. Autoclave for 15 min at 15 psi pressure–121°C.

Preparation of Medium: To 70.0mL of cooled sterile base solution, aseptically add 20.0mL of sterile horse serum, 10.0mL of sterile fresh yeast extract solution, 1.0mL of sterile penicillin solution, 0.5mL of sterile CVA enrichment, 0.5mL of sterile L-cysteine·HCl·H_2O solution, 0.4mL of sterile urea solution and 0.1mL of sterile Phenol Red solution. Mix thoroughly. Aseptically distribute into sterile tubes or flasks.

Use: For the isolation and cultivation of *Ureaplasma urealyticum* from clinical specimens.

Staphylococcus–Streptococcus Selective Medium

Composition per 1060mL:

Columbia blood agar base 1.0L
Horse blood, defibrinated 50.0mL
Antibiotic inhibitor 10.0mL

pH 7.3 ± 0.2 at 25°C

Columbia Blood Agar Base:

Composition per liter:

Special peptone 23.0g
Agar 10.0g
NaCl 5.0g
Starch 1.0g

Source: Columbia blood agar base is available as a premixed powder from Oxoid Unipath.

Preparation of Columbia Blood Agar Base: Add components to distilled/deionized water and bring volume to 1.0L. Mix thoroughly. Gently heat and bring to boiling. Autoclave for 15 min at 15 psi pressure–121°C. Cool to 45° to 50°C.

Antibiotic Inhibitor:

Composition per 10mL:

Nalidixic acid 0.015g
Colistin sulfate 0.010g
Ethanol (95% solution) 10.0mL

Preparation of Antibiotic Inhibitor: Add components to 10.0mL ethanol. Mix thoroughly. Filter sterilize.

Preparation of Medium: To 1.0L of cooled sterile Columbia blood agar base aseptically add sterile horse blood and sterile antibiotic inhibitor. Mix thoroughly. Pour into sterile Petri dishes or distribute into sterile tubes.

Use: For the selective isolation of *Staphylococcus aureus* and streptococci from clinical specimens or foods.

Starch Agar with Bromcresol Purple

Composition per liter:

Agar 15.0g
Cornstarch 10.0g
Meat peptone 10.0g
Bromcresol Purple solution 1.2mL

pH 6.8 ± 0.2 at 25°C

Bromcresol Purple Solution:

Composition per 10mL:

Bromcresol Purple 0.16g
Ethanol (95% solution) 10.0mL

Preparation of Bromcresol Purple Solution: Add Bromcresol Purple to 10.0mL of 95% ethanol. Mix thoroughly.

Preparation of Medium: Add components to distilled/deionized water and bring volume to 1.0L. Mix thoroughly. Gently heat and bring to boiling. Distribute into tubes or flasks. Autoclave for 15 min at 15 psi pressure–121°C. Pour into sterile Petri dishes or leave in tubes.

Use: For the differentiation of *Gardnerella vaginalis (Haemophilus vaginalis, Corynebacterium vaginale)* from other microorganisms found in the genitourinary tract with the exception of some strains of *Streptococcus* and *Lactobacillus*. Differentiation is based on starch hydrolysis. Bacteria that can hydrolyze starch appear as colonies surrounded by a yellow zone.

Starch Agar with Bromcresol Purple

Composition per liter:

Solution 1 .... 200.0mL
Solution 2 .... 20.0mL

pH 7.8 ± 0.2 at 25°C

Solution 1:

Composition per 200mL:

Heart infusion agar 5.0mL
Bromcresol Purple solution 0.2mL

Preparation of Solution 1: Add components to distilled/deionized water and bring volume to 200.0mL. Mix thoroughly. Gently heat while stirring and bring to boiling.

Heart Infusion Agar:

Composition per liter:

Beef heart, solids from infusion 500.0g
Agar 15.0g
Tryptose 10.0g
NaCl 5.0g

Preparation of Heart Infusion Agar: Add components to distilled/deionized water and bring volume to 1.0L. Mix thoroughly. Gently heat and bring to boiling.

Bromcresol Purple Solution:

Composition per 10mL:

Bromcresol Purple 0.16g
Ethanol (95% solution) 10.0mL

Preparation of Bromcresol Purple Solution: Add Bromcresol Purple to 10.0mL of ethanol. Mix thoroughly.

Solution 2:

Composition per 20mL:

Starch 0.4g

Preparation of Solution 2: Add starch to distilled/deionized water and bring volume to 20.0mL. Mix thoroughly. Gently heat while stirring and bring to boiling.

Preparation of Medium: Combine solution 1 and solution 2. Mix thoroughly. Autoclave for 15 min at 15 psi pressure–121°C. Pour into sterile Petri dishes or distribute into sterile tubes.

Use: For the differentiation of *Gardnerella vaginalis* (*Haemophilus vaginals, Corynebacterium vaginale*) from other microorganisms found in the genitourinary tract with the exception of some strains of *Streptococcus* and *Lactobacillus*. Differentiation is based on starch hydrolysis. Bacteria that can hydrolyze starch appear as colonies surrounded by a yellow zone.

Starch Hydrolysis Agar

Composition per liter:

Beef heart, infusion from 500.0g
Soluble starch 20.0g
Agar 15.0g
Tryptose 10.0g
NaCl 5.0g

pH 7.4 ± 0.2 at 25°C

Preparation of Medium: Add components to distilled/deionized water and bring volume to 1.0L. Mix thoroughly. Gently heat and bring to boiling. Distribute into tubes or flasks. Autoclave for 15 min at 15 psi pressure–121°C. Pour into sterile Petri dishes or leave in tubes.

Use: For the cultivation and differentiation of a variety of microorganisms based on amylase production. After incubation, starch hydrolysis is determined by the addition of Gram's or Lugol's iodine solution. Organisms that produce amylase appear as colonies surrounded by a clear zone.

Stock Culture Agar

Composition per liter:

Beef heart infusion 500.0g
Gelatin 10.0g
Proteose peptone 10.0g
Agar 7.5g
Casein 5.0g
Na_2HPO_4 .... 4.0g
Sodium citrate 3.0g
Glucose 0.5g

pH 7.5 ± 0.2 at 25°C

Source: This medium is available as a premixed powder from Difco Laboratories.

Preparation of Medium: Add components to cold distilled/deionized water and bring volume to 1.0L. Mix thoroughly. Gently heat while stirring and bring to boiling. Distribute into tubes or flasks. Autoclave for 15 min at 15 psi pressure–121°C. Pour into sterile Petri dishes or leave in tubes.

Use: For the maintenance of pathogenic and nonpathogenic bacteria, especially streptococci.

Stock Culture Agar with L-Asparagine

Composition per liter:

Beef heart infusion 500.0g
Gelatin 10.0g
Proteose peptone 10.0g
Agar 7.5g
Casein 5.0g
Na_2HPO_4 .... 4.0g
Sodium citrate 3.0g
L-Asparagine 1.0g
Glucose 0.5g

pH 7.5 ± 0.2 at 25°C

Preparation of Medium: Add components to cold distilled/deionized water and bring volume to 1.0L. Mix thoroughly. Gently heat while stirring and bring to boiling. Distribute into tubes or flasks. Autoclave for 15 min at 15 psi pressure–121°C. Pour into sterile Petri dishes or leave in tubes.

Use: For the maintenance of pathogenic and nonpathogenic bacteria, especiallly streptococci.

Strep ID Quad Plate

Composition per liter:

Quadrant I 5.0mL
Quadrant II 5.0mL
Quadrant III 5.0mL
Quadrant IV 5.0mL

Source: Available as a prepared medium from BBL Microbiology Systems.

Quadrant I:

Composition per 5.0mL:

Bacitracin 0.5mg
TSA II agar 5.0mL

Quadrant II:

Composition per 5.0mL:

TSA II agar 5.0mL
Sheep blood, defibrinated 0.25mL

Quadrant III:

Composition per 5.0mL:

Bile esculin agar 5.0mL

Quadrant IV:

Composition per 5.0mL:

Blood agar base with 6.5% NaCl 5.0mL

Use: For the differentiation and presumptive identification of streptococci. The Strep (*Streptococcus)* ID (Identification) Quad Plate is a four-sectored plate each containing a different medium.

Streptococcus Agar

Composition per liter:

Glucose 20.0g
Pancreatic digest of casein 20.0g
Agar 15.0g
K_2HPO_4 .... 2.0g
$MgSO_4 \cdot 7H_2O$ 0.1g

pH 6.8 ± 0.2 at 25°C

Preparation of Medium: Add components to distilled/deionized water and bring volume to 1.0L. Mix thoroughly. Gently heat and bring to boiling. Distribute into tubes or flasks. Autoclave for 15 min at 15 psi pressure–121°C. Pour into sterile Petri dishes or leave in tubes.

Use: For the cultivation and maintenance of *Streptococcus* species.

Streptococcus Blood Agar, Selective

Composition per liter:

Agar 15.0g
Pancreatic digest of casein 10.0g
Beef extract 6.7g
Nucleic acid 6.0g
NaCl 5.0g
Sheep blood, defibrinated 50.0mL
Maltose solution 10.0mL
Antibiotic inhibitor 10.0mL

pH 7.3 ± 0.2 at 25°C

Maltose Solution:

Composition per 10mL:

Maltose 0.25–5.0g

Preparation of Maltose Solution: Add maltose to distilled/deionized water and bring volume to 10.0mL. Mix thoroughly. Filter sterilize.

Antibiotic Inhibitor:

Composition per 10mL:

Polymyxin B sulfate 0.020g
Neomycin sulfate 0.010g

Preparation of Antibiotic Inhibitor: Add components to distilled/deionized water and bring volume to 10.0mL. Mix thoroughly. Filter sterilize.

Preparation of Medium: Add components—except sheep blood, maltose solution, and antibiotic inhibitor—to distilled/deionized water and bring volume to 930.0mL. Mix thoroughly. Gently heat and bring to boiling. Autoclave for 15 min at 15 psi pressure–121°C. Cool to 45° to 50°C. Aseptically add sterile sheep blood, sterile maltose solution, and sterile antibiotic inhibitor. Mix thoroughly. Pour into sterile Petri dishes or distribute into sterile tubes.

Use: For the isolation and cultivation of Group A hemolytic *Streptococcus* species from the human respiratory tract.

Streptococcus Medium

Composition per liter:

Agar 15.0g
Glucose 4.0g
K_2HPO_4 .... 3.8g
Pancreatic digest of casein 2.5g
Yeast extract 2.5g

pH 7.6 ± 0.2 at 25°C

Preparation of Medium: Add components to distilled/deionized water and bring volume to 1.0L. Mix thoroughly. Gently heat and bring to boiling. Distribute into tubes or flasks. Autoclave for 15 min at 15 psi pressure–121°C. Pour into sterile Petri dishes or leave in tubes.

Use: For the cultivation and maintenance of *Enterococcus faecalis*.

Streptococcus mutans Medium

Composition per 100mL:

Pancreatic digest of casein 2.0g
Mannitol 0.5g
NaCl 0.25g
Lactoalbumin 0.25g
Agar 0.075g
L-Cystine 0.05g

Sodium thioglycollate 0.05g
Thallium acetate 0.025g
Crystal Violet 0.1mg
Bromcresol Purple (0.04% solution)............. 15.0mL
pH 7.1 ± 0.2 at 25°C

Caution: Thallium salts are toxic.

Preparation of Medium: Add components to distilled/deionized water and bring volume to 1.0L. Mix thoroughly. Gently heat and bring to boiling. Distribute into tubes in 5.0mL volumes. Autoclave for 15 min at 15 psi pressure–121°C.

Use: For the selective isolation and cultivation of *Streptococcus mutans*. Bacteria that turn the medium yellow are presuptive for *S. mutans*.

Streptococcus Selective Medium

Composition per liter:

Special peptone 23.0g
Agar...10.0g
NaCl 5.0g
Starch 1.0g
Horse blood, defibrinated............................. 50.0mL
Antibiotic inhibitor....................................... 10.0mL
pH 7.3± 0.2 at 25°C

Source: This medium is available as a premixed powder from Oxoid Unipath.

Antibiotic Inhibitor:

Composition per 10mL:

Colistin sulfate 10.0mg
Oxolinic acid 5.0mg

Preparation of Antibiotic Inhibitor: Add components to distilled/deionized water and bring volume to 10.0mL. Mix thoroughly. Filter sterilize.

Preparation of Medium: Add components, except horse blood and antibiotic inhibitor, to distilled/deionized water and bring volume to 940.0mL. Mix thoroughly. Gently heat and bring to boiling. Autoclave for 15 min at 15 psi pressure–121°C. Cool to 45° to 50°C. Aseptically add sterile horse blood, and sterile antibiotic inhibitor. Mix thoroughly. Pour into sterile Petri dishes or distribute into sterile tubes.

Use: For the selective isolation of streptococci from clinical specimens or foodstuffs.

Streptosel™ Agar

Composition per liter:

Pancreatic digest of casein...............................15.0g
Agar...12.0g
Glucose 5.0g
Papaic digest of soybean meal 5.0g
NaCl 4.0g
Sodium citrate 1.0g
L-Cystine 0.2g
NaN_3..0.2g
Na_2SO_3..0.2g
Crystal Violet 0.2mg
pH 7.4 ± 0.2 at 25°C

Source: This medium is available as a premixed powder from BBL Microbiology Systems.

Preparation of Medium: Add components to distilled/deionized water and bring volume to 1.0L. Mix thoroughly. Gently heat and bring to boiling. Distribute into tubes or flasks. If medium is used the same day, do not autoclve. Pour into sterile Petri dishes or leave in tubes. If medium is to be stored, autoclave for 15 min at 13 psi pressure–118°C. Pour into sterile Petri dishes or leave in tubes.

Use: For the selective isolation, cultivation and enumeration of streptococci from specimens containing a mixed flora.

Streptosel™ Broth

Composition per liter:

Pancreatic digest of casein...............................15.0g
Glucose 5.0g
Papaic digest of soybean meal 5.0g
NaCl 4.0g
Sodium citrate 1.0g
L-Cystine 0.2g
Na_2SO_3..0.2g
NaN_3..0.2g
Crystal Violet 0.2mg
pH 7.4 ± 0.2 at 25°C

Source: This medium is available as a premixed powder from BBL Microbiology Systems.

Preparation of Medium: Add components to distilled/deionized water and bring volume to 1.0L. Mix thoroughly. Distribute into tubes or flasks. Autoclave for 15 min at 13 psi pressure–118°C.

Use: For the selective isolation and cultivation of streptococci from specimens containing a mixed flora.

Use: For the selective isolation and cultivation of *Brochothrix thermosphacta*.

Stuart *Leptospira* Broth, Modified

Composition per liter:

NaCl 1.93g
Na_2HPO_4...0.66g

NH_4Cl....0.34g
$MgCl_2·6H_2O$....0.19g
L-Asparagine....0.13g
KH_2PO_4....0.08g
Glycerol....5.0mL
Rabbit serum, inactivated at 56°C, 30 min....100.0mL
pH 7.4 ± 0.2 at 25°C

Preparation of Medium: Add each component, except rabbit serum, to distilled/deionized water in separate flasks and bring each volume to 100.0mL. Mix thoroughly. Combine the seven solutions, except the rabbit serum. Mix thoroughly. Gently heat and bring to boiling. Autoclave for 15 min at 15 psi pressure–121°C. Cool to 45° to 50°C. Aseptically add sterile rabbit serum. Mix thoroughly. Aseptically distribute into sterile tubes or flasks.

Use: For the cultivation of *Leptospira* species.

Stuart Medium Base

Composition per 1100mL:

NaCl....1.8g
Na_2HPO_4....0.67g
$MgCl_2·6H_2O$....0.41g
NH_4Cl....0.27g
Asparagine....0.13g
KH_2PO_4....0.09g
Phenol Red....0.01g
Glycerol....5.0mL
Leptospira enrichment (Difco)....100.0mL
pH 7.6 ± 0.2 at 25°C

Source: This medium is available as a premixed powder from Difco Laboratories. *Leptospira* enrichment contains rabbit serum and hemoglobin and is available from Difco Laboratories.

Preparation of Medium: Add components, except glycerol and *Leptospira* enrichment, to distilled/deionized water and bring volume to 995.0mL. Mix thoroughly. Add glycerol. Mix thoroughly. Autoclave for 15 min at 15 psi pressure–121°C. Cool to 45° to 50°C. Aseptically add *Leptospira* enrichment. Mix thoroughly. Aseptically distribute into sterile screw-capped tubes in 10.0mL volumes.

Use: For the cultivation of *Leptospira* species.

Stuart Transport Medium

Composition per liter:

Sodium glycerophosphate....10.0g
Sodium thioglycollate....1.0g
$CaCl_2·2H_2O$....0.1g
Methylene Blue....2.0mg
pH 7.4 ± 0.2 at 25°C

Preparation of Medium: Add components to distilled/deionized water and bring volume to 1.0L. Mix thoroughly. Gently heat and bring to boiling. Distribute into 7.0mL screw-capped tubes. Fill tubes to capacity. Autoclave for 15 min at 15 psi pressure–121°C.

Use: For the preservation of *Neisseria* species and other fastidious organisms during their transport.

Stuart Transport Medium, Modified

Composition per liter:

Sodium glycerophosphate....10.0g
Agar....5.0g
Cysteine·HCl·H_2O....0.5g
Sodium thioglycollate....0.5g
$CaCl_2·2H_2O$....0.1g
Methylene Blue....1.0mg
pH 7.4 ± 0.2 at 25°C

Source: This medium is available as a premixed powder from Oxoid Unipath.

Preparation of Medium: Add components to distilled/deionized water and bring volume to 1.0L. Mix thoroughly. Gently heat and bring to boiling. Distribute into 7.0mL screw-capped tubes. Fill tubes to capacity. Autoclave for 15 min at 15 psi pressure–121°C.

Use: For the preservation of *Neisseria* species and other fastidious organisms during their transport.

Sucrose Phosphate Glutamate Transport Medium

Composition per liter:

Sucrose....75.0g
K_2HPO_4....0.52g
Na_2HPO_4....1.22g
Glutamic acid....0.72g
Bovine serum....50.0mL
Antibiotic inhibitor....10.0mL
pH 7.4–7.6 at 25°C

Antibiotic Inhibitor:

Composition per 10mL:

Vancomycin....0.1g
Streptomycin....0.050g
Nystatin....25,000U

Preparation of Antibiotic Inhibitor: Add components to distilled/deionized water and bring volume to 10.0mL. Mix thoroughly. Filter sterilize.

Preparation of Medium: Add components, except bovine serum and antibiotic inhibitor, to distilled/deionized water and bring volume to 940.0mL. Mix thoroughly. Gently heat and bring to boiling.

Adjust pH to 7.4–7.6. Autoclave for 15 min at 15 psi pressure–121°C. Cool to 45° to 50°C. Aseptically add sterile bovine serum and sterile antibiotic inhibitor. Mix thoroughly. Aseptically distribute into sterile tubes or flasks.

Use: For the maintenance of *Chlamydia* species during transport.

Sucrose Phosphate Transport Medium

Composition per liter:

Sucrose..68.5g
K_2HPO_4..2.1g
KH_2PO_4..1.1g
Bovine serum..50.0mL
Antibiotic inhibitor..................................10.0mL

pH 7.0 ± 0.2 at 25°C

Antibiotic Inhibitor:

Composition per 10mL:

Vancomycin..0.1g
Streptomycin..0.050g
Nystatin..25,000U

Preparation of Antibiotic Inhibitor: Add components to distilled/deionized water and bring volume to 10.0mL. Mix thoroughly. Filter sterilize.

Preparation of Medium: Add components, except bovine serum and antibiotic inhibitor, to distilled/deionized water and bring volume to 940.0mL. Mix thoroughly. Gently heat and bring to boiling. Adjust pH to 7.0. Autoclave for 15 min at 15 psi pressure–121°C. Cool to 45° to 50°C. Aseptically add sterile bovine serum and sterile antibiotic inhibitor. Mix thoroughly. Aseptically distribute into sterile tubes or flasks.

Use: For the maintenance of *Chlamydia* species during transport.

Sucrose Teepol Tellurite Agar (STT Agar)

Composition per liter:

Agar..20.0g
Beef extract..1.0g
Peptone...1.0g
Sucrose...1.0g
NaCl..0.5g
Bromthymol Blue (0.2% solution)..................2.5mL
Tellurite solution..2.5mL
Sodium lauryl sulfate
(Teepol—0.1% solution)..........................0.2mL

pH 8.0 ± 0.2 at 25°C

Tellurite Solution:

Composition per 100mL:

K_2TeO_3..0.05g

Preparation of Tellurite Solution: Add the K_2TeO_3 to distilled/deionized water and bring the volume to 100.0mL. Mix thoroughly. Filter sterilize. Use freshly prepared solution.

Caution: Potassium tellurite is toxic.

Preparation of Medium: Add components to distilled/deionized water and bring volume to 1.0L. Mix thoroughly. Gently heat and bring to boiling. Do not autoclave. Pour into sterile Petri dishes.

Use: For the selective isolation, cultivation and differentiation of *Vibrio* species based on their ability to ferment sucrose. *Vibrio cholerae* appears as flat yellow colonies. *Vibrio parahaemolyticus* appears as levated green-yellow mucoid colonies.

SXT Blood Agar

Composition per liter:

Pancreatic digest of casein................................14.5g
Agar...14.0g
NaCl..5.0g
Papaic digest of soybean meal.............................5.0g
Growth factor, BBL...1.5g
Sulfamethoxazole...0.024g
Trimethoprim...1.25mg
Sheep blood, defibrinated.................................50.0mL

pH 7.3 ± 0.2 at 25°C

Source: This medium is available as a premixed powder from BBL Microbiology Systems.

Preparation of Medium: Add components, except defibrinated sheep blood, to distilled/deionized water and bring volume to 950.0mL. Mix thoroughly. Gently heat and bring to boiling. Autoclave for 15 min at 15 psi pressure–121°C. Cool to 45° to 50°C. Aseptically add 50.0mL defibrinated sheep blood. Mix thoroughly. Pour into sterile Petri dishes or distribute into sterile tubes.

Use: For the selective isolation of Lancefield Group A and Group B streptococci from throat cultures and other clinical specimens.

TB Nitrate Reduction Broth

Composition per 100mL:

$Na_2HPO_4 \cdot 12H_2O$ 0.485g
KH_2PO_4 .... 0.117g
$NaNO_3$ 0.085g
pH 7.0 ± 0.2 at 25°C

Preparation of Medium: Add components to distilled/deionized water and bring volume to 100.0mL. Mix thoroughly. Distribute into tubes or flasks. Autoclave for 15 min at 15 psi pressure–121°C.

Use: For the differentiation of *Mycobacterium* species based on nitrate reduction. After growth of cells in appropriate medium, nitrate reduction is determined by making a suspension of cells in TB nitrate reduction broth and adding hydrochloric acid, sulfanilamide and *N*-naphylenendiamine. Nitrate reduction turns the medium pink. *M. tuberculosis* reduces nitrate and turns the medium deep pink within 1 min. *M. bovis* does not reduce nitrate and does not change the medium.

TCBS Agar (Thiosulfate Citrate Bile Salt Sucrose Agar)

Composition per liter:

Sucrose 20.0g
Agar 14.0g
NaCl 10.0g
Sodium citrate 10.0g
$Na_2S_2O_3$ 10.0g
Yeast extract 5.0g
Pancreatic digest of casein 5.0g
Peptic digest of animal tissue 5.0g
Oxgall 5.0g
Sodium cholate 3.0g
Ferric citrate 1.0g
Thymol Blue 0.04g
Bromthymol Blue 0.04g
pH 8.6 ± 0.2 at 25°C

Source: This medium is available as a premixed powder from BBL Microbiology Systems and Difco Laboratories.

Preparation of Medium: Add components to distilled/deionized water and bring volume to 1.0L. Mix thoroughly. Gently heat while stirring and bring to boiling. Do not autoclave. Cool to 45° to 50°C. Pour into sterile Petri dishes or distribute into sterile tubes.

Use: For the selective isolation of *Vibrio cholerae* and *Vibrio parahaemolyticus* from a variety of clinical specimens and in epidemiological investigations.

TCH Medium (Thiophene 2 Carboxylic Acid Hydrazide Medium)

Composition per 1105mL:

Thiophene-2-carboxylic acid hydrazide 1.1mg
Middlebrook 7H10 Agar Base 1.0L
OADC Enrichment 100.0mL
Glycerol 5.0mL
pH 6.6 ± 0.2 at 25°C

Middlebrook 7H10 Agar Base:
Composition per liter:

Agar 15.0g
Na_2HPO_4 .... 1.5g
KH_2PO_4 .... 1.5g
$(NH_4)_2SO_4$ 0.5g
L-Glutamic Acid 0.5g
Sodium citrate 0.4g
Ferric ammonium citrate 0.04g
$MgSO_4 \cdot 7H_2O$ 0.025g
$ZnSO_4 \cdot 7H_2O$ 1.0mg
$CuSO_4 \cdot 5H_2O$ 1.0mg
Pyridoxine 1.0mg
Biotin 0.5mg
$CaCl_2 \cdot 2H_2O$ 0.5mg
Malachite Green 0.25mg

Preparation of Middlebrook 7H10 Agar Base: Add glycerol to 900.0mL of distilled/deionized water and add remaining components. Mix thoroughly. Gently heat and bring to boiling.

Middlebrook OADC Enrichment:
Composition per 100mL:

Bovine albumin fraction V 5.0g
Glucose 2.0g
NaCl 0.85g
Oleic acid 0.05g
Catalase 4.0mg

Source: Available as a prepared enrichment from Difco Laboratories.

Preparation of Middlebrook OADC Enrichment: Add components to distilled/deionized water and bring volume to 100.0mL. Mix thoroughly. Filter sterilize.

Preparation for Medium: Combine components. Mix thoroughly. Distribute into tubes or flasks. Autoclave for 15 min at 15 psi pressure–121°C. Pour into sterile Petri dishes or leave in tubes.

Use: For the differentiation of *Mycobacterium* species. *M. bovis* is inhibited by TCH. *M. tuberculosis* and other mycobacteria are generally resistant to low concentrations of TCH. This distinguishes *M. bovis* from other slow-growing mycobacteria.

Tergitol 7 Agar

Composition per liter:

Lactose	20.0g
Agar	13.0g
Peptone	10.0g
Yeast extract	6.0g
Meat extract	5.0g
Tergitol-7	0.1g
Bromothymol Blue	0.05g
TTC solution	5.0mL

pH 7.2 ± 0.2 at 25°C

Source: This medium is available as a premixed powder from Oxoid Unipath.

TTC Solution:

Composition per 100mL:

Triphenyltetrazolium chloride	0.05g

Preparation of TTC Solution: Add triphenyltetrazolium chloride to distilled/deionized water and bring volume to 100.0mL. Mix thoroughly. Filter sterilize.

Preparation of Medium: Add components to distilled/deionized water and bring volume to 995.0mL. Mix thoroughly. Gently heat and bring to boiling. Autoclave for 15 min at 15 psi pressure–121°C. Cool to 50°C. Aseptically add 5.0mL of sterile TTC solution. Mix thoroughly. Pour into sterile Petri dishes or distribute into sterile tubes.

Use: For the detection and enumeration of coliforms. Lactose-fermenting bacteria appear as yellow colonies. Non-lactose fermenting bacteria appear as blue colonies.

Tergitol 7 Agar

Composition per liter:

Agar	15.0g
Lactose	10.0g
Yeast extract	3.0g
Pancreatic digest of casein	2.5g
Peptic digest of animal tissue	2.5g
Tergitol 7	0.1g
Bromthymol Blue	25.0mg
TTC solution	3.0mL

pH 6.9 ± 0.2 at 25°C

Source: This medium is available as a premixed powder from BBL Microbiology Systems and Difco Laboratories.

TTC Solution:

Composition per 100mL:

Triphenyl tetrazolium chloride	1.0g

Preparation of TTC Solution: Add triphenyltetrazolium chloride to distilled/deionized water and bring volume to 100.0mL. Mix thoroughly. Filter sterilize.

Preparation of Medium: Add components to distilled/deionized water and bring volume to 997.0mL. Mix thoroughly. Gently heat and bring to boiling. Autoclave for 15 min at 15 psi pressure–121°C. Cool to 50°C. Aseptically add 3.0mL of sterile TTC solution. Mix thoroughly. Pour into sterile Petri dishes or distribute into sterile tubes.

Use: For the selective isolation and differentiation of of coliform bacteria based on lactose fermentation. Lactose-fermenting bacteria appear as yellow colonies. Non-lactose fermenting bacteria appear as blue colonies.

Tergitol 7 Agar H

Composition per liter:

Agar	15.0g
Lactose	10.0g
Yeast extract	3.0g
Pancreatic digest of casein	2.5g
Peptic digest of animal tissue	2.5g
Ferric ammonium citrate	0.5g
$Na_2S_2O_3$	0.5g
Tergitol 7	0.1g
Bromthymol Blue	0.025g

pH 7.2 ± 0.2

Preparation of Medium: Add components to distilled/deionized water and bring volume to 1.0L. Mix thoroughly. Gently heat and bring to boiling. Distribute into tubes or flasks. Autoclave for 15 min at 15 psi pressure–121°C. Pour into sterile Petri dishes or leave in tubes.

Use: For the selective isolation and differentiation of enteric bacteria from urine.

Tergitol 7 Broth

Composition per liter:

Lactose	10.0g
Yeast extract	3.0g
Pancreatic digest of casein	2.5g
Peptic digest of animal tissue	2.5g
Tergitol 7	0.1g
Bromthymol Blue	25.0mg
TTC solution	3.0mL

pH 6.9 ± 0.2 at 25°C

Source: This medium is available as a premixed powder from BBL Microbiology Systems and Difco Laboratories.

TTC Solution:

Composition per 100mL:

Triphenyl tetrazolium chloride 1.0g

Preparation of TTC Solution: Add tri-phenyltetrazolium chloride to distilled/deionized water and bring volume to 100.0mL. Mix thoroughly. Filter sterilize.

Preparation of Medium: Add components to distilled/deionized water and bring volume to 997.0mL. Mix thoroughly. Gently heat while stirring and bring to boiling. Autoclave for 15 min at 15 psi pressure–121°C. Cool to 25°C. Aseptically add 3.0mL of sterile TTC solution. Mix thoroughly.

Use: For the isolation and cultivation of coliforms, *Salmonella* species and other enteric bacteria.

Tetrathionate Broth

Composition per liter:

$Na_2S_2O_3$ 40.7g
$CaCO_3$ 25.0g
NaCl 4.5g
Peptone 4.5g
Yeast extract 1.8g
Beef extract 0.9g
Iodine solution 20.0mL

Source: This medium is available as a premixed powder from Oxoid Unipath.

Iodine Solution:

Composition per 20mL:

Iodine 6.0g
KI 5.0g

Preparation of Iodine Solution: Add iodine and KI to distilled/deionized water and bring volume to 20.0mL. Mix thoroughly.

Preparation of Medium: Add components, except iodine solution, to distilled/deionized water and bring volume to 980.0mL. Mix thoroughly. Gently heat and bring to boiling. Do not autoclave. Cool to 40°C. Add 20.0mL of iodine solution. Mix thoroughly. Distribute into tubes in 10.0mL volumes. Use medium the same day it is prepared.

Use: For the selective isolation and enrichment of *Salmonella* species from feces and other specimens.

Tetrathionate Broth (TT Broth)

Composition per liter:

$Na_2S_2O_3$ 30.0g
$CaCO_3$ 10.0g
Proteose peptone 5.0g
Bile salts 1.0g
Iodine solution 20.0mL

pH 8.4± 0.2 at 25°C

Source: This medium is available as a premixed powder from Difco Laboratories and BBL Microbiology Systems.

Iodine Solution:

Composition per 20mL:

Iodine 6.0g
KI 5.0g

Preparation of Iodine Solution: Add iodine and KI to distilled/deionized water and bring volume to 20.0mL. Mix thoroughly.

Preparation of Medium: Add components, except iodine solution, to distilled/deionized water and bring volume to 980.0mL. Mix thoroughly. Gently heat and bring to boiling. Do not autoclave. Cool to 40°C. Add 20.0mL of iodine solution. Mix thoroughly. Distribute into tubes in 10.0mL volumes. Use medium the same day it is prepared.

Use: For the selective isolation and enrichment of *Salmonella* species from infectious material.

Tetrathionate Broth (m–Tetrathionate Broth) (m–TT Broth)

Composition per liter:

$Na_2S_2O_3$ 30.0g
$CaCO_3$ 10.0g
Pancreatic digest of casein 2.5g
Peptic digest of animal tissue 2.5g
Iodine-iodide solution 20.0mL

pH 8.0 ± 0.2 at 25°C

Iodine-Iodide Solution:

Composition per 20mL:

Iodine 6.0g
KI 5.0g

Preparation of Iodine-Iodide Solution: Add iodine and KI to distilled/deionized water and bring volume to 20.0mL. Mix thoroughly.

Preparation of Medium: Add components, except iodine-iodide solution, to distilled/deionized water and bring volume to 980.0mL. Mix thoroughly. Gently heat and bring to boiling. Do not autoclave. Cool to 40°C. Add 20.0mL of iodine-iodide solution. Mix thoroughly. Distribute into tubes in 10.0mL volumes. Use medium the same day it is prepared.

Use: For the selective isolation in the membrane filter method of *Salmonella* species from feces, urine, and other specimens.

Tetrathionate Broth (m–Tetrathionate Broth)

Composition per liter:

$Na_2S_2O_3$	30.0g
Proteose peptone	5.0g
Bile salts	1.0g
Iodine solution	20.0mL

pH 8.0 ± 0.2 at 25°C

Source: This medium is available as a premixed powder from Difco Laboratories.

Iodine Solution:

Composition per 20mL:

Iodine	6.0g
KI	5.0g

Preparation of Iodine Solution: Add iodine and KI to distilled/deionized water and bring volume to 20.0mL. Mix thoroughly.

Preparation of Medium: Add components, except iodine solution, to distilled/deionized water and bring volume to 980.0mL. Mix thoroughly. Gently heat and bring to boiling. Do not autoclave. Cool to 40°C. Add 20.0mL of iodine solution. Mix thoroughly. Use medium the same day it is prepared.

Use: For the enrichment of *Salmonella* species in the membrane filter method prior to placing the filter on selective media such as Brilliant Green broth.

Tetrathionate Broth, Hajna (TT Broth, Hajna)

Composition per liter:

$Na_2S_2O_3$	38.0g
$CaCO_3$	25.0g
Casein/meat peptone (50/50)	18.0g
NaCl	5.0g
D–Mannitol	2.5g
Yeast extract	2.0g
Glucose	0.5g
Sodium deoxycholate	0.5g
Brilliant Green	0.01g
Iodine solution	40.0mL

pH 7.5–7.8 at 25°C

Source: This medium is available as a premixed powder from Difco Laboratories.

Iodine Solution:

Composition per 40mL:

KI	8.0g
Iodine	5.0g

Preparation of Iodine Solution: Add iodine and KI to distilled/deionized water and bring volume to 40.0mL. Mix thoroughly.

Preparation of Medium: Add components, except iodine solution, to distilled/deionized water and bring volume to 960.0mL. Mix thoroughly. Gently heat and bring to boiling. Do not autoclave. Cool to 40°C. Add 40.0mL of iodine solution. Mix thoroughly. Distribute into tubes in 10.0mL volumes. Use medium the same day it is prepared.

Use: For the isolation of *Salmonella* species from feces, urine, and other specimens.

Tetrathionate Broth, USA (TT Broth, USA)

$Na_2S_2O_3$	30.0g
$CaCO_3$	10.0g
Casein peptone	2.5g
Meat peptone	2.5g
Bile salts	1.0g

Source: This medium is available as a premixed powder from Oxoid Unipath.

Iodine-Iodide Solution:

Composition per 20mL:

Iodine	6.0g
KI	5.0g

Preparation of Iodine-Iodide Solution: Add iodine and KI to distilled/deionized water and bring volume to 20.0mL. Mix thoroughly.

Preparation of Medium: Add components, except iodine solution, to distilled/deionized water and bring volume to 980.0mL. Mix thoroughly. Gently heat and bring to boiling. Do not autoclave. Cool to 40°C. Add 20.0mL of iodine solution. Mix thoroughly. Distribute into tubes in 10.0mL volumes. Use medium the same day it is prepared.

Use: For the selective enrichment of *Salmonella* species from feces, urine, and other specimens.

Tetrathionate Broth with Novobiocin

Composition per liter:

$Na_2S_2O_3$	38.0g
$CaCO_3$	25.0g
Casein/meat peptone (50/50)	18.0g
NaCl	5.0g
Yeast extract	2.0g
D–Mannitol	0.5g
Glucose	0.5g
Sodium deoxycholate	0.5g
Brilliant Green	0.01g
Novobiocin	4.0mg
Iodine solution	40.0mL

pH 7.5–7.8 at 25°C

Iodine Solution:

Composition per 40mL:

KI 8.0g
Iodine 5.0g

Preparation of Iodine Solution: Add iodine and KI to distilled/deionized water and bring volume to 20.0mL. Mix thoroughly.

Preparation of Medium: Add components, except iodine solution, to distilled/deionized water and bring volume to 960.0mL. Mix thoroughly. Gently heat and bring to boiling. Do not autoclave. Cool to 40°C. Add 40.0mL of iodine solution. Mix thoroughly. Distribute into tubes in 10.0mL volumes. Use medium the same day it is prepared.

Use: For the isolation of *Salmonella* species from feces, urine, and other specimens. Novobiocin suppresses the growth of *Proteus* species.

Tetrathionate Crystal Violet Enhancement Broth

Composition per liter:

Potassium tetrathionate 20.0g
Casein/meat peptone (50/50) 8.6g
NaCl 6.4g
Crystal Violet 0.005g
pH 6.5 ± 0.2 at 25°C

Preparation of Medium: Add components to distilled/deionized water and bring volume to 1.0L. Mix thoroughly. Distribute into tubes or flasks. Autoclave for 15 min at 15 psi pressure–121°C.

Use:For the isolation of *Salmonella* species from feces, urine, and other specimens.

Tetrazolium Tolerance Agar (TTC Agar)

Composition per liter:

Pancreatic digest of casein 15.0g
Agar 15.0g
Triphenyltetrazolium chloride 10.0g
Papaic digest of soybean meal 5.0g
NaCl 5.0g
pH 7.3 ± 0.2 at 25°C

Preparation of Medium: Add components to distilled/deionized water and bring volume to 1.0L. Mix thoroughly. Gently heat and bring to boiling. Distribute into tubes or flasks. Autoclave for 15 min at 15 psi pressure–121°C. Do not overheat. Pour into sterile Petri dishes or leave in tubes.

Use: For the differentiation of bacteria based upon the ability to tolerate and grow in the presence of tetrazolium.Enter*ococcus faecalis* (enterococci) rapidly reduce tetrazolium.

Thayer–Martin Agar, Modified (MTM II) (Modified Thayer–Martin Agar)

Composition per liter:

Agar 12.0g
Hemoglobin 10.0g
Pancreatic digest of casein 7.5g
Selected meat peptone 7.5g
NaCl 5.0g
K_2HPO_4 .... 4.0g
Cornstarch 1.0g
KH_2PO_4 .... 1.0g
C-N-V-T inhibitor 10.0mL
Supplement solution 10.0mL
pH 7.2 ± 0.2 at 25°C

CNVT Inhibitor:

Composition per 10mL:

Colistin sulfate 7.5mg
Trimethoprim lactate 5.0mg
Vancomycin 3.0mg
Nystatin 12,500U

Preparation of CNVT Inhibitor: Add components to distilled/deionized water and bring volume to 10.0mL. Mix thoroughly. Filter sterilize.

Supplement Solution:

Composition per liter:

Glucose 100.0g
L-Cysteine·HCl 25.9g
L-Glutamine 10.0g
L-Cystine 1.1g
Adenine 1.0g
Nicotinamide adenine dinucleotide 0.25g
Vitamin B_{12} 0.1g
Thiamine pyrophosphate 0.1g
Guanine·HCl 0.03g
$Fe(NO_3)_3·6H_2O$ 0.02g
p-Aminobenzoic acid 0.013g
Thiamine·HCl 3.0mg

Source: The supplement solution IsoVitaleX® enrichment is available from BBL Microbiology Laboratories. This enrichment may be replaced by supplement VX from Difco Laboratories.

Preparation of Supplement Solution: Add components to distilled/deionized water and bring volume to 1.0L. Mix thoroughly. Filter sterilize.

Preparation of Medium: Add components, except C-N-V-T inhibitor and supplement solution, to distilled/deionized water and bring volume to

990.0mL. Mix thoroughly. Gently heat and bring to boiling. Distribute into tubes or flasks. Autoclave for 15 min at 15 psi pressure–121°C. Cool to 45° to 50°C. Aseptically add 10.0mL of sterile C-N-V-T inhibitor and 10.0mL of sterile Isosupplement solution. Mix thoroughly. Pour into sterile Petri dishes or distribute into sterile tubes.

Use: For the isolation of *Neisseria gonorrhoeae* and other *Neisseria* species from specimens containing mixed flora of bacteria and fungi.

Thayer–Martin Medium

Composition per liter:

GC agar base 740.0mL
Hemoglobin solution 250.0mL
Vitox supplement 10.0mL
pH 7.3 ± 0.2 at 25°C

GC Agar Base:

Composition per 740mL:

Special peptone 15.0g
Agar 10.0g
NaCl 5.0g
K_2HPO_4 .... 4.0g
Cornstarch 1.0g
KH_2PO_4 .... 1.0g
pH 7.2 ± 0.2 at 25°C

Preparation of GC Agar Base: Add components of GC medium base and the hemoglobin to distilled/deionized water and bring volume to 740.0mL. Mix thoroughly. Gently heat until boiling. Autoclave for 15 min at 15 psi pressure–121°C. Cool to 45° to 50°C.

Hemoglobin Solution:

Composition per 250mL:

Hemoglobin 5.0g

Preparation of Hemoglobin Solution: Add hemoglobin to distilled/deionized water and bring volume to 250.0mL. Mix thoroughly. Autoclave for 15 min at 15 psi pressure–121°C. Cool to 45° to 50°C.

Vitox Supplement:

Composition per 10.0mL:

Glucose 2.0g
L-Cysteine·HCl 0.518g
L-Glutamine 0.2g
L-Cystine 0.022g
Adenine sulfate 0.010g
Nicotinamide adenine dinucleotide 5.0mg
Cocarboxylase 2.0mg
Guanine·HCl 0.6mg
$Fe(NO_3)_3 \cdot 6H_2O$ 0.4mg
p-Aminobenzoic acid 0.26mg
Vitamin B_{12} 0.2mg
Thiamine·HCl 0.06mg

Preparation of Vitox Supplement: Add components to distilled/deionized water and bring volume to 10.0mL. Mix thoroughly. Filter sterilize.

Preparation of Medium: To 740.0mL of cooled sterile GC agar base, aseptically add 250.0mL of sterile hemoglobin solution and 10.0mL of sterile Vitox supplement. Mix thoroughly. Pour into sterile Petri dishes or distribute into sterile tubes.

Use: For the isolation and cultivation of fastidious microorganisms, especially *Neisseria gonorrhoeae* and related species.

Thayer–Martin Medium

Composition per liter:

Hemoglobin 10.0g
GC medium base 980.0mL
CNVT inhibitor 10.0mL
Supplement B (Difco) 10.0mL
pH 7.3 ± 0.2 at 25°C

Source: Available as a prepared medium in tubes from Difco Laboratories.

GC Medium Base:

Composition per 980mL:

Proteose Peptone No. 3 .... 15.0g
Agar 10.0g
NaCl 5.0g
K_2HPO_4 .... 4.0g
Cornstarch 1.0g
KH_2PO_4 .... 1.0g
pH 7.2 ± 0.2 at 25°C

Preparation of GC Medium Base: Add components of GC medium base and the hemoglobin to distilled/deionized water and bring volume to 1.0L. Mix thoroughly. Gently heat until boiling. Autoclave for 15 min at 15 psi pressure–121°C. Cool to 45° to 50°C.

CNVT Inhibitor:

Composition per 10mL:

Colistin sulfate 7.5mg
Trimethoprim lactate 5.0mg
Vancomycin 3.0mg
Nystatin 12,500U

Preparation of CNVT Inhibitor: Add components to distilled/deionized water and bring volume to 10.0mL. Mix thoroughly. Filter sterilize.

Preparation of Medium: To 980.0mL of cooled sterile GC medium base, aseptically add 10.0mL of sterile CNVT inhibitor and 10.0mL of sterile supplement B. Mix thoroughly. Pour into sterile Petri dishes or distribute into sterile tubes.

Use: For the isolation and cultivation of fastidious microorganisms, especially *Neisseria gonorrhoeae* and related species.

Thayer–Martin Medium, Modified (Modified Thayer–Martin Agar)

Composition per liter:

GC agar base 720.0mL
Hemoglobin solution 250.0mL
GC supplement 30.0mL
pH 7.3 ± 0.2 at 25°C

GC Agar Base:
Composition per 720mL:

Special peptone 15.0g
Agar 10.0g
NaCl 5.0g
K_2HPO_4 .... 4.0g
Cornstarch 1.0g
KH_2PO_4 .... 1.0g
pH 7.2 ± 0.2 at 25°C

Preparation of GC Agar Base: Add components of GC medium base and the hemoglobin to distilled/deionized water and bring volume to 720.0mL. Mix thoroughly. Gently heat until boiling. Autoclave for 15 min at 15 psi pressure–121°C. Cool to 45° to 50°C.

Hemoglobin Solution:
Composition per 250mL:

Hemoglobin 5.0g

Preparation of Hemoglobin Solution: Add hemoglobin to distilled/deionized water and bring volume to 250.0mL. Mix thoroughly. Autoclave for 15 min at 15 psi pressure–121°C. Cool to 45° to 50°C.

GC Supplement:
Composition per 30.0mL:

Yeast autolysate 10.0g
Glucose 1.5g
$NaHCO_3$ 0.15g
Colistin sulfate 7.5mg
Trimethoprim lactate 5.0mg
Vancomycin 3.0mg
Nystatin 12,500U

Preparation of GC Supplement: Add components to distilled/deionized water and bring volume to 30.0mL. Mix thoroughly. Filter sterilize.

Preparation of Medium: To 720.0mL of cooled sterile GC agar base, aseptically add 250.0mL of sterile hemoglobin solution, and 30.0mL of sterile GC supplement. Mix thoroughly. Pour into sterile Petri dishes or distribute into sterile tubes.

Use: For the selective isolation and cultivation of fastidious microorganisms, especially *Neisseria gonorrhoeae* and related species.

Thayer–Martin Medium, Modified (Modified Thayer–Martin Agar)

Composition per liter:

GC agar base 730.0mL
Hemoglobin solution 250.0mL
Vitox supplement 10.0mL
VCNT antibiotic solution 10.0mL
pH 7.3 ± 0.2 at 25°C

GC Agar Base:
Composition per 730mL:

Special peptone 15.0g
Agar 10.0g
NaCl 5.0g
K_2HPO_4 .... 4.0g
Cornstarch 1.0g
KH_2PO_4 .... 1.0g
pH 7.2 ± 0.2 at 25°C

Preparation of GC Agar Base: Add components of GC medium base and the hemoglobin to distilled/deionized water and bring volume to 730.0mL. Mix thoroughly. Gently heat until boiling. Autoclave for 15 min at 15 psi pressure–121°C. Cool to 45° to 50°C.

Hemoglobin Solution:
Composition per 250mL:

Hemoglobin 5.0g

Preparation of Hemoglobin Solution: Add hemoglobin to distilled/deionized water and bring volume to 250.0mL. Mix thoroughly. Autoclave for 15 min at 15 psi pressure–121°C. Cool to 45° to 50°C.

Vitox Supplement:
Composition per 10.0mL:

Glucose 2.0g
L-Cysteine·HCl 0.518g
L-Glutamine 0.2g
L-Cystine 0.022g
Adenine sulfate 0.010g
Nicotinamide adenine dinucleotide 5.0mg
Cocarboxylase 2.0mg
Guanine·HCl 0.6mg
$Fe(NO_3)_3 \cdot 6H_2O$ 0.4mg
p-Aminobenzoic acid 0.26mg
Vitamin B_{12} 0.2mg
Thiamine·HCl 0.06mg

Preparation of Vitox Supplement: Add components to distilled/deionized water and bring volume to 10.0mL. Mix thoroughly. Filter sterilize.

VCNT Antibiotic Solution:
Composition per 10mL:

Colistin methane sulfonate 7.5mg
Trimethoprim lactate 5.0mg

Vancomycin 3.0mg
Nystatin 12,500U

Preparation of VCNT Antibiotic Solution: Add components to distilled/deionized water and bring volume to 10.0mL. Mix thoroughly. Filter sterilize.

Preparation of Medium: To 730.0mL of cooled sterile GC agar base, aseptically add 250.0mL of sterile hemoglobin solution, 10.0mL of sterile Vitox supplement, and 10.0mL of VCNT antibiotic solution. Mix thoroughly. Pour into sterile Petri dishes or distribute into sterile tubes.

Use: For the selective isolation and cultivation of fastidious microorganisms, especially *Neisseria gonorrhoeae* and related species.

Thayer–Martin Medium, Selective

Composition per liter:

GC agar base 730.0mL
Hemoglobin solution 250.0mL
Vitox supplement 10.0mL
VCN antibiotic solution 10.0mL

pH 7.3 ± 0.2 at 25°C

GC Agar Base:
Composition per 730mL:

Special peptone 15.0g
Agar 10.0g
NaCl 5.0g
K_2HPO_4 .... 4.0g
Cornstarch 1.0g
KH_2PO_4 .... 1.0g

pH 7.2 ± 0.2 at 25°C

Preparation of GC Agar Base: Add components of GC medium base and the hemoglobin to distilled/deionized water and bring volume to 730.0mL. Mix thoroughly. Gently heat until boiling. Autoclave for 15 min at 15 psi pressure–121°C. Cool to 45° to 50°C.

Hemoglobin Solution:
Composition per 250mL:

Hemoglobin 5.0g

Preparation of Hemoglobin Solution: Add hemoglobin to distilled/deionized water and bring volume to 250.0mL. Mix thoroughly. Autoclave for 15 min at 15 psi pressure–121°C. Cool to 45° to 50°C.

Vitox Supplement:
Composition per 10.0mL:

Glucose 2.0g
L-Cysteine·HCl 0.518g
L-Glutamine 0.2g
L-Cystine 0.022g
Adenine sulfate 0.010g
Nicotinamide adenine dinucleotide 5.0mg
Cocarboxylase 2.0mg
Guanine·HCl 0.6mg
$Fe(NO_3)_3 \cdot 6H_2O$ 0.4mg
p-Aminobenzoic acid 0.26mg
Vitamin B_{12} 0.2mg
Thiamine·HCl 0.06mg

Preparation of Vitox Supplement: Add components to distilled/deionized water and bring volume to 10.0mL. Mix thoroughly. Filter sterilize.

VCN Antibiotic Solution:
Composition per 10mL:

Colistin methane sulfonate 7.5mg
Vancomycin 3.0mg
Nystatin 12,500U

Preparation of VCN Antibiotic Solution: Add components to distilled/deionized water and bring volume to 10.0mL. Mix thoroughly. Filter sterilize.

Preparation of Medium: To 730.0mL of cooled sterile GC agar base, aseptically add 250.0mL of sterile hemoglobin solution , 10.0mL of sterile Vitox supplement, and 10.0mL of VCN antibiotic solution. Mix thoroughly. Pour into sterile Petri dishes or distribute into sterile tubes.

Use: For the selective isolation and cultivation of fastidious microorganisms, especially *Neisseria gonorrhoeae* and related species.

Thayer–Martin Selective Agar

Composition per liter:

Agar 12.0g
Hemoglobin 10.0g
Pancreatic digest of casein 7.5g
Selected meat peptone 7.5g
NaCl 5.0g
K_2HPO_4 .... 4.0g
Cornstarch 1.0g
KH_2PO_4 .... 1.0g
Supplement solution 10.0mL
V-C-N inhibitor 10.0mL

pH 7.2 ± 0.2 at 25°C

Source: This medium is available as a premixed powder from BBL Microbiology Systems.

Supplement Solution:
Composition per liter:

Glucose 100.0g
L-Cysteine·HCl 25.9g
L-Glutamine 10.0g
L-Cystine 1.1g
Adenine 1.0g

Nicotinamide adenine dinucleotide	0.25g
Vitamin B_{12}	0.1g
Thiamine pyrophosphate	0.1g
Guanine·HCl	0.03g
$Fe(NO_3)_3 \cdot 6H_2O$	0.02g
p-Aminobenzoic acid	0.013g
Thiamine·HCl	3.0mg

Source: The supplement solution IsoVitaleX® enrichment is available from BBL Microbiology Laboratories. This enrichment may be replaced by supplement VX from Difco Laboratories.

Preparation of Supplement Solution: Add components to distilled/deionized water and bring volume to 1.0L. Mix thoroughly. Filter sterilize.

VCN Inhibitor:

Composition per 10mL:

Colistin	7.5mg
Vancomycin	3.0mg
Nystatin	12,500U

Preparation of VCN Inhibitor: Add components to distilled/deionized water and bring volume to 10.0mL. Mix thoroughly. Filter sterilize.

Preparation of Medium: Add components, except supplement solution and V-C-N inhibitor, to distilled/deionized water and bring volume to 980.0mL. Mix thoroughly. Gently heat and bring to boiling. Autoclave for 15 min at 15 psi pressure–121°C. Cool to 45° to 50°C. Aseptically add sterile V-C-N inhibitor and sterile supplement solution. Mix thoroughly. Pour into sterile Petri dishes or distribute into sterile tubes.

Use: For the selective isolation of *Neisseria gonorrhoeae* and *Neisseria meningitidis* from specimens containing mixed flora of bacteria and fungi.

Thiogel® Medium

Composition per liter:

Gelatin	50.0g
Pancreatic digest of casein	17.0g
Glucose	6.0g
Papaic digest of soybean meal	3.0g
NaCl	2.5g
Sodium thioglycollate	0.5g
Agar	0.7g
Na_2SO_3	0.1g
L-Cystine	0.25g

pH 7.0 ± 0.2 at 25°C

Source: This medium is available as a premixed powder from BBL Microbiology Systems.

Preparation of Medium: Add components to distilled/deionized water preheated to 50°C and bring volume to 1.0L. Mix thoroughly. Let stand for 5 min. Gently heat while stirring and bring to boiling. Distribute into tubes filling them half full. Autoclave for 15 min at 13 psi pressure–118°C. Pour into sterile Petri dishes or leave in tubes.

Use: For the differentiation of microorganisms based on their ability to liquefy gelatin.

Thioglycollate Bile Broth

Composition per 1050mL:

Pancreatic digest of casein	15.0g
Glucose	5.5g
Yeast extract	5.0g
NaCl	2.5g
Agar	0.75g
L-Cystine	0.5g
Sodium thioglycollate	0.5g
Bile solution	50.0mL

pH 7.1 ± 0.2 at 25°C

Bile Solution:

Composition per 100mL:

Oxgall	40.0g
Sodium deoxycholate	2.0g

Preparation of Bile Solution: Add components to distilled/deionized water and bring volume to 100.0mL. Mix thoroughly. Filter sterilize.

Preparation of Medium: Add components, except bile solution, to distilled/deionized water and bring volume to 1.0L. Mix thoroughly. Gently heat and bring to boiling. Distribute into tubes in 10.0mL volumes. Autoclave for 15 min at 15 psi pressure–121°C. Cool to 45° to 50°C. Aseptically add 0.5mL of sterile bile solution to each tube. Mix thoroughly.

Use: For the cultivation of *Bacteroides fragilis* and *Clostridium perfringens* from clinical specimens.

Thioglycollate Broth USP, Alternative

Composition per liter:

Pancreatic digest of casein	15.0g
Glucose	5.5g
Yeast extract	5.0g
NaCl	2.5g
L-Cystine	0.5g
Sodium thioglycollate	0.5g

pH 7.1 ± 0.2 at 25°C

Source: This medium is available as a premixed powder from Oxoid Unipath.

Preparation of Medium: Add components to distilled/deionized water and bring volume to 1.0L. Mix

thoroughly. Distribute into tubes or flasks. Autoclave for 15 min at 15 psi pressure–121°C. Prepare freshly or boil and cool the medium just before use.

Use: For the cultivation of both aerobic and anaerobic organisms in the performance of sterility tests of turbid or viscous specimens.

Thioglycollate Gelatin Medium

Composition per liter:

Gelatin....................50.0g
Pancreatic digest of casein....................15.0g
Yeast extract....................5.0g
NaCl....................2.5g
Glucose....................2.0g
Agar....................0.75g
L-Cystine....................0.25g
Na_2SO_3....................0.1g
Thioglycollic acid....................0.3mL
pH 7.0 ± 0.2 at 25°C

Source: This medium is available as a premixed powder from Difco Laboratories.

Preparation of Medium: Add components to distilled/deionized water and bring volume to 1.0L. Mix thoroughly. Gently heat and bring to 50°C. Let stand 5 min. Gently heat and bring to boiling. Distribute into tubes or flasks. Autoclave for 15 min at 15 psi pressure–121°C.

Use: For the determination of gelatin liquefaction by aerobes, microaerophiles and anaerobes without special incubation.

Thioglycollate Medium, Enriched (THIO Medium) (Thioglycollate Medium with Vitamin K_1 and Hemin)

Composition per liter:

Thioglycollate medium without indicator....................1.0L
Hemin solution....................0.5mL
Vitamin K_1 solution....................0.1mL
pH 7.0 ± 0.2 at 25°C

Thioglycollate Medium without Indicator:
Composition per liter:

Pancreatic digest of casein....................17.0g
Glucose....................6.0g
Papaic digest of soybean meal....................3.0g
NaCl....................2.5g
Agar....................0.7g
Sodium thioglycollate....................0.5g
L-Cystine....................0.25g
Na_2SO_3....................0.1g

Source: Thioglycollate Medium without Indicator is available as a premixed powder from Oxoid Unipath and BBL Microbiology Systems.

Preparation of Thioglycollate Medium without Indicator: Add components to distilled/deionized water and bring volume to 1.0L. Mix thoroughly.

Vitamin K_1 Solution:
Composition per 100mL:

Vitamin K_1....................1.0g

Preparation of Vitamin K_1 Solution: Add vitamin K_1 to 99.0mL of absolute ethanol. Mix thoroughly.

Hemin Solution:
Composition per 100mL:

Hemin....................1.0g
NaOH (1*N* solution)....................20.0mL

Preparation of Hemin Solution: Add hemin to 20.0mL of 1*N* NaOH solution. Mix thoroughly. Bring volume to 100.0mL with distilled/deionized water.

Preparation of Medium: Add 0.5mL of hemin solution and 0.1mL of vitamin K_1 solution to 1.0L of thioglycollate medium without indicator. Mix thoroughly. Distribute into screw-capped tubes or flasks. Autoclave for 15 min at 15 psi pressure–121°C. Cool tubes or flasks under 85% N_2 + 10% H_2 + 5% CO_2. Tighten caps.

Use: For the isolation, cultivation and identification of a wide variety of obligate anaerobic bacteria.

Thioglycollate Medium, USP

Composition per liter:

Pancreatic digest of casein....................15.0g
Glucose....................5.5g
Yeast extract....................5.0g
NaCl....................2.5g
Agar....................0.5g
L-cystine....................0.5g
Sodium thioglycollate....................0.5g
Resazurin....................1.0mg
pH 7.1 ± 0.2 at 25°C

Source: This medium is available as a premixed powder from Oxoid Unipath.

Preparation of Medium: Add components to distilled/deionized water and bring volume to 1.0L. Mix thoroughly. Gently heat and bring to boiling. Distribute into tubes or flasks. Autoclave for 15 min at 15 psi pressure–121°C.

Use: For the cultivation of both aerobic and anaerobic organisms in the performance of sterility tests.

Thioglycollate Medium with 20% Bile (THIO + Bile Medium)

Composition per liter:

Oxgall	20.0g
Thioglycollate medium without indicator	1.0L
Hemin solution	0.5mL
Vitamin K_1 solution	0.1mL

pH 7.0 ± 0.2 at 25°C

Thioglycollate Medium without Indicator:
Composition per liter:

Pancreatic digest of casein	17.0g
Glucose	6.0g
Papaic digest of soybean meal	3.0g
NaCl	2.5g
Agar	0.7g
Sodium thioglycollate	0.5g
L-Cystine	0.25g
Na_2SO_3	0.1g

Preparation of Thioglycollate Medium Without Indicator: Add components to distilled/deionized water and bring volume to 1.0L. Mix thoroughly.

Vitamin K_1 Solution:
Composition per 100mL:

Vitamin K_1	1.0g

Preparation of Vitamin K_1 Solution: Add vitamin K_1 to 99.0mL of absolute ethanol. Mix thoroughly.

Hemin Solution:
Composition per 100mL:

Hemin	1.0g
NaOH (1*N* solution)	20.0mL

Preparation of Hemin Solution: Add hemin to 20.0mL of 1*N* NaOH solution. Mix thoroughly. Bring volume to 100.0mL with distilled/deionized water.

Preparation of Medium: Add 0.5mL of hemin solution, 0.1mL of vitamin K_1 solution, and oxgall to 1.0L of Thioglycolate Medium without Indicator. Mix thoroughly. Distribute into screw-capped tubes or flasks. Autoclave for 15 min at 15 psi pressure–121°C. Cool tubes or flasks under 85% N_2 + 10% H_2 + 5% CO_2. Tighten caps.

Use: For the isolation, cultivation and identification of a variety of obligate anaerobic bile-tolerant bacteria.

Thioglycollate Medium without Glucose

Composition per liter:

Pancreatic digest of casein	15.0g
Yeast extract	5.0g
NaCl	2.5g
Agar	0.75g
L-Cystine	0.25g
Methylene Blue	2.0mg
Thioglycollic acid	0.3mL

pH 7.2 ± 0.2 at 25°C

Source: This medium is available as a premixed powder from Difco Laboratories.

Preparation of Medium: Add components to distilled/deionized water and bring volume to 1.0L. Mix thoroughly. Gently heat and bring to boiling. Distribute into tubes or flasks. Autoclave for 15 min at 15 psi pressure–121°C. If medium becomes oxidized before use (Methylene Blue turns blue) heat in a boiling water bath to expel absorbed O_2. Cool to 25°C.

Use: For the cultivation of anaerobic, microaerophilic and aerobic microorganisms. For use in sterility testing of a variety of specimens.

Thioglycollate Medium without Glucose and Indicator

Composition per liter:

Pancreatic digest of casein	15.0g
Yeast extract	5.0g
NaCl	2.5g
Agar	0.75g
L-Cystine	0.25g
Thioglycollic acid	0.3mL

pH 7.2 ± 0.2 at 25°C

Source: This medium is available as a premixed powder from Difco Laboratories.

Preparation of Medium: Add components to distilled/deionized water and bring volume to 1.0L. Mix thoroughly. Gently heat and bring to boiling. Distribute into tubes or flasks. Autoclave for 15 min at 15 psi pressure–121°C. If medium becomes oxidized before use, heat in a boiling water bath to expel absorbed O_2. Cool to 25°C.

Use: For the cultivation of anaerobic, microaerophilic and aerobic microorganisms. For use in sterility testing of a variety of specimens.

Thioglycollate Medium without Indicator

Composition per liter:

Pancreatic digest of casein	15.0g
Yeast extract	5.0g
Glucose	5.0g
NaCl	2.5g

Agar....0.75g
Sodium thioglycollate....0.5g
L-Cystine....0.25g
pH 7.2 ± 0.2 at 25°C

Source: This medium is available as a premixed powder from Difco Laboratories.

Preparation of Medium: Add components to distilled/deionized water and bring volume to 1.0L. Mix thoroughly. Gently heat and bring to boiling. Distribute into tubes or flasks. Autoclave for 15 min at 15 psi pressure–121°C. If medium becomes oxidized before use, heat in a boiling water bath to expel absorbed O_2. Cool to 25°C.

Use: For the cultivation of anaerobic, microaerophilic and aerobic microorganisms. For use in sterility testing of a variety of specimens.

Thioglycollate Medium without Indicator

Composition per liter:
Pancreatic digest of casein....17.0g
Glucose....6.0g
Papaic digest of soybean meal....3.0g
NaCl....2.5g
Agar....0.7g
Sodium thioglycollate....0.5g
L-Cystine....0.25g
Na_2SO_3....0.1g
pH 7.0 ± 0.2 at 25°C

Source: This medium is available as a premixed powder from Oxoid Unipath.

Preparation of Medium: Add components to distilled/deionized water and bring volume to 1.0L. Mix thoroughly. Distribute into tubes or flasks. Autoclave for 15 min at 15 psi pressure–121°C. Prepare freshly or boil and cool the medium just before use.

Use: For the growth of aerobic and anaerobic microorganisms in diagnostic bacteriology.

Thioglycollate Medium without Indicator–135C

Composition per liter:
Pancreatic digest of casein....17.0g
Glucose....6.0g
Papaic digest of soybean meal....3.0g
NaCl....2.5g
Agar....0.7g
Sodium thioglycollate....0.5g
Na_2SO_3....0.1g
L-Cystine....0.25g
pH 7.0 ± 0.2 at 25°C

Source: This medium is available as a premixed powder from BBL Microbiology Systems.

Preparation of Medium: Add components to distilled/deionized water and bring volume to 1.0L. Mix thoroughly. Gently heat while stirring and bring to boiling. Distribute into tubes or flasks filling them half full. For maintenance of cultures, a small quantity of $CaCO_3$ may be added to tubes before adding medium. Autoclave for 15 min at 13 psi pressure–118°C. Prepare freshly or boil and cool the medium just before use. Store prepared medium at 2° to 8°C in the dark.

Use: For the isolation and cultivation of a wide variety of microorganisms, particularly obligate anaerobes, from clinical specimens and other materials.

Thioglycollate Medium without Indicator with Hemin (Thioglycollate Medium, Supplemented)

Composition per liter:
Pancreatic digest of casein....17.0g
$CaCO_3$, chips or powder....10.0g
Glucose....6.0g
Papaic digest of soybean meal....3.0g
NaCl....2.50g
Agar....0.70g
Sodium thioglycollate....0.50g
L-Cystine....0.25g
Na_2SO_3....0.10g
Hemin....5.0mg
Na_2CO_3 solution....10.0mL
Vitamin K_1 solution....10.0mL
pH 7.2 ± 0.2 at 25°C

Na_2CO_3 Solution:
Composition per 10mL:
Na_2CO_3....1.0g

Preparation of Na_2CO_3 Solution: Add Na_2CO_3 to distilled/deionized water and bring volume to 10.0mL. Mix thoroughly. Filter sterilize.

Vitamin K_1 Solution:
Composition per 100mL:
Vitamin K_1....1.0g
Ethanol, absolute....99.0mL

Preparation of Vitamin K_1 Solution: Add vitamin K_1 to 99.0mL of absolute ethanol. Mix thoroughly.

Preparation of Medium: Add components, except $CaCO_3$, Na_2CO_3 solution and vitamin K_1 solution, to distilled/deionized water and bring volume to 990.0mL. Mix thoroughly. Gently heat and bring to boiling. Add 0.1g of $CaCO_3$ chips or powder to each of 100 test tubes. Distribute broth into the same tubes in 10.0mL volumes. Autoclave for 15 min at 15 psi pressure–121°C. Cool to 45° to 50°C. Aseptically add 0.1mL of sterile Na_2CO_3 solution and 0.1mL of sterile vitamin K_1 solution to each tube. Mix thoroughly.

Use: For the cultivation of a wide variety of obligate anaerobes.

Thiol Broth

Composition per liter:

Proteose peptone, No. 3	10.0g
Thiol complex (Difco)	8.0g
Yeast extract	5.0g
NaCl	5.0g
Glucose	1.0g
p-Aminobenzoic acid	0.05g

pH 7.1 ± 0.2 at 25°C

Source: This medium is available as a premixed powder from Difco Laboratories.

Preparation of Medium: Add components to distilled/deionized water and bring volume to 1.0L. Mix thoroughly. Gently heat and bring to boiling. Distribute into tubes or flasks. For neutralization of penicillin, distribute medium into tubes to a depth of 60 mm. For neutralization of streptomycin, distribute medium into tubes in shallow layers. Autoclave for 15 min at 15 psi pressure–121°C.

Use: For the cultivation of bacteria from body fluids and other materials containing penicillin, streptomycin or sulfonamides. Also used for the cultivation and maintenance of *Bifidobacterium* species.

Thiol Medium

Composition per liter:

Proteose peptone, No. 3	10.0g
Thiol complex	8.0g
Yeast extract	5.0g
NaCl	5.0g
Glucose	1.0g
Agar	1.0g
p-Aminobenzoic acid	0.05g

pH 7.1 ± 0.2 at 25°C

Source: This medium is available as a premixed powder from Difco Laboratories.

Preparation of Medium: Add components to distilled/deionized water and bring volume to 1.0L. Mix thoroughly. Gently heat and bring to boiling. Distribute into tubes or flasks. For neutralization of penicillin, distribute medium into tubes to a depth of 60 mm. For neutralization of streptomycin, distribute medium into tubes in shallow layers. Autoclave for 15 min at 15 psi pressure–121°C.

Use: For the cultivation of bacteria from body fluids and other materials containing penicillin, streptomycin or sulfonamides. Also used for the cultivation and maintenance of *Bifidobacterium* species.

Tinsdale Agar

Composition per 1100mL:

Proteose peptone	20.0g
Agar	15.0g
NaCl	5.0g
Yeast extract	5.0g
L-Cystine	0.24g
Tinsdale supplement	150.0mL

pH 7.4 ± 0.2 at 25°C

Source: This medium is available as a premixed powder from Difco Laboratories and Oxoid Unipath.

Tinsdale Supplement:

Composition per100mL:

$Na_2S_2O_3$	0.43g
K_2TeO_3	0.35g
Serum	100.0mL

Preparation of Tinsdale Supplement: Add $Na_2S_2O_3$ and K_2TeO_3 to serum. Mix thoroughly. Filter sterilize.

Caution: Potassium tellurite is toxic.

Preparation of Medium: Add components, except Tinsdale supplement, to distilled/deionized water and bring volume to 1.0L. Mix thoroughly. Gently heat and bring to boiling. Autoclave for 15 min at 15 psi pressure–121°C. Cool to 50° to 55°C. Aseptically add 100.0mL of sterile Tinsdale supplement. Mix thoroughly. Pour into sterile Petri dishes or distribute into sterile tubes.

Use: For the primary isolation and identification of *Corynebacterium diphtheriae.*

Tissue Culture Amino Acids, HeLa 100X (TC Amino Acids, HeLa 100X)

Composition per liter:

L-Lysine	0.029g
L-Isoleucine	0.026g

L-Leucine....0.026g
L-Threonine....0.023g
L-Valine....0.023g
L-Tyrosine....0.018g
L-Arginine....0.017g
L-Phenylalanine....0.016g
L-Cystine....0.012g
L-Histidine....7.8mg
L-Methionine....7.5mg
L-Tryptophan....4.1mg
pH 7.2–7.4 at 25°C

Preparation of Tissue Culture Amino Acids, HeLa 100X: Add components to distilled/deionized water and bring volume to 1.0L. Mix thoroughly. Adjust pH to 7.2–7.4. Filter sterilize.

Use: For the preparation of Eagle HeLa Medium for tissue culture procedures and virus studies.

Tissue Culture Amino Acids, Minimal Eagle 50X (TC Amino Acids, Minimal Eagle 50X)

Composition per liter:

L-Arginine....0.1g
L-Lysine....0.058g
L-Isoleucine....0.052g
L-Leucine....0.052g
L-Threonine....0.048g
L-Valine....0.046g
L-Tyrosine....0.036g
L-Phenylalanine....0.032g
L-Histidine....0.031g
L-Cystine....0.024g
L-Methionine....0.015g
L-Tryptophan....0.010g
pH 7.2–7.4 at 25°C

Preparation of Tissue Culture Amino Acids, Minimal Eagle 50X: Add components to distilled/deionized water and bring volume to 1.0L. Mix thoroughly. Adjust pH to 7.2–7.4. Filter sterilize.

Use: For the preparation of TC Minimal Medium Eagle for tissue culture procedures and virus studies.

Tissue Culture Dulbecco Solution (TC Dulbecco Solution)

Composition per liter:

NaCl....8.0g
Na_2HPO_4....1.15g
KH_2PO_4....0.2g
KCl....0.2g
$CaCl_2 \cdot 2H_2O$....0.1g
$MgCl_2 \cdot 6H_2O$....0.1g
pH 7.2–7.4 at 25°C

Preparation of Tissue Culture Dulbecco Solution: Add components to distilled/deionized water and bring volume to 1.0L. Mix thoroughly. Adjust pH to 7.2–7.4. Filter sterilize.

Use: For use in tissue culture and virus preparations.

Tissue Culture Earle Solution (TC Earle Solution)

Composition per 1002mL:

NaCl....6.8g
$NaHCO_3$....2.2g
Glucose....1.0g
KCl....0.4g
$CaCl_2 \cdot 2H_2O$....0.2g
NaH_2PO_4....0.125g
$MgSO_4 \cdot 7H_2O$....0.1g
Phenol Red (1% solution)....2.0mL
pH 7.2–7.4 at 25°C

Preparation of Tissue Culture Earle Solution: Add components, except phenol red, to distilled/deionized water and bring volume to 1.0L. Mix thoroughly. Add 2.0mL of phenol red solution. Adjust pH to 7.2–7.4. Filter sterilize.

Use: For use in tissue culture and virus preparations.

Tissue Culture Hanks Solution (TC Hanks Solution)

Composition per liter:

NaCl....8.0g
Glucose....1.0g
KCl....0.4g
$NaHCO_3$....0.35g
$CaCl_2 \cdot 2H_2O$....0.14g
$MgCl_2 \cdot 6H_2O$....0.1g
$MgSO_4 \cdot 7H_2O$....0.1g
Na_2HPO_4....0.06g
KH_2PO_4....0.06g
Phenol Red....0.02g
pH 7.2–7.4 at 25°C

Source: This medium is available as a premixed solution from Difco Laboratories.

Preparation of Tissue Culture Hanks Solution: Add components to distilled/deionized water and bring volume to 1.0L. Mix thoroughly. Adjust pH to 7.2–7.4. Filter sterilize.

Use: For use in tissue culture procedures.

Tissue Culture Medium 199 (TC Medium 199)

Composition per 1050mL:

NaCl 8.0g
Glucose 1.0g
KCl 0.4g
$NaHCO_3$ 0.35g
DL-Glutamic acid 0.15g
$CaCl_2 \cdot 2H_2O$ 0.14g
DL-Leucine 0.12g
L-Glutamine 0.1g
$MgSO_4 \cdot 7H_2O$ 0.1g
L-Arginine 0.07g
L-Lysine 0.07g
DL-Aspartic acid 0.06g
Na_2HPO_4 .... 0.06g
KH_2PO_4 .... 0.06g
DL-Threonine 0.06g
DL-Alanine 0.05g
Glycine 0.05g
DL-Phenylalanine 0.05g
DL-Serine 0.05g
Sodium acetate 0.05g
DL-Valine 0.05g
DL-Isoleucine 0.04g
L-Proline 0.04g
L-Tyrosine 0.04g
DL-Methionine 0.03g
L-Cystine 0.02g
L-Histidine 0.02g
Phenol Red 0.02g
DL-Tryptophan 0.02g
Adenine 0.01g
L-Hydroxyproline 0.01g
Tween™ 80 .... 5.0mg
Adenosine triphosphate 1.0mg
Choline 0.5mg
Deoxyribose 0.5mg
Ribose 0.5mg
Guanine 0.3mg
Hypoxanthine 0.3mg
Thymine 0.3mg
Uracil 0.3mg
Xanthine 0.3mg
Adenylic acid 0.2mg
Cholesterol 0.2mg
Calciferol 0.1mg
$Fe(NO_3)_3 \cdot 9H_2O$ 0.1mg
L-Cysteine 0.1mg
Vitamin A 0.1mg
α-Tocopherol phosphate 0.01mg
Biotin 0.01mg
Calcium pantothenate 0.01mg
Folic acid 0.01mg
Menadione 0.01mg
Riboflavin 0.01mg
Thiamine·HCl 0.01mg
p-Aminobenzoic acid 0.05mg
Ascorbic acid 0.05mg
Glutathione 0.05mg
Inositol 0.05mg
Niacin 0.025mg
Niacinamide 0.025mg
Pyridoxine·HCl 0.025mg
Pyridoxal·HCl 0.025mg
Serum 50.0–100.0mL

pH 7.2–7.4 at 25°C

Preparation of Medium: Add components, except serum, to distilled/deionized water and bring volume to 1.0L. Mix thoroughly. Adjust pH to 7.2–7.4 with 10% Na_2CO_3 solution. Filter sterilize. Aseptically add 50.0–100.0mL of sterile serum. Human serum, bovine serum, horse serum or fetal calf serum may be used. Mix thoroughly. If desired antibacterial inhibitors may be added. Aseptically add 500,000U penicillin and 0.5g streptomycin to 1050.0mL of the complete medium.

Use: For cultivation of a wide variety of cell lines in tissue culture. It is especially useful for detection, titering and identification of viruses in tissue culture cells.

Tissue Culture Medium Eagle, HeLa (TC Medium Eagle, HeLa)

Composition per 1056mL:

NaCl 5.85g
$NaHCO_3$ 1.68g
Glucose 0.9g
KCl 0.373g
NaH_2PO_4 .... 0.138g
$MgCl_2 \cdot 6H_2O$ 0.120g
$CaCl_2 \cdot 2H_2O$ 0.11g
L-Lysine 0.0269g
L-Isoleucine 0.0262g
L-Leucine 0.0262g
L-Threonine 0.0238g
L-Valine 0.0234g
L-Tyrosine 0.0181g
L-Arginine 0.0174g
L-Phenylalanine 0.0165g
L-Cystine 0.012g
L-Histidine 7.8mg
L-Methionine 7.5mg
Phenol Red 5.0mg
L-Tryptophan 4.1mg
Folic acid 0.44mg
Thiamine·HCl 0.34mg

Biotin 0.24mg
Pantothenic acid 0.22mg
Pyridoxal·HCl 0.2mg
Choline chloride 0.14mg
Nicotinamide 0.12mg
Riboflavin 0.04mg
Serum 50.0mL–100.0mL
Glutamine solution 6.0mL
pH 7.2–7.4 at 25°C

Glutamine Solution:
Composition per 100mL:
L-Glutamine 5.0g
NaCl (0.85% solution) 100.0mL

Preparation of Glutamine Solution: Add the glutamine to the 0.85% NaCl solution. Mix thoroughly. Filter sterilize.

Preparation of Medium: Add components, except glutamine and serum, to distilled/deionized water and bring volume to 1.0L. Mix thoroughly. Adjust pH to 7.2–7.4. Filter sterilize. Aseptically add 6.0mL of sterile glutamine solution and 50.0–100.0mL of sterile serum. Human serum, bovine serum, horse serum or fetal calf serum may be used. Mix thoroughly.

Use: For the cultivation and maintenance of HeLa and other cell lines in tissue culture, and for studying the cytopathogenicity of viral agents.

Tissue Culture Medium Eagle with Earle Balanced Salt Solution (TC Medium Eagle with Earle BSS)

Composition per 1056mL:
NaCl 6.8g
$NaHCO_3$ 2.2g
Glucose 1.0g
KCl 0.4g
$CaCl_2 \cdot 2H_2O$ 0.2g
NaH_2PO_4 .... 0.125g
$MgSO_4 \cdot 7H_2O$ 0.1g
L-Isoleucine 0.026g
L-Leucine 0.026g
L-Lysine 0.026g
L-Threonine 0.024g
L-Valine 0.0235g
L-Tyrosine 0.018g
L-Arginine 0.0174g
L-Phenylalanine 0.0165g
L-Cystine 0.012g
L-Histidine 8.0mg
L-Methionine 7.5mg
Phenol Red 5.0mg
L-Tryptophan 4.0mg
Inositol 1.8mg
Biotin 1.0mg
Calcium pantothenate 1.0mg
Choline chloride 1.0mg
Folic acid 1.0mg
Nicotinamide 1.0mg
Pyridoxal·HCl 1.0mg
Thiamine·HCl 1.0mg
Riboflavin 0.1mg
Serum 50.0mL–100.0mL
Glutamine solution 6.0mL
pH 7.2–7.4 at 25°C

Glutamine Solution:
Composition per 100mL:
L-Glutamine 5.0g
NaCl (0.85% solution) 100.0mL

Preparation of Glutamine Solution: Add the glutamine to the 0.85% NaCl solution. Mix thoroughly. Filter sterilize.

Preparation of Medium: Add components, except glutamine and serum, to distilled/deionized water and bring volume to 1.0L. Mix thoroughly. Adjust pH to 7.2–7.4. Filter sterilize. Aseptically add 6.0mL of sterile glutamine solution and 50.0–100.0mL of sterile serum. Human serum, bovine serum, horse serum or fetal calf serum may be used. Mix thoroughly.

Use: For the cultivation of HeLa, KB and other tissue culture cell lines.

Tissue Culture Medium Eagle with Hanks Balanced Salt Solution (TC Medium Eagle with Hanks BSS)

Composition per 1056mL:
NaCl 8.0g
Glucose 1.0g
KCl 0.4g
$CaCl_2 \cdot 2H_2O$ 0.14g
$MgSO_4 \cdot 7H_2O$ 0.1g
KH_2PO_4 .... 0.06g
Na_2HPO_4 .... 0.05g
L-Isoleucine 0.026g
L-Leucine 0.026g
L-Lysine 0.026g
L-Threonine 0.024g
L-Valine 0.0235g
L-Tyrosine 0.018g
L-Arginine 0.0174g
L-Phenylalanine 0.0165g
L-Cystine 0.012g

L-Histidine....................8.0mg
L-Methionine....................7.5mg
Phenol Red....................5.0mg
L-Tryptophan....................4.0mg
Inositol....................1.8mg
Biotin....................1.0mg
Folic acid....................1.0mg
Calcium pantothenate....................1.0mg
Choline chloride....................1.0mg
Nicotinamide....................1.0mg
Pyridoxal·HCl....................1.0mg
Thiamine·HCl....................1.0mg
Riboflavin....................0.1mg
Serum....................50.0mL–100.0mL
Glutamine solution....................6.0mL

pH 7.2–7.4 at 25°C

Glutamine Solution:

Composition per 100mL:

L-Glutamine....................5.0g
NaCl (0.85% solution)....................100.0mL

Preparation of Glutamine Solution: Add the glutamine to the 0.85% NaCl solution. Mix thoroughly. Filter sterilize.

Preparation of Medium: Add components, except glutamine and serum, to distilled/deionized water and bring volume to 1.0L. Mix thoroughly. Adjust pH to 7.2–7.4. Filter sterilize. Aseptically add 6.0mL of sterile glutamine solution and 50.0–100.0mL of sterile serum. Human serum, bovine serum, horse serum or fetal calf serum may be used. Mix thoroughly.

Use: For use as a base in the preparation of liquid media used for the cultivation of tissue culture cell lines.

Tissue Culture Medium Ham F10 (TC Medium Ham F10)

Composition per 1050mL:

NaCl....................7.4g
Glucose....................1.1g
Na_2HPO_4....................0.290g
KCl....................0.285g
L-Arginine....................0.211g
$MgSO_4 \cdot 7H_2O$....................0.153g
L-Glutamine....................0.1462g
Sodium pyruvate....................0.110g
KH_2PO_4....................0.083g
$CaCl_2 \cdot 2H_2O$....................0.044g
L-Cystine....................0.0315g
L-Lysine....................0.0293g
L-Histidine....................0.021g
L-Asparagine....................0.015g
L-Glutamic acid....................0.0147g
L-Aspartic acid....................0.0133g
L-Leucine....................0.0131g
L-Proline....................0.0115g
L-Serine....................0.0105g
L-Alanine....................8.91mg
Glycine....................7.51mg
L-Phenylalanine....................4.96mg
L-Methionine....................4.48mg
Hypoxanthine....................4.0mg
L-Threonine....................3.57mg
L-Valine....................3.5mg
L-Isoleucine....................2.6mg
L-Tyrosine....................1.81mg
Cyanocobalamin....................1.3mg
Folic acid....................1.3mg
Phenol Red....................1.2mg
Thiamine·HCl....................1.0mg
$FeSO_4 \cdot 7H_2O$....................0.83mg
Calcium pantothenate....................0.7mg
Thymidine....................0.7mg
Choline chloride....................0.69mg
Niacinamide....................0.6mg
L-Tryptophan....................0.6mg
i-Inositol....................0.54mg
Riboflavin....................0.37mg
Lipoic acid....................0.2mg
Pyridoxine·HCl....................0.2mg
$ZnSO_4 \cdot 7H_2O$....................0.028mg
Biotin....................0.024mg
$CuSO_4 \cdot 5H_2O$....................2.5µg
Fetal calf serum....................50.0–100.0mL

pH 7.2–7.4 at 25°C

Preparation of Medium: Add components, except fetal calf serum, to distilled/deionized water and bring volume to 1.0L. Mix thoroughly. Adjust pH to 7.2–7.4 with 10% Na_2CO_3 solution. Filter sterilize. Aseptically add 50.0–100.0mL of sterile fetal calf serum. Mix thoroughly.

Use: For cultivation of a wide variety of cell lines in tissue culture.

Tissue Culture Medium NCTC 109 (TC Medium NCTC 109)

Composition per 1050mL:

NaCl....................6.8g
$NaHCO_3$....................2.2g
Glucose....................1.0g
KCl....................0.4g
L-Cysteine....................0.26g
$CaCl_2 \cdot 2H_2O$....................0.20g
NaH_2PO_4....................0.14g
L-Glutamine....................0.14g

$MgSO_4 \cdot 7H_2O$ 0.10g
Sodium acetate 0.05g
Ascorbic acid 0.05g
L-Alanine 0.03g
L-Lysine 0.03g
L-Arginine 0.026g
L-Valine 0.025g
L-Leucine 0.020g
Phenol Red 0.020g
L-Histidine 0.019g
L-Threonine 0.019g
L-Isoleucine 0.018g
L-Tryptophan 0.017.g
L-Phenylalanine 0.017g
L-Tyrosine 0.016g
Glycine 0.014g
Tween™ 80 .... 0.012g
L-Serine 0.011g
L-Cystine 0.010g
Glutathione 0.010g
Cyanocobalamin 0.010g
Deoxycytidine 0.010g
Deoxyguanosine 0.010g
Deoxyadenosine 0.010g
Thymidine 0.010g
L-Aspartic acid 9.91mg
L-Glutamic acid 8.26mg
L-Arginine 8.09mg
L-Ornithine 7.38mg
Nicotinamide adenine dinucleotide 7.0mg
L-Proline 6.13mg
L-α-N-Butyric acid 5.51mg
L-Methionine 4.44mg
L-Taurine 4.18mg
L-Hydroxyproline 4.09mg
D-Glucosamine 3.2mg
Coenzyme A 2.5mg
Glucuronolactone 1.8mg
Sodium glucuronate 1.8mg
Choline chloride 1.25mg
Cocarboxylase 1.0mg
Flavin adenine dinucleotide 1.0mg
Uridine triphosphate 1.0mg
Nicotinamide adenine dinucleotide phosphate 1.0mg
Vitamin A 0.25mg
Calciferol 0.25mg
i-Inositol 0.125mg
p-Aminobenzoic acid 0.125mg
5-Methylcytosine 0.1mg
Pyridoxine·HCl 0.0625mg
Pyridoxal·HCl 0.0625mg
Niacin 0.0625mg
Niacinamide 0.0625mg
Biotin 0.025mg
Folic acid 0.025mg
Menadione 0.025mg
Pantothenate 0.025mg
Riboflavin 0.025mg
Thiamine·HCl 0.025mg
α-Tocopherol phosphate 0.025mg
Serum 50.0–100.0mL

pH 7.2–7.4 at 25°C

Preparation of Medium: Add components, except serum, to distilled/deionized water and bring volume to 1.0L. Mix thoroughly. Adjust pH to 7.2–7.4 with 10% Na_2CO_3 solution. Filter sterilize. Aseptically add 50.0–100.0mL of sterile serum. Human serum, bovine serum, horse serum or fetal calf serum may be used. Mix thoroughly.

Use: For cultivation of a wide variety of cell lines in tissue culture.

Tissue Culture Medium RPMI #1640 (TC Medium RPMI #1640)

Composition per liter:

NaCl 6.46g
Glucose 2.0g
$NaHCO_3$ 2.0g
NaH_2PO_4 .... 1.512g
KCl 0.4g
L-Glutamine 0.3g
L-Arginine 0.2g
Calcium nitrate 0.1g
$MgSO_4 \cdot 7H_2O$ 0.1g
L-Asparagine 0.05g
L-Cystine 0.05g
L-Isoleucine 0.05g
L-Leucine 0.05g
L-Lysine HCl 0.04g
Inositol 0.035g
L-Serine 0.030g
Hydroxy-L-Proline 0.020g
L-Aspartic acid 0.020g
L-Glutamic acid 0.020g
L-Proline 0.020g
L-Threonine 0.020g
L-Tyrosine 0.020g
L-Valine 0.020g
L-Histidine 0.015g
L-Methionine 0.015g
L-Phenylalanine 0.015g
Glycine 0.01g
L-Tryptophan 5.0mg
Phenol Red 5.0mg
Choline chloride 3.0mg

p-Aminobenzoic acid 1.0mg
Folic acid 1.0mg
Glutathione 1.0mg
Nicotinamide 1.0mg
Pyridoxine HCl 1.0mg
Thiamine HCl 1.0mg
Calcium pantothenate 0.25mg
Biotin 0.20mg
Riboflavin 0.20mg
Vitamin B_{12} 5.0mg
Serum 50.0–100.0mL

pH 7.2–7.4 at 25°C

Source: This medium is available as a premixed powder and solution from Difco Laboratories.

Preparation of Medium: Add components, except serum, to distilled/deionized water and bring volume to 1.0L. Mix thoroughly. Adjust pH to 7.2–7.4 with 10% Na_2CO_3 solution. Filter sterilize. Aseptically add 50.0–100.0mL of sterile serum. Human serum, bovine serum, horse serum or fetal calf serum may be used. Mix thoroughly.

Use: For cultivation of a wide variety of cell lines in tissue culture.

Tissue Culture Minimal Medium Eagle

Composition per liter:

Sterile salt solution 944.0mL
TC amino acids, minimal Eagle 50X 20.0mL
TC $NaHCO_3$, 10% 20.0mL
TC vitamins minimal Eagle 100X 10.0mL
TC glutamine, 5% 6.0mL

pH 7.2–7.4 at 25°C

Sterile Salt Solution:

Composition per 944mL:

NaCl 6.8g
Glucose 1.0g
KCl 0.4g
$CaCl_2$ 0.2g
$MgCl_2$ 0.2g
NaH_2PO_4 .... 0.15g

Preparation of Sterile Salt Solution: Add components to distilled/deionized water and bring volume to 944.0mL. Mix thoroughly. Filter sterilize.

Tissue Culture Amino Acids, Minimal Eagle 50X:

Composition per liter:

L-Arginine 0.1g
L-Lysine 0.06g
L-Isoleucine 0.05g
L-Leucine 0.05g
L-Threonine 0.05g
L-Valine 0.05g
L-Tyrosine 0.04g
L-Phenylalanine 0.03g
L-Histidine 0.03g
L-Cystine 0.02g
L-Methionine 0.02g
L-Tryptophan 0.01g

Preparation of Tissue Culture Amino Acids, Minimal Eagle 50X: Add components to distilled/deionized water and bring volume to 1.0L. Mix thoroughly. Adjust pH to 7.2–7.4. Filter sterilize.

TC $NaHCO_3$, 10%:

Composition per 100mL:

$NaHCO_3$ 10.0g

Preparation of TC $NaHCO_3$, 10%: Add $NaHCO_3$ to distilled/deionized water and bring volume to 100.0mL. Mix thoroughly. Filter sterilize.

TC Vitamins, Minimal Eagle 100X:

Composition per liter:

Inositol 2.0mg
Calcium pantothenate 1.0mg
Choline chloride 1.0mg
Folic acid 1.0mg
Nicotinamide 1.0mg
Pyridoxal 1.0mg
Thiamine·HCl 1.0mg
Riboflavin 0.1mg

Preparation of TC Vitamins, Minimal Eagle 100X: Add components to distilled/deionized water and bring volume to 1.0L. Mix thoroughly. Filter sterilize.

TC Glutamine, 5%:

Composition per 100mL:

L-Glutamine 5.0g
NaCl (0.85% solution) 100.0mL

Preparation of TC Glutamine, 5%: Add the glutamine to the 0.85% NaCl solution. Mix thoroughly. Filter sterilize.

Preparation of Medium: Aseptically combine 944.0mL of sterile salt solution, 20.0mL of sterile TC amino acids, minimal Eagle 50X, 20.0mL of sterile TC $NaHCO_3$, 10%, 10.0mL of sterile TC vitamins minimal Eagle 100X, and 6.0mL of sterile TC glutamine, 5%. Mix thoroughly. Adjust pH to 7.2–7.4.

Use: For the cultivation of mammalian cells in monolayer or suspension for tissue culture procedures and virus preparation.

Tissue Culture Minimal Medium Eagle Spinner Modified (TC Minimal Medium Eagle Spinner Modified MEM–S)

Composition per 1056mL:

NaH_2PO_4	1.35g
NaCl	6.8g
$NaHCO_3$	2.2g
Glucose	1.0g
KCl	0.4g
$CaCl_2 \cdot 2H_2O$	0.2g
NaH_2PO_4	0.125g
$MgSO_4 \cdot 7H_2O$	0.1g
L-Isoleucine	0.026g
L-Leucine	0.026g
L-Lysine	0.026g
L-Threonine	0.024g
L-Valine	0.0235g
L-Tyrosine	0.018g
L-Arginine	0.0174g
L-Phenylalanine	0.0165g
L-Cystine	0.012g
L-Histidine	8.0mg
L-Methionine	7.5mg
Phenol Red	5.0mg
L-Tryptophan	4.0mg
Inositol	1.8mg
Biotin	1.0mg
Calcium pantothenate	1.0mg
Choline chloride	1.0mg
Folic acid	1.0mg
Nicotinamide	1.0mg
Pyridoxal·HCl	1.0mg
Thiamine·HCl	1.0mg
Riboflavin	0.1mg
Serum	50.0mL–100.0mL
Glutamine solution	6.0mL

pH 7.2–7.4 at 25°C

Glutamine Solution:

Composition per 100mL:

L-Glutamine	5.0g
NaCl (0.85% solution)	100.0mL

Preparation of Glutamine Solution: Add the glutamine to the 0.85% NaCl solution. Mix thoroughly. Filter sterilize.

Preparation of Medium: Add components, except glutamine and serum, to distilled/deionized water and bring volume to 1.0L. Mix thoroughly. Adjust pH to 7.2–7.4 with 10% Na_2CO_3 solution. Filter sterilize. Aseptically add 6.0mL of sterile glutamine solution and 50.0–100.0mL of sterile serum. Human serum, bovine serum, horse serum or fetal calf serum may be used. Mix thoroughly.

Use: For the cultivation of mammalian cells in suspension.

Tissue Culture Minimal Medium Eagle with Earle Balanced Salts Solution (TC Minimal Medium Eagle with Earle BSS)

Composition per 1056mL:

NaCl	6.8g
Glucose	1.0g
KCl	0.4g
$CaCl_2 \cdot 2H_2O$	0.2g
$MgCl_2 \cdot 6H_2O$	0.2g
NaH_2PO_4	0.15g
L-Arginine	0.10g
L-Lysine	0.06g
L-Isoleucine	0.05g
L-Leucine	0.05g
L-Threonine	0.05g
L-Valine	0.05g
L-Tyrosine	0.04g
L-Phenylalanine	0.03g
L-Histidine	0.03g
L-Cystine	0.02g
L-Methionine	0.02g
L-Tryptophan	0.01g
i-Inositol	2.0mg
Calcium pantothenate	1.0mg
Choline chloride	1.0mg
Folic acid	1.0mg
Nicotinamide	1.0mg
Pyridoxal	1.0mg
Thiamine·HCl	1.0mg
Riboflavin	0.1mg
Serum	50.0mL–100.0mL
Glutamine solution	6.0mL

pH 7.2–7.4 at 25°C

Glutamine Solution:

Composition per 100mL:

L-Glutamine	5.0g
NaCl (0.85% solution)	100.0mL

Preparation of Glutamine Solution: Add the glutamine to the 0.85% NaCl solution. Mix thoroughly. Filter sterilize.

Preparation of Medium: Add components, except glutamine and serum, to distilled/deionized water and bring volume to 1.0L. Mix thoroughly. Adjust pH to 7.2–7.4 with 10% Na_2CO_3 solution. Filter sterilize. Aseptically add 6.0mL of sterile glutamine solution and 50.0–100.0mL of sterile serum. Human serum, bovine serum, horse serum or fetal calf serum may be used. Mix thoroughly. To grow cells in a monolayer, aseptically add 2.0mL of a sterile 10% $CaCl_2 \cdot 2H_2O$ solution. To grow cells in suspension, omit the $CaCl_2 \cdot 2H_2O$ solution.

Use: For preparation of Eagle's Minimal Medium for the cultivation of cells in monolayer or suspension in tissue culture.

Tissue Culture Tyrode Solution (TC Tyrode Solution)

Composition per 1002mL:

NaCl	8.0g
Glucose	1.0g
$NaHCO_3$	1.0g
$CaCl_2 \cdot 2H_2O$	0.2g
KCl	0.2g
$MgCl_2 \cdot 6H_2O$	0.1g
NaH_2PO_4	0.05g
Phenol Red (1% solution)	2.0mL

pH 7.2–7.4 at 25°C

Preparation of Tissue Culture Tyrode Solution: Add components, except Phenol Red, to distilled/deionized water and bring volume to 1.0L. Mix thoroughly. Add 2.0mL of Phenol Red solution. Adjust pH to 7.2–7.4. Filter sterilize.

Use: For use in tissue culture procedures.

Tissue Culture Vitamins Minimal Eagle, 100X (TC Vitamins Minimal Eagle, 100X)

Composition per liter:

Inositol	2.0mg
Calcium pantothenate	1.0mg
Choline chloride	1.0mg
Folic acid	1.0mg
Nicotinamide	1.0mg
Pyridoxal	1.0mg
Thiamine·HCl	1.0mg
Riboflavin	0.1mg

pH 7.2–7.4 at 25°C

Preparation of TC Vitamins, Minimal Eagle 100X: Add components to distilled/deionized water and bring volume to 1.0L. Mix thoroughly. Filter sterilize.

Use: For the preparation of TC Minimal Medium Eagle used in tissue culture procedures.

TOC Agar (Tween™ 80 Oxgall Caffeic acid Agar)

Composition per liter:

Agar	20.0g
Oxgall	10.0g
Caffeic acid	0.3g
Tween™ 80	10.0mL

Source: Available as a prepared medium from BBL Microbiology Systems.

Preparation of Medium: Add components to distilled/deionized water and bring volume to 1.0L. Mix thoroughly. Gently heat and bring to boiling. Autoclave for 15 min at 15 psi pressure–121°C. Pour into sterile Petri dishes.

Use: For the differentiation and identification of *Candida albicans* and *Cryptococcus neoformans. C. albicans* produces germ tubes and chlamydospores when grown on this medium. *C. neoformans* appears as tan to brown colonies.

Todd–Hewitt Broth

Composition per liter:

Beef heart, infusion from	500.0g
Neopeptone	20.0g
Na_2CO_3	2.5g
Glucose	2.0g
NaCl	2.0g
Na_2HPO_4	0.4g

pH 7.8 ± 0.2 at 25°C

Source: This medium is available as a premixed powder from Difco Laboratories.

Preparation of Medium: Add components to distilled/deionized water and bring volume to 1.0L. Mix thoroughly. Distribute into tubes or flasks. Autoclave for 15 min at 15 psi pressure–121°C.

Use: For the cultivation of Group A streptococci used in serological typing, and for the cultivation of a variety of pathogenic microorganisms.

Todd–Hewitt Broth

Composition per liter:

Pancreatic digest of casein	20.0g
Infusion from 450g fat-free minced meat	10.0g
Glucose	2.0g
$NaHCO_3$	2.0g
NaCl	2.0g
Na_2HPO_4	0.4g

pH 7.8 ± 0.2 at 25°C

Source: This medium is available as a premixed powder from Oxoid Unipath.

Preparation of Medium: Add components to distilled/deionized water and bring volume to 1.0L. Mix thoroughly. Distribute into tubes or flasks. Autoclave for 10 min at 10 psi pressure–115°C.

Use: For the cultivation of Group A streptococci used in serological typing, and for the cultivation of a variety of pathogenic microorganisms.

Todd–Hewitt Broth (ATCC Medium 235)

Composition per liter:

Peptone....................20.0g
Beef heart, solids from infusion....................3.1g
Na_2CO_3....................2.5g
Glucose....................2.0g
NaCl....................2.0g
Na_2HPO_4....................0.4g
pH 7.8 ± 0.2 at 25°C

Source: This medium is available as a premixed powder from BBL Microbiology Systems.

Preparation of Medium: Add components to distilled/deionized water and bring volume to 1.0L. Mix thoroughly. Distribute into tubes or flasks. Autoclave for 15 min at 15 psi pressure–121°C.

Use: For the cultivation of Group A streptococci used in serological typing, and for the cultivation of a variety of pathogenic microorganisms.

Todd–Hewitt Broth, Modified

Composition per liter:

Neopeptone....................20.0g
Glucose....................2.0g
$NaHCO_3$....................2.0g
NaCl....................2.0g
Na_2HPO_4....................0.4g
Beef heart infusion....................1.0L
pH 7.8 ± 0.2 at 25°C

Preparation of Medium: Add components to distilled/deionized water and bring volume to 1.0L. Mix thoroughly. Distribute into tubes or flasks. Autoclave for 10 min at 10 psi pressure–115°C.

Use: For cultivation of streptococci for serological identification.

Toluidine Blue DNA Agar

Composition per liter:

Agar....................10.0g
NaCl....................10.0g
Tris(hydroxymethyl)aminomethane buffer....................6.1g
Deoxyribonucleic acid (DNA)....................0.3g
Toluidine Blue O....................0.083g
$CaCl_2$, anhydrous....................1.1mg
pH 7.3 ± 0.2 at 25°C

Preparation of Medium: Add components, except Toluidine Blue O, to distilled/deionized water and bring volume to 1.0L. Mix thoroughly. Gently heat and bring to boiling. Add Toluidine Blue O. Mix thoroughly. Medium does not have to be sterilized if used immediately. Pour into sterile Petri dishes or distibute into sterile tubes. Allow tubes to cool in a slanted position.

Use: For the cultivation and differentiation of bacteria based on their production of deoxyribonuclease (DNase). Bacteria that produce DNase turn the medium pink.

Transgrow Medium

Composition per liter:

GC agar base....................730.0mL
Hemoglobin solution....................250.0mL
Vitox supplement....................10.0mL
VCN antibiotic solution....................10.0mL
pH 7.3 ± 0.2 at 25°C

GC Agar Base:
Composition per 730mL:

Special peptone....................15.0g
Agar....................20.0g
NaCl....................5.0g
K_2HPO_4....................4.0g
Cornstarch....................1.0g
KH_2PO_4....................1.0g
pH 7.2 ± 0.2 at 25°C

Preparation of GC Agar Base: Add components of GC medium base and the hemoglobin to distilled/deionized water and bring volume to 730.0mL. Mix thoroughly. Gently heat until boiling. Autoclave for 15 min at 15 psi pressure–121°C. Cool to 45° to 50°C.

Hemoglobin Solution:
Composition per 250mL:

Hemoglobin....................5.0g

Preparation of Hemoglobin Solution: Add hemoglobin to distilled/deionized water and bring volume to 250.0mL. Mix thoroughly. Autoclave for 15 min at 15 psi pressure–121°C. Cool to 45° to 50°C.

Vitox Supplement:
Composition per 10.0mL:

Glucose....................2.0g
L-Cysteine·HCl....................0.518g
L-Glutamine....................0.2g

L-Cystine	0.022g
Adenine sulfate	0.010g
Nicotinamide adenine dinucleotide	5.0mg
Cocarboxylase	2.0mg
Guanine·HCl	0.6mg
$Fe(NO_3)_3 \cdot 6H_2O$	0.4mg
p-Aminobenzoic acid	0.26mg
Vitamin B_{12}	0.2mg
Thiamine·HCl	0.06mg

Preparation of Vitox Supplement: Add components to distilled/deionized water and bring volume to 10.0mL. Mix thoroughly. Filter sterilize.

VCN Antibiotic Solution:
Composition per 10mL:

Colistin methane sulfonate	7.5mg
Vancomycin	3.0mg
Nystatin	12,500U

Preparation of VCN Antibiotic Solution: Add components to distilled/deionized water and bring volume to 10.0mL. Mix thoroughly. Filter sterilize.

Preparation of Medium: To 730.0mL of cooled sterile GC agar base, aseptically add 250.0mL of sterile hemoglobin solution , 10.0mL of sterile Vitox supplement, and 10.0mL of VCN antibiotic solution. Mix thoroughly. Pour into sterile Petri dishes or distribute into sterile tubes.

Use: For the cultivation and transport of fastidious microorganisms, especially *Neisseria* species.

Transgrow Medium

Composition per liter:

GC medium base	730.0mL
Hemoglobin solution	250.0mL
Supplement B (Difco)	10.0mL
VCNT antibiotic solution	10.0mL

pH 7.3 ± 0.2 at 25°C

GC Medium Base:
Composition per 730mL:

Proteose peptone No. 3	15.0g
Agar	20.0g
NaCl	5.0g
K_2HPO_4	4.0g
Glucose	1.5g
Cornstarch	1.0g
KH_2PO_4	1.0g

pH 7.2 ± 0.2 at 25°C

Preparation of GC Medium Base: Add components to distilled/deionized water and bring volume to 730.0mL. Mix thoroughly. Gently heat until boiling. Autoclave for 15 min at 15 psi pressure–121°C. Cool to 45° to 50°C.

Hemoglobin Solution:
Composition per 250mL:

Hemoglobin	10.0g

Preparation of Hemoglobin Solution: Add hemoglobin to distilled/deionized water and bring volume to 250.0mL. Mix thoroughly. Autoclave for 15 min at 15 psi pressure–121°C. Cool to 45° to 50°C.

Supplement B:
Composition per 10mL:

Supplement B contains yeast concentrate, glutamine, coenzyme, cocarboxylase, hematin and growth factors.

Preparation of Supplement B: Add components to distilled/deionized water and bring volume to 10.0mL. Mix thoroughly. Filter sterilize.

VCNT Antibiotic Solution:
Composition per 10mL:

Colistin methane sulfonate	7.5mg
Trimethoprim lactate	5.0mg
Vancomycin	3.0mg
Nystatin	12,500U

Preparation of VCNT Antibiotic Solution: Add components to distilled/deionized water and bring volume to 10.0mL. Mix thoroughly. Filter sterilize.

Preparation of Medium: To 730.0mL of cooled sterile GC medium base, aseptically add 250.0mL of sterile hemoglobin solution , 10.0mL of sterile supplement B, and 10.0mL of VCNT antibiotic solution. Mix thoroughly. Pour into sterile Petri dishes or distribute into sterile tubes.

Use: For the cultivation and transport of fastidious microorganisms, especially *Neisseria* species.

Transgrow Medium with Trimethoprim

Composition per liter:

Agar	20.0g
Hemoglobin	10.0g
Pancreatic digest of casein	7.5g
Selected meat peptone	7.5g
NaCl	5.0g
K_2HPO_4	4.0g
Glucose	1.5g
Cornstarch	1.0g
KH_2PO_4	1.0g
Supplement solution	10.0mL
V-C-N-T inhibitor	10.0mL

pH 6.7 ± 0.2 at 25°C

Source: Available as a prepared medium from BBL Microbiology Systems.

Supplement Solution:
Composition per liter:

Glucose 100.0g
L-Cysteine·HCl 25.9g
L-Glutamine 10.0g
L-Cystine 1.1g
Adenine 1.0g
Nicotinamide adenine dinucleotide 0.25g
Vitamin B_{12} 0.1g
Thiamine pyrophosphate 0.1g
Guanine·HCl 0.03g
$Fe(NO_3)_3 \cdot 6H_2O$ 0.02g
p-Aminobenzoic acid 0.013g
Thiamine·HCl 3.0mg

Source: The supplement solution IsoVitaleX® enrichment is available from BBL Microbiology Laboratories. This enrichment may be replaced by supplement VX from Difco Laboratories.

Preparation of Supplement Solution: Add components to distilled/deionized water and bring volume to 1.0L. Mix thoroughly. Filter sterilize.

VCNT Inhibitor:
Composition per 10mL:

Colistin 7.5mg
Trimethoprim lactate 5.0mg
Vancomycin 3.0mg
Nystatin 12,500U

Preparation of VCNT Inhibitor: Add components to distilled/deionized water and bring volume to 10.0mL. Mix thoroughly. Filter sterilize.

Preparation of Medium: Add components, except supplement solution and V-C-N-T inhibitor, to distilled/deionized water and bring volume to 980.0mL. Mix thoroughly. Gently heat and bring to boiling. Autoclave for 15 min at 15 psi pressure–121°C. Cool to 45° to 50°C under 5–30% CO_2. Aseptically add 10.0mL of sterile supplement solution and 10.0mL of sterile V-C-N-T inhibitor. Mix thoroughly. Aseptically distribute under 5–30% CO_2 into sterile screw-capped tubes.

Use: For the transportation and recovery of pathogenic *Neisseria* species.

Transgrow Medium without Trimethoprim

Composition per liter:

Agar 20.0g
Hemoglobin 10.0g
Pancreatic digest of casein 7.5g
Selected meat peptone 7.5g
NaCl 5.0g
K_2HPO_4 .... 4.0g
Glucose 1.5g
Cornstarch 1.0g
KH_2PO_4 .... 1.0g
Supplement solution 10.0mL
V-C-N inhibitor 10.0mL

pH 6.7 ± 0.2 at 25°C

Source: Available as a prepared medium from BBL Microbiology Systems.

Supplemement Solution:
Composition per liter:

Glucose 100.0g
L-Cysteine·HCl 25.9g
L-Glutamine 10.0g
L-Cystine 1.1g
Adenine 1.0g
Nicotinamide adenine dinucleotide 0.25g
Vitamin B_{12} 0.1g
Thiamine pyrophosphate 0.1g
Guanine·HCl 0.03g
$Fe(NO_3)_3 \cdot 6H_2O$ 0.02g
p-Aminobenzoic acid 0.013g
Thiamine·HCl 3.0mg

Source: The supplement solution IsoVitaleX® enrichment is available from BBL Microbiology Laboratories. This enrichment may be replaced by supplement VX from Difco Laboratories.

Preparation of Supplement Solution: Add components to distilled/deionized water and bring volume to 1.0L. Mix thoroughly. Filter sterilize.

VCN Inhibitor:
Composition per 10mL:

Colistin 7.5mg
Vancomycin 3.0mg
Nystatin 12,500U

Preparation of VCN Inhibitor: Add components to distilled/deionized water and bring volume to 10.0mL. Mix thoroughly. Filter sterilize.

Preparation of Medium: Add components, except supplement solution and V-C-N inhibitor, to distilled/deionized water and bring volume to 980.0mL. Mix thoroughly. Gently heat and bring to boiling. Autoclave for 15 min at 15 psi pressure–121°C. Cool to 45° to 50°C under 5–30% CO_2. Aseptically add 10.0mL of sterile supplement solution and 10.0mL of sterile V-C-N inhibitor. Mix thoroughly. Aseptically distribute under 5–30% CO_2 into sterile screw-capped tubes.

Use: For the transportation and recovery of pathogenic *Neisseria* species.

Transport Medium

Composition per liter:

Sodium glycerophosphate 10.0g
Agar 3.0g
Sodium thioglycollate 1.0g
$CaCl_2 \cdot 2H_2O$ 0.1g
Methylene Blue 2.0mg
pH 7.3 ± 0.2 at 25°C

Source: This medium is available as a premixed powder from BBL Microbiology Systems.

Preparation of Medium: Add components to distilled/deionized water and bring volume to 1.0L. Mix thoroughly. Gently heat while stirring and bring to boiling. Distribute into screw-capped tubes or vials. Fill tubes nearly to capacity. Leave only enough space so that when a small swab is introduced the tube does not overflow. Autoclave for 10 min at 15 psi pressure–121°C. Tighten caps on tubes.

Use: For the transportation of swab specimens for the recovery of a wide variety of microorganisms, including *Neisseria gonorrhoeae*.

Transport Medium Stuart

Composition per liter:

Sodium glycerophosphate 10.0g
Agar 3.0g
Sodium thioglycollate 0.9g
$CaCl_2 \cdot 2H_2O$ 0.1g
Methylene Blue 2.0mg
pH 7.3 ± 0.2 at 25°C

Source: This medium is available as a premixed powder from Difco Laboratories.

Preparation of Medium: Add components to distilled/deionized water and bring volume to 1.0L. Mix thoroughly. Gently heat while stirring and bring to boiling. Distribute into screw-capped tubes or vials. Fill tubes nearly to capacity. Leave only enough space so that when a small swab is introduced the tube does not overflow. Autoclave for 10 min at 15 psi pressure–121°C. Tighten caps on tubes.

Use: For the transportation of swab specimens for the recovery of a wide variety of microorganisms, including *Neisseria gonorrhoeae*.

Trichomonas Selective Medium

Composition per liter:

Liver digest 25.0g
NaCl 6.5g
Glucose 5.0g
Agar 1.0g
Horse serum 80.0mL
Antibiotic inhibitor 10.0mL
pH 6.4 ± 0.2 at 25°C

Horse Serum:
Composition per 80mL:

Horse serum 80.0mL

Preparation of Horse Serum: Gently heat sterile horse serum to 56°C for 30 min. Aseptically adjust pH to 6.0 with 0.1*N* HCl. Use immediately.

Antibiotic Inhibitor:
Composition per 10mL:

Chloramphenicol 100.0mg

Preparation of Antibiotic Inhibitor: Add components to distilled/deionized water and bring volume to 10.0mL. Mix thoroughly. Filter sterilize.

Preparation of Medium: Add components, except horse serum, to distilled/deionized water and bring volume to 910.0mL. Mix thoroughly. Gently heat and bring to boiling. Autoclave for 15 min at 15 psi pressure–121°C. Cool to 45° to 50°C. Aseptically add 80.0mL of freshly prepared sterile horse serum and 10.0mL of sterile antibiotic inhibitor. Mix thoroughly. Aseptically distribute into sterile tubes or flasks.

Use: For the cultivation of *Trichomonas vaginalis* from specimens with a mixed bacterial flora.

Trichophyton Agar 1

Composition per liter:

Glucose 40.0g
Agar 15.0g
Vitamin assay casamino acids 2.5g
KH_2PO_4 .... 1.8g
$MgSO_4 \cdot 7H_2O$ 0.1g
pH 6.8 ± 0.2 at 25°C

Source: This medium is available as a premixed powder from Difco Laboratories.

Preparation of Medium: Add components to distilled/deionized water and bring volume to 1.0L. Mix thoroughly. Gently heat and bring to boiling. Distribute into tubes. Autoclave for 15 min at 15 psi pressure–121°C. Allow tubes to cool in a slanted position.

Use: For the differentiation of the *Trichophyton* species.

Trichophyton Agar 2

Composition per liter:

Glucose 40.0g
Agar 15.0g
Vitamin assay casamino acids 2.5g
KH_2PO_4 .... 1.8g

$MgSO_4 \cdot 7H_2O$ 0.1g
Inositol 50.0mg

pH 6.8 ± 0.2 at 25°C

Source: This medium is available as a premixed powder from Difco Laboratories.

Preparation of Medium: Add components to distilled/deionized water and bring volume to 1.0L. Mix thoroughly. Gently heat and bring to boiling. Distribute into tubes. Autoclave for 15 min at 15 psi pressure–121°C. Allow tubes to cool in a slanted position.

Use: For the differentiation of the *Trichophyton* species.

Trichophyton Agar 3

Composition per liter:

Glucose 40.0g
Agar 15.0g
Vitamin assay casamino acids 2.5g
KH_2PO_4 .... 1.8g
$MgSO_4 \cdot 7H_2O$ 0.1g
Inositol 0.05g
Thiamine·HCl 0.2mg

pH 6.8 ± 0.2 at 25°C

Source: This medium is available as a premixed powder from Difco Laboratories.

Preparation of Medium: Add components to distilled/deionized water and bring volume to 1.0L. Mix thoroughly. Gently heat and bring to boiling. Distribute into tubes. Autoclave for 15 min at 15 psi pressure–121°C. Allow tubes to cool in a slanted position.

Use: For the differentiation of the *Trichophyton* species.

Trichophyton Agar 4

Composition per liter:

Glucose 40.0g
Agar 15.0g
Vitaminassay casamino acids 2.5g
KH_2PO_4 .... 1.8g
$MgSO_4 \cdot 7H_2O$ 0.1g
Thiamine·HCl USP 200.0μg

pH 6.8 ± 0.2 at 25°C

Source: This medium is available as a premixed powder from Difco Laboratories.

Preparation of Medium: Add components to distilled/deionized water and bring volume to 1.0L. Mix thoroughly. Gently heat and bring to boiling. Distribute into tubes. Autoclave for 15 min at 15 psi pressure–121°C. Allow tubes to cool in a slanted position.

Use: For the differentiation of the *Trichophyton* species.

Trichophyton Agar 5

Composition per liter:

Glucose 40.0g
Agar 15.0g
Vitamin Assay Casamino Acids 2.5g
KH_2PO_4 .... 1.8g
$MgSO_4 \cdot 7H_2O$ 0.1g
Nicotinic acid 2.0mg

pH 6.8 ± 0.2 at 25°C

Source: This medium is available as a premixed powder from Difco Laboratories.

Preparation of Medium: Add components to distilled/deionized water and bring volume to 1.0L. Mix thoroughly. Gently heat and bring to boiling. Distribute into tubes. Autoclave for 15 min at 15 psi pressure–121°C. Allow tubes to cool in a slanted position.

Use: For the differentiation of the *Trichophyton* species.

Trichophyton Agar 6

Composition per liter:

Glucose 40.0g
Agar 15.0g
KH_2PO_4 .... 1.8g
Ammonium nitrate 1.5g
$MgSO_4 \cdot 7H_2O$ 0.1g

pH 6.8 ± 0.2 at 25°C

Source: This medium is available as a premixed powder from Difco Laboratories.

Preparation of Medium: Add components to distilled/deionized water and bring volume to 1.0L. Mix thoroughly. Gently heat and bring to boiling. Distribute into tubes. Autoclave for 15 min at 15 psi pressure–121°C. Allow tubes to cool in a slanted position.

Use: For the differentiation of the *Trichophyton* species.

Trichophyton Agar 7

Composition per liter:

Glucose 40.0g
Agar 15.0g
KH_2PO_4 .... 1.8g
Ammonium nitrate 1.5g
$MgSO_4 \cdot 7H_2O$ 0.1g
Histidine·HCl 0.03g

pH 6.8 ± 0.2 at 25°C

Source: This medium is available as a premixed powder from Difco Laboratories.

Preparation of Medium: Add components to distilled/deionized water and bring volume to 1.0L. Mix thoroughly. Gently heat and bring to boiling. Distribute into tubes. Autoclave for 15 min at 15 psi pressure–121°C. Allow tubes to cool in a slanted position.

Use: For the differentiation of the *Trichophyton* species.

Triple Sugar Iron Agar (TSI Agar)

Composition per liter:

Peptone	20.0g
Agar	12.0g
Lactose	10.0g
Sucrose	10.0g
NaCl	5.0g
Beef extract	3.0g
Yeast extract	3.0g
Glucose	1.0g
Ferric citrate	0.3g
$Na_2S_2O_3$	0.3g
Phenol Red	0.025g

pH 7.4 ± 0.2 at 25°C

Source: This medium is available as a premixed powder from Difco Laboratories and Oxoid Unipath.

Preparation of Medium: Add components to distilled/deionized water and bring volume to 1.0L. Mix thoroughly. Gently heat and bring to boiling. Distribute into tubes or flasks. Autoclave for 15 min at 15 psi pressure–121°C. Allow tubes to cool in a slanted position to form a 1.0 inch butt.

Use: For the differentiation of members of the Enterobacteriaceae based on their fermentation of lactose, sucrose and glucose and the production of H_2S.

Triple Sugar Iron Agar (TSI Agar)

Composition per liter:

Agar	13.0g
Pancreatic digest of casein	10.0g
Peptic digest of animal tissue	10.0g
Lactose	10.0g
Sucrose	10.0g
NaCl	5.0g
Glucose	1.0g
$Fe(NH_4)_2(SO_4)_2{\cdot}6H_2O$	0.2g
$Na_2S_2O_3$	0.2g
Phenol Red	0.025g

pH 7.3 ± 0.2 at 25°C

Source: This medium is available as a premixed powder from BBL Microbiology Systems.

Preparation of Medium: Add components to distilled/deionized water and bring volume to 1.0L. Mix thoroughly. Gently heat and bring to boiling. Distribute into tubes or flasks. Autoclave for 15 min at 15 psi pressure–121°C. Allow tubes to cool in a slanted position to form a 1.0 inch butt.

Use: For the differentiation of members of the Enterobacteriaceae based on their fermentation of lactose, sucrose and glucose and the production of H_2S.

Tryptic Digest Broth

Composition per liter:

Tryptic digest of beef heart	10.0g
NaCl	5.0g
Glucose	1.0g

pH 7.6 ± 0.2 at 25°C

Source: This medium is available as a premixed powder from Difco Laboratories.

Preparation of Medium: Add components to distilled/deionized water and bring volume to 1.0L. Mix thoroughly. Distribute into tubes or flasks. Autoclave for 15 min at 15 psi pressure–121°C.

Use: For use as a base medium to which enrichments are added for cultivation of fastidious microorganisms.

Tryptic Nitrate Medium

Composition per liter:

Tryptose	20.0g
Na_2HPO_4	2.0g
Agar	1.0g
Glucose	1.0g
KNO_3	1.0g

pH 7.6 ± 0.2 at 25°C

Source: This medium is available as a premixed powder from Difco Laboratories.

Preparation of Medium: Add components to distilled/deionized water and bring volume to 1.0L. Mix thoroughly. Gently heat and bring to boiling. Distribute into tubes in 10.0mL volumes. Autoclave for 15 min at 15 psi pressure–121°C.

Use: For the cultivation and differentiation of *Pseudomonas* and related genera. For the differentiation of bacteria based on their reduction of nitrate to nitrite. After incubation of the bacterium in tryptic nitrate medium for 18–24 hr, sulfanillic acid and α-naphthol reagents are added. Nitrate reduction is indicated by the development of a red to violet color.

Trypticase™ Agar Base with Carbohydrate

Composition per liter:

Pancreatic digest of casein 20.0g
Carbohydrate 5.0g
Agar 3.5g
Phenol Red 0.02g

pH 7.4 ± 0.2 at 25°C

Preparation of Medium: Add components to distilled/deionized water and bring volume to 1.0L. Mix thoroughly. Gently heat and bring to boiling. Distribute into tubes. Autoclave for 15 min at 13 psi pressure–118°C. Do not overheat. Pour into sterile Petri dishes or leave in tubes.

Use: For differentiation of microorganisms based on their motility and fermentation reactions. Fermentation of carbohydrate changes the medium yellow.

Trypticase™ Soy Agar

Composition per liter:

Pancreatic digest of casein 17.0g
Agar 15.0g
NaCl 5.0g
Papaic digest of soybean meal 3.0g
K_2HPO_4 ....2.5g
Glucose 2.5g

pH 7.3 ± 0.2 at 25°C

Preparation of Medium: Add components to distilled/deionized water and bring volume to 1.0L. Mix thoroughly. Gently heat and bring to boiling. Distribute into tubes or flasks. Autoclave for 15 min at 15 psi pressure–121°C. Pour into sterile Petri dishes or leave in tubes.

Use: For the cultivation and maintenance of a wide variety of heterotrophic microorganisms.

Trypticase™ Soy Agar (Tryptic Soy Agar) (Soybean Casein Digest Agar) (ATCC Medium 77)

Composition per liter:

Pancreatic digest of casein 15.0g
Agar 15.0g
Papaic digest of soybean meal 5.0g
NaCl 5.0g

pH 7.3 ± 0.2 at 25°C

Source: This medium is available as a premixed powder from BBL Microbiology Systems and Difco Laboratories.

Preparation of Medium: Add components to distilled/deionized water and bring volume to 1.0L. Mix thoroughly. Gently heat and bring to boiling. Distribute into tubes or flasks. Autoclave for 15 min at 15 psi pressure–121°C. Do not overheat. Pour into sterile Petri dishes or leave in tubes.

Use: For the isolation and cultivation of a wide variety of fastidious as well as nonfastidious microorganisms.

Trypticase™ Soy Agar with Sheep Blood (Tryptic Soy Blood Agar) (TSA Blood Agar)

Composition per liter:

Pancreatic digest of casein 15.0g
Agar 15.0g
Papaic digest of soybean meal 5.0g
NaCl 5.0g
Sheep blood, defibrinated....50.0mL

pH 7.3 ± 0.2 at 25°C

Preparation of Medium: Add components, except sheep blood, to distilled/deionized water and bring volume to 950.0mL. Mix thoroughly. Gently heat and bring to boiling. Autoclave for 15 min at 15 psi pressure–121°C. Cool to 45° to 50°C. Aseptically add sterile sheep blood. Mix thoroughly. Pour into sterile Petri dishes in 17.0mL volumes or distribute into sterile tubes.

Use: For the cultivation of a wide variety of fastidious microorganisms. For the observation of hemolytic reactions of a variety of bacteria. It may be used to perform the CAMP test for the presumptive identification of group B streptococci (*Streptococcus agalactiae*).

Trypticase™ Soy Agar with Sheep Blood and Gentamicin (TSA II™ with Sheep Blood and Gentamicin)

Composition per liter:

Pancreatic digest of casein 14.5g
Agar 14.0g
Papaic digest of soybean meal 5.0g
NaCl 5.0g
Growth factors (BBL) 1.5g
Sheep blood, defibrinated....50.0mL
Gentamicin solution 10.0mL

pH 7.3 ± 0.2 at 25°C

Source: This medium is available as a premixed powder from BBL Microbiology Systems.

Gentamicin Solution:
Composition per 10mL:
Gentamicin 2.5mg

Preparation of Gentamicin Solution: Add gentamicin to distilled/deionized water and bring volume to 10.0mL. Mix thoroughly. Filter sterilize.

Preparation of Medium: Add components, except sheep blood and gentamicin solution, to distilled/deionized water and bring volume to 940.0mL. Mix thoroughly. Gently heat and bring to boiling. Autoclave for 15 min at 15 psi pressure–121°C. Cool to 45° to 50°C. Aseptically add sterile sheep blood and sterile gentamicin solution. Mix thoroughly. Pour into sterile Petri dishes or distribute into sterile tubes.

Use: For the isolation of *Streptococcus pneumoniae* from a variety of clinical specimens.

Trypticase™ Soy Agar with Sheep Blood, Formate and Fumarate

Composition per liter:
Pancreatic digest of casein 14.5g
Agar 14.0g
Papaic digest of soybean meal 5.0g
NaCl 5.0g
Sucrose 2.0g
Growth factors (BBL) 1.5g
Sheep blood, defibrinated 50.0mL
Formate-fumarate solution 13.0mL
pH 7.3 ± 0.2 at 25°C

Formate-Fumarate Solution:
Composition per 100.0mL:
Sodium formate 6.0g
Fumaric acid 6.0g

Preparation of Formate-Fumarate Solution: Add components to distilled/deionized water and bring volume to 100.0mL. Mix thoroughly. Adjust pH to 7.0. Filter sterilize.

Preparation of Medium: Add components, except sheep blood, to distilled/deionized water and bring volume to 950.0mL. Mix thoroughly. Gently heat and bring to boiling. Autoclave for 15 min at 15 psi pressure–121°C. Cool to 45° to 50°C. Aseptically add sterile sheep blood. Mix thoroughly. Pour into sterile Petri dishes. Prior to inoculation aseptically spread 0.2mL of sterile formate-fumarate solution on each plate.

Use: For the isolation of *Streptococcus pneumoniae* from a variety of clinical specimens.

Trypticase™ Soy Agar with Sheep Blood, Sucrose and Tetracycline

Composition per liter:
Pancreatic digest of casein 14.5g
Agar 14.0g
Papaic digest of soybean meal 5.0g
NaCl 5.0g
Sucrose 2.0g
Growth factors (BBL) 1.5g
Sheep blood, defibrinated 50.0mL
Tetracycline solution 10.0mL
pH 7.3 ± 0.2 at 25°C

Tetracycline Solution:
Composition per 10mL:
Tetracycline 0.5mg

Preparation of Tetracycline Solution: Add tetracycline to distilled/deionized water and bring volume to 10.0mL. Mix thoroughly. Filter sterilize.

Preparation of Medium: Add components, except sheep blood and tetracycline solution, to distilled/deionized water and bring volume to 940.0mL. Mix thoroughly. Gently heat and bring to boiling. Autoclave for 15 min at 15 psi pressure–121°C. Cool to 45° to 50°C. Aseptically add sterile sheep blood and sterile tetracycline solution. Mix thoroughly. Pour into sterile Petri dishes or distribute into sterile tubes.

Use: For the isolation of *Streptococcus pneumoniae* from a variety of clinical specimens.

Trypticase™ Soy Broth (Soybean Casein Digest Broth, USP)

Composition per liter:
Pancreatic digest of casein 17.0g
NaCl 5.0g
Papaic digest of soybean meal 3.0g
K_2HPO_4 .... 2.5g
Glucose 2.5g
pH 7.3 ± 0.2 at 25°C

Source: This medium is available as a premixed powder from BBL Microbiology Systems and Difco Laboratories.

Preparation of Medium: Add components to distilled/deionized water and bring volume to 1.0L. Mix thoroughly. Distribute into tubes or flasks. Autoclave for 15 min at 15 psi pressure–121°C.

Use: For the cultivation of a wide variety of fastidious and nonfastidious microorganisms from clinical specimens and in epidemiological investigations. Also used for the rapid estimation of the bacteriological quality of water.

Trypticase™ Soy Broth with 0.1% Agar

Composition per liter:

Pancreatic digest of casein 17.0g
NaCl 5.0g
Papaic digest of soybean meal 3.0g
K_2HPO_4 .... 2.5g
Glucose 2.5g
Agar 1.0g

pH 7.3 ± 0.2 at 25°C

Source: This medium is available as a premixed powder from BBL Microbiology Systems.

Preparation of Medium: Add components to distilled/deionized water and bring volume to 1.0L. Mix thoroughly. Distribute into tubes or flasks. Autoclave for 15 min at 15 psi pressure–121°C.

Use: For the cultivation of anaerobic microorganisms. For the cultivation of microorganisms isolated of root canals and other clinical specimens.

Trypticase™ Soy Broth with 10mM Glucose

Composition per liter:

Pancreatic digest of casein 17.0g
NaCl 5.0g
Papaic digest of soybean meal 3.0g
K_2HPO_4 .... 2.5g
Glucose 1.8g

pH 7.3 ± 0.2 at 25°C

Preparation of Medium: Add components to distilled/deionized water and bring volume to 1.0L. Mix thoroughly. Distribute into tubes or flasks. Autoclave for 15 min at 15 psi pressure–121°C.

Use: For the cultivation of a variety of fastidious and nonfastidious microorganisms from clinical specimens and in epidemiological investigations.

Trypticase™ Tellurite Agar Base

Composition per liter:

Agar 20.0g
Pancreatic digest of casein 10.0g
Peptic digest of animal tissue 10.0g
NaCl 5.0g
Glucose 2.0g
Serum 50.0mL
Chapman tellurite solution 10.0mL

pH 7.5 ± 0.2 at 25°C

Source: This medium is available as a premixed powder from BBL Microbiology Systems.

Chapman Tellurite Solution:
Composition per 100mL:

K_2TeO_3 .... 1.0g

Preparation of Chapman Tellurite Solution: Add K_2TeO_3 to distilled/deionized water and bring volume to 100.0mL. Mix thoroughly. Filter sterilize.

Caution: Potassium tellurite is toxic.

Preparation of Medium: Add components, except serum and Chapman tellurite solution, to distilled/deionized water and bring volume to 940.0mL. Mix thoroughly. Gently heat and bring to boiling. Autoclave for 15 min at 15 psi pressure–121°C. Cool to 45° to 50°C. Aseptically add sterile serum and sterile Chapman tellurite solution. Sheep serum, rabbit serum or human serum may be used. Mix thoroughly. Pour into sterile Petri dishes.

Use: For the selective isolation of microorganisms from clinical specimens, especially from the nose, throat and vagina.

Tryptone Phosphate Broth

Composition per liter:

Pancreatic digest of casein 20.0g
NaCl 5.0g
K_2HPO_4 .... 2.0g
KH_2PO_4 .... 2.0g
Tween™ 80 .... 15.0mL

pH 7.0 ± 0.2 at 25°C

Preparation of Medium: Add components to distilled/deionized water and bring volume to 1.0L. Mix thoroughly. Distribute into tubes or flasks. Autoclave for 15 min at 15 psi pressure–121°C.

Use: For the cultivation of enteropathogenic *Escherichia coli.*

Tryptone Soya Broth

Composition per liter:

Pancreatic digest of casein 17.0g
NaCl 5.0g
Pancreatic digest of soybean meal 3.0g
K_2HPO_4 .... 2.5g
Glucose 2.5g

pH 7.3 ± 0.2 at 25°C

Source: This medium is available as a premixed powder from Oxoid Unipath.

Preparation of Medium: Add components to distilled/deionized water and bring volume to 1.0L. Mix thoroughly. Distribute into tubes or flasks. Autoclave for 15 min at 15 psi pressure–121°C.

Use: For the cultivation of a wide variety of microorganisms.

Tryptone Water Broth (Tryptone Broth)

Composition per liter:

Pancreatic digest of casein 10.0g
NaCl 5.0g

pH 7.5 ± 0.2 at 25°C

Source: This medium is available as a premixed powder from Oxoid Unipath.

Preparation of Medium: Dissolve 15.0g in 1 liter of distilled water and distribute into final containers. Sterilize by autoclaving at 121°C for 15 minutes.

Use: For the cultivation of production of indole by microorganisms.

Tryptone Yeast Extract Agar

Composition per liter:

Pancreatic digest of casein 10.0g
Agar 2.0g
Yeast extract 1.0g
Bromcresol Purple 0.04g
Carbohydrate solution 100.0mL

pH 7.0 ± 0.2 at 25°C

Carbohydrate Solution:
Composition per 100mL:

Carbohydrate 10.0g

Preparation of Carbohydrate Solution: Add carbohydrate to distilled/deionized water and bring volume to 100.0mL. Glucose or mannitol may be used. Mix thoroughly. Filter sterilize.

Preparation of Medium: Add components, except carbohydrate solution, to distilled/deionized water and bring volume to 900.0mL. Mix thoroughly. Adjust pH to 7.0. Gently heat and bring to boiling. Distribute into tubes in 13.5mL volumes. Autoclave for 20 min at 10 psi pressure–115°C. Cool to 45° to 50°C. Aseptically add 1.5mL of carbohydrate solution to each tube. Mix thoroughly. Solidify agar quickly by placing tubes in ice water.

Use: For the cultivation and differentiation of *Staphylococcus aureus* based on glucose and mannitol fermentation. Bacteria that ferment the added carbohydrate turn the medium yellow.

Tryptophan 1% Solution (Trypticase™ 1% Solution) (Tryptone 1% Solution)

Composition per liter:

Pancreatic digest of casein 10.0g

pH 7.0 ± 0.2 at 25°C

Source: Available as a premixed powder from BBL Microbiology Systems and Difco Laboratories.

Preparation of Medium: Add components to distilled/deionized water and bring volume to 1.0L. Mix thoroughly. Distribute into tubes or flasks. Autoclave for 15 min at 15 psi pressure–121°C.

Use: For the differentiation of bacteria, especially members of the Enterobacteaceae, based on their production of indole.

Tryptophan Broth

Composition per 100mL:

L-Tryptophan 0.5g
NaCl 0.5g
KH_2PO_4 .... 0.25g

pH 7.4 ± 0.2 at 25°C

Preparation of Medium: Add components to distilled/deionized water and bring volume to 100.0mL. Mix thoroughly. Adjust pH to 7.4. Filter sterilize. Aseptically distribute in 1.0mL volumes into sterile screw-capped tubes.

Use: For the cultivation of *Flavobacterium* species and a variety of other bacteria. Also used to differentiate bacteria based on indole production. Indole is determined by the addition of modified Kovacs reagent to cultures which have incubated for 18–24 hr. Formation of a red color in the upper layer indicates indole formation.

Tryptose Agar

Composition per liter:

Agar 15.0g
Pancreatic digest of casein 10.0g
Peptic digest of animal tissue 10.0g
NaCl 5.0g
Glucose 1.0g

pH 7.2 ± 0.2 at 25°C

Source: This medium is available as a premixed powder from Difco Laboratories.

Preparation of Medium: Add components to distilled/deionized water and bring volume to 1.0L. Mix thoroughly. Gently heat and bring to boiling. Distribute into tubes or flasks. Autoclave for 15 min at 15 psi pressure–121°C. Pour into sterile Petri dishes or leave in tubes.

Use: For the cultivation and maintenance of fastidious aerobic and facultative microorganisms.

Tryptose Agar with Citrate

Composition per liter:

Agar	15.0g
Pancreatic digest of casein	10.0g
Peptic digest of animal tissue	10.0g
Sodium citrate	10.0g
NaCl	5.0g
Glucose	1.0g

pH 7.2 ± 0.2 at 25°C

Preparation of Medium: Add components to distilled/deionized water and bring volume to 1.0L. Mix thoroughly. Gently heat and bring to boiling. Distribute into tubes or flasks. Autoclave for 15 min at 15 psi pressure–121°C. Pour into sterile Petri dishes or leave in tubes.

Use: For the cultivation and maintenance of fastidious aerobic and facultative microorganisms including *Brucella* species and streptococci.

Tryptose Agar with Thiamine

Composition per liter:

Agar	15.0g
Pancreatic digest of casein	10.0g
Peptic digest of animal tissue	10.0g
NaCl	5.0g
Glucose	1.0g
Thiamine·HCl	5.0mg

pH 7.2 ± 0.2 at 25°C

Preparation of Medium: Add components to distilled/deionized water and bring volume to 1.0L. Mix thoroughly. Gently heat and bring to boiling. Distribute into tubes or flasks. Autoclave for 15 min at 15 psi pressure–121°C. Pour into sterile Petri dishes or leave in tubes.

Use: For the cultivation and maintenance of fastidious aerobic and facultative microorganisms including *Brucella* species and streptococci.

Tryptose Blood Agar

Composition per liter:

Agar	12.0g
Tryptose	10.0g
NaCl	5.0g
Beef extract	3.0g
Sheep blood, defibrinated	70.0mL

pH 7.2 ± 0.2 at 25°C

Source: This medium is available as a premixed powder from Oxoid Unipath.

Preparation of Medium: Add components, except sheep blood, to distilled/deionized water and bring volume to 930.0mL. Mix thoroughly. Gently heat and bring to boiling. Autoclave for 15 min at 15 psi pressure–121°C. Cool to 45° to 50°C. Aseptically add sterile sheep blood. Mix thoroughly. Pour into sterile Petri dishes in 17.0mL volumes or distribute into sterile tubes.

Use: A general purpose medium for the cultivation and maintenance of a wide variety of microorganisms, including fastidious bacteria.

Tryptose Broth

Composition per liter:

Pancreatic digest of casein	10.0g
Peptic digest of animal tissue	10.0g
NaCl	5.0g
Glucose	1.0g

pH 7.2 ± 0.2 at 25°C

Source: This medium is available as a premixed powder from Difco Laboratories.

Preparation of Medium: Add components to distilled/deionized water and bring volume to 1.0L. Mix thoroughly. Distribute into tubes or flasks. Autoclave for 15 min at 15 psi pressure–121°C.

Use: For the cultivation of fastidious aerobic and facultative microorganisms including streptococci.

Tryptose Broth

Composition per liter:

Pancreatic digest of casein	10.0g
Peptic digest of animal tissue	10.0g
NaCl	5.0g
Glucose	1.0g
Thiamine·HCl	5.0mg

pH 7.2 ± 0.2 at 25°C

Source: This medium is available as a premixed powder from BBL Microbiology Systems.

Preparation of Medium: Add components to distilled/deionized water and bring volume to 1.0L. Mix thoroughly. Distribute into tubes or flasks. Autoclave for 15 min at 15 psi pressure–121°C.

Use: For the cultivation of fastidious aerobic and facultative microorganisms.

Tryptose Broth with Citrate

Composition per liter:

Pancreatic digest of casein 10.0g
Peptic digest of animal tissue 10.0g
Sodium citrate 10.0g
NaCl 5.0g
Glucose 1.0g
Thiamine·HCl 5.0mg

pH 7.2 ± 0.2 at 25°C

Preparation of Medium: Add components to distilled/deionized water and bring volume to 1.0L. Mix thoroughly. Distribute into tubes or flasks. Autoclave for 15 min at 15 psi pressure–121°C.

Use: For the isolation and cultivation of a variety of fastidious aerobic microorganisms especially *Brucella* species from clinical sources and dairy products.

Tryptose Phosphate Broth

Composition per liter:

Tryptose 20.0g
NaCl 5.0g
Na_2HPO_4 .... 2.5g
Glucose 2.0g

pH 7.3 ± 0.2 at 25°C

Source: This medium is available as a premixed powder from Difco Laboratories and Oxoid Unipath.

Preparation of Medium: Add components to distilled/deionized water and bring volume to 1.0L. Mix thoroughly. Distribute into tubes or flasks. Autoclave for 15 min at 15 psi pressure–121°C. Prior to the inoculation of anaerobic microorganisms, place tubes of sterile medium in a 100°C bath for 15 min and cool undisturbed.

Use: For the cultivation of a variety of fastidious and nonfastidious bacteria.

Tryptose Phosphate Broth, Modified

Composition per liter:

Enzymatic digest of casein 20.0g
NaCl 5.0g
Na_2HPO_4 .... 2.5g
Glucose 2.0g

pH 7.3 ± 0.2 at 25°C

Source: This medium is available as a premixed powder from BBL Microbiology Systems.

Preparation of Medium: Add components to distilled/deionized water and bring volume to 1.0L. Mix thoroughly. Distribute into tubes or flasks. Autoclave for 15 min at 15 psi pressure–121°C. Prior to the inoculation of anaerobic microorganisms, place tubes of sterile medium in a 100°C bath for 15 min and cool undisturbed.

Use: For the cultivation of a variety of fastidious microorganisms including pneumococci, streptococci and meningococci.

Tween™ 80 Hydrolysis Broth

Composition per liter:

Na_2HPO_4 .... 5.79g
NaH_2PO_4 .... 3.53g
Neutral Red 0.02g
Tween™ 80 .... 5.00mL

pH 7.0 ± 0.2 at 25°C

Preparation of Medium: Add components to distilled/deionized water and bring volume to 1.0L. Mix thoroughly. Distribute into tubes or flasks. Autoclave for 15 min at 15 psi pressure–121°C.

Use: For the differentiation of *Mycobacterium* species. Strains that hydrolyze Tween™ 80 within 5 days turn the medium pink to red.

Tyrosine Agar

Composition per liter:

Solution 1 .... 900.0mL
Solution 2 .... 100.0mL

pH 7.0 ± 0.2 at 25°C

Solution 1:

Composition per 900mL:

Agar 15.0g
Pancreatic digest of gelatin 5.0g
Beef extract 3.0g

Preparation of Solution 1: Add components to distilled/deionized water and bring volume to 900.0mL. Mix thoroughly. Gently heat and bring to boiling.

Solution 2:

Composition per 100mL:

Tyrosine 5.0g

Preparation of Solution 2: Add tyrosine to distilled/deionized water and bring volume to 100.0mL. Mix thoroughly. Gently heat and bring to boiling.

Preparation of Medium: Combine solutions 1 and 2. Mix thoroughly. Distribute into tubes or flasks. Autoclave for 15 min at 15 psi pressure–121°C. Pour into sterile Petri dishes or leave in tubes.

Use: For the differentiation of aerobic *Actinomycete* species. Clearing around a colony indicates utilization of tyrosine. *Streptomyces* and *Actinomadura* species utilize tyrosine. *Nocaridia asteroides, N. caviae, N. braziliensis* and *Mycobacterium fortuitum* do not utilize tyrosine.

U Agar Plates (*Ureaplasma* Agar Plates) (MES Agar)

Composition per 100.2mL:

Base agar..65.0mL
Horse serum...20.0mL
Yeast dialysate...10.0mL
MES (2-N-Morpholinoethane
sulfonic acid) buffer solution....................3.0mL
Penicillin solution..2.0mL
Urea solution..0.2mL

pH 5.5 ± 0.2 at 25°C

Base Agar:

Composition per liter:

Papaic digest of soybean meal.........................20.0g
Agarose...10.0g
NaCl...5.0g
Phenol Red (2% solution)...............................1.0mL

Preparation of Base Agar: Add components to distilled/deionized water and bring volume to 1.0L. Mix thoroughly. Gently heat and bring to boiling. Adjust pH to 7.3. Autoclave for 15 min at 15 psi pressure–121°C. Cool to 45° to 50°C.

Yeast Dialysate:

Composition per 10mL:

Yeast, active dried 450.0g

Preparation of Yeast Dialysate: Add active, dried yeast to distilled/deionized water and bring volume to 1250.0mL. Gently heat and bring to 40°C. Autoclave for 15 min at 15 psi pressure–121°C. Put into dialysis tubing. Dialyze against 1.0L of distilled/deionized water for 2 days at 4°C. Discard tubing and its contents. Autoclave dialysate for 15 min at 15 psi pressure–121°C. Store at –20°C.

MES Buffer Solution:

Composition per 100mL:

MES (2-N-Morpholinoethane
sulfonic acid) buffer................................19.52g

Preparation of MES Buffer Solution: Add MES buffer to distilled/deionized water and bring volume to 100.0mL. Mix thoroughly. Adjust pH to 5.5. Filter sterilize.

Penicillin Solution:

Composition per 10mL:

Penicillin...100,000U

Preparation of Penicillin Solution: Add penicillin to distilled/deionized water and bring volume to 10.0mL. Mix thoroughly. Filter sterilize.

Urea Solution:

Composition per 100mL:

Urea..6.0g

Preparation of Urea Solution: Add urea to distilled/deionized water and bring volume to 100.0mL. Mix thoroughly. Filter sterilize.

Preparation of Medium: To 65.0mL of cooled, sterile base agar, aseptically add 10.0mL of sterile yeast dialysate, 20.0mL of horse serum, 2.0mL of sterile penicillin solution, 3.0mL of sterile MES buffer solution, and 0.2mL of sterile urea solution. Mix thoroughly. Pour into 10mm × 35mm Petri dishes in 5.0mL volumes. Allow plates to stand overnight at 25°C to remove excess surface moisture.

Use: For the isolation and cultivation of *Ureaplasma* species.

U Broth (*Ureaplasma* Broth)

Composition per 99.5mL:

Base agar..65.0mL
Horse serum...20.0mL
Yeast dialysate...10.0mL
Penicillin solution..2.0mL
MES (2-N-Morpholinoethane
sulfonic acid) buffer solution....................1.0mL
Na_2SO_3 solution..1.0mL
Urea solution..0.5mL

pH 5.5 ± 0.2 at 25°C

Base Agar:

Composition per liter:

Papaic digest of soybean meal.........................20.0g
Agarose...10.0g
NaCl...5.0g
Phenol Red (2% solution)...............................1.0mL

Preparation of Base Agar: Add components to distilled/deionized water and bring volume to 1.0L. Mix thoroughly. Gently heat and bring to boiling. Adjust pH to 7.3. Autoclave for 15 min at 15 psi pressure–121°C. Cool to 45° to 50°C.

Yeast Dialysate:

Composition per 10mL:

Yeast, active dried 450.0g

Preparation of Yeast Dialysate: Add active, dried yeast to distilled/deionized water and bring volume to 1250.0mL. Gently heat and bring to 40°C. Autoclave for 15 min at 15 psi pressure–121°C. Put into dialysis tubing. Dialyze against 1.0L of distilled/deionized water for 2 days at 4°C. Discard tubing and its contents. Autoclave dialysate for 15 min at 15 psi pressure–121°C. Store at –20°C.

Penicillin Solution:

Composition per 10mL:

Penicillin...100,000U

Preparation of Penicillin Solution: Add penicillin to distilled/deionized water and bring volume to 10.0mL. Mix thoroughly. Filter sterilize.

MES Buffer Solution:

Composition per 100mL:

MES (2-N-Morpholinoethane sulfonic acid) buffer 19.52g

Preparation of MES Buffer Solution: Add MES buffer to distilled/deionized water and bring volume to 100.0mL. Mix thoroughly. Adjust pH to 5.5. Filter sterilize.

Na_2SO_3 Solution:

Composition per 10mL:

Na_2SO_3 .... 0.126g

Preparation of Na_2SO_3 Solution: Add Na_2SO_3 to distilled/deionized water and bring volume to 10.0mL. Mix thoroughly. Filter sterilize.

Urea Solution:

Composition per 100mL:

Urea 6.0g

Preparation of Urea Solution: Add urea to distilled/deionized water and bring volume to 100.0mL. Mix thoroughly. Filter sterilize.

Preparation of Medium: To 65.0mL of cooled, sterile base agar, aseptically add 10.0mL of sterile yeast dialysate, 20.0mL of horse serum, 2.0mL of sterile penicillin solution, 1.0mL of sterile MES buffer solution, 1.0mL of sterile Na_2SO_3 solution, and 0.5mL of sterile urea solution. Mix thoroughly. Pour into 10mm × 35mm Petri dishes in 5.0mL volumes. Allow plates to stand overnight at 25°C to remove excess surface moisture. Use within 48 hr.

Use: For the isolation and cultivation of *Ureaplasma urealyticum.*

U9 Broth (Urease Color Test Medium)

Composition per 101.6mL:

U9 base 95.0mL
Horse serum, unheated 5.0mL
Penicillin G solution 1.0mL
Urea solution 0.5mL
Phenol Red solution 0.1mL

pH 6.0 ± 0.2 at 25°C

U9 Base:

Composition per 100mL:

NaCl 0.63g
Pancreatic digest of casein 0.425g
Papaic digest of soybean meal 0.075g
K_2HPO_4 .... 0.063g
Glucose 0.063g
KH_2PO_4 .... 0.02g

Preparation of U9 Base: Add components to distilled/deionized water and bring volume to 100.0mL. Mix thoroughly. Adjust pH to 5.5 with 1*N* HCl. Autoclave for 15 min at 15 psi pressure–121°C. Cool to 45° to 50°C.

Penicillin G Solution:

Composition per 10mL:

Penicillin G .... 0.63 g

Preparation of Penicillin G Solution: Add Penicillin G to distilled/deionized water and bring volume to 10.0mL. Mix thoroughly. Filter sterilize.

Urea Solution:

Composition per 30mL:

Urea 3.0g

Preparation of Urea Solution: Add urea to distilled/deionized water and bring volume to 30.0mL. Mix thoroughly. Filter sterilize.

Phenol Red Solution:

Composition per 10mL:

Phenol Red 0.1g

Preparation of Phenol Red Solution: Add Phenol Red to distilled/deionized water and bring volume to 10.0mL. Mix thoroughly. Filter sterilize.

Preparation of Medium: To 95.0mL of cooled, sterile U9 base, aseptically add 5.0mL of sterile horse serum, 1.0mL of sterile penicillin G solution, 0.5mL of sterile urea solution, and 0.1mL of sterile Phenol Red solution. Mix thoroughly. Aseptically distribute into sterile tubes or flasks.

Use: For the isolation and identification of mycoplasmas from clinical specimens, especially *Ureaplasma urealyticum. U. urealyticum* is the only species of the *Mycoplasma* group known to contain urease. Bacteria with urease activity turn the medium dark pink.

U9 Broth with Amphotericin B

Composition per 101.6mL:

U9 base 95.0mL
Horse serum, unheated 5.0mL
Antibiotic solution 1.0mL
Urea solution 0.5mL
Phenol Red solution 0.1mL

pH 6.0 ± 0.2 at 25°C

U9 Base:

Composition per 100mL:

NaCl 0.63g
Pancreatic digest of casein 0.425g

Papaic digest of soybean meal....................0.075g
K_2HPO_4....................0.063g
Glucose....................0.063g
KH_2PO_4....................0.02g

Preparation of U9 Base: Add components to distilled/deionized water and bring volume to 100.0mL. Mix thoroughly. Adjust pH to 5.5 with 1*N* HCl. Autoclave for 15 min at 15 psi pressure–121°C. Cool to 45° to 50°C.

Antibiotic Solution:
Composition per 10mL:
Penicillin G....................0.63g
Amphotericin B....................2.5mg

Preparation of Antibiotic Solution: Add Penicillin G and Amphotericin B to distilled/deionized water and bring volume to 10.0mL. Mix thoroughly. Filter sterilize.

Urea Solution:
Composition per30mL:
Urea....................3.0g

Preparation of Urea Solution: Add urea to distilled/deionized water and bring volume to 30.0mL. Mix thoroughly. Filter sterilize.

Phenol Red Solution:
Composition per 10mL:
Phenol Red....................0.1g

Preparation of Phenol Red Solution: Add Phenol Red to distilled/deionized water and bring volume to 10.0mL. Mix thoroughly. Filter sterilize.

Preparation of Medium: To 95.0mL of cooled, sterile U9 base, aseptically add 5.0mL of sterile horse serum, 1.0mL of sterile antibiotic solution, 0.5mL of sterile urea solution, and 0.1mL of sterile Phenol Red solution. Mix thoroughly. Aseptically distribute into sterile tubes or flasks.

Use: For the isolation and identification of mycoplasmas from clinical specimens, especially *Ureaplasma urealyticum. U. urealyticum* is the only species of the *Mycoplasma* group known to contain urease. Bacteria with urease activity turn the medium dark pink.

U9B Broth

Composition per 102.1mL:
U9 base....................95.0mL
Horse serum, unheated....................5.0mL
Penicillin G solution....................1.0mL
Urea solution....................0.5mL
Cysteine·HCl·H_2O solution....................0.5mL
Phenol Red solution....................0.1mL
pH 6.0 ± 0.2 at 25°C

U9 Base:
Composition per 100mL:
NaCl....................0.63g
Pancreatic digest of casein....................0.425g
Papaic digest of soybean meal....................0.075g
K_2HPO_4....................0.063g
Glucose....................0.063g
KH_2PO_4....................0.02g

Preparation of U9 Base: Add components to distilled/deionized water and bring volume to 100.0mL. Mix thoroughly. Adjust pH to 5.5 with 1*N* HCl. Autoclave for 15 min at 15 psi pressure–121°C. Cool to 45° to 50°C.

Penicillin G Solution:
Composition per 10mL:
Penicillin G....................0.63 g

Preparation of Penicillin G Solution: Add Penicillin G to distilled/deionized water and bring volume to 10.0mL. Mix thoroughly. Filter sterilize.

Urea Solution:
Composition per 30mL:
Urea....................3.0g

Preparation of Urea Solution: Add urea to distilled/deionized water and bring volume to 30.0mL. Mix thoroughly. Filter sterilize.

Cysteine·HCl·H_2O Solution:
Composition per50mL:
Cysteine·HCl·H_2O....................1.0g

Preparation of Cysteine·HCl·H_2O Solution: Add urea to distilled/deionized water and bring volume to 50.0mL. Mix thoroughly. Filter sterilize.

Phenol Red Solution:
Composition per 10mL:
Phenol Red....................0.1g

Preparation of Phenol Red Solution: Add Phenol Red to distilled/deionized water and bring volume to 10.0mL. Mix thoroughly. Filter sterilize.

Preparation of Medium: To 95.0mL of cooled, sterile U9 base, aseptically add 5.0mL of sterile horse serum, 1.0mL of sterile penicillin G solution, 0.5mL of sterile urea solution, 0.5mL of sterile cysteine·H-Cl·H_2O solution and 0.1mL of sterile Phenol Red solution. Mix thoroughly. Aseptically distribute into sterile tubes or flasks.

Use: For the isolation and identification of mycoplasmas from clinical specimens, especially *Ureaplasma urealyticum. U. urealyticum* is the only species of the *Mycoplasma* group known to contain urease. Bacteria with urease activity turn the medium dark pink.

U9C Broth

Composition per 102mL:

U9C base	90.0mL
Horse serum, unheated	10.0mL
Penicillin G solution	1.0mL
Urea solution	0.3mL
Cysteine·HCl·H_2O solution	0.5mL
GHL tripeptide solution	0.1mL
Phenol Red solution	0.1mL

pH 6.0 ± 0.2 at 25°C

U9C Base:

Composition per 100mL:

NaCl	0.85g
Pancreatic digest of casein	0.25g
Papaic digest of soybean meal	0.15g
K_2HPO_4	0.12g
Glucose	0.12g
$MgCl_2 \cdot 6H_2O$	0.2g
Yeast extract	0.1g
KH_2PO_4	0.02g

Preparation of U9C Base: Add components to distilled/deionized water and bring volume to 100.0mL. Mix thoroughly. Adjust pH to 5.5 with 2*N* HCl. Autoclave for 15 min at 15 psi pressure–121°C. Cool to 45° to 50°C.

Penicillin G Solution:

Composition per 10mL:

Penicillin G	0.63 g

Preparation of Penicillin G Solution: Add Penicillin G to distilled/deionized water and bring volume to 10.0mL. Mix thoroughly. Filter sterilize.

Urea Solution:

Composition per 30mL:

Urea	3.0g

Preparation of Urea Solution: Add urea to distilled/deionized water and bring volume to 30.0mL. Mix thoroughly. Filter sterilize.

Cysteine·HCl·H_2O Solution:

Composition per 50mL:

Cysteine·HCl·H_2O	1.0g

Preparation of Cysteine·HCl·H_2O Solution: Add urea to distilled/deionized water and bring volume to 50.0mL. Mix thoroughly. Filter sterilize.

GHL Tripeptide Solution:

Composition per 10mL:

GHL tripeptide	0.2mg

Preparation of GHL Tripeptide Solution: Add GHL tripeptide (glycyl-L-histidyl-L-lysine acetate) to distilled/deionized water and bring volume to 10.0mL. Mix thoroughly. Filter sterilize.

Phenol Red Solution:

Composition per 10mL:

Phenol Red	0.1g

Preparation of Phenol Red Solution: Add Phenol Red to distilled/deionized water and bring volume to 10.0mL. Mix thoroughly. Filter sterilize.

Preparation of Medium: To 90.0mL of cooled, sterile U9C base, aseptically add 10.0mL of sterile horse serum, 1.0mL of sterile penicillin G solution, 0.3mL of sterile urea solution, 0.5mL of sterile cysteine·HCl·H_2O solution, 0.1mL of sterile GHL tripeptide solution and 0.1mL of sterile Phenol Red solution. Mix thoroughly. Aseptically distribute into sterile tubes or flasks.

Use: For the isolation and identification of mycoplasmas from clinical specimens, especially *Ureaplasma urealyticum. U. urealyticum* is the only species of the *Mycoplasma* group known to contain urease. Bacteria with urease activity turn the medium dark pink.

Urea Agar (Urease Test Agar) (Urea Agar Base, Christensen)

Composition per liter:

Urea	20.0g
Agar	15.0g
NaCl	5.0g
KH_2PO_4	2.0g
Peptone	1.0g
Glucose	1.0g
Phenol Red	0.012g

pH 6.8 ± 0.2 at 25°C

Source: This medium is available as a premixed powder from BBL Microbiology Systems and Difco Laboratories.

Preparation of Medium: Add components, except agar, to distilled/deionized water and bring volume to 100.0mL. Mix thoroughly. Filter sterilize. Add agar to distilled/deionized water and bring volume to 900.0mL. Mix thoroughly. Gently heat and bring to boiling. Autoclave for 15 min at 15 psi pressure–121°C. Cool to 50°C. Aseptically add the 100.0mL of sterile basal medium. Mix thoroughly. Distribute into sterile tubes. Allow tubes to solidify in a slanted position.

Use: For the differentiation of a variety of microorganisms, especially members of the Enterobacteriaceae, aerobic actinomycetes, streptococci and nonfermenting Gram-negative bacteria, on the basis of urease production.

Urea Agar Base

Composition per liter:

Agar	15.0g
NaCl	5.0g
Na_2HPO_4	1.2g
Peptone	1.0g
Glucose	1.0g
KH_2PO_4	0.8g
Phenol Red	0.012g
Urea solution	50.0mL

pH 6.8 ± 0.2 at 25°C

Source: This medium is available as a premixed powder from Oxoid Unipath.

Urea Solution:

Composition per 100mL:

Urea	40.0g

Preparation of Urea Solution: Add urea to distilled/deionized water and bring volume to 100.0mL. Mix thoroughly. Filter sterilize.

Preparation of Medium: Add components, except urea solution, to distilled/deionized water and bring volume to 950.0mL. Mix thoroughly. Gently heat and bring to boiling. Autoclave for 20 min at 10 psi pressure–115°C. Cool to 50°C. Aseptically add 50.0mL of sterile urea solution. Mix thoroughly. Pour into sterile Petri dishes or distribute into sterile tubes. Allow tubes to solidify in a slanted position.

Use: For the detection of *Proteus* species based on rapid urease activity and the identification of other members of the Enterobacteriaceae based on urease activity. Urease positive bacteria turn the medium pink.

Urea Broth 10B for *Ureaplasma urealyticum*

Composition per 100.5mL:

PPLO broth without Crystal Violet	70.0mL
Horse serum, unheated	20.0mL
Fresh yeast extract solution	10.0mL
L-Cysteine·HCl·H_2O	0.5mL
CVA enrichment	0.5mL
Urea solution	0.4mL
Phenol Red	0.1mL

PPLO Broth without Crystal Violet:

Composition per 900mL:

Beef heart, solids from infusion	16.1g
Peptone	3.25g
NaCl	1.61g

Preparation of PPLO Broth without Crystal Violet: Add components to distilled/deionized water and bring volume to 900mL. Adjust pH to 5.5 with 2*N* HCl. Autoclave for 15 min at 15 psi pressure–121°C. Cool to 37°C.

Fresh Yeast Extract Solution:

Composition per 100mL:

Baker's yeast live, pressed, starch-free	25.0g

Preparation of Fresh Yeast Extract Solution: Add the live Baker's yeast to 100.0mL of distilled/deionized water. Autoclave for 90 min at 15 psi pressure–121°C. Allow to stand. Remove supernatant solution. Adjust pH to 6.6–6.8.

Cysteine·HCl·H_2O Solution:

Composition per 50mL:

Cysteine·HCl·H_2O	1.0g

Preparation of Cysteine·HCl·H_2O Solution: Add cysteine·HCl·H_2O to distilled/deionized water and bring volume to 50.0mL. Mix thoroughly. Filter sterilize.

CVA Enrichment:

Composition per liter:

Glucose	100.0g
L-Cysteine·HCl·H_2O	25.9g
L-Glutamine	10.0g
Adenine	1.0g
L-Cystine·2HCl	1.0g
Nicotinamide adenine dinucleotide	0.25g
Cocarboxylase	0.1g
Guanine·HCl	0.03g
$Fe(NO_3)_3$	0.02g
Vitamin B_{12}	0.01g
p-Aminobenzoic acid	0.013g
Thiamine·HCl	3.0mg

Preparation of CVA Enrichment: Add components to distilled/deionized water and bring volume to 1.0L. Mix thoroughly. Filter sterilize.

Urea Solution:

Composition per 30mL:

Urea	3.0g

Preparation of Urea Solution: Add urea to distilled/deionized water and bring volume to 30.0mL. Mix thoroughly. Filter sterilize.

Preparation of Medium: Aseptically combine the components, except the PPLO broth without Crystal Violet. Aseptically add this mixture to the cooled, sterile PPLO broth without Crystal Violet. Mix thoroughly. Aseptically distribute into sterile tubes or flasks.

Use: For the cultivation and maintenance of *Ureaplasma urealyticum* and other *Ureaplasma* species. Urease positive bacteria turn the medium to peach orange.

Urea Broth Base

Composition per liter:

NaCl	5.0g
Na_2HPO_4	1.2g
Peptone	1.0g
Glucose	1.0g
KH_2PO_4	0.8g
Phenol Red	0.012g
Urea solution	50.0mL

pH 6.8 ± 0.2 at 25°C

Source: This medium is available as a premixed powder from Oxoid Unipath.

Urea Solution:

Composition per 100mL:

Urea	40.0g

Preparation of Urea Solution: Add urea to distilled/deionized water and bring volume to 100.0mL. Mix thoroughly. Filter sterilize.

Preparation of Medium: Add components, except urea solution, to distilled/deionized water and bring volume to 950.0mL. Mix thoroughly. Autoclave for 20 min at 10 psi pressure–115°C. Cool to 50°C. Aseptically add 50.0mL of sterile urea solution. Mix thoroughly. Aseptically distribute into sterile tubes or flasks.

Use: For the differentiation of members of the Enterobacteriaceae based on urease production. Urease positive bacteria turn the medium pink.

Urea R Broth (Urea Rapid Broth)

Composition per liter:

Urea	20.0g
Yeast extract	0.1g
Na_2HPO_4	0.095g
KH_2PO_4	0.091g
Phenol Red	0.01g

pH 6.9 ± 0.2 at 25°C

Source: Available as a prepared medium from Difco Laboratories.

Preparation of Medium: Add components to distilled/deionized water and bring volume to 1.0L. Mix thoroughly. Filter sterilize. Aseptically distribute into sterile tubes or flasks.

Use: For the differentiation of members of the Enterobacteriaceae based on rapid detection of urease activity. Urease positive bacteria turn the medium cerise.

Urea Semisolid Medium

Composition per liter:

Solution A	400.0mL
Solution B	50.0mL

Solution A:

Composition per 400mL:

Pancreatic digest of casein	6.0g
Yeast extract	2.0g
NaCl	1.0g
Yeast extract	0.8g
Agar	0.3g
L-Cystine	0.1g
Thioglycollic acid	0.12mL

pH 7.2 ± 0.2 at 25°C

Preparation of Solution A: Add components to distilled/deionized water and bring volume to 400.0mL. Mix thoroughly. Gently heat and bring to boiling. Autoclave for 15 min at 15 psi pressure–121°C. Cool to 60°C.

Solution B:

Composition per 50mL:

Urea	8.0g
Na_2HPO_4	3.8g
KH_2PO_4	3.64g
Yeast extract	0.04g
Phenol Red	4.0mg

Preparation of Solution B: Add components to distilled/deionized water and bring volume to 50.0mL. Mix thoroughly. Filter sterilize.

Preparation of Medium: Aseptically combine 400.0mL of sterile solution A and 50.0mL of sterile solution B. Mix thoroughly. Aseptically distribute into sterile screw-capped tubes in 7.0mL volumes. Pass the tubes into an anaerobic chamber containing 85% N_2 + 10% H_2 + 5% CO_2 for 60 min. Close screw caps tightly.

Use: For the cultivation and differentiation of anaerobic bacteria based on their production of urease. Bacteria that produce urease turn the medium bright red.

Urea Test Broth

Composition per liter:

Urea	20.0g
Na_2HPO_4	9.5g
KH_2PO_4	9.1g
Yeast extract	0.1g
Phenol Red	0.01g
Urea solution	100.0mL

Urea Solution:

Composition per 100mL:

Urea	20.0g

Preparation of Urea Solution: Add urea to distilled/deionized water and bring volume to 100.0mL. Mix thoroughly. Filter sterilize.

Preparation of Medium: Add components, except urea solution, to distilled/deionized water and bring volume to 900.0mL. Mix thoroughly. Autoclave for 15 min at 15 psi pressure–121°C. Cool to 45° to 50°C. Aseptically add sterile urea solution. Mix thoroughly. Aseptically distribute into sterile tubes in 3.0mL volumes.

Use: For the cultivation and differentiation of members of the Enterobacteriaceae and aerobic actinomycetes based on their production of urease. Bacteria that produce urease turn the medium bright red.

Urea Test Broth

Composition per 99.6mL:

H broth base	85.0mL
Horse serum	10.0mL
Penicillin solution	2.0mL
MES (2-N-Morpholinoethane sulfonic acid) buffer solution	1.0mL
Na_2SO_3 solution	1.0mL
Urea solution	0.5mL
Phenol Red solution	0.1mL

pH 7.2 ± 0.2 at 25°C

H Broth Base:

Composition per liter:

NaCl	5.0g
Pancreatic digest of casein	5.0g
Peptone	5.0g
Beef extract	3.0g
K_2HPO_4	2.5g
Glucose	1.0g

Preparation of H Broth Base: Add components to distilled/deionized water and bring volume to 1.0L. Mix thoroughly. Gently heat and bring to boiling. Distribute into tubes in 4.0mL volumes. Autoclave for 15 min at 10 psi pressure–115°C. Cool to 45° to 50°C.

Penicillin Solution:

Composition per 10mL:

Penicillin	100,000U

Preparation of Penicillin Solution: Add penicillin to distilled/deionized water and bring volume to 10.0mL. Mix thoroughly. Filter sterilize.

MES Buffer Solution:

Composition per 100mL:

MES (2-N-Morpholinoethane sulfonic acid) buffer	19.52g

Preparation of MES Buffer Solution: Add MES buffer to distilled/deionized water and bring volume to 100.0mL. Mix thoroughly. Adjust pH to 5.5. Filter sterilize.

Na_2SO_3 Solution:

Composition per 10mL:

Na_2SO_3	0.126g

Preparation of Na_2SO_3 Solution: Add Na_2SO_3 to distilled/deionized water and bring volume to 10.0mL. Mix thoroughly. Filter sterilize.

Urea Solution:

Composition per 100mL:

Urea	6.0g

Preparation of Urea Solution: Add urea to distilled/deionized water and bring volume to 100.0mL. Mix thoroughly. Filter sterilize.

Phenol Red Solution:

Composition per 10mL:

Phenol Red	0.1g

Preparation of Phenol Red Solution: Add Phenol Red to distilled/deionized water and bring volume to 10.0mL. Mix thoroughly. Filter sterilize.

Preparation of Medium: To 85.0mL of cooled sterile H broth base, aseptically add 10.0mL of sterile horse serum, 2.0mL of sterile penicillin solution, 1.0mL of MES buffer solution, 1.0mL of Na_2SO_3 solution, 0.5mL of urea solution, and 0.1 mL of sterile Phenol Red solution. Mix thoroughly. Aseptically distribute into test tubes in 3.0mL volumes.

Use: For the cultivation and differentiation of *Ureaplasma* species based on their production of urease.

Urease Test Broth (Urea Broth)

Composition per liter:

Urea	20.0g
Na_2HPO_4	9.5g
KH_2PO_4	9.1g
Yeast extract	0.1g
Phenol Red	0.01g

pH 6.8 ± 0.2 at 25°C

Source: This medium is available as a premixed powder from BBL Microbiology Systems and Difco Laboratories.

Preparation of Medium: Add components to distilled/deionized water and bring volume to 1.0L. Mix thoroughly. Filter sterilize. Aseptically distribute into sterile tubes or flasks.

Use: For the differentiation of organisms, especially the Enterobacteriaceae, on the basis of urease production. Urease positive bacteria turn the medium pink.

UVM *Listeria* Enrichment Broth (University of Vermont *Listeria* Enrichment Broth)

Composition per liter:

NaCl	20.0g
Na_2HPO_4	9.6g
Pancreatic digest of casein	5.0g
Peptic digest of animal tissue	5.0g
Beef extract	5.0g
Yeast extract	5.0g
KH_2PO_4	1.35g
Esculin	1.0g
Nalidixic acid	40.0mg
Acriflavine·HCl	12.0mg

pH 7.2 ± 0.2 at 25°C

Preparation of Medium: Add components to distilled/deionized water and bring volume to 1.0L. Mix thoroughly. Distribute into tubes or flasks. Autoclave for 15 min at 15 psi pressure–121°C.

Use: For the selective isolation of *Listeria monocytogenes*.

UVM Modified *Listeria* Enrichment Broth (University of Vermont Modified *Listeria* Enrichment Broth)

Composition per liter:

NaCl	20.0g
Na_2HPO_4	9.6g
Pancreatic digest of casein	5.0g
Peptic digest of animal tissue	5.0g
Beef extract	5.0g
Yeast extract	5.0g
KH_2PO_4	1.35g
Esculin	1.0g
Nalidixic acid	20.0mg
Acriflavine·HCl	12.0mg

pH 7.2 ± 0.2 at 25°C

Source: This medium is available as a premixed powder from BBL Microbiology Systems.

Preparation of Medium: Add components to distilled/deionized water and bring volume to 1.0L. Mix thoroughly. Distribute into tubes or flasks. Autoclave for 15 min at 15 psi pressure–121°C.

Use: For the selective isolation of *Listeria monocytogenes*.

V Agar

Composition per liter:

Agar	13.5g
Pancreatic digest of casein	12.0g
Peptone	10.0g
Peptic digest of animal tissue	5.0g
NaCl	5.0g
Beef extract	3.0g
Yeast extract	3.0g
Cornstarch	1.0g
Human blood, anticoagulated	50.0mL

pH 7.4 ± 0.2 at 25°C

Source: This medium is available as a prepared medium from BBL Microbiology Systems.

Preparation of Medium: Add components, except human blood, to distilled/deionized water and bring volume to 950.0mL. Mix thoroughly. Gently heat and bring to boiling. Distribute into tubes or flasks. Autoclave for 15 min at 15 psi pressure–121°C. Cool to 50°C. Aseptically add 50.0mL of human blood. Mix thoroughly. Pour into sterile Petri dishes or leave in tubes.

Use: For the isolation and differentiation of *Gardnerella vaginalis* from clinical specimens. Plates are incubated under an atmosphere with 3-10% CO_2. *G. vaginalis* appears as small white colonies with diffuse β-hemolysis.

V–8™ Agar

Composition per liter:

Agar	20.0g
Yeast powder	10.0g
V-8 canned vegetable juice	500.0mL

pH 6.9± 0.2 at 25°C

Preparation of Medium: Add components to distilled/deionized water and bring volume to 1.0L. Mix thoroughly. Gently heat and bring to boiling. Distribute into tubes or flasks. Autoclave for 15 min at 15 psi pressure–121°C. Pour into sterile Petri dishes or leave in tubes.

Use: For the isolation and cultivation of yeasts and especially for ascospore formation.

Veal Infusion Agar

Composition per liter:

Agar	15.0g
Veal, infusion from	10.0g
Pancreatic digest of casein	5.0g
Peptic digest of animal tissue	5.0g
NaCl	5.0g

pH 7.4 ± 0.2 at 25°C

Source: This medium is available as a premixed powder from BBL Microbiology Systems.

Preparation of Medium: Add components to distilled/deionized water and bring volume to 1.0L. Mix

thoroughly. Gently heat and bring to boiling. Distribute into tubes or flasks. Autoclave for 15 min at 15 psi pressure–121°C. Pour into sterile Petri dishes or leave in tubes.

Use: For the cultivation and maintenance of a variety of microorganisms. Can be used for cultivation of fastidious microorganisms when enriched with blood or serum.

Veal Infusion Broth

Composition per liter:

Veal, infusion from 10.0g
Pancreatic digest of casein 5.0g
Peptic digest of animal tissue 5.0g
NaCl 5.0g

pH 7.4 ± 0.2 at 25°C

Source: This medium is available as a premixed powder from BBL Microbiology Systems.

Preparation of Medium: Add components to distilled/deionized water and bring volume to 1.0L. Mix thoroughly. Gently heat and bring to boiling. Distribute into tubes or flasks. Autoclave for 15 min at 15 psi pressure–121°C. Use freshly prepared solution.

Use: For the cultivation of streptococci and other microorganisms.

Veal Infusion Broth with Horse Serum

Composition per liter:

Veal, infusion from 500.0g
Pancreatic digest of casein 5.0g
Peptic digest of animal tissue 5.0g
NaCl 5.0g
Horse serum, heat-inactivated 100.0mL

pH 7.4 ± 0.2 at 25°C

Preparation of Medium: Add components, except horse serum, to distilled/deionized water and bring volume to 900.0mL. Mix thoroughly. Gently heat and bring to boiling. Autoclave for 15 min at 15 psi pressure–121°C.Cool to 50°C. Aseptically add 100.0mL horse serum. Mix thoroughly. Aseptically distribute into tubes or flasks. Use freshly prepared solution or boil without mixing prior to use.

Use: For the cultivation and maintenance of *Streptococcus pyogenes*.

Viral Transport Medium (VTM)

Composition per 104.1mL:

Bovine serum albumin 0.5g
Veal infusion broth 100.0mL
Phenol Red 0.4mL
Amphotericin B solution 2.0mL
Gentamicin 1.0mL
Vancomycin 0.2mL

pH 7.4 ± 0.2 at 25°C

Veal Infusion Broth:
Composition per liter:

Veal, infusion from 500.0g
NaCl 5.0g
Pancreatic digest of casein 5.0g
Peptic digest of animal tissue 5.0g

Preparation of Veal Infusion Broth: Add components to distilled/deionized water and bring volume to 1.0L. Mix thoroughly. Distribute into tubes or flasks. Autoclave for 15 min at 15 psi pressure–121°C. Use freshly prepared solution.

Amphotericin B Solution:
Composition per 10mL:

Amphotericin B 2.5g

Preparation of Amphotericin B Solution: Add amphotericin B to distilled/deionized water and bring volume to 10.0mL. Mix thoroughly. Filter sterilize.

Gentamicin Solution:
Composition per 10mL:

Gentamicin 0.5g

Preparation of Gentamicin Solution: Add gentamicin to distilled/deionized water and bring volume to 10.0mL. Mix thoroughly. Filter sterilize.

Vancomycin Solution:
Composition per 10mL:

Vancomycin 0.5g

Preparation of Vancomycin Solution: Add vancomycin to distilled/deionized water and bring volume to 10.0mL. Mix thoroughly. Filter sterilize.

Preparation of Medium: To 100.0mL of sterile veal infusion broth, aseptically add bovine serum albumin, Phenol Red, amphotericin B solution, gentamicin solution and vancomycin solution. Mix thoroughly. Dispense 2.0mL of medium into serum vials. Store at 4°C and use for up to 2 months.

Use: For the maintenance and transport of specimens submitted for viral culture.

Vogel and Johnson Agar

Composition per liter:

Agar 16.0g
Pancreatic digest of casein 10.0g
D-Mannitol 10.0g
Glycine 10.0g

Yeast extract 5.0g
K_2HPO_4 .... 5.0g
LiCl 5.0g
Phenol Red 0.025g
K_2TeO_3 solution 20.0mL
pH 7.2 ± 0.2 at 25°C

Source: This medium is available as a premixed powder from BBL Microbiology Systems, Difco Laboratories and Oxoid Unipath.

K_2TeO_3 Solution:
Composition per 100mL:
K_2TeO_3 .... 1.0g

Preparation of K_2TeO_3 Solution: Add K_2TeO_3 to distilled/deionized water and bring volume to 100.0mL. Mix thoroughly. Filter sterilize.

Caution: Potassium tellurite is toxic.

Preparation of Medium: Add components, except K_2TeO_3 solution, to distilled/deionized water and bring volume to 980.0mL. Mix thoroughly. Gently heat and bring to boiling. Autoclave for 15 min at 15 psi pressure–121°C. Cool to 45° to 50°C. Aseptically add 20.0mL of sterile K_2TeO_3 solution. Mix thoroughly. Pour into sterile Petri dishes or distribute into sterile tubes.

Use: For the detection of coagulase-positive *Staphylococcus aureus.*

Wilkins–Chalgren Agar

Composition per liter:
Agar 15.0g
Gelatin peptone 10.0g
Pancreatic digest of casein 10.0g
NaCl 5.0g
Yeast extract 5.0g
Glucose 1.0g
L-Arginine 1.0g
Sodium pyruvate 1.0g
Hemin 5.0mg
Vitamin K_1 (Menadione) 0.5mg
pH 7.1 ± 0.2 at 25°C

Source: This medium is available as a premixed powder from Difco Laboratories.

Preparation of Medium: Add components to distilled/deionized water and bring volume to 1.0L. Mix thoroughly. Gently heat and bring to boiling. Distribute into tubes or flasks. Autoclave for 15 min at 15 psi pressure–121°C. Cool to 50–55°C. Add antibiotic to be assayed; varying concentrations of antibiotics are used. Mix thoroughly. Pour into sterile Petri dishes or leave in tubes.

Use: For the cultivation and maintenance of anaerobic bacteria. For standardized antimicrobic susceptibility testing to determine the minimum inhibitory concentrations of antimicrobics for anaerobic bacteria.

Wilkins–Chalgren Anaerobe Agar

Composition per liter:
Agar 10.0g
Pancreatic digest of casein 10.0g
Gelatin peptone 10.0g
NaCl 5.0g
Yeast extract 5.0g
Glucose 1.0g
L-Arginine 1.0g
Sodium pyruvate 1.0g
Haemin 5.0mg
Menadione 0.5mg
Defibrinated blood 50.0mL
Tween™ 80 .... 1.0mL
pH 7.1 ± 0.2 at 25°C

Source: This medium is available as a premixed powder from Oxoid Unipath.

Preparation of Medium: Add components, except defibrinated blood, to distilled/deionized water and bring volume to 950.0mL. Mix thoroughly. Gently heat and bring to boiling. Distribute into tubes or flasks. Autoclave for 15 min at 15 psi pressure–121°C. Cool to 50–55°C. Aseptically add 50.0mL of defibrinated blood. Mix thoroughly. Pour into sterile Petri dishes or leave in tubes.

Use: For the cultivation of nonsporulating anaerobes.

Wilkins–Chalgren Anaerobe Agar with GN Supplement

Composition per liter:
Agar 10.0g
Pancreatic digest of casein 10.0g
Gelatin peptone 10.0g
NaCl 5.0g
Yeast extract 5.0g
Glucose 1.0g
L-Arginine 1.0g
Sodium pyruvate 1.0g
Hemin 5.0mg
Menadione 0.5mg
Defibrinated blood 50.0mL
G-N anaerobe selective supplement .... 20.0 mL
pH 7.1 ± 0.2 at 25°C

Source: This medium is available as a premixed powder from Oxoid Unipath.

GN Anaerobe Selective Supplement
Composition per 20mL:
Nalidixic acid 10.0mg
Hemin 5.0mg

Sodium succinate 2.5mg
Vancomycin 2.5mg
Menadione 0.5mg

Preparation of GN Anaerobe Selective Supplement: Add components to distilled/deionized water and bring volume to 20.0mL. Mix thoroughly. Filter sterilize.

Preparation of Medium: Add components, except defibrinated blood and GN anaerobe selective supplement, to distilled/deionized water and bring volume to 900.0mL. Mix thoroughly. Gently heat and bring to boiling. Distribute into tubes or flasks. Autoclave for 15 min at 15 psi pressure–121°C. Cool to 50–55°C. Aseptically add 20.0mL of G-N anaerobe selective supplement and 50.0mL of defibrinated blood. Bring volume to 1.0L with distilled/deionized water. Mix thoroughly. Pour into sterile Petri dishes or leave in tubes.

Use: For the selective isolation of Gram-negative anaerobes.

Wilkins–Chalgren Anaerobe Agar with NS Supplement

Composition per liter:

Agar 10.0g
Pancreatic digest of casein 10.0g
Gelatin peptone 10.0g
NaCl 5.0g
Yeast extract 5.0g
Glucose 1.0g
L-Arginine 1.0g
Sodium pyruvate 1.0g
Hemin 5.0mg
Menadione 0.5mg
Defibrinated blood 50.0mL
N-S anaerobe selective supplement .... 20.0 mL
Tween™ 80 .... 1.0mL

pH 7.1 ± 0.2 at 25°C

Source: This medium is available as a premixed powder from Oxoid Unipath.

NS Anaerobe Selective Supplement:
Composition per 20mL:

Sodium pyruvate 1.0g
Nalidixic acid 0.01g
Hemin 5.0mg
Menadione 0.5mg

Preparation of NS Anaerobe Selective Supplement: Add components to distilled/deionized water and bring volume to 20.0mL. Mix thoroughly. Filter sterilize.

Preparation of Medium: Add components, except defibrinated blood and N-S anaerobe selective supplement, to distilled/deionized water and bring volume to 900.0mL. Mix thoroughly. Gently heat and bring to boiling. Distribute into tubes or flasks. Autoclave for 15 min at 15 psi pressure–121°C. Cool to 50–55°C. Aseptically add 20.0mL of N-S anaerobe selective supplement and 50.0mL of defibrinated blood. Bring volume to 1.0L with distilled/deionized water. Mix thoroughly. Pour into sterile Petri dishes or leave in tubes.

Use: For the selective isolation of nonsporulating anaerobes.

Wilkins–Chalgren Anaerobe Broth (Anaerobe Broth, MIC)

Composition per liter:

Pancreatic digest of casein 10.0g
Gelatin peptone 10.0g
NaCl 5.0g
Yeast extract 5.0g
Glucose 1.0g
L-Arginine 1.0g
Sodium pyruvate 1.0g
Hemin 5.0mg
Menadione 0.5mg

pH 7.1 ± 0.2 at 25°C

Source: This medium is available as a premixed powder from Difco Laboratories and Oxoid Unipath.

Preparation of Medium: Add components to distilled/deionized water and bring volume to 1.0L. Mix thoroughly. Distribute into tubes or flasks. Autoclave for 15 min at 15 psi pressure–121°C.

Use: For the cultivation and antimicrobial susceptibility (MIC) testing of anaerobic bacteria.

Wolin–Bevis Medium

Composition per liter:

Agar 20.0g
$(NH_4)_2SO_4$ 1.0g
KH_2HPO_4 .... 1.0g
Glucose 0.25g
L-Histidine·HCl 0.25g
Tween™ 80 (polysorbate 80) 3.0mL

pH 5.4 ± 0.2 at 25°C

Preparation of Medium: Add components to distilled/deionized water and bring volume to 1.0L. Mix thoroughly. Gently heat and bring to boiling. Distribute into tubes or flasks. Autoclave for 15 min at 15 psi pressure–121°C. Pour into sterile Petri dishes or leave in tubes.

Use: For the enhanced production of chlamydospores by *Candida albicans.*

Xanthine Agar

Composition per liter:

Solution 1 .... 900.0mL
Solution 2 .... 100.0mL

pH 7.0 ± 0.2 at 25°C

Solution 1:
Composition per 900mL:

Agar 15.0g
Pancreatic digest of gelatin 5.0g
Beef extract 3.0g

Preparation of Solution 1: Add components to distilled/deionized water and bring volume to 900.0mL. Mix thoroughly. Gently heat and bring to boiling.

Solution 2:
Composition per 100mL:

Xanthine 4.0g

Preparation of Solution 2: Add xanthine to distilled/deionized water and bring volume to 100.0mL. Mix thoroughly. Gently heat and bring to boiling.

Preparation of Medium: Combine solutions 1 and 2. Mix thoroughly. Distribute into tubes or flasks. Autoclave for 15 min at 15 psi pressure–121°C. Pour into sterile Petri dishes or leave in tubes.

Use: For the differentiation of aerobic actinomycete species. Clearing around a colony indicates utilization of xanthine. *Streptomyces* species utilize xanthine; most *Nocardia* and *Actinomadura* species do not utilize xanthine.

XL Agar Base (Xylose Lysine Agar Base)

Composition per liter:

Agar 13.5g
Lactose 7.5g
Sucrose 7.5g
L-Lysine 5.0g
NaCl 5.0g
Xylose 3.5g
Yeast extract 3.0g
Phenol Red 0.08g
Thiosulfate-citrate solution 20.0mL

pH 7.5 ± 0.2 at 25°C

Source: This medium is available as a premixed powder from BBL Microbiology Systems.

Thiosulfate-Citrate Solution:
Composition per 100mL:

$Na_2S_2O_3$ 34.0g
Ferric ammonium citrate 4.0g

Preparation of Thiosulfate-Citrate Solution: Add components to distilled/deionized water and bring volume to 100.0mL. Mix thoroughly.

Preparation of Medium: Add components, except thiosulfate-citrate solution, to distilled/deionized water and bring volume to 980.0mL. Mix thoroughly. Gently heat while stirring and bring to boiling. Distribute into tubes or flasks. Autoclave for 10 min at 14 psi pressure–118°C. Cool to 55°C. Aseptically add 20.0 mL of the sterile thiosulfate-citrate solution. Mix thoroughly. Pour into sterile Petri dishes or leave in tubes.

Use: For the isolation, cultivation, and differentiation of enteric pathogens. Nonfermenting xylose/lactose/sucrose bacteria appear as red colonies. Xylose-fermenting lysine-decarboxylating bacteria appear as red colonies. Xylose-fermenting lysine-non-decarboxylating bacteria appear as opaque yellow colonies. Lactose or sucrose-fermenting bacteria appear as yellow colonies.

XL Agar Base

Composition per liter:

Agar .... 15 g
Lactose 7.5g
Sucrose 7.5g
L-Lysine 5.0g
NaCl 5.0g
Xylose 3.75g
Yeast extract 3.0g
Phenol Red 0.08g
Thiosulfate-citrate solution 20.0mL

pH 7.4 ± 0.2 at 25°C

Source: This medium is available as a premixed powder from Difco Laboratories.

Thiosulfate-Citrate Solution:
Composition per 100mL:

$Na_2S_2O_3$ 34.0g
Ferric ammonium citrate 4.0g

Preparation of Thiosulfate-Citrate Solution: Add components to distilled/deionized water and bring volume to 100.0mL. Mix thoroughly.

Preparation of Medium: Add components, except thiosulfate-citrate solution, to distilled/deionized water and bring volume to 980.0mL. Mix thoroughly. Gently heat while stirring and bring to boiling. Distribute into tubes or flasks. Autoclave for 10 min at 14 psi pressure–118°C. Cool to 55°C. Aseptically add 20.0 mL of the sterile thiosulfate-citrate solution. Mix thoroughly. Pour into sterile Petri dishes or leave in tubes.

Use: For the isolation, cultivation, and differentiation of enteric pathogens. Nonfermenting xylose/lactose/sucrose bacteria appear as red colonies. Xylose-fermenting lysine-decarboxylating bacteria appear as red colonies. Xylose-fermenting lysine-non-decarboxylating bacteria appear as opaque yellow colonies. Lactose or sucrose-fermenting bacteria appear as yellow colonies.

XLD Agar (Xylose Lysine Deoxycholate Agar)

Composition per liter:

Agar	13.5g
Lactose	7.5g
Sucrose	7.5g
$Na_2S_2O_3$	6.8g
L-Lysine	5.0g
NaCl	5.0g
Xylose	3.5g
Yeast extract	3.0g
Sodium desoxycholate	2.5g
Ferric ammonium citrate	0.8g
Phenol Red	0.08g

pH 7.5 ± 0.2 at 25°C

Source: This medium is available as a premixed powder from BBL Microbiology Systems, Difco Laboratories and Oxoid Unipath.

Preparation of Medium: Add components to distilled/deionized water and bring volume to 1.0L. Mix thoroughly. Gently heat and bring to boiling. Do not overheat. Distribute into tubes or flasks. Autoclave for 15 min at 15 psi pressure–121°C. Pour into sterile Petri dishes or leave in tubes. Plates should be poured as soon as possible to avoid precipitation.

Use: For isolation and differentiation of enteric pathogens, especially *Shigella* and *Providencia* species. Nonfermenting xylose/lactose/sucrose bacteria appear as red colonies. Xylose-fermenting lysine-decarboxylating bacteria appear as red colonies. Xylose-fermenting lysine-non-decarboxylating bacteria appear as opaque yellow colonies. Lactose or sucrose-fermenting bacteria appear as yellow colonies.

Xylose Sodium Deoxycholate Citrate Agar

Composition per liter:

Agar	12.0g
Xylose	10.0g
Sodium citrate	5.0g
$Na_2S_2O_3 \cdot 5H_2O$	5.0g
Beef extract	5.0g
Peptone	5.0g
NaCl	2.5g
Sodium deoxycholate	2.5g
Ferric ammonium citrate	1.0g
Neutral Red (1% solution)	2.5mL

pH 7.5 ± 0.2 at 25°C

Preparation of Medium: Add components to distilled/deionized water and bring volume to 1.0L. Mix thoroughly. Gently heat and bring to boiling for 20 sec. Do not autoclave. Cool to 45° to 50°C. Pour into sterile Petri dishes.

Use: For the cultivation of *Salmonella* species and some *Shigella* species.

Y 1 Adrenal Cell Growth Medium

Composition per 101mL:

Ham's F-10 Medium	90.0mL
Fetal bovine serum	10.0mL
Penicillin-streptomycin solution	1.0mL

pH 7.0 ± 0.2 at 25°C

Ham's F-10 Medium:
Composition per liter:

NaCl	7.4g
$NaHCO_3$	1.2g
Glucose	1.1g
$NaH_2PO_4 \cdot H_2O$	0.29g
KCl	0.28g
L-Arginine·HCl	0.21g
L-Glutamine	0.15g
$MgSO_4 \cdot 7H_2O$	0.15g
Sodium pyruvate	0.11g
KH_2PO_4	0.08g
$CaCl_2 \cdot 2H_2O$	0.04g
L-Cystine·2HCl	0.04g
L-Histidine·HCl·H_2O	0.02g
L-Lysine·HCl	0.02g
L-Asparagine-H_2O	0.01g
L-Aspartic Acid	0.01g
L-Glutamic acid	0.01g
L-Leucine	0.01g
L-Proline	0.01g
L-Serine	0.01g
L-Alanine	8.9mg
Glycine	7.5mg
D-Phenylalanine	5.0mg
L-Methionine	4.5mg
Hypoxanthine	4.1mg
L-Threonine	3.6mg
L-Valine	3.5mg
L-Isoleucine	2.6mg

L-Tyrosine 1.8mg
Vitamin B_{12} 1.4mg
Folic acid 1.3mg
Phenol Red 1.2mg
Thiamine·HCl 1.0mg
$FeSO_4·7H_2O$ 0.8mg
Choline chloride 0.7mg
D-Calcium pantothenate 0.7mg
Thymidine 0.7mg
Niacinamide 0.6mg
L-Tryptophan 0.6mg
Isoinositol 0.5mg
Riboflavin 0.4mg
Lipoic acid 0.2mg
Pyridoxine·HCl 0.2mg
$ZnSO_4·7H_2O$ 0.03mg
Biotin 0.02mg
$CuSO_4·5H_2O$ 3.0μg

Preparation of Ham's F-10 Medium: Add components to distilled/deionized water and bring volume to 1.0L. Mix thoroughly.

Penicillin-Streptomycin Solution:
Composition per 100mL:
Streptomycin 0.5g
Penicillin G 500,000U

Preparation of Penicillin-Streptomycin Solution: Add components to distilled/deionized water and bring volume to 100.0mL. Mix thoroughly. Filter sterilize.

Preparation of Medium: Aseptically combine components. Filter sterilize. Store at 4-5°C.

Use: For the cultivation of Y-1 mouse adrenal tissue culture cells used for the detection of heat-labile toxin (LT) produced by enterotoxigenic strains of *Escherichia coli*. LT causes the conversion of elongated fibroblast-like cells into round, refractile cells.

Y 1 Adrenal Cell Growth Medium

Composition per 580mL:
Ham's F-10 Medium 500.0mL
Fetal bovine serum 75.0mL
Penicillin-streptomycin solution 5.0mL
pH 7.0 ± 0.2 at 25°C

Ham's F-10 Medium:
Composition per liter:
NaCl 7.4g
$NaHCO_3$ 1.2g
Glucose 1.1g
$NaH_2PO_4·H_2O$ 0.29g
KCl 0.28g
L-Arginine·HCl 0.21g
L-Glutamine 0.15g
$MgSO_4·7H_2O$ 0.15g
Sodium pyruvate 0.11g
KH_2PO_4 .... 0.08g
$CaCl_2·2H_2O$ 0.04g
L-Cystine·2HCl 0.04g
L-Histidine·HCl·H_2O 0.02g
L-Lysine·HCl 0.02g
L-Asparagine-H_2O 0.01g
L-Aspartic Acid 0.01g
L-Glutamic acid 0.01g
L-Leucine 0.01g
L-Proline 0.01g
L-Serine 0.01g
L-Alanine 8.9mg
Glycine 7.5mg
D-Phenylalanine 5.0mg
L-Methionine 4.5mg
Hypoxanthine 4.1mg
L-Threonine 3.6mg
L-Valine 3.5mg
L-Isoleucine 2.6mg
L-Tyrosine 1.8mg
Vitamin B_{12} 1.4mg
Folic acid 1.3mg
Phenol Red 1.2mg
Thiamine·HCl 1.0mg
$FeSO_4·7H_2O$ 0.8mg
Choline chloride 0.7mg
D-Calcium pantothenate 0.7mg
Thymidine 0.7mg
Niacinamide 0.6mg
L-Tryptophan 0.6mg
Isoinositol 0.5mg
Riboflavin 0.4mg
Lipoic acid 0.2mg
Pyridoxine·HCl 0.2mg
$ZnSO_4·7H_2O$ 0.03mg
Biotin 0.02mg
$CuSO_4·5H_2O$ 3.0μg

Preparation of Ham's F-10 Medium: Add components to distilled/deionized water and bring volume to 1.0L. Mix thoroughly.

Penicillin-Streptomycin Solution:
Composition per 100mL:
Streptomycin 0.5g
Penicillin G 500,000U

Preparation of Penicillin-Streptomycin Solution: Add components to distilled/deionized water and bring volume to 100.0mL. Mix thoroughly. Filter sterilize.

Preparation of Medium: Aseptically combine components. Filter sterilize. Store at 4° to 5°C.

Use: For the cultivation of Y-1 mouse adrenal tissue culture cells used for the detection of cholera enterotoxin (CT) produced by enterotoxigenic strains of *Vibrio cholerae* or *Vibrio mimicus*. CT causes the conversion of elongated fibroblast-like cells into round, refractile cells.

Yeast Dextrose Agar

Composition per liter:

Agar....................15.0g
Glucose....................10.0g
Yeast extract....................10.0g

pH 7.0 ± 0.2 at 25°C

Preparation of Medium: Add components to distilled/deionized water and bring volume to 1.0L. Mix thoroughly. Gently heat and bring to boiling. Adjust pH to 7.0. Distribute into tubes or flasks. Autoclave for 15 min at 15 psi pressure–121°C.

Use: For the cultivation of a variety of heterotrophic microorganisms.

Yeast Extract Agar

Composition per liter:

Agar....................20.0g
Yeast extract....................1.0g
Buffer solution....................2.0mL

pH 6.0 ± 0.2 at 25°C

Buffer Solution:

Composition per 400mL:

KH_2PO_4....................60.0g
Na_2HPO_4....................40.0g

Preparation of Buffer Solution: Add 40.0g Na_2HPO_4 to 300.0mL of distilled/deionized water. Mix thoroughly. Add 60.0g of KH_2PO_4. Mix thoroughly. Adjust pH to 6.0.

Preparation of Medium: Add components to distilled/deionized water and bring volume to 1.0L. Mix thoroughly. Autoclave for 15 min at 15 psi pressure–121°C. Pour into sterile Petri dishes.

Use: For identification of *Histoplasma capsulatum*, *Blastomyces dermatitidis* and *Coccidioides immitis;* also for the cultivation of filamentous fungi.

Yeast Extract Agar

Composition per liter:

Agar....................15.0g
Proteose peptone....................10.0g
NaCl....................5.0g
Yeast extract....................3.0g

Preparation of Medium: Add components to distilled/deionized water and bring volume to 1.0L. Mix thoroughly. Gently heat and bring to boiling. Distribute into tubes or flasks. Autoclave for 15 min at 15 psi pressure–121°C. Pour into sterile Petri dishes or leave in tubes.

Use: For the cultivation of a variety of heterotrophic microorganisms, including filamentous fungi.

Yeast Extract Phosphate Agar (YEP Agar) (Smith's Medium)

Composition per liter:

Agar....................20.0g
Yeast extract....................1.0g
KH_2PO_4....................0.3g
Na_2HPO_4....................0.2g
Phenol Red....................1.0mg

Source: This medium is available as a premixed powder from BBL Microbiology Systems.

Preparation of Medium: Add components to distilled/deionized water and bring volume to 1.0L. Mix thoroughly. Gently heat and bring to boiling. Distribute into tubes or flasks. Autoclave for 15 min at 15 psi pressure–121°C. Pour into sterile Petri dishes or leave in tubes.

Use: For the isolation of dimorphic pathogenic fungi from clinical specimens. Promotes and retains the spore-forming abilities of molds, especially *Histoplasma capsulatum* and *Blastomyces dermtitidis.*

Yeast Extract Phosphate Medium with Chloramphenicol

Composition per liter:

Yeast extract....................1.0g
Chloramphenicol....................50.0mg
Phosphate buffer stock solution....................2.0mL
NH_4OH (sterile, concentrated)....................0.05mL
Agar....................20.0g

Phosphate Buffer Stock Solution:

Composition per 300mL:

Na_2HPO_4....................40.0g
KH_2PO_4....................60.0

Preparation of Phosphate Buffer Stock Solution: Add components to distilled/deionized water and bring volume to 300.0mL. Mix thoroughly.

Preparation of Medium: Add components except NH_4OH to distilled/deionized water and bring volume to 1.0L. Mix thoroughly. Autoclave for 15 min at 15 psi pressure–121°C. Aseptically dispense into sterile Petri dishes. After inoculation with specimens, such as sputum or other contaminated fluids, onto the agar medium, add one drop of cold NH_4OH to the edge of the medium surface and allow to diffuse throughout the medium.

Use: For the isolation of *Histoplasma capsulatum* and *Blastomyces dermatitidis* from contaminated specimens. The ammonium hydroxide inhibits the growth of most yeasts, saprophytic filamentous fungi, and bacteria.

Yeast Fermentation Broth

Composition per liter:

Carbohydrate	10g
Pancreatic digest of gelatin	7.5g
Yeast extract	5.5g
Bromcresol Purple	16.0mg

Source: This medium is available as a premixed powder from BBL Microbiology Systems.

Preparation of Medium: Add components to distilled/deionized water and bring volume to 1.0L. Mix thoroughly. Distribute into test tubes, each containing an inverted Durham tube. Autoclave for 15 min at 15 psi pressure–121°C.

Use: For fermentation tests of specific carbohydrates used in the characterization and identification of yeasts. Gas accumulation in the Durham tube and a color change of the medium to yellow indicates carbohydrate fermentation.

Yeast Fermentation Medium

Composition per liter:

Peptone	7.5g
Yeast extract	4.5g
Bromthymol Blue (1.6% solution)	1.0mL
Carbohydrate solution	1.0mL

Carbohydrate Solution:

Composition per 10mL:

Carbohydrate	0.6g

Preparation of Carbohydrate Solution: Add carbohydrate to distilled/deionized water and bring volume to 10.0mL. Glucose, maltose, lactose, galactose or trehalose may be used. If raffinose is used, prepare a 12% solution. Mix thoroughly. Filter sterilize.

Preparation of Medium: Add components, except carbohydrate solution, to distilled/deionized water and bring volume to 1.0L. Mix thoroughly. Gently heat and bring to boiling. Distribute in 2.0mL volumes into test tubes that contain an inverted Durham tube. Autoclave for 15 min at 15 psi pressure–121°C. Cool to 45° to 50°C. Aseptically add 1.0mL of sterile carbohydrate solution. Mix thoroughly.

Use: For the cultivation and differentiation of yeast based on carbohydrate fermentation patterns. Yeast that can ferment a specific carbohydrate turn the medium yellow.

Yeast Morphology Agar

Composition per liter:

$(NH_4)_2SO_4$	32.5g
Agar	18.0g
Glucose	10.0g
Asparagine	1.5g
KH_2PO_4	1.0g
$MgSO_4 \cdot 7H_2O$	0.5g
NaCl	0.5g
$CaCl_2 \cdot 2H_2O$	0.1g
DL-Methionine	0.02g
DL-Tryptophan	0.02g
L-Histidine·HCl	0.01g
Inositol	2.0mg
H_3BO_3	0.5mg
Calcium pantothenate	0.4mg
$MgSO_4 \cdot 7H_2O$	0.4mg
Niacin	0.4mg
Pyridoxine·HCl	0.4mg
Thiamine·HCl	0.4mg
$ZnSO_4 \cdot 7H_2O$	0.4mg
p-Aminobenzoic acid	0.2mg
$FeCl_3$	0.2mg
Riboflavin	0.2mg
$Na_2MoO_4 \cdot 4H_2O$	0.2mg
KI	0.1mg
$CuSO_4 \cdot 5H_2O$	0.04mg
Biotin	2.0μg
Folic acid	2.0μg

pH 5.6 ± 0.2 at 25°C

Source: This medium is available as a premixed powder from Difco Laboratories.

Preparation of Medium: Add components to distilled/deionized water and bring volume to 1.0L. Mix thoroughly. Gently heat and bring to boiling. Distribute into tubes or flasks. Autoclave for 15 min at 15 psi pressure–121°C.

Use: For agar dilution and diffusion disk testing with 5-flucytosine. For the microbiological assay of flucytosine using *Candida kefyr* ATCC 28838 or *Saccharomyces cerevisiae* ATCC 36375 as the indicator microorganisms.

Yeast Nitrogen Base

Composition per liter:

$(NH_4)_2SO_4$	5.0g
KH_2PO_4	1.0g

$MgSO_4 \cdot 7H_2O$ 0.5g
NaCl 0.1g
$CaCl_2 \cdot 2H_2O$ 0.1g
DL-Methionine 0.02g
DL-Tryptophan 0.02g
L-Histidine·HCl 0.01g
Inositol 2.0mg
KI 0.1mg
H_3BO_3 .... 0.5mg
$ZnSO_4 \cdot 7H_2O$ 0.4mg
$MnSO_4 \cdot 4H_2O$ 0.4mg
Thiamine·HCl 0.4mg
Pyroxidine·HCl 0.4mg
Niacin 0.4mg
Calcium pantothenate 0.4mg
p-Aminobenzoic acid 0.2mg
Riboflavin 0.2mg
$FeCl_3$ 0.2mg
$Na_2MoO_4 \cdot 4H_2O$ 0.2mg
$CuSO_4 \cdot 5H_2O$ 0.04mg
Folic acid 2.0μg
Biotin 2.0μg

pH 5.5 ± 0.2 at 25°C

Source: This medium is available as a premixed powder from BBL Microbiology Systems and Difco Laboratories.

Preparation of Medium: Add components to distilled/deionized water and bring volume to 1.0L. Mix thoroughly. Distribute into tubes or flasks. Autoclave for 15 min at 15 psi pressure–121°C. Alternately for carbon assimilation tests, prepare a 10× concentrated solution by adding components to distilled/deionized water and bring volume to 100.0mL. Mix thoroughly. Distribute into tubes or flasks. Autoclave for 15 min at 15 psi pressure–121°C. Prepare a carbohydrate solution by adding 0.5g of carbohydrate to 90.0mL of distilled/deionized water. Mix thoroughly. Filter sterilize. Aseptically add 0.5mL of the 10× concentrated solution to 4.5mL of the filter sterilized carbohydrate solution. Mix thoroughly.

Use: For carbohydrate assimilation tests in the characterization and identification of yeasts.

Yeast Nitrogen Base with Carbohydrate

Composition per liter:

Carbohydrate 5.0g
$(NH_4)_2SO_4$ 5.0g
KH_2PO_4 .... 1.0g
$MgSO_4 \cdot 7H_2O$ 0.5g
NaCl 0.1g
$CaCl_2 \cdot 2H_2O$ 0.1g
DL-Methionine 0.02g
DL-Tryptophan 0.02g
L-Histidine·HCl 0.01g
Inositol 2.0mg
KI 0.1mg
H_3BO_3 .... 0.5mg
$ZnSO_4 \cdot 7H_2O$ 0.4mg
$MnSO_4 \cdot 4H_2O$ 0.4mg
Thiamine·HCl 0.4mg
Pyroxidine·HCl 0.4mg
Niacin 0.4mg
Calcium pantothenate 0.4mg
p-Aminobenzoic acid 0.2mg
Riboflavin 0.2mg
$FeCl_3$ 0.2mg
$Na_2MoO_4 \cdot 4H_2O$ 0.2mg
$CuSO_4 \cdot 5H_2O$ 0.04mg
Folic acid 2.0μg
Biotin 2.0μg

pH 5.6 ± 0.2 at 25°C

Preparation of Medium: Add components to distilled/deionized water and bring volume to 1.0L. Mix thoroughly. Filter sterilize. Aseptically distribute into tubes or flasks.

Use: For carbohydrate assimilation tests in the characterization and identification of yeasts.

Yeast Nitrogen Base, 10X with Asparagine and Glucose

Composition per liter:

Glucose 10.0g
$(NH_4)_2SO_4$ 5.0g
L-Asparagine 1.5g
KH_2PO_4 .... 1.0g
$MgSO_4 \cdot 7H_2O$ 0.5g
NaCl 0.1g
$CaCl_2 \cdot 2H_2O$ 0.1g
DL-Methionine 0.02g
DL-Tryptophan 0.02g
L-Histidine·HCl 0.01g
Inositol 2.0mg
H_3BO_3 .... 0.5mg
$ZnSO_4 \cdot 7H_2O$ 0.4mg
$MnSO_4 \cdot 4H_2O$ 0.4mg
Thiamine·HCl 0.4mg
Pyridoxine·HCl 0.4mg
Niacin 0.4mg
Calcium pantothenate 0.4mg
p-Aminobenzoic acid 0.2mg
Riboflavin 0.2mg
$FeCl_3$ 0.2mg
$Na_2MoO_4 \cdot 4H_2O$ 0.2mg
KI 0.1mg
$CuSO_4 \cdot 5H_2O$ 0.04mg

Folic Acid.. 2.0μg
Biotin 2.0μg

pH 5.6 ± 0.2 at 25°C

Preparation of Medium: Add components to distilled/deionized water and bring volume to 1.0L. Dilute 100.0mL of 10× medium with 900.0mL of distilled/deionized water. Mix thoroughly. Filter sterilize. Aseptically distribute into sterile tubes or flasks.

Use: For susceptibility testing of yeasts and filamentous fungi.

Yersinia Selective Agar Base

Composition per liter:

Mannitol 20.0g
Peptone...17.0g
Agar...12.5g
Proteose peptone 3.0g
Yeast extract...2.0g
Sodium pyruvate 2.0g
NaCl 1.0g
Sodium desoxycholate 0.5g
$MgSO_4 \cdot 7H_2O$ 0.01g
Neutral Red 0.03g
Crystal Violet 1.0mg
Selective supplement 6.0mL

pH 7.4 ± 0.2 at 25°C

Source: This medium is available as a premixed powder from Difco Laboratories and Oxoid Unipath.

Selective Supplement:
Composition per 6mL:

Cefsulodin 15.0mg
Irgasan 4.0mg
Novobiocin.. 2.5mg
Ethanol 2.0mL

Preparation of Selective Supplement: Aseptically add components to 4.0mL distilled/deionized water and 2.0mL ethanol. Mix thoroughly.

Preparation of Medium: Add components to distilled/deionized water and bring volume to 1.0L. Mix thoroughly. Gently heat and bring to boiling. Distribute into tubes or flasks. Autoclave for 15 min at 15 psi pressure–121°C. Cool to 50°C. Aseptically add selective supplement. Mix thoroughly. Pour into sterile Petri dishes or leave in tubes.

Use: For the isolation and enumeration of *Yersinia enterocolitica* from food and clinical specimens.

Index

D

E

H

M

T

U

V